2nd International Conference on Sustainability in Architectural Design and Urbanism (ICSADU 2018)

IOP Conference Series: Earth and Environmental Science
Volume 213

Semarang, Indonesia
29 August 2018

ISBN: 978-1-5108-7955-3
ISSN: 1755-1307

Printed from e-media with permission by:

Curran Associates, Inc.
57 Morehouse Lane
Red Hook, NY 12571

Some format issues inherent in the e-media version may also appear in this print version.

Institute of Physics
Dirac House, Temple Back
Bristol BS1 6BE UK

Phone: 44 1 17 929 7481
Fax: 44 1 17 920 0979

techtracking@iop.org

Additional copies of this publication are available from:

Curran Associates, Inc.
57 Morehouse Lane
Red Hook, NY 12571 USA
Phone: 845-758-0400
Fax: 845-758-2633
Email: curran@proceedings.com
Web: www.proceedings.com

TABLE OF CONTENTS

BEYOND SUSTAINABILITY: RECONSIDERING THE HEALING QUALITIES OF THE BUILT ENVIRONMENT 1
Almantas Samalavicius

PLAYABILITY FOR CITIES 9
Saehoon Kim

3D INTERACTIVE VISUALIZATION SCENARIO TO IMPROVE PUBLIC PARTICIPATION IN THE VISUAL COMFORT ASSESSMENT FOR THE PEDESTRIAN IN PROVINCIAL STREET, MALANG, INDONESIA 16
Herry Santosa, Jenny Ernawati, Lisa Dwi Wulandari

ENERGY OPTIMUM BUILDING WALL FROM A MIXTURE OF PAPERCRETE AND LOCAL SOIL IN TROPICAL HUMID AREA 25
Vincentius Totok Noerwasito

WATER RESOURCE AS AXIS-MUNDI: AN EFFORT TO PRESERVE WATER RESOURCE SUSTAINABILITY 32
Ema Yunita Titisari, Antariksa, Lisa Dwi Wulandari, Surjono

CAUSAL FACTORS IN RESILIENCE OF OLD TAMBAK LOROK FISHING PORT IN SEMARANG 41
Satriya Wahyu Firmandhani, Masyiana Arifah Alfia Riza

SCENARIOS OF VILLAGE ALLIANCES PROMOTING LOCAL DEVELOPMENT 49
Listen Prima

CULTURAL SIGNIFICANCE OF THE PLACE: THE KEY FACTOR IN URBAN HERITAGE PROGRAMME 59
Yohannes Firzal, Irham Temas

SPATIAL CONFLICT IN URBAN KAMPONG DEVELOPMENT A CASE STUDY IN KAMPONG PENDRIKAN, SEMARANG 67
Bangun Ir Harsritanto, Wulani Enggar Sari, Rizqi Jamaluddin, Ratih Widiastuti, Rona Fika Jamila

THE COMMUNITY ROLE IN GREEN AREA SUSTAINABILITY AS A MODEL OF ENERGY-EFFICIENT BUILDINGS IN THE HUMID TROPICAL REGION 73
Sri Yuliani, Wiwik Setyaningsih

THE SUNDANESE AND JAVANESE HOUSE LOCAL WISDOM TO RESPOND TO THE CLIMATE CHANGES 81
Handajani Asriningpuri, Estuti Rochimah

TEXTILE TOURISM IMAGE AS AN IDENTITY OF CIGONDEWAH IN BANDUNG CITY 96
Karto Wijaya, Asep Yudi Permana

RESIDENTIAL SPACE TRANSFORMATION AS THE LEGITIMACY SPACE A CASE STUDY: MAGERSARI NDALEM SASANAMULYA BALUWARTI SUNANATE PALACE OF SURAKARTA 106
Avi Marlina

THE MORPHOLOGICAL CHANGES FACTORS IN KOTAGEDE FRINGE AREAS 118
Bambang Setioko, Bangun Ir Harsritanto, Mustika K Wardhani

INVESTIGATION OF URBAN PLACES IN SEOUL DIGITAL INDUSTRIAL COMPLEX (G-VALLEY)... 125
Suha Ryu, Saehoon Kim

WHAT IS THE SUSTAINABLE METHOD ENOUGH FOR OUR BUILT ENVIRONMENT? 132
A. Djoko Istiadji, Gagoek Hardiman, Prasasto Satwiko

INVESTIGATING PROBLEMS OF VEHICLE PARKING IN A LOW-RISE RESIDENTIAL AREA IN NANGOK-DONG, SEOUL 142
Seungwoo Seo, Saehoon Kim

SPATIAL CONFIGURATION OF OPTIMIZATION MODEL AT FORT ORANJE AREA IN TERNATE CITY TO CREATE THERMAL COMFORT 148
Asri A. Muhammad, Firdawaty Marasabessy, Arif Kusumawanto, Ardhya Nareswari

REINVIGORATING LOCAL BUILDING ENVELOPE FOR TRANSLATED MODERN BUILDING AND CHANGED ENVIRONMENT.. 160
Fx Teddy Badai Samodra, Irvansyah, Collinthia Erwindi

EMPOWERING TRADITIONAL HOUSE FOR FUTURE NOISE CONTROL BASED ON SOUND FREQUENCY PROPAGATION 168
Fx Teddy Badai Samodra, Irvansyah, Collinthia Erwindi

THE TRANSFORMATION OF GEGERKALONG GIRANG AREA, BANDUNG CITY: AMID EDUCATIVE AND RELIGIOUS AREAS.. 177
Asep Yudi Permana, Indah Susanti, Karto Wijaya

DRIVING FACTORS OF DEVELOPERS IN DEVELOPING APARTMENT IN DEPOK......................... 186
Yusup Setiadi, Hendrajaya Isnaeni

UNDERSTANDING GENIUS LOCI TO SUSTAIN UME BANGKA'S TRADITIONAL ARCHITECTURE BASED ON INTANGIBLE MATERIAL CULTURE 194
Kemas Ridwan Kurniawan, Elita Nuraeny

THE SPATIAL AND SETTLEMENT PATTERN IN MABI HAMLET, BELANTIH VILLAGE, KINTAMANI: CONSERVATION OF TANGIBLE AND INTANGIBLE OF BALINESE TRADITIONAL ARCHITECTURE IN BALI AGA VILLAGES 202
Ni Made Yudantini

INDOOR THERMAL ENVIRONMENT IN TROPICAL ARCHIPELAGO CITY 210
Mustamin Rahim, Baharuddin Hamzah

QUESTIONING URBAN SYMBOLISM IN THE OLD CITY OF SEMARANG... 217
Yulia Nurliani Lukito, Putri Widya Pratama

TYPOLOGY OF BAJO TRIBE SETTLEMENT IN NORTH MALUKU 225
Mustamin Rahim, Ardi Basri, Hendra Fauzi

HYBRID ARCHITECTURE AND CONTESTED SPACE OF COLONIAL INTERACTION: COLONIAL EXHIBITIONS AT PASAR GAMBIR OF BATAVIA AND JAARMARKT OF SURABAYA 232
Yulia Nurliani Lukito, Fildza Miranda

PREFACE

Praise to God Almighty, who has endowed us with knowledge and ability to complete this publication.

The 2nd International Conference on Sustainability in Architectural Design and Urbanism (ICSADU) is a forum that provides a space for academics, practitioners, students, bureaucrats to convey their thoughts, ideas, views related to sustainable development issues in the field of architecture and urbanism. Furthermore, the forum provides a place to publish papers or scientific articles into acceptable and reputable journals. The ultimate aim of this forum is to create community and networks in the field of sustainable urban-space and architecture. Following the success of ICSADU 2017, on this occasion, the highlight theme is "empowering nature and culture toward a better quality of life." Participants came from Indonesia, Japan, Korea, Thailand, Malaysia, Germany & Lithuania. The committee accepted one hundred twenty five abstracts, including five from keynote speakers and sixty five selected papers that were presented in the conference on August 29, 2018.

Being a world issue for more than three decades, 'sustainable development' has evolved as a concept and movement receiving much support as well as mockery. Various obstacles were tackled with the lack of synchronization of economic, socio-cultural and ecological development. The concept of sustainable development has been applied by traditional societies, such as traditional villages of Bali and Toraja in Indonesia. On the other hand, the concept of garden city, the smart building, intelligent urbanism concept are several examples of the application in modern society. The field of architectural design has been applied from design-preparation of development, selection of material, technology of development, occupancy and post-occupancy stage. Those indicates that the works made are quite massive in the area of architecture and urbanism. Furthermore, it requires hard efforts to empower existing potentials in order to achieve a better life continuously, particularly in developing countries that face several challenges of globalization, demographic, urbanization, land scarcity, environmental quality degradation, global warming threats, natural and man-made disasters.

Socio-cultural harmonies, survival also maintenance of cultural-historical heritage are urgent parts of strengthening identity and fostering social harmony. Many architectural values are built on considering socio-cultural harmonies, which become important capitals of human being development in the future. Nevertheless, they are often threatened by the pressures of modernization and capitalization. Some are discussed in this book based on cases in Bali, Yogjakarta, Malang, Semarang, Surabaya, Jakarta, Seoul, Bangkok, Moskow, Lisabon and many more cities.

This book also discusses several issues, such as: smart building energy consumption, smart cities as parts of the balance with environment, an adaptation to the climate by planning and design, environmentally friendly materials, the use of IT to measure the city compactness, as well as a tool of observation methods, smart transportation, smart buildings and smart cities will facilitate various aspects of human life. However, maintaining harmony in the natural environment is more important than any material benefit, because this will reduce fund in the life cycle project, and also avoid the social disintegration.

On behalf of committee, I would like to thank those who have actively contributed to the preparation stage, the process of managing the manuscript, the presentation and discussion sessions of the conference, the preparation and publication of papers into the aforementioned journal.

1. All the Management Diponegoro University, Rector, Vice Rector, LPPM, WCU team, Dean of Faculty of Engineering, Chairman of Department of Architecture.
2. The reviewers team, who consisted of experts from various universities, including Ritsumeikan University Japan, The Ryukyus University Japan, Seoul National University Korea, Silpakorn University Thailand, King Mongkut University Thailand, UNDIP, ITS, UNS.

IOP Conf. Series: Earth and Environmental Science **213** (2018) 011001 doi:10.1088/1755-1315/213/1/011001

3. All Participants, who came from Japan, South Korea, Malaysia, Thailand, Germany, Lithuania, Indonesia especially from ITB, UI, UNS, UGM, ITS, UNHAS, UNIBRA, UPI, UNIKA, and many more.
4. Team of IOP Conference Series Earth of Environmental Sciences.
5. All of parties, readers and, that we cannot mention one by one.

It is my fondest hope that this publication can contribute to the development of knowledge particularly in the field of architecture as well as urbanism.

Semarang, November 2018

Chair Person

Dr. Atiek Suprapti

ICSADU 2018

Organizing Committee

Dr. Ir. Atiek Suprapti, MT

Satriya Wahyu Firmandhani, ST, MT

Masyiana Arifah Alfia Riza, ST, MT

Mustika Kusumaning Wardhani, ST

Edward Endrianto Pandelaki, ST, MT, PhD

Dr. Ir. R Siti Rukayah, MT

Dr. Eng. Bangun I R H, ST, MT

Chely Novia Bramiana, ST, BBE, M.Sc

Ratih Widiastuti, ST, MT

Ir. Mohammad Ischak, MT

Augi Sekatia, ST.,MT

Kania Kinasih, ST

Rini Lestari Widiastuti, S.Kom

Sri Wiyono

Marwan

Scientific Committee

Prof. Alamantas Samalavičius (Vilnius Gediminas Technical University, Lithuania)

Assc. Prof. Dr. Saehoon-Kim (Seoul National University, South Korea)

Prof. Dr. Jun-Ichiro Giorgos Tsutsumi (University of the Ryukyus, Japan)

Prof. Dr. Happy Ratna Santosa (Institut Teknologi Sepuluh November)

Prof. Dr. Gunawan Tjahjono (Universitas Indonesia)

Prof. Dr. Widjaja Martokusumo (Institut Teknlogi Bandung)

Prof. Dr. Bambang Setioko (Universitas Diponegoro)

Prof. Dr. Nany Yuliastuti (Universitas Diponegoro)

Prof. Dr. Edi Purwanto (Universitas Diponegoro)

Dr. Agung Budi Sardjono (Universitas Diponegoro)

Dr. Atiek Suprapti (Universitas Diponegoro)

IOP Conf. Series: Earth and Environmental Science **213** (2018) 011002 doi:10.1088/1755-1315/213/1/011002

This Conference is sponsored by:

Department of Architecture

Faculty of Engineering

Universitas Diponegoro

Lembaga Penelitian dan Pengabdian Masyarakat

Universitas Diponegoro Research Centre

IOP Conf. Series: Earth and Environmental Science **213** (2018) 011003 doi:10.1088/1755-1315/213/1/011003

ICSADU 2018 Photographs

2nd ICSADU
International Conference on Sustainability in Architectural Design and Urbanism
Semarang, August 29th, 2018
Empowering Culture and Nature toward a Better
Keynote Spe
Prof. Dr. Ir. E
Prof. Dr. Jur
Prof. Almant
Prof. Dr. Kin
Dr. Acharaw

Semarang, August 29
Empowering Culture and Nature toward a Better

IOP Conf. Series: Earth and Environmental Science **213** (2018) 011003 doi:10.1088/1755-1315/213/1/011003

IOP Conf. Series: Earth and Environmental Science **213** (2018) 011003 doi:10.1088/1755-1315/213/1/011003

IOP Conf. Series: Earth and Environmental Science **213** (2018) 011003 doi:10.1088/1755-1315/213/1/011003

IOP Conf. Series: Earth and Environmental Science **213** (2018) 011003 doi:10.1088/1755-1315/213/1/011003

IOP Conf. Series: Earth and Environmental Science **213** (2018) 011003 doi:10.1088/1755-1315/213/1/011003

Peer review statement

All papers published in this volume of *IOP Conference Series: Earth and Environmental Science* have been peer reviewed through processes administered by the proceedings Editors. Reviews were conducted by expert referees to the professional and scientific standards expected of a proceedings journal published by IOP Publishing.

Beyond sustainability: reconsidering the healing qualities of the built environment

Almantas Samalavičius
Vilnius Gediminas Technical University, Vilnius, Lithuania.

Corresponding author e-mail: almantsam@yahoo.com

Abstract. Sustainability, resilience and other concepts have recently emerged as a response to the most urgent problems of our time including the ones of the built environment. Throughout the last century little attention if any at all was paid to the adaptability of the built environment to nature and the needs of human beings. The ideology that shaped Modernism was hostile to the psychological qualities of the built environment and instead cared about shaping new aesthetics and advanced technology in building industry. The ascent of Modernism during the last century among other things was associated with dissemination of certain building doctrines and dogmas that neglected environmental issues and played their role in discrediting traditional ways of building architectural structures as well as traditional urbanism. As a result the built environment lost its adaptability to natural environment as well as its human qualities. Thus a new approach to building and rethinking of the legacy of Modernism as well as current practices is urgently needed. Though the rise of environmental consciousness gave an impetus to reconsider present quality of built environment, more steps in this direction are required. Reconsideration of architectural and urban design practices could and should be triggered by the new findings in psychology, neuroscience as well as re-reading of pioneering work in architectural and urban theory done by Christopher Alexander and his co-workers and more recently revisited by Nikos Salingaros. Reasoning that the development of technology in itself can solve the problems of the built environment is deeply flawed and thus should be transcended.

Keywords: urbanism, sustainability, resilience, healing environment, design patterns.

1. Introduction

There is more and more evidence that the present course of human activities on earth is hardly sustainable, thus efforts have been and continue to be made to make it more liveable, moreover so that knowledge about the scope as well as reasons of contemporary environmental crisis has essentially expanded during recent decades. The need for a new vocabulary and new practices have been expressed by the introduction and wide dissemination of such important notions and concepts as sustainability, resilience and the like. The built environment is directly connected to these concerns about the future quality of human life in various parts of the globe especially as problems of climate change have become a pressing global issue. Renowned American theologian, philosopher and environmentalist John B. Cobb whom has contributed significantly to the concept of sustainability in the early stage of its conceptualization has timely and persuasively insisted that "The finitude of our planet requires us to work toward a human society that accepts limits and seeks a decent life for all within the. Such a society should live in a balance with other species and primarily on the renewable resources of the planet. It should use non-renewable resources only at a rate that is agreed upon in the light of technological progress in safe substitution of more plentiful resources." [1] As energy and economy analyst Richard Heinberg has suggested while discussing the problems of economic growth that is adopted as present global paradigm, "most of the basic elements in our current way of life will have to adapt or become unsupportable." [2] Meanwhile forecasting the future of humanity in the mid-21st century Jorgen Randers – one of the authors of original *The Limits to Growth* study commissioned by the Club of Rome (1972) has recently argued in his new book that by 2052, "the new paradigm – 'sustainable being based on renewable energy' – will be exerting increasing influence on policy making. Not only because of the ominous threat from the approaching climate disaster in the second half of the twenty-first century, but also because the energy sector by then will have completed one-half of the transfer from a fossil to a solar base" [3] The common awareness about the situation of the earth in this respect (environmental damage, climate change, greenhouse effects, etc.) has largely expanded, meanwhile the notions and strategies of sustainability and resilience has been incorporated into both architectural and urban theory as well as practice [4] Studies in the analysis of the built environment and environmental quality have been abundant and there is hardly any need to name any of them.

However, one must admit that despite new (and promising) notions, concepts and encouraging slogans we are still dealing with the burden of the last century and modernist ideology that has largely shaped the urban tissue of contemporary cities and their architecture. Despite of the advent of post-modernity that questioned the legacy of Modernism, its essential assumptions are still largely dominating the practices of architectural and urban design and are rarely subjected to rigorous questioning. More often than not calls for sustainability and resilience mask the unwillingness to essentially revise the legacy of Modernism in the sphere of architectural and urban design, even some efforts are made to introduce sustainable practices, the building industry largely follows the route taken during the ascent of Modernism during the last century. That is why the revision of Modernist architectural and urban design is necessary and consideration of alternative approaches is not just meaningful activity but is of utmost necessity while dealing with complex issues of the built environment.

The aim of this paper is to discuss some of the reasons why discourse of sustainability and resilience often misses its target despite its ambitious and far-reaching goals. It suggests and outlines the need to go beyond conceptualizations of Modernism and its practices while trying to find appropriate patterns for design our living environment while taking into account adaptability of architectural and urban design and psychological comfort provided or not provided by the built structures. At the same time attempts are made to revisit and reconsider some of the theoretical concepts of architecture and urban design that could be especially helpful in rethinking the forms of architectural and urban design largely conditioned by the theory and practice of Modernism and the general course of the twentieth century's design. Focus on the pioneering work by Christopher Alexander and its recent revisiting made by Nikos Salingaros is expected to provide a new perspective while discussing the healing character of the built environment that is often missing in contemporary design practices.

2. Methods

The article is of theoretical and inevitably interpretative origin, thus it employs the qualitative methods peculiar to the theoretical critical analysis, i. e. those of critical analysis of sources, historical analysis of discourse in this particular field as well as critical interpretation of written sources.

IOP Conf. Series: Earth and Environmental Science **213** (2018) 012001 doi:10.1088/1755-1315/213/1/012001

3. Discussion

3.1. The Mental Legacy of Modernism and Its Ambiguities

Discussing the legacy of Modernism is still a somewhat taboo subject in Western architectural and urban scholarship that continues to be deeply committed to its key assumptions despite of the obvious fact that the architecture of Modernism was not based on ideas that are currently needed to establish a framework urgently needed for sustainable and resilient building practices. As architect and urban designer Leon Krier has recently argued, "Modernist architecture is an architectural Esperanto dependent on the synthetic building materials and climate control machinery. The problem is not so much its nature and existence but its pretension to be the only legitimate language of modernity, declaring traditional architecture and techniques to be "historical" and thus "outdated" and "anachronistic." [5] Architect and architectural critic Kenneth Frampton who obviously has a different opinion about architectural Modernism than more "Orthodox" thinker like Krier, nevertheless has argued that whatever one might think of its roots, "modernizing barbarism" as a part of global contemporary building production is "appalling". [6] Some other authors, like Michael W. Mehaffy and Nikos Salingaros have recently demonstrated that the very term "modern" in itself crates a phenomenon that they call "a false dichotomy of values" [7] According to both researchers, this category is largely responsible for giving up the accumulated principles throughout human history and instead offering a pursuit of a "false utopian promise of progress" [8] Furthermore, the reign of Modernist aesthetics and general view is associated with the making of an industrialized world that imposed its own models on urban (as well as architectural) structures. [9] The general failure of modernist architectural practice has been examined in detail by Malcolm Miles, who has provided a well-researched documentation on how as large number of the world's iconic buildings failed to meet requirements both in terms of function and aesthetics. [10] One of numerous notorious examples of famous, iconic and often paradigmatic modernist architectural buildings that were destined to fail despite of ambitions of their designers is the Unite d'Habitation in Marseilles, France designed by Le Corbusier and uncritically presented as exemplary piece of modern architecture in a large number of overviews and anthologies of the last century. [11] Despite of growing dissatisfaction with modernist legacy in architectural design and urban planning expressed by the people who are their users and reputable academic researchers, the myths of novelty continue to haunt imagination of most contemporary architects. The damage done by modernist urban planning and urban design have been discussed by a variety of authors, including architectural historian Vincent Scully who was extremely critical of its attitude to the old, historical cities [12] as well as such perceptive urban historians like Spiro Kostof [13] and more lately Peter Hall. [14] At the same time, architectural Modernism remained hostile to traditional building techniques and aesthetics as it was focused on the ideas of progress and technological future. Because of modernist design practices traditional ways of building have been either eliminated or marginalized in most parts of the world and homogenization of the built environment was offered both a s a global remedy and as a visible image of modernizing ideology. Thought architectural design in many ways recently started to move beyond simplistic Modernism in many aspects, its essential assumptions, especially contempt and ignorance of the past still persist in a architectural as well as urban design as well as in criticism and architectural theory. There is an urgent need of transcending the modernist ideology and practices in architectural and urban design in order to take a course toward more mentally and psychologically rewarding built environment than that one that was constructed following short-sighted visions of Modernism and its expropriation by the global building industry.

3.2. The Need for a Healing Environment and its Foundations

Opinions about the flaws of the built environment constructed under the imp licit or explicit influence of Modernist architectural ideology are abundant. Most users often find out that the buildings that were supposed to be functional and beneficial to users embody the qualities that contradict their initial goals. Not much happens even if occasionally dwellers or users of particular modernist buildings are interviewed about their attitudes: the building industry continues to follow the patterns that have been compromised long ago. It is often (and rightly) argued that the built environment in itself becomes a source of negative emotions and feeling and provides a setting for these to proliferate and take over. For example, Arnold Berleant - one of the pioneers who developed still largely novel the field of urban aesthetics maintains that "Architectural accomplishment is not universal, to be sure, and uninspired building continues to oppress most of the increasing population of an increasingly urbanized world.' [15] More lately another leading philosopher of architecture, Alberto Perez-Gomez concluded that "An environment that becomes increasingly devoid of qualities, reduced to a set of coordinates in a global positioning device, for instance, tends to exacerbate our contemporary psychopathologies – our sense of despair in view of the "meaninglessness of existence", contributing to debilitating nihilism. In other words, while the qualities of space remain directly assessable through our senses and our emotional consciousness, the technological world tends to deny the cognitive value of such perception and in fact reinforces the opposite assumption…" [16] Even giving up a complicated discussion about what really makes the legacy of architectural and urban Modernism and which trends in contemporary architecture might be viewed as its offspring and which as distortions imposed by international building industry, it is possible to speak about architectural structures and urban tissue in terms of healing environment and adaptive design that requires a paradigm shift as suggested by mathematician and urban and architectural theorist Nikos Salingaros. In this sense it is possible to speak about the present paradigm of industrial design based on a 'copying' rather than 'generating' form. [17] In the new framework of thinking that involves reconsideration of industrial design pattern that dominated throughout the last century, it is important to distinguish a more sensitive design meant for organism as opposed to that one that was mechanical and machine-like [18] it is well known that rhetoric of Le Corbusier was based on interpreting urban elements, like houses, streets, etc, as 'machines' for moving or living. This paradigm od architectural design is no longer feasible and thus should be transcended. This task can be accomplished by using the insights from contemporary science, taking into consideration such different phenomena: complex adaptive systems, the network city, geometry of resilience, fractals, self-organization, evidence-based and biophilic design, computation and evolution of patterns, etc. [19] The new approach toward design is needed as it is openly admitted that even if there had been progress in technology's influence over architectural work, the current principles of sustainable design 'remain far from holistic' [20] Moreover, already several decades ago some architectural theorists pursuing the course of phenomenology research came up with a questions about why such a large part of modern built environment has no appeal to human emotions, like for e.g. Juhani Pallasmaa in his essay "Geometry of Feeling', published as early as 1985 [21] or Joseph Rykwert. [22] Though introduction of phenomenology into architectural studies started with the writings of Norwegian architectural theorist Christian Norberg-Schulz [23] and phenomenological approach analyzing the role of human senses in interactions with natural and built environment was further developed by Pallasmaa and others, more lately some other options of understanding sensitive relations between human being and their environments were supplied by the findings of evolutional psychology and their application into the field of architectural studies. More lately there had been a lot of interest in appropriating the findings of psychology in architectural and urban studies. A lot of evidence has been provided on how people react to their immediate environment, on how important details and diversity are for visual perception making the students of architecture to doubt a number of modernist myths [24] Thus psychological qualities of the built environment prove to be very important and in many ways they determine whether people feel at home in interior and exterior spaces or experience feeling of estrangement, fair, desperation and even leading to the loss of meaning and eventual depression. Thus knowledge accumulated by a number of scientific disciplines, including evolutional psychology provides a firm basis to suggest that is no longer possible to rely solely on the modernist aesthetics, ignoring the implications of human psychological interaction with their living environment that architectural modernism has largely ignored and continues to do so. The findings of disciplines concerned with space and human senses have guided researchers into distinguishing between light, somatic, thermal and acoustic spaces and discussing the benefits of multisensory design. [25] The development of biophilic design and its theoretical foundations also reflects the common tendency to transcend the legacy of Modernism that is taking root in various parts of the world.

3.3. Beyond Industrial Design: New Directions and the Virtue of Looking Backward

Architectural modernism has always associated itself with the vanguard in the development of human culture, promoted aesthetics novelty and yet, often failed to grasp the implications of knowledge provided by other disciplines. Architectural modernism itself was rooted in denial of tradition and continuity: it often claimed that it is necessary to break the shackles of the past in order to create a new brave living environment. However, after more than a hundred years have passed since this aesthetic ideology came into being, it is possible to conclude that most of its promises were not delivered. Instead it borrowed itself to industrial mode of building that proliferated all over the globe and largely homogenized the contemporary built environment. To cut a long story short, architectural modernism seems to have created more problems all over the world homogenizing the built environment and imposing simplified and often simplistic forms than it was able to offer aesthetics and/or technical solutions of complex and complicated problems of human habitat. Besides, instead of understanding the mental function of architecture as ways "to enrich, articulate and strengthen the relations with the world, and ultimately our awareness of ourselves' [26] it often chose an opposite direction – to create an environment that blocked the relations with the world, disregarded the human senses, by-passed or deliberately ignored architectural and cultural traditions and silenced communities. As was already mentioned, some trends in architectural thinking attempted to overcome this inbred ambiguity of architectural Modernism and development of phenomenological view might be seen as one of the responses to its dogmatism and narrowness of vision. There were, however, other options as well. Some forty years ago mathematician, architect and architectural theorist Christopher Alexander developed his concepts of "pattern language" [27] and "timeless way of building" [28] offering a way to get over the tendencies of architectural and urban design of the last century. In *Timeless Way of Building* Christopher Alexander has made an interesting remark on why patterns are needed. According to Alexander "in period when languages are no longer widely shared, when people have been robbed of their intuitions by specialists, when they no longer even know the simplest patterns that were once implicit in their habits, it becomes necessary to make patterns explicit precisely and scientifically, so that they can be shared in a new way – explicitly, instead of implicitly – and discussed in public." [29] Alexander maintains that timelessness is important element that is capable of describing the environment that was created by indigenous vernacular builders in various localities of the globe; besides he believes traditional builders were able to grasp the meaning of nature and while constructing their dwelling followed the course parallel to nature's creation. Eventually such reasoning provided the architectural theorist to claim a certain way of building - a timeless one. His concept of timelessness of building has been occasionally questioned and even opposed, especially in terms of overemphasized sustainable character indigenous buildings allegedly possessed For e.g. Nezar Alsayyad and Gabriel Arboleda have argued that this kind of view neglecting particular historical and regional varieties and peculiarities might be interpreted as mythological one as "in many cases, he buildings and settlement practices that may have been sustainable in the past are no longer so because the natural and built environment in most of the world's traditional communities have been profoundly affected by diverse factors. [30] However, the discussion whether Alexander's view really belongs to the realm of mythology (if any at all) does not seem grounded enough as nowhere in *Timeless Way of Building* or *Pattern Language* Alexander suggests that all and every indigenous building practice has been sustainable or resilient. Instead, drawing on numerous building traditions and historical periods he offered a possible way to develop a humane approach toward building environment. Moreover, his concept of "patterns" goes beyond indigenous architectural traditions and might be considered as being associated with human activity in the sphere of building in different historical periods under different circumstances. Accordingly the "pattern language" should be treated not as a theoretical hypothesis but as instrumental and theoretically valid building guidelines that have ben substantially researcher by Alexander and his associates. Moreover, it is difficult or almost impossible to ignore Alexander's insight about the universal character of the building patterns, since their existence can be well attested not only by theoretical speculations, but by both collective and individual experience. Among many other things he rightly argued that any single architectural structure should not be taken as an isolated element as it exists in the contexts of other buildings – a warning that modern architectural practice ignored and continues to ignore. The reason of this ignorance is most likely, the persistence of the ideology of modernism that stubbornly refuses to acknowledge that other paths are possible to create a psychologically comfortable, healing and rewarding contemporary built environment other than Modernism has envisioned. While critically re-examining the modernist paradigm of architectural and urban design, it might be highly beneficial to explore the possibilities set forward by Alexander's pattern language abandoning biases and prejudices associated with alternatives *to* architectural Modernism rather than *inside* Modernism.

While addressing the members of the Royal Institute of British Architects back in 1984 philosopher and historian Ivan Illich drew a striking comparison between the art of vernacular dwelling and the practice of modern housing, suggesting that unlike inherent human ability to dwell peculiar to different cultures "housing provides cubicles in which residents are housed. Such housing is planned, built and equipped for them. To be allowed to dwell minimally in one's own housing constitutes a special privilege: only the rich may move a door or drive a nail into a wall. Thus the vernacular space of dwelling is replaced by the homogeneous space of the garage. Settlements look the same from Taiwan to Ohio and from Lima to Peking. Everywhere you find the same garage for the human – shelves to store the work-force overnight, handy for the means of transportation. Inhabitants dwelling in spaces they fashioned have been replaced by residents sheltered in buildings produced for them, duly registered as consumers of housing protected by the Tenants' or the Credit receivers' Act." 31 This remark seems to be still valid, despite the proliferation of new building technologies and innovations, as housing continues to be understood in terms of homogeneous garage though often wrapped up in ecological discourse. Though taking a different approach (less anthropological, yet more practical) toward dwelling, Christopher Alexander's studies offer a way of understanding the disappearing vernacular values and using them meaningfully in contemporary architectural practice. Thus his pioneering work in studying the patterns of living deserves more attention than it had received so far.

4. Conclusion

The legacy of Modernism in architectural and urban design has proved to be burdensome as this aesthetic ideology has institutionalized a systematic assault on traditions and traditional modes of architectural practice and urban design. The intellectual agenda se by Le Corbusier and hordes of his associates and disciples and eventually adopted by the building industry in the west and east alike have compromised, damaged or even erased traditional building practices labelling them as anti-Modern and thus having no adequate place in the sphere of architectural and urban design. However, the promises of architectural Modernism have failed to face the challenges of contemporary environmental crisis, climate change and its avatars, neither architectural Modernism offered any meaningful alternatives. Currently emerging design projects oriented toward sustainability and resilience often fail to meet their goals despite their high expectations partially because Modernist concepts continue privilege novelty for the sake of novelty and promote only certain types of aesthetics that fits the narrow scale of expression drafted by its ideology and espouses contempt for the practices and concepts of the past. There have been attempts to overcome the phase of liminality and the legacy of architectural Modernism by revisiting "timeless" building traditions and working out design patterns that can potentially be applied universally in order to create more satisfactory built environment than one that was constructed throughout the last century not only in terms of sustainability and resilience but also as psychologically comfortable, inspiring and assuring environment strengthening the humanness of human beings and their psychic balance with surroundings. The contribution of Christopher Alexander and his collaborators is one of the possible and plausible ways of creating this kind of the built environment. Further inquiry into design patterns developed by Alexander and his team a few decades ago as well as critical revision and reconsideration of his concept of 'timeless way of building' provides an ample space for further research in this direction that might help to create more positive qualities of the built environment going beyond the narrow and restricted framework of architectural Modernism and opening new vistas to human dwelling on the planet plagued by the lack of balance between natural and built environment and its debilitating effect on the human mind.

5. References

[1] Cobb J B 2007 (1992) Sustainability: Economics, Ecology and Justice Eugene, Oregon: Wipf and Stock, 7
[2] Heinberg R 2011 The End of Growth; Adapting to Our New Economic Reality, Gabriola Island: New Society Publishers, 113
[3] Randers J 2012 2052: A Global Forecast for the Next Forty Years. White River Junction, Vermont: Chelsey Green Publishing, 205
[4] Gauzin-Muller D 2002 Sustainable Architecture and Urbanism. Basel: Birkhauser, 12-31
[5] Krier L 2017 *"Modernity and Its Discontents"* in Samalavicius, A Ed. Rethinking Modernism and the Built Environment, Newcastle upon Tyne: Cambridge Scholars Publishing, 78.
[6] Frampton K 2017 "*The Impossibility of Building Cities*" in Samalavicius A. Ed. Rethinking Modernism and the Built Environment, Newcastle upon Tyne: Cambridge Scholars Publishing, 165
[7] Mehaffy MW and Salingaros N 2016 The End of the Modern World, htttp://www.orthodoxytoday.org/articles7/SalingarosModernism.php
[8] Ibid.
[9] Salingaros N 2017 Design Patterns and Living Architecture, Portland, Oregon: Sustasis Press
[10] Miles M 2009 Exploding the Myths of Modern Architecture. London: Francis Lincoln Limited
[11] Miles M 2015 A critical appraisal of the design, construction and influence of the Unite d'Habitation, Marseilles, France, *Journal of Architecture and Urbanism*, Vol 39, No 2, 103-115
[12] Scully V 1994 "Foreword" in Katz P Ed. The New Urbanism: Toward an Architecture of Community. New York: McGraw Hill
[13] Kostof S 2005 The City Assembled: The Elements of Urban Form throughout History, New York: Thames and Hudson
[14] Hall P 2002 Cities of Tomorrow, London: Blackwell
[15] Berleant, A 1992 The Aesthetics of Environment, Philadelphia; Temple University Press,149-150
[16] Gomez-Perez A 2016 Attunement: Architectural Meaning after the Crisis of Modern Science, Cambridge, Mass.: The MIT Press, 109
[17] Salingaros N 2017 Design Patterns and Living Architecture, Portland: Oregon: Sustasis Press,
[18] Ibid, 11.
[19] Salingaros N, Mehaffy MW 2015 Design for a Living Planet, Portland, Oregon: Sustasis Press
[20] Frampton K 2011 "*Urbanization and Its Discontents*" in Lee S. Ed. Aesthetics of Sustainable Architecture, Rotterdam: 010 Publishers, 99
[21] Pallasmaa, J 2005 Encounters: Architectural Essays, Helsinki; Rakennustieto Oy
[22] Rykwert J 1996 The Dancing Column, Cambridge, Mass.: The MIT Press
[23] Norberg-Schulz, C. 1979 Genius Loci: Towards a Phenomenology of Architecture, New York: Rizzoli
[24] Susman A and Hollander JB 2015 Cognitive Architecture, New York: Routledge
[25] Erwine B 2017 Creating Sensory Spaces: The Architecture of the Invisible, New York and London: Routledge

[26] Pallasmaa J 2007 "An Archipelago of Authenticity" in Caiico G. Ed. Architecture, Ethics and the Personhood of Place, Hanover and London: University Press of New England
[27] Alexander C et al. 1977 A Pattern Language. New York: Oxford University Press
[28] Alexander C 1979 Timeless Way of Building, New York: Oxford University Press
[29] Ibid, 246
[30] Alsayyad N and Arboleda G 2011 "The Sustainable indigenous Vernacular: Interrogating a Myth" in Lee S. Ed. Aesthetics of Sustainable Architecture, Rotterdam: 010 Publishers
[31] Illich, I 1992 In the Mirror of the Past, London: Maryon Boyars, 57-58

IOP Conf. Series: Earth and Environmental Science **213** (2018) 012002 doi:10.1088/1755-1315/213/1/012002

Playability for Cities

Saehoon Kim
Graduate School of Environmental Studies and Environmental Planning Institute, Seoul National University, Seoul, South Korea

Corresponding e-mail: skim5@snu.ac.kr

Abstract. Urban design and planning can contribute to the making of enjoyable civic environments. Urban places beloved by the citizens do not only exhibit visual comfort and convenient urban structure, but also provide a feeling of satisfaction and psychological rewards. In the article, urban places like the entertainment zones in the United States and playable places in South Korea are described. A sense of liking, desires for social activities, and immersive environments (mental involvement in urban environment) associated with play and participatory experience are discussed as part of the playable features embedded in cities.

Keywords: Playability, Pleasure, Entertainment, Urban design

1. Introduction

In the 1980s and 90s, Korean society underwent changes that were difficult to reverse in many respects. Outside the country, the era of the Cold War and ideological confrontation began to recede, at least slowly. Within the Korean peninsula, a number of social problems have emerged after the 1980s Olympics and the economic boom. As political democratization took place, it was more urgent to find the identity of the individuals—raising the question of "who am I" and "what do I like" —rather than to fabricate a set of communal values at a regional or a national level. At the same time, a new kind of play culture and a desire for pleasing experience spread into the cities. For instance, in the early 1990s, the 'karaoke room' popped up all over the country in a short period of time and was established as a place for entertainment and social gathering. During the period, various 'love hotels' and 'decadent cafes' mushroomed. At least for some people, they became a place for taboo behaviors and deviation from routine everyday lives.

Cities evolve in response to the emergence of social demands and marketable place-based services. Among them, play and pleasure constitute one of the strong driving forces transforming the urban environment. From a broader perspective, the "quality of life" in cities is not only associated with affordable services and quality living environments, but also is affected by the manner that urban citizens are capable of enjoying the city. As previously proposed in the hedonic treadmill theory, an individual has to keep walking on a treadmill to at least stay in the same place. In the same manner, a sense of happiness and pleasure in cities is accomplished only when the urban environments transform (or 'walk') according to the change in social expectation and individuals' desire. Urban citizens, including us and our nearby neighbors, continue to pursue happiness and pleasure in the city. Once they achieve it, wheels must move again to achieve a new kind of joy and excitement.

IOP Conf. Series: Earth and Environmental Science **213** (2018) 012002 doi:10.1088/1755-1315/213/1/012002

In some cities around the world, desire for deviation from the norms has shaped the urban identity. For instance, one can experience a burst of Latin music and liquor shops in the streets of Calle Obispo in Havana, Cuba, which dates back to the late 16th century. Entertainment shows and gambling characterize the experience in Las Vegas, which is often referred to as the capital of the world's entertainment and casino. Angeles in Philippines—sometimes called the Sin City—is filled with decadent night-time shops, places for prostitution and illegal drugs. As the wheels of pleasure roll, various places infiltrate the cities, sometimes legitimately and in other times illegally.

1.1. The Concept of Playable City

Pleasurable experience in cities goes beyond the ambiguous state hanging between the legal and the illegal lines. The concept of "playable city" began to capture the imagination of cities and activists around the world, offering a new insight of connecting people and culture. Kids' play in a neighborhood park, night-time walking or bicycling along a trail, or even sharing a co-working space with other start-ups constitute playable experiences that attract a large number of urban citizens to more cohesive and participatory environments.

The basic assumption is that some urban problems like loneliness, isolation, lack of liveliness, and social inactiveness should be addressed by communities' collective action. Here, "play" does not have to be a frivolous experiment or a surprise event as spurred with the temporary installation of a 300-foot water slide in Bristol. Instead, any immersive activities that motivate individual responses or group actions in an exploratory manner are part of the components of the playable city. Through this, urban citizens can stay away from everyday routines narrowly focusing on a functional means to an end and learn to relate to other people in a relaxed, cooperative, and sometimes creative atmosphere.

But transforming cities toward a more playable future is not always a straightforward vision. This is especially true when the efforts of transformation are planned through the cooperation of different social actors and local authorities. This is partly because social perception towards playability in cities vary substantially by people and is often undervalued. Thus, in some places, private entrepreneurs play a prominent role, ranging from renting a small room in a commercial area and converting it into an entertainment place for business purposes. The entertainment rooms scattered across Korean cities illustrate such an effort.

1.2. Entertainment Rooms in Korea

At least over the past 20 years, a number of small-scaled entertainment rooms sprout in Korean cities. The rooms provided diversified programs, such as singing, drinking, resting, studying, relaxing, and being entertained, among others. The origins of the rooms are different, but they offer a wide variety of drinks and beverages, songs and dances, rest and entertainment, and social encounters with strangers or different people. Some showed up in red-light districts, while others were established in a mixed-use residential neighborhood. Until recently, the rooms are continuing their vitality with customized services to the customer's tastes. Frequent renovation of the facility was undertaken to have a new theme or enhanced experience.

Some entertainment rooms caused a social controversy, and much of that controversy is still ongoing. For example, protests against the development permission of love hotels in the vicinity of a residential area and a school in Ilsan, Korea has started in the late 1990s. The subsequent denial of property tax payments by the residents continued until quite recently. Nevertheless, for at least some building uses, the regulation has been relaxed or sometimes converted into a legitimate use, as shown in karaoke. In the early 1990s, karaoke was first introduced in Busan, and then spread across the country with the emergence of related businesses by companies like ASSA, Taejin Media (now TJ Media), and Geum Young. As a result of the popularity of the karaoke, the government enacted the "Act on the Regulation of Customs" in 1992, and banned night-time operations and the access of minors to the karaoke. However, in the late 1990s, the regulation was stigmatized as a policy to unnecessarily restrict the freedom of enjoying entertainment. As the ban on business hours was abolished and the admission of youths was permitted at least until the midnight hours, the controversy over the harmfulness of karaoke did not last more than ten years.

Figure 1. Commercial street in Seoul filled with entertainment rooms

It is also noteworthy how deviant behaviors violating informal social norms are accepted in cities. For example, talking to an unfamiliar guest or asking for a service that is not on the menu is allowed in some places. In other places, abnormal behaviors like yelling, high-spirited shouting, violent actions, and drunkenness are accepted to be less unpleasant. French philosopher Michel Foucault labeled the "heterotopias of deviation" for a space in which individual behaviors are accepted at an appropriate level even though it falls outside the typical norm of a society. Foucault exemplified spaces like a prison and a psychiatric hospital, where those who committed deviations are accommodated. But if relatively less serious deviations or spontaneous pursuits of pleasure are included, small-scaled entertainment rooms in Korean cities are good examples of the heterotopias of deviation (Figure 1).

1.3. Entertainment Zones in the United States

Professor Daniel Campo at the Morgan State University conducted a study on the "Entertainment Zone" in the US, where heterogeneous behaviors of drinking, dancing, entertaining, and gambling take place at night. Although not well known to the people outside the country, the entertainment zones are located across many cities in the US, such as in the Alston district of Boston, Chicago's Rush Street, or in the corner of the Philadelphia Old Town. The zones often "occupy the margins of downtowns in former commercial and industrial areas, underutilized retain corridors or underdeveloped waterfronts" and were progressively formed without being master planned [1].

The district is formed with the agglomeration of pubs, clubs, restaurants, and motels in urban blocks that have lost the original function. The uses settle down in a run-down, mixed-use neighborhood to avoid high rents in the city center. Buildings inside the block are relatively old and small with a width of about 6-15 m. Commercial activities take place along the street but the whole block is often comprised of a number of different uses rather than being monopolized by a single business. According to Campo and Ryan (2008), this kind of entertainment district is a rare place for experiencing authentic urbanity, although it is difficult to achieve commercial successes unlike the casinos in Las Vegas or the Disney World in Orlando. In the US, a large portion of actual living or working space is located away from urban centers and often around the suburbs. Thus, spending time in an entertainment zone exposes urban citizens to immerse in the atmosphere of congestion, noise, intense interactions, and crowdedness, enabling temporary escape from the suburbia (Figure 2).

Figure 2. Entertainment zone along Water Street in Milwaukee, United States (Source: [1])

The implication is that while enjoying in a contemporary shopping mall like the Jamsil Lotte World or Yeouido IFC Mall in Seoul is one type of urban entertainment, a neighborhood with a mixture of small-sized commercial, manufacturing, shopping, and amusement spaces also serves as a vital urban experience. The entertainment zone is a messy but fresh site that individuals can find a place with great playability. In this way, neighborhoods in Seoul like Mullae-dong, Haebang-chon, Seo-chon, Seongsu-dong, or areas near the Cheonggyecheon exemplify a Korean entertainment zone, where places of amusement, entertainment, and consumption are mixed with residential, industrial and educational uses. The spaces are not organized according to one narrative or a theme. Rather, it is a mix of unpredictable elements, and this mixture can vary to some extent depending on the will of the operators and the consumers. It does not mean that a carefully designed place by an architect is less playable. But it is necessary to think about how to accommodate the elements of pleasure and playability in the cities.

2. Discussion

Outside the entertainment zones, virtually all of the urban spaces are potential candidates for playable urban activities. For instance, an outdoor space for active physical activities or a street suitable for biking can be part of the pleasant urban places. Although there may be differences depending on the individuals, the activities represent a state that a high level of liking and wanting induce both physically and mentally healthy activities.

I believe that a well-designed urban space can serve as a powerful trigger for enjoyment in relation to the motivations of 'liking' and 'wanting.' From the perspective of liking, let's look around the space that urban citizens visit and enjoy with others. Some places are frequently used for specific purposes even though a sense of liking is lacking. It is difficult to judge whether a sense of 'liking' prevails in some place by simply checking the frequency of uses or the pedestrian movement. This is because the frequency of use is not always proportional to the degree of entertainment (especially liking). Additionally, a space aimed only at stimulating sensorial pleasure or unexpected surprise is not sustainable from the perspective of entertainment. Some people will be excited by the first encounter of such a place, but the next experience will fail unless far more intense stimulus is prepared. An urban space loved by many people for a long time exhibits a signal of both 'liking' and 'wanting', which is an emotion that is instantly experienced (= liking) and a feeling of long-term satisfaction and reward (= wanting), respectively.

Figure 3. A playable urban environment of Gamcheon Maeul in Busan, South Korea

Unfortunately, it is not well known how people experience psychological rewards and satisfaction in an urban space. This is particularly prominent in a neighborhood experiencing decreases in adaptability and cohesive environment of the neighborhood [2]. Here, some hypothetical ideas are proposed as follows. Compared to a space where the whole area is planned and operated by a single person or an organization, a space where collective values and perspectives of many people permeate is better. A space that is open to change, adaptation, and having a room for layered accumulation is better. A space with ambiguous boundary between the nature and the artificial may provoke a sense of discovery and emotional attachment. A space that allows people to experience a different person or a new cultural value is better (Figure 3).

In contemporary cities, visionary ideas about the walkability, the bikeability, and the age-friendliness of cities are proposed. Many meaningful attempts have been made under the name of the walkable city or bikeable design. But the approaches often focused on the quality of convenience and functional usability. For instance, the walkable city is defined in studies like Southworth (2005) as having an environment that "invites walking by means of a richly connected path network" and "provides access to the everyday places people want to go…safe and comfortable with streets that are easy to across [3]." But we need to look back at the essential meaning of pleasure in cities. Professor Bohyeon Lee from Ewha Womans University, for example, pointed out that the term "walkability" is a compound word of "walk" + "usability."

Based on the definition, attempts of making the walkable and bike-friendly city are overloaded with the notion of improving usability and efficiency of mobility infrastructure, such as building a safe pathway or a bicycle road, and installation of street furniture and storage facilities. Without doubt, they can make walking and bicycle riding safer and more convenient. However, beyond this usefulness, the notion of 'playability' or 'gamification' of walking/biking in cities needs to be better

IOP Conf. Series: Earth and Environmental Science **213** (2018) 012002 doi:10.1088/1755-1315/213/1/012002

highlighted. How to amplify a sense of playability in our city should be discussed along with physically expanding the walkways and providing street furniture.

For example, how do you feel the sense of accomplishment in cities by walking/riding a bicycle, how to share the emotion of activeness with others, how to perform an unusual mission with biking, and how to explore new places through a team play are examples that the notion of playability in cities can be accounted for. An example of traffic mimes in Bogota, Colombia sheds light on the playable method for managing the city (Figure 4). In the city, traffic fatality rates became extremely high and the roads were chaotic. About 420 mimes were hired to control traffic on the roads by mocking reckless drivers and performing the frustration experienced by pedestrians moving through the traffic.

Figure 4. Traffic mimes in Bogota, Colombia (Source: https://solutions.thischangeseverything.org)

3. Conclusion

The article discussed many aspects of pleasure and playable experience in contemporary cities. Overconfidence should be avoided as the feature of playability will not be a silver bullet that will cure all forms of urban problems and the inactiveness of cities. However, playable cities provide an insightful vision for revitalizing an urban area.

Moving on to a far broader topic of the quality of life and sustainability, the following idea is proposed. Previous discussions on urban sustainability have largely been limited to a few narrow areas, such as conserving energy and reducing carbon dioxide emissions. But I believe that much more debate is needed about the pleasures in the city - especially a sense of likes, desires, playfulness, and immersion (including mental involvement in urban environments) associated with play, experience, and commitment. A city that consumes less energy and less CO_2 comprises one of many important aspects of a sustainable city. But cities should provoke a sense of "liking" and "wanting" across a number of citizens to meet the diverse needs of urban society. In other words, a city that seeks to increase its eco-friendliness while seriously limiting a sense of playability is difficult to sustain. Matthias Sauerbruch, a German architect, proposed that cities that enforce environmental sustainability are not fun.

As narrow-minded environmentalism should be avoided, efforts should be made about how to reshape the urban spaces more pleasant and playable. Urban design professionals have to understand diverse aspects of playability in cities. Here, the term pleasure does not only mean a sensual appeal to some people. The spectrum of pleasure is very wide, such as a high level of playfulness, an accessible enjoyment to a multitude of people, spatial flexibility accommodating changes and collective memories over time, and an appropriate level of anonymity allowing for activities by different generations.

4. Acknowledgments

The keynote speech article was made based on the translation of the chapter of a book titled *Searching for good cities through cities* (Dosieseo Dosireul Chajda, Chapter 9) written by the author [4]. This work was supported by the National Research Foundation of Korea(NRF) grant funded by the Korea government(MSIP) (NRF-2017R1A1A1A05001205).

5. References

[1] Campo D and Ryan B 2008 The entertainment zone: Unplanned nightlife and the revitalization of the American downtown *Journal of Urban Design*. **13**(3) 291-315.

[2] Lee J, Won S and Kim S 2016. Describing changes in the built environment of shrinking cities: Case study of Incheon, South Korea *Journal of Urban Planning and Development*. **142**(2) 05015010.

[3] Southworth M 2005 Designing the walkable city *Journal of Urban Planning and Development*. **131**(4) 246-257.

[4] Kim S 2017 *Searching for good cities through cities (Dosieseo Dosireul Chajda)* Seoul: Hansup.

3D Interactive Visualization Scenario to Improve Public Participation in the Visual Comfort Assessment for the Pedestrian in Provincial Street, Malang, Indonesia

Herry Santosa[1], Jenny Ernawati[2], Lisa Dwi Wulandari[2]
[1] Art and Architecture Design Laboratory, Architecture Department, Faculty of Engineering, Brawijaya University, Jl. MT.Haryono 167 Malang-65145, Indonesia
[2] Human Settlement and Urban Design Laboratory, Architecture Department, Faculty of Engineering, Brawijaya University, Jl. MT.Haryono 167 Malang-65145, Indonesia

Corresponding e-mail: herrysantosa@ub.ac.id

Abstract. The visual comfort of pedestrians in the street corridor is essential to encourage a continuously satisfying experience of a visual aesthetic leading to a better quality of urban spaces, which is significant for making cityscape, especially in the streetscape. Valuable visual attributes of the pedestrian spaces are indicated by the presence of visual comfort for pedestrians as they pass through the pedestrian space. The creation of visual comfort in the pedestrian spaces also encourages positive spaces along the road corridor, generating an excellent space utilization, which eventually led to many positive appreciations of the space environment.Creating high-quality urban places triggers a more sustainable urban development by increasing pedestrian satisfaction in visual comfort through empowering public engagement in online decision-making system. Thus, public participatory is urgently needed to evaluate the spatial and visual comfort of provincial streetscapes. The disturbance of visual comfort for pedestrians often occurs in the main road corridors, especially the provincial streets in Malang, which has a strategic location for the expansion of commercial functions. This study evaluated a visual comfort of pedestrians in the provincial streets. It employed the 3D modeling method combined with the interactive simulation through the use of a multimedia authoring platform software to develop the 3D Interactive Visualization Scenario for online participation. The result of this study was the scenario of 3D visualization interactivity development to improve the effectiveness of the decision-making system in the assessment of pedestrian visual comfortalong the provincial street corridor.

Keywords: Public Engagement, Visual Comfort,Interactive Simulation,Online Decision-Making System, Provincial Street

1. Introduction

Urban spaces play a prominent role in accommodating citizen's activities around the city, which is very important to provide safe, comfortable, and proper urban spaces for the community. Pedestrian ways alongside urban streetscapes become crucial to form urban spaces. Consequently, the visual and spatial comfort of the pedestrian way should be well-maintained to create a better quality of urban spaces, which is significant for making cityscape, especially in the streetscape. Creating high-quality urban places triggers a more sustainable urban development by increasing pedestrian satisfaction in spatial and visual comfort and later on reducing vehicle use. The creation of good urban images will encourage people to understand and establish the valuable meaning of urban spaces [9].

Malang is one of the most significant cities in East Java Province that ever had the best urban planning among other cities in the colonial era in Indonesia. Currently, Malang hasrapid growth of trade and service facilities along with an increase in the influx of immigrants, especially students who want to study in Malang. Over time, this condition led to the development of the city that has become uncontrolled with the rapid growth of buildings, citizens, and vehicles. Therefore, this situation creates the lack of control of good governance. There are five provincial streetscapes are functioning as inter-city lane connecting Malang and other cities, which become the inter-city transport mobilization mode. Thus, these street corridors have become a strategic route causing the uncontrolled rapid growth of commercial area alongside the streets and lead to the occurrence of unsatisfying spatial and visual comfort for pedestrians.

The evaluation and control ofthe spatial and visual quality of pedestrian comfort need to perform the spatial and visual judgments process through public participatory in the decision-making process. The conventional method of public participation sorely less able to cope with the rapid development of urban space, so the community needs to gain appreciation in its involvement in urban spatial planning through easy-to-understand media with varying levels of community knowledge and skills [8]. Currently, the development of 3D computer simulation techniques has reached the level of visualization that can approach the real world. This condition encourages a change in public participation from a 2-dimensional paper-based drawing model to a 3-dimensional visualization computer-based model, which is capable of delivering high accuracy and complexity of data, present the speed of various 3D simulations, as well as the speed of design change [1,4]. Moreover, the dynamic presentations of the 3D simulation are capable of presenting the variety of future environment and can generate an active perception of space experience that is very similar to the real world [2].

In fact, the coupling method of delivering computer and Virtual Reality technology with a public assessment by the Semantic Differential method has been familiarly used to examine the quality of urban landscape comfort. Some researchers have proved that the use of a coupling method is very useful in supporting the visual and spatial streetscape quality assessment [6,8,14,19]. Moreover, the prototype development of the 3D Interactive Simulation System has been conducted in Kayutangan, a historical street corridor in Malang, through a series of continuous research stages. It involves public preference of visual quality in the historiccommercial streetscape, the construction of low polygon 3D modeling, the development of 3D interactive visualization, and the development of user interface design for 3D interactivity in the virtual environment [11,12].

1.1. 3D Visualization stages

3D visualization development for the provincial streetscapes requires at least six essential stages that consist of 3D simulation scenario, data collection, 3D modeling, geometry optimization, texture application, and the entity object [12]. 3D simulation scenario is the fundamental step in preparing the basic concept of creating a 3D visualization. There is four essential considerationfor constructing 3D visualization. First, the 3D simulation scenarioperforms estimating of the final form of 3D models provincial streetscapes, including building mass number of objects and the number of streetscape elements. Second, the concept of level of detail geometry is well into account the number of polygons in any 3D object creation models.Third, interactive concepts related to the number of entities on each 3D model object. Finally, the concept of navigation that determines the navigation control system for handling interactions on the VE. It related to the current estimate of the movement of the user views on space exploration simulation.

1.2. 3D Interactive Simulation

3D interactive simulation development as a tool in the decision-making process in urban visual landscape planning demands the importance of observation and navigation activities in the passivelevel to interactivelevel in the Virtual Environment [12,18]. A passive interaction of spaceassistedfor the evaluation activities in a predefined animation of 3D simulation modeling. However, users do not serve the freedom to control the direction of movement,andthere is a lack of interaction simulation of 3D objects [4].A passiveobservation does appear to be sufficient for creating a sense of place and maybe perfectly adequate for virtual tourism experiences. The method of an active interaction perform as well as an active exploration, and it should be initiated and established for built environment evaluation.On the other hand, an active interaction deliverythe ability to take control ofnavigation of a virtual model that might lead to a deeper and richer understandingof the modeled space, since people tend to experience the real world throughself-determined movement through it[2]. Some researchers have conducted a study related to the use of 3D interactive simulation combined with the development of user interface design for building consensus through public participation [3,5, 7, 15, 17].

1.3. 3D spatial multimedia

Recently, the development of 3D spatial multimedia as a decision-making support system in the planning process began to flourish and widely used in urban planning that associated with the public participation [16]. The use of spatial multimedia system intended as a medium of interaction between the planning concepts of the built environment with the community as a user. The development of 3D spatial multimedia systeminvolving the scripting language to construct a 3D interactive simulation in web interface design [12]. One of the 3D spatial multimedia application authoring platforms is Adobe Director software. Adobe Director Software has Lingo script as a scripting language for 3D behaviorinteractionwithin3D Virtual Environment.The utilization of behavioral scripts in Adobe Director Software can enrich the interactive techniques in 3D simulation and increase user participation in the utilization of the system as one tool in modeling decision-making. Moreover, the development of spatial multimedia systems inquires an understanding the strategies of good building consensus simultaneously.

2. Methods

This study is the second phase of the advancement of 3D interactive simulation construction for Landscape Visual Planning System in the streetscape in Malang, which is the continuation of the first stage research investigating people preferences through semantic differential method. This study aims to construct a 3D interactive simulation scenario of a Decision Support Systemfor visual quality evaluation in the commercial streetscape. This study method involved three main activities based on 3D street corridor modeling development, development of three interactive levels of 3D street corridor, and development of 3D spatial multimedia scenario based on the web. This study also emphasizes the elaboration of 3D behavior scriptto support adequate user interactivity for empowering public engagement in online decision-making system.

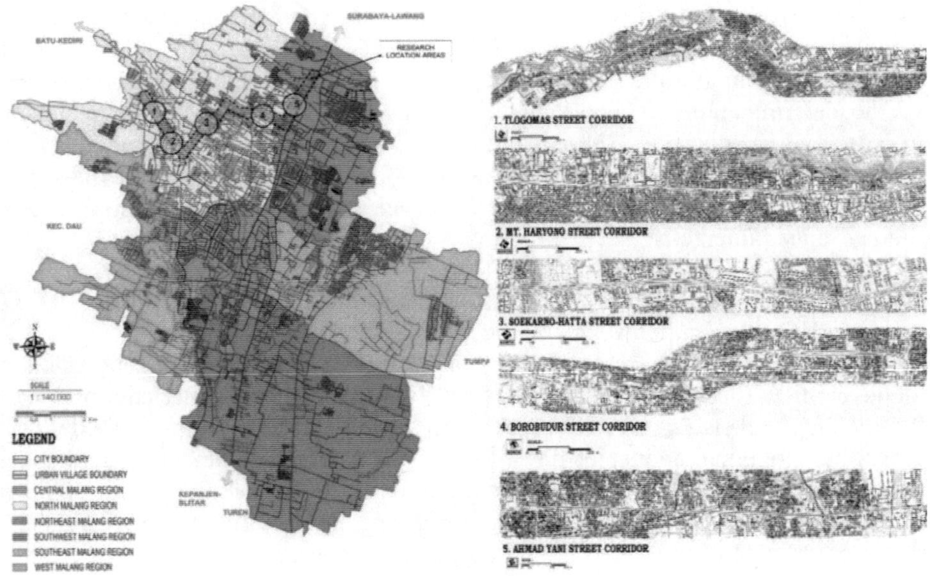

Figure 1. Map of the five provincial street corridors

2.1. Research location

The research of system developmentfocusedattention on five provincial street corridors in Malangthat crossed the north area of Malang, as presented in Figure 1. These streetscape locatedin the strategic areas which became the prominentaccess connecting the north with the western region of Malang,acting as promising commercial area and an activeroute for trading activities and services.

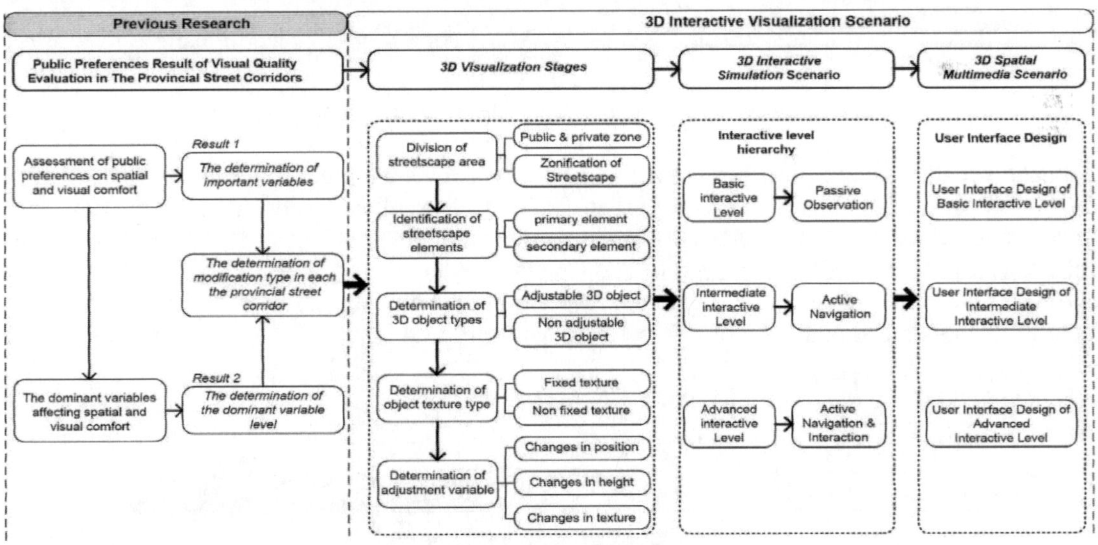

Figure 2. Research framework

2.2. Research framework

As illustratedin Figure 2, this research is in the subsequential phase of the preceding of multiple staged-projects which are addressed to construct a 3D interactive simulation decision support system in Provincial streets in Malang [13]. The previous research conductedthe developing a conceptual framework of Decision Support System (DSS) for the assessment of visual and spatial comfort of the urban streetscapebased on public preferences in several procedures. First, conducting a Semantic Differential method through public preferences in the provincial street corridor. Second,classifying the variables into the positive and negative zone. Third, separating the streetscape elements into primary, secondary and tertiary, based on the result of multiple linear regression analysis of the public preferences. Fourth, determining the modification type in the provincial street corridor according to the result of the determination of important variable and dominant variable level to analyze the modification criteria of each corridor.Subsequently, this study performedthree crucial stages covering 3D visualization stages, 3D interactive simulation scenario, and 3D spatial multimedia scenario.These primary stages produce three interactive hierarchy level covering basic interactive, intermediate interactive, and advanced interactive [12].

3. Discussion

3.1. Development of 3D visualization stages

A 3D modelingprocess is the most crucial stage in the 3D visualization concepts. This stage will determine the level of detailed geometry model,and the depth of virtual reality environments. Steps to making 3D modeling consist of four types, namely modeling topographic modeling, building modeling, landscape modeling, and street furniture modeling. Each 3D object model has a degree of difficulty and different challenges and requires considerable time to complete it correctly.Optimizing the geometry is undoubtedly needed to ease the rendering process in the active rendering mode. Nonetheless, the optimizing of geometry should consider the degree of urgency of the elementalform of the building details. Eventually, the 3D visualization concepts coveringfiveessential scenarios of 3D modelingconstruction todirect the development of the 3D simulation as shown in figure 3.

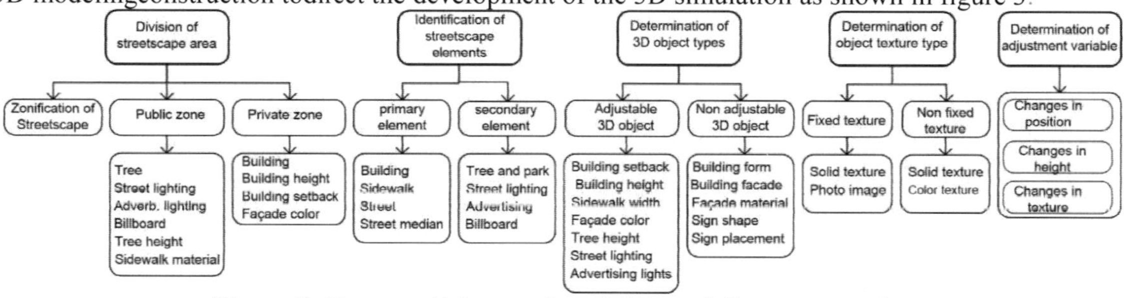

Figure 3. Fiveessential scenarios of 3D modeling construction

The process of 3D visualization developments requires a level of precision and high accuracy, to produce a scene that can represent the 3D urban features of excellent quality in the world of virtual reality.On another hand, it is essential to organize the number of object entity in each of the 3D objects. This setting is intended for the benefit of interactive levels in Adobe Director Software. If the level of user interaction developed through the interaction of each element in each building, then each element of the building should be broken down into some entities.

3.2. Development of 3D Interactive Simulation Scenario

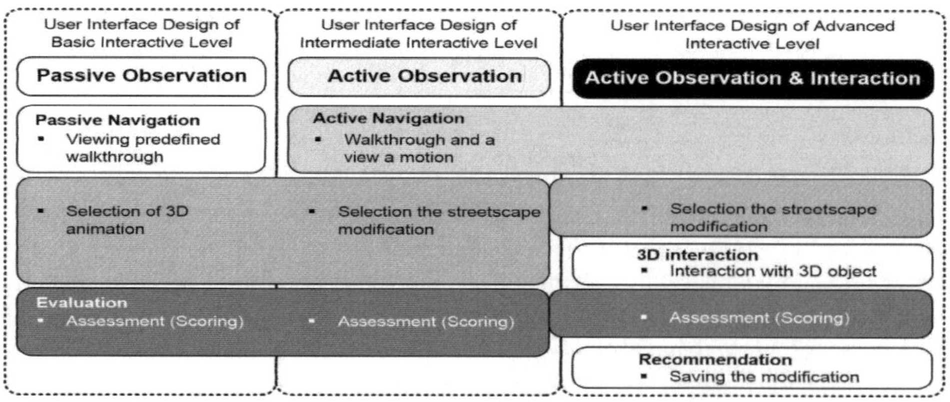

Figure 4. A scenario of 3D interactive simulation

IOP Conf. Series: Earth and Environmental Science **213** (2018) 012003 doi:10.1088/1755-1315/213/1/012003

The three types of 3D simulation were categorized by the capability of the simulation to interact with the user at three levels of interaction[12], as follows:

1. Basic interactive level
   The basic interactive level is the first type with less user-interactivity. The 3D simulation delivered a passive observation. It offers a walkthrough animation of variousstreetscape compositions, and users observed the model during the predefined animation. Users could give the assessment orscoring.
2. Intermediate interactive level
   The intermediate interactive level is the second typewith moderate user-interactivity. The 3D simulation performed the active-observation. It offers a VR of various streetscape compositions. Users perform a walkthrough and affect their motion in the virtual environment through navigation aids. Users could give the assessment or scoring.
3. Advanced interactive level
   The advanced interactive level is the third typewith high user-interactivity. The 3D simulation carried out the active-observation and 3D interactive. It offers a VR for various streetscape compositions. Users perform a walkthrough, affect their motion, and interact with 3D objects in the virtual environment through navigation aids, and some provided control panels. The user can save the results of modifications to a file for the benefit of decision-making.

3.3. Utilization of 3D behavior on interactive 3D simulation

Lingo is Director's scripting language which provides the capability of Adobe Director Software on the level of complexity, navigation, and a more in-depthinteractive. Lingo script language is also available in the form of the instant Lingo script, or it can be obtained from commercial providers Lingo scripts, such as Chrome Lib, as well as from an independent creator Lingo script.

Table 1. An example of 3D behavior script in advanced interactive level

Types of 3D Behavior	Public Zone	Private Zone
Navigation behavior	*The navigation behavior settings are set through the camera view using instant behavior*	
Visibility behaviour		
1. showing the model	on mouseUp me member("street0").model("Billboard2").visibility=#front end	on mouseUp me member("street0").model("Building1").visibility=#front end
2. hiding the model	on mouseUp me member("street0").model("Billboard2").visibility=#none end	on mouseUp me member("street0").model("Building1").visibility=#none end
Scaling behaviour		
1. elevating the model	on mouseUp me member("street0").model("tree").transform.scale(1,1,1.5) end	on mouseUp me member("world").model("Building1").transform.scale(1,1,1.5) end
2. shorten the model	on mouseUp me member("street0").model("tree").transform.scale(1,1,0.5) end	on mouseUp me member("world").model("Building1").transform.scale(1,1,0.5) end
Position behaviour		
1. widening the building setback		on mouseUp me member("world").model("Building1").worldPosition = vector(18, 100, 0) end
2. restoring to its original position		on mouseUp me member("world").model("Building1").worldPosition = vector(0, 0, 0) end
Texturing behaviour	on mouseUp me member("street0").shader("Shader12").textureList[1] = member("street0").texture("paving1") member("street0").shader("Shader12").textureTransform.scale = vector(0.5,0.5,1)	
Color behavior		
1. white color		on mouseUp me member("Building1").shader("Shader7").diffuse = rgb(255,255,255) end
2. gray color		on mouseUp me member("Building1").shader("Shader7").diffuse = rgb(96,96,96) end
3. black color		on mouseUp me member("Building1").shader("Shader7").diffuse = rgb(0,0,0) end
Reset model behavioor	on mouseUp me member("street0").resetWorld() go to frame 5 end	on mouseUp me member("world").resetworld() member("Building1").resetworld() go to frame 5 end

Save model behavior		on mouseUp me
		fileio = new xtra("fileio")
		myFile = fileio.displaysave("Save w3d As Building1" , ".w3d")
		if myFile <> void then
		member("world").savew3d(myFile)
		end if
		fileio = void
		end

3D behavior script is the most imperative scripting language in Adobe Director Software. The script is advantageously in the development of interactive 3D simulations. Interactive simulation allows user involvement in making a change or adjustment to the model or a 3D object.At the advanced interactive level, the 3D behavior scripting arrangement created two categories of interactive space zones, namely the public zone,and private zone (table 1). Public zone is part of the street corridor space attached to the public interest, including street, street furniture, street lighting, billboard, signboard, parks,and trees. Meanwhile, the private zone is part of the street corridor space attached to private interests, namely building the facade, building setback, building height, and commercial advertising. Furthermore, the types of 3D behavior scripting contained in the public zone consist of five types of scripting: navigation behavior, visibility behavior, scaling behavior, texturing behavior, and reset behavior models. While the 3D behavior scripting contained in the private zone includes seven types of scripting which are navigation behavior, visibility behavior, scaling behavior, position behavior, color behavior, reset behavior model, and save model behavior.

3.4. Spatial multimedia

Design scenario ofthe spatialmultimedia conformed to thethree levels of 3D interactive simulationscenario that consisted of less user-interactivity level up to high user-interactivity level.Accordingly, the concept of spatial multimedia generatedthree user interface design.The first type of user interface design offers three activities consisted of passive observation, selection of choices, and evaluation. A passiveobservationrelated to the viewing activity of predefined activity walkthrough animation. A selection of choices provided a panels images for selecting the type of 3D animation on a wide range of streetscape compositions. The second type offers three activities consisted of active-observation, selection of choices, and evaluation. Active observationconducted an exploration activity of the 3D simulation streetscape using the navigation control. Capturing the sceneconductedthe recording of a picture of the scene from each 3D simulation intended as a follow-up evaluation in the future discussion. On the other hand, the third type offers three activities consisted of active-observation, interaction with objects, and evaluation. 3D object-interactionconducted an object editing activity through the button panels and identified by the naming of a 3D object.Whilethe evaluation consisted of three activities, namely capturing images from the scene, saving 3D models, and scoring. Figure 5 displayed an example of a user interface design scenario at the advanced interactive level.

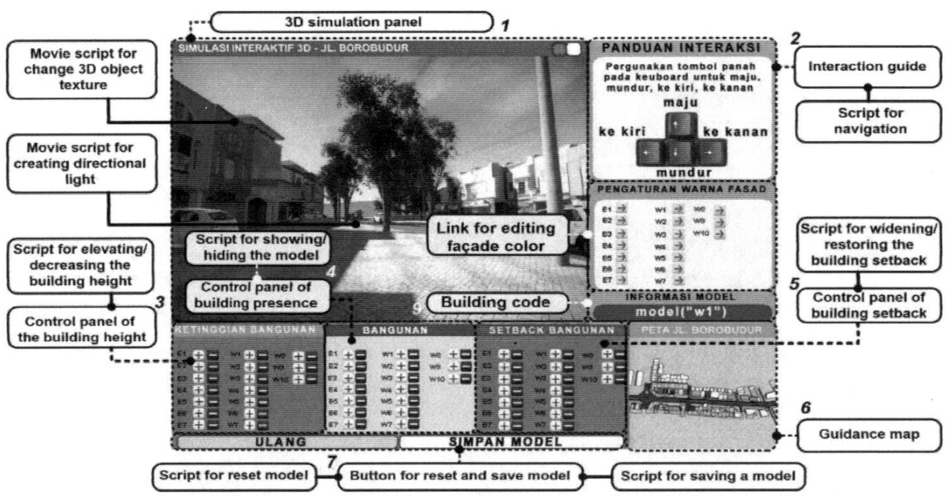

Figure 5. A scenario of user interface design at the advanced interactive level

4. Conclusion

The advancement of a 3D interactive simulation scenario of Decision Support System (DSS) for visual quality evaluation in the provincial street corridors cover 3D visualization stages, 3D interactive simulation scenario, and 3D spatial multimedia scenario. The creation of three interactive levels ranging from less user-interactivity, medium user-interactivity to high user-interactivity aimed to accommodate the diversity of knowledge and skills of the community in the system operation and the amount of time available to the community. This diversity of user-interactivity levels enhances active participation as well as online building consensus learning. Based on the experience in the pilot workshop of system implementation in the case of other street corridors [12], the obstacles and challenges to be aware of consist of four aspects: understanding of the system, operation of the system, quality of elements, and user interface composition design. First, a thorough and comprehensive understanding of the system requires optimal manual guidance. Second, people still feel unfamiliar and difficult in running the interactivity on the system. Third, the image used as navigational aids must have high quality. Fourth, the design of the user interface layout should be clear and easy to understand. Accordingly, the public requires sufficient adaptation to the comprehension of the system as a whole.

The outcome of this study indicates that the utilization and the development of 3D behavior scripts in Adobe Director Software can enrich the interactive techniques in 3D simulation and simultaneously capable of increasinguser participation in the utilization of the system as one tool in online decision-making.In the following stage, the final authoring of the 3D Interactive Visualization Scenario for online participation stand in needsgradual tests and dissemination to establish system reliability.

5. Acknowledgments
The authors acknowledged the support from the Ministry of Research and Technology and the Higher Education Republic of Indonesia. The support is under the research grant PTUPT UB of The year 2018 Contract Number 338.123/UN10.C10/PN/2018.

6. References
[1] Al-Kodmany K 2000 Visualization Tools and Methods for Participatory Planning and Design *Journal of Urban Technology***8** 2 pp 1-37
[2] Conniff A,T Craig, R Laing, CGalán-Díaz2010 A comparison of active-navigation and passive observation of desktop models of future built environments *Design Studies* Elsevier Ltd. **31** 5 pp 419-438
[3] Honjo T, E. Lim 2001 Visualization of Landscape by VRML System, *Journal of Landscape and Urban Planning***55**pp 175–183
[4] Howard T,Gaborit N 2007 Using Virtual Environment Technology to Improve Public Participation in Urban Planning Process. *Journal Urban Planning*Dev. **133** 4 pp233–241
[5] Kawakami M, Z Shen 2006 Study on Decision Support System for District Planning in Public Participation-A case study in Kanazawa City, Japan, Van Leeuwen JP and H J P Timmermans (eds.) *Progress in Design & Decision Support Systems in Architecture and Urban Planning, Eindhoven: Eindhoven University of Technology*, ISBN-10: 90-386-1756-9, ISBN-13: 978-90-386-1756-5 pp 203-218
[6] Koba N, T.Kishimoto 2009 Impression Evaluation of Streetscape by Super High-Rise Buildings Set-Back in High-Rise Part *Journal of Architecture and Planning AIJ* **74** 645 pp 2435-2442
[7] Koga M, SIkaruga, KTadamura, AOhgai, MMatsuo 2008 Study on Image Sharing Support Method in Landscape Planning Using Information Technology *Journal of Architecture and Planning AIJ* **73**633pp 2409–2416
[8] Koizumi K, T.Kishimoto, N.Koba 2009 Streetscape Analysis Focusing on Unity of the Unity of The Streetscape and Identity of Building Façade in Yokohama Motomachi Street *Journal of Architecture and Planning AIJ***74** 636 pp 393-400
[9] Laing R, J. Scott 2011 3D High-definition Scanning: Data Recording and Virtual Modeling of The Built Heritage *Journal of Building Appraisal***6**pp 201–211
[10] Nasar Jack L 1998 *The evaluativeimage of the city*, (London& New Delhi: Sage publications, Thousand Oaks)
[11] Santosa H, S Ikaruga, T Kobayashi 2013 Visual Evaluation of Urban Commercial Streetscape Through Building Owners Judgment: Case Study in Malang city, Indonesia *Journal of Architecture and Planning***78** 691 pp 1995-2005

[12] Santosa H, S Ikaruga, T Kobayashi 2014 Development of Landscape Support System Using Interactive 3D Visualization *Journal of Architecture and Planning***79** 706 pp 1995-2005

[13] Santosa H, Ernawati J, Wulandari L D 2018 Visual Quality Evaluation of Urban Commercial Streetscape for The Development of Landscape Visual Planning System in Provincial Street Corridors in Malang, Indonesia *Friendly City 4 'From Research to Implementation For Better Sustainability' IOP Conf. Series: Earth and Environmental Science* **126** (2018) 012202

[14] Seta S, N. Matsumoto, J.Tanabe 2003 A Study of Usefulness of Fourier Transform Method for Evaluating Streetscapes *Journal of Architecture and Planning AIJ* 569 pp 85-92

[15] Shen Z, M. Kawakami 2010 An Online Visualization Tool for Internet-based Local Townscape Design*Computers, Environment,and Urban Systems***34** 2pp104-116

[16] Stefano Burigat, Luca Chittaro 2007 Navigation in 3D virtual environments: Effects of user experience and location-pointing navigation aids, International *Journal of Human-Computer Studies*Elsevier Ltd.**65** 11 pp 945-958

[17] Takiguchi H, T Arima, T Sakai, S Hagishima2003 Multimedia Computer System Supporting Park-Planning Workshop *Journal of Architecture and Planning***574**pp129-135

[18] Wu H, He Z, Gong J 2010A Virtual Globe-based 3D Visualization and Interactive Framework for Public Participation in Urban Planning Processes*Computers, Environment and Urban Systems***34**pp 291–298

[19] Yoshimura H, Y. Tsukamoto 2010 Study of Streetscape and Building Frontage in Suburban Roadside Areas *Journal of Architecture and Planning AIJ***75** 649, pp751-758

Energy Optimum Building Wall From a Mixture of Papercrete and Local Soil in Tropical Humid Area

Vincentius Totok Noerwasito
Architecture Department, Institut Teknologi Sepuluh Nopember, Surabaya, Indonesia

Corresponding e-mail: *totoknoerwasito@yahoo.com*

Abstract. Paper waste is today's problem in urban areas. It is important to utilize papercrete waste as the raw material of building materials. This research explored the relationship between papercrete wall and the energy produced.The problem of this research was to obtain a simple house wall optimally against overheating and embodied energy. The wall materials used papercrete and local soil. The research location was in Surabaya. The method of wall production was mixing paper and local soil and adhesive i.e. cement that was compacted and molded. The research variable was walls thickness and the measurement of embodied energy was unit value of embodied energy of material. The value of overheating was obtained by using Archipak simulation program on building simulation model with dimension 6 x 6 m. Energy optimization was carried out on variables. The results revealed wall of papercrete and local soil in simulation building model; wall thickness of 8 cm was energy optimal wall to embody energy and overheating. That was one of the principles of green design.

Keywords: papercrete, soil, simulation, embodied energy, cooling energy

1. Introduction

The world's current problems are waste increase and natural resource depletion. Those problemsare closely related to the development of public consumption. This paper focused on paper waste. The problem of paper waste will be reduced if the waste is utilized as the basic materials of building materials. Waste processing can be done by reuse or recycle. To date researchers mostly investigated the development of research on waste processing into building materials [1]

The focus of this research is to discuss the building wall using the basic materials of paper waste. In the field of wall building construction of paper called papercrete [2]. The composition of the papercrete wall used consists of paper as raw materials plus local soil. Addition of soil increases the resistance to climate and material density [3]. The adhesive materialsare cement and lime. The block-making process of the papercrete wall uses a compacting process and is dried in a natural way. This is similar to the process of making compressed earth blocks [4].

The purpose of the study was to investigate the potential of this papercerete wall to the energy it produced. The wall of the building in question was a wall of buildings in simple house. The energy discussed here is cooling energy and embodied energy. The research problem is how the condition of simple house wall is right to produce optimum energy against overheating and embodied energy.

The method used simulation and optimization by using model of simple building type 36 which had a width of 6 m; width 6 m and height 3 m. This building is widely developed in Indonesia. This study was conducted in Surabaya. The variable used wall thickness of building. Optimization of embodied energy with climate needs to be done [5]. In this research, energy optimization was done on overheating and embodied energy in the model.

The result of this research is simple wall thickness design using paper composition and soil that is energy efficient. Besides, the results of this study show the use of papercrete wall not only solve the problem of paper waste and problem in the field of construction alone but also on the development of the energy field. So using papercrete and soil as a wall material is part of the concept of green design and green material.

2.Methods

This paper conducted a study of wall construction materials consisting of papercrete and local soil. The composition adhesive used cement. This was carried out so that the composition had hardness and density. Activities undertaken were: simulation model of building design that using walls of papercrete and local soil. The building had efficient energy to overheating and embodied energy. While embodied energy was energy used building materials for the production process from the search of natural resources to be building materials. Here is the research thinking flow (Figure 1).

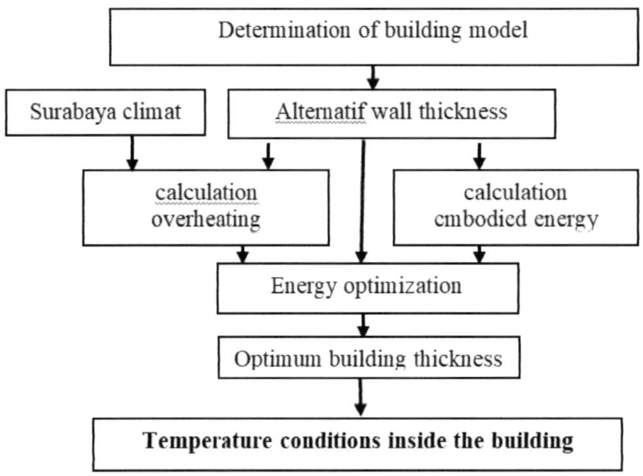

Figure1. the thought line of the study

The dimensions of the building model were determined according to the size of the modest house. So the dimension of simulation building model used with size 6 x 6 x 3 m (see figure 2). Alternative wall design was done to obtain optimal temperature conditions. For that alternative of different wall thickness was made. Thickness of the wall was a variable of research. The condition of the wall can be seen in Figure 3. While the wall plan variables can be seen in table 1.

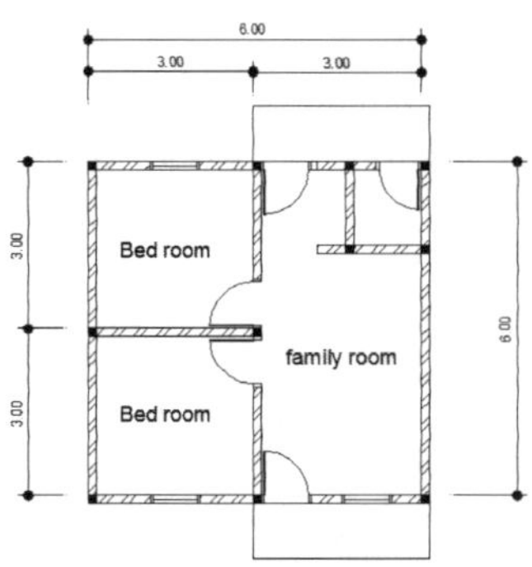

Figure 2. Plan of model building type 36

IOP Conf. Series: Earth and Environmental Science **213** (2018) 012004 doi:10.1088/1755-1315/213/1/012004

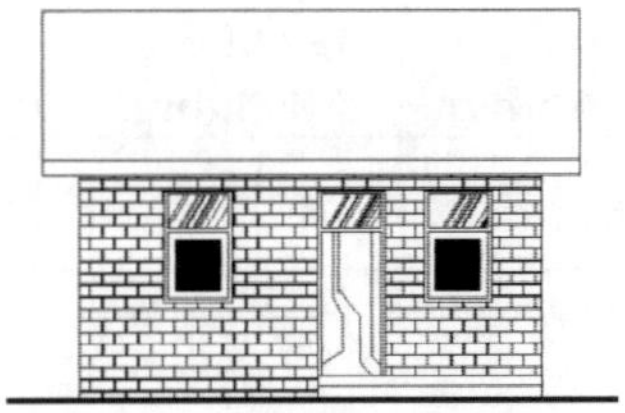

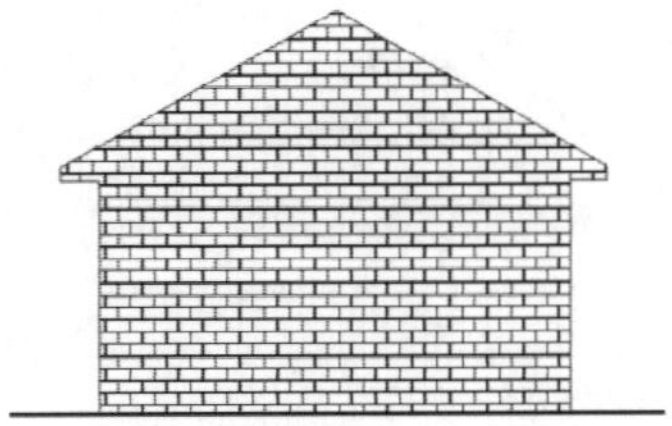

Front elevation of model Back elevation of model

Figure 3. Plan and elevation of model

Calculation of overheating used archipak program. As the variable was wall thickness that was 8 cm, 11cm and 16 cm and they had code W8; W11 and W16 (see table 1). Embodied energy calculation was done by multiplying the width or volume of the building material with the embodied energy value per unit of each building material.

3. Discussion

3.1. Overheating

Overheating describes uncomfortable thermal conditions above and below comfortable thermal conditions. The determination and calculation of overheating requires data about thermal properties of materials. The thermal properties obtained by using Archipak software can be seen in table 1.

Table 1. Thermal property of papercrete mix soil walls

No	Wall Thickness	U-value	Admitance	Time lag	Decrement factor
1	8 cm	5,53	5,76	1	0,97
2	11 cm	5,52	5,92	1,35	0,94
3	16cm	5,51	6,24	1,89	0,88

Results of calculations with Archipak

Table 2 shows the cooling energy and overheating of papercrete and soil for a year from each model with different wall thickness. The location of the calculation was in Surabaya. The model with the thickest wall had the smallest cooling energy and smallest overheated. While model with the thinnest walls have the greatest cooling energy and greatest overheated. In terms of cooling energy and overheated, models with thick walls were the best wall, because the energy was lower.

Table 2. Cooling energy and overheated

No	code	Wall thickness	Cooling kWh/years	Overheating K.h/years
1	W8	8 cm	18029	10037
2	W11	11 cm	17652	10285
3	W16	16cm	16604	9354

Results of calculations with Archipak

Table 3 shows the maximum and minimum energy in Surabaya. In this table the minimum energy value was in July. While the maximum energy value was in October. July was the coldest month, while October was the hottest month.

Table 3. Energy maximum and minimum of model

No	Code	minimum			maksimum		
		Cooling kWh/bln	Overheated K.h/bln	month	Cooling kWh/bln	Overheated K.h/bln	month
1	W8	1324	579	July	1721	1182	October
2	W11	1293	588	July	1682	1207	October
3	W16	1232	506	July	1595	1127	October

Results of calculations with Archipak

In this study energy that used as main energy was overheating. Figure 4 shows that the highest overheatingwas in the W11 building model, i.e.building with wall thickness of 11 cm where as the lowest overheating was found in the building model W16. The building had a wall with a thickness of 16 cm. Comparison of cooling load between building W8 with W11 was not significant that was2%. The comparison between lowest and highest cooling loads in all three buildings was 931 Kh/years or 10%.This shows that the difference in overheating of W8 dan W11of walls was not significant.

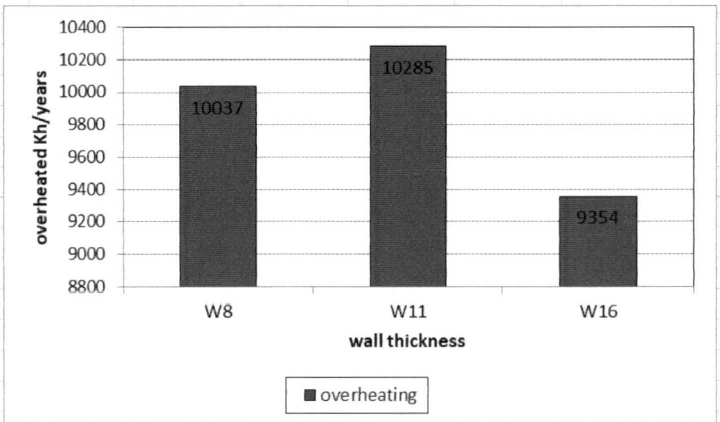

Results of calculations with Archipak

Figure 4. Cooling energy on differents wall

3.2. Embodied Energy

The process of production materials from raw materials into finished materials is called embodied energy [7]. The main building material used in this research was the composition of the papercrete and the soil around. Each material had a different embodied energy. There were two types of embodied energy in this research that was embodied energy total and embodied energy per m^2 of building. Materials used for the building model with embodied energy per unit can be seen in Table 4.

Table 4 embodied energy per unit

Materials	Embodied energy perunit
Papercrete and soil	32855 MJ/m^3
Roof	251 MJ/m^2
floor	5250 MJ/m^3
Wood door	388Mj/m^2

Source: Lawson [9]

Table 4 shows that the values per unit of embodied energy in each material were different. The units were based on volume, weight and area of materials. Each building model had different embodied energy depending on the volume or weight of the material used. The volume of material and building area could reduce the value of embodied energy [8].

Compositions of the papercrete and soil had influenced value of embodied energy in the building model. This was due to the weight and volume of both materials is relatively large. The embodied energy value of each building model was calculated based on the multiplication of the area or volume of the building material with the embodied energy per unit value. The result of calculation of total Embodied Energy from each building can be seen in Figure 5. The building had the highest thickness wall had the highest total embodied energy i.e. W16 which had wall thickness 16 cm.

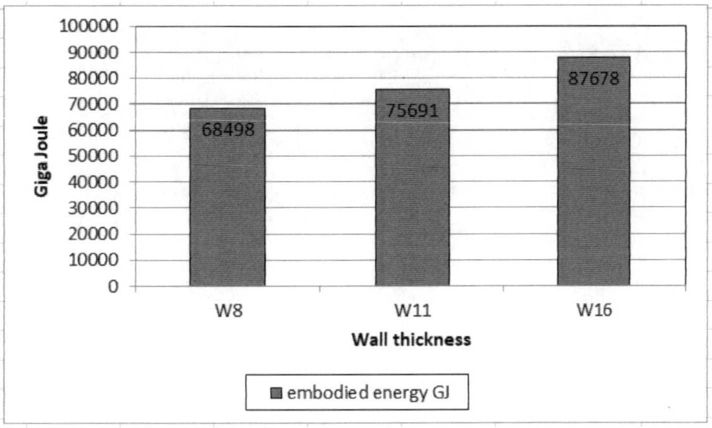

Figure 5. Embodied total energy on different walls

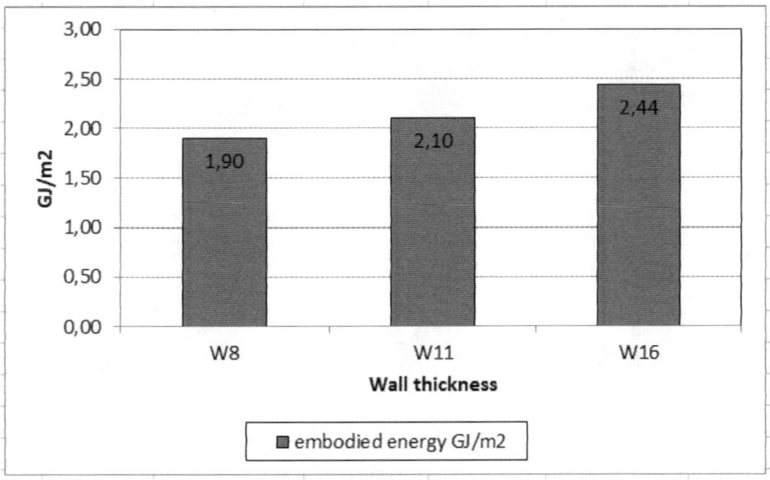

Figure 6. Embodied energy per m^2 on different wall

Figure 6 shows value of the embodied energy per m^2 of the building. Buildings with thinner walls contained the lowest embodied energy per m^2. This condition was equal to the value of total embodied energy (figure 4). Where, the lowest embodied energy was owned by the building with the thinnest wall (W8). The cause of same condition between total embodied energy and embodied energy per m^2 because both had the same building area. In this case a good embodied energy was owned by building model with a wall thickness of 8 cm (W8).

3.3. Optimization

The condition of each building model had different energy value between cooling energy with embodied energy. There were building models that had low embodied energy but they had high cooling energy. Besides that there were also models that had high-embodied energy but cooling energy was low. Therefore, it is necessary to optimize the all energy in each building model. The energy optimization required data about yearly cooling load and embodied energy per year. The value of cooling energy and embodied energy can be seen in Figures 3 and 6.

Energy optimization was carried out by looking at the energy position of the building model against the value of cooling energy and embodied energy. The optimization process used the percentage graph of energy. The percentage comparison of energy can be seen in table 5.

Table 5. Embodied energy and overheated model

Model	Embodied energy	overheated
	GJ/m2	Kh/years
W8	1,90	1182
W11	2,10	1207
W16	2,44	1127
average	2,15	1172

Figure 7 shows an application of energy in the graph. There was a combined difference of embodied energy and overheating in each model. The W8 and W11 models were low embodied energy models. Furthermore, the W8 and W11 models were chosen most optimally. Criteria to determine the condition in the room were embodied energy. The lower embodied energy of the room became more efficient in materials. Therefore, among the lower W8 and W11 models the embodied energy was W8 (see Figure 7). However, in reality the embodied energy W11 was higher than W8 namely 10,5%. This shows that the wall with the composition of mixture of papercrete and soil with 8 cm thick was the optimal wall in overheating and embodied energy.

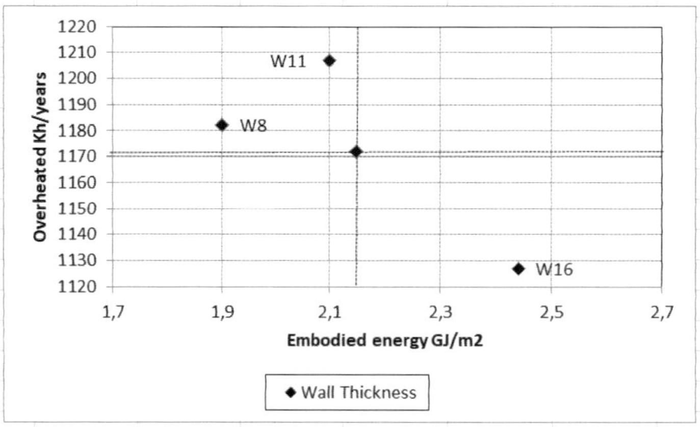

Figure 7. Optimization of overheated and embodied energy

3.4. Indoor Temperature

Based on table 3 shows overheating conditions of W8 (wall with 8 cm thick) in the hottest and coldest month i.e. July and October. Condition of indoor temperature in those months can be seen in figure 8. In October, the hottest indoor temperature was at 4pm, at 35.1 C^{O}. The coldest indoor temperature was at 6 pm at 26.8 C^{O}. The hottest temperature in July was 32.6 C^{O} occurred at 4 pm. The coldest temperature occurred at 7 pm at a temperature of 24.9 C^{O}.

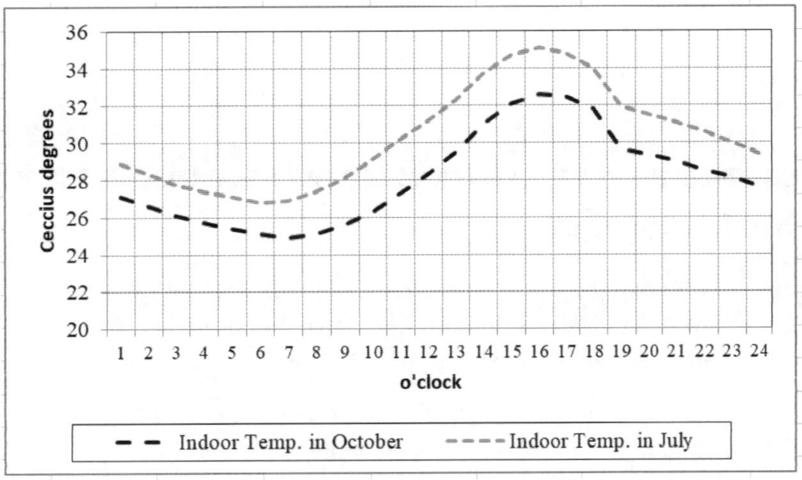

Figure 8. Indoor temperature conditions of wall with 8 cm in July and October

4. Conclusion

In this study optimum results in overheating and embodied energy were not on the wall papercrete only but from the composition of the mixture of concrete paper with soil.

Thickness of the wall from composition of mixture papercrete and soil affected the value of material thermal properties. Cooling energy building was affected by thickness of the wall and climatic conditions around.

The wall that had the optimum energy in this study was a wall with a thickness of 8 cm. The walls had relatively low overheating and the lowest embodied energy compared to wall thickness of 11 cm and 16 cm. Optimum energy was not always present in buildings with walls that had the lowest embodied energy and the lowest cooling energy, but they must be balanced.

The highest indoor temperature occurred in October and the lowest indoor temperature minimum occurred in July. In those months the maximum indoor temperature occurred at 4pm.

The papercrete mix soil wall absorbed the water so it needed to be protected on the bottom of the wall. Its top was quite protected with cantilever of building.

5. References

[1] Myriam Marie Delcasse al 2017. A Papercrete brick consists of recycled material and therefore cost is low compared to conventional bricks, Int. *Journal of Engineering Research and Application* www.ijera.com ISSN : 2248-9622, Vol. 7, Issue 3, (Part -6) March, pp.09-14.

[2] Garrett McElroy, Charis Thompson. Kayla Williams 2010. Make It Complete With Papercrete. Engineering 215 Intro to Design. Haiti.

[3] Bashar S. Mohammed 2009.Papercrete as Infill Materials for Composite Wall System. *European Journal of Scientific Research.*EuroJournals Publishing, ISSN 1450-216X Vol.34 No.4.pp.455-462

[4] Rigassi Vincent 1995. Blocs de terre comprime Vol 1 Manuel de Production. CRA-Terre EAG. Grenoble, France.

[5] Sattary, A and Thorpe 2012, D. Optimizing embodied energy of building construction through bioclimatic principles In: Smith, S.D (Ed) Procs 28th Annual ARCOM Conference, 3-5 September, Edinburgh, UK, *Association of Researchers in Construction Management*, pp1401-1411.

[6] Tracy Mumma 1995. Reducing the Embodied Energy of Buildings. Construction. Home Energy Magazine Online January/February.

[7] Krishnakedar, S. Gumaste 2006.Embodied Energy Computations in Buildings, *Advances in Energy Research.* Pp 404-409.

[8] Jil Tushar Sheth, Saransh Joshi 2015. Paper Crete: A Sustainable Building Material. Strategic *Technologies of Complex Environmental Issues-A Sustainable Approach.* p85-90. `

Water Resource as *Axis-Mundi*: an Effort to Preserve Water Resource Sustainability

Ema Yunita Titisari, Antariksa, Lisa Dwi Wulandari, Surjono
University of Brawijaya, Malang, Indonesia,

Corresponding e-mail: ema.yunita@gmail.com

Abstract. Sumberawan Site is an entity of water resource and an old *stupa*. Its existence is corroborated by myths, folklore, and legends about *amerta* water and paradise garden. Rituals, myths, and sacred artefacts are clues that lead to the *axis-mundi* concept. It is characterized by a sacred space as the central of the environment. It is unique that Sumberawan Site is not located in the village centre, but the activities and symbolic attributes lead to its function as axis-mundi. This possibility was explored by phenomenological paradigm. The study used two approaches: phenomenology and non-verbal communication approach. Field observations and in-depth interviews were conducted to collect data. Observational objects included time, space, and activities. In-depth interviews were conducted on key informants to explore their experience. The data were analysed inductively through five stages of phenomenological analysis. Space attributes were recorded, drawn, and interpreted by non-verbal communication approach to support phenomenological interpretation. The results reveal that (1) Sumberawan Site is an axis-mundi which means *source of life*; (2) This concept relates to water resources in the context of agrarian societies' culture; (3) In the settlement setting, it is located 'outside' the village; (4) The concept of life source is vertical-transcendental, and directly connected to the horizontal-social concept; (5) The sacred zone is bound by time. The findings of this study indicate that efforts to preserve water resources as a source of livelihood of agrarian societies are not only conducted by physical effort, but also directly linked to vertical-transcendental and social-horizontal values.

Keywords: centre, *stupa*, water resource, sustainability, local wisdom

1. Introduction

The awareness of agrarian-traditional society towards water needs has created special culture and tradition. This tradition is in the form of ritual, corroborated by myth and folklore[1], toponyms, artifacts, and space attributes [2]which lead to the creation of water source sacrality. The Javanese people believe that the water source becomes the abode of the *danhyang*, the spirits of the ancestors and other spirits [3]. Belief about a place affects behavior and policies over the place, even creating attachment to the place [4].

The activities at Sumberawan water source show sacred character (meditation, prayer, traditional-religious ceremony, and bath ritual) in addition to profane activities.The belief in transcendental energy in water and the place has developed since before the construction of the temple (the temple was built around the 14th century). This place is referred to as *Kasurangganan* (angel's garden of paradise) in the Book of *Negara Kertagama*. It is said that the water is *amerta* water. People believe

IOP Conf. Series: Earth and Environmental Science **213** (2018) 012005 doi:10.1088/1755-1315/213/1/012005

that the kings of Singosari retired to this place after descending from the throne in preparation for entering the world of saint[5].Sacred activities at Sumberawan water sources led to the concept of *axis mundi*.

Axis mundi is a central point as well as a vertical-transcendental bridge that connects the middle world ofhuman with the upper wolrd of gods and the middle world with the beneath world of evil [2,6–8]. The space around it is sacred[6,9]. This point is the expression of human promordial yearning to always be near the center of the world (God). The markers are sacred artifacts. *Axis-mundi* is the origin and return point; the point of birth as well as the point of death (back to God).At times marked as a turn of the period, traditional societies commemorate the rebirth process (*cosmogony*) at that point with the aim of transforming*chaostic* into *cosmos* (regular/equilibrium/harmonious) as at the beginning of creation[6].In Javanese society, the ancestors became the intermediary of the transcendental relationship of man-God[10].

Studies on traditional settlements indicate that *axis-mundi* becomes the point of reference in shaping the structure of domestic space for the creation of security, safety, social-family relations, and power classificationkekuasaan [11–15]. Generally, in the context of settlements, *axis-mundi* positioned as the central point. It lies in the middle of residential area and forms cosmic axes according to the sacred-profane direction[7,16–18]. In the wider context, mountain is an*axis-mundi*. It isrepresented on sacred artifacts, building, or building elements[16,19,20]. The symbols of *axis-mundi* can also be stones [14], trees [8], compositions of open space with place of worship and palaces [12,14,21], caves, small pit in the floor, pyramids [22], totem poles, temple or other religious buildings [4,6,23–29]. Water can be function as axis-mundi,(8) i.e. whirlpool or rough water (in a river or lake). It connects Middle Earth with Beneath Earth. But, as we know, no research has shown that water sources can also act as *axis-mundi*.This will be the novelty of this research.

The possibility of water source as axis-mundi is interesting to study. This relates to the concept of sustainability of natural resources (water resources) in traditional societies, especially agrarian societies.Although located far outside the residential area, but Sumberawan water source phenomenon leads to function as *axis-mundi*. Concerning the concept of sustainable development on conservation and management of natural resources (water resources), the function of water resources as *axis-mundi* in agrarian society is interesting to study.This research attempts to explain how the concept of *axis-mundi* on Sumberawan water source according to agrarian society of Toyomerto Village and the pilgrims.If the water source is *axis mundi*, then it is possible that beliefs and prayers that are implemented at once in behavior, policies, and actions towards water sources are one way of maintaining water quality and sustainability. Those who study about prayers state that prayers, feelings, thoughts, words, and beliefs must come true [30]. The limit for this research is on the expression of the axis mundi concept of Java's agrarian society (in this case the people of Toyomerto Village), which may be related to water sources.

2. Methods

This research employed phenomenology approach[31,32]andnon-verbal communication approach[33]. These methods wereselected in accordance with the purpose of research, namely: to determine whether Sumberawan water source has a meaning of axis mundi. Phenomenological approach aims to describe the meaning of a phenomenon, while the non-verbal communication approach aims to determine the meaning of a place based on the attributes of space and its settings.

The data were collected directly from the field through interviews with key informants or participants in natural setting [34]. The focus of the interview was on what participants experience and how they experience it. Participants were selected purposively from the ritual actors in Sumberawan. Data analysis was carried out inductively in five stages of analysis, namely: identifying phenomena, reducing information to important statements or important citations, integrating them into themes, developing textural descriptions (what is experienced) and structural descriptions (how participants experience it), then combining textures and structural descriptions to convey essence or essential invariant structure of the phenomenon under study [31]. In this process, the researchers conductedbracketting or epoche so that the exploration of the participants' experience was not disturbed by the subjectivity of the researcher's experience. Non-verbal communication approach method was mainly used in the exploration process of structural description related to the setting of space or place. The focus of the observations is on fixed, semi-fixed, and non-fixed feature elements [33].

3. Discussion

3.1. Key Themes

Participants were selected from those who performed ritual and bathing activities at Sumberawan Site. Of the 25 key participants, 78 important statements about their experience at Sumberawan Site emerged. The meaning of the statements was formulated and subsequently grouped into 6 themes, namely:

1. Water that heals
2. The holy water of source of life
3. Garden of paradise (*Kasuranggganan*)
4. Place with a magical atmosphere
5. Reminders of human life
6. Buddhist symbol

Almost all participants stated that Sumberawan's water could heal a sick body, refresh, and make the people become ageless. They believed the water of Sumberawan as *amerta water* (water of life). Participants obtained information about the efficacy of water from generation to generation or from people they knew, both in the form of water mythology, folklore, and news about those who recovered after bathing and drinking Sumberawan water. The stories motivated their arrival to use Sumberawan water as an intermediary for healing, health, and beauty. Although there wereparticipants who believed that people must go through a special ritualin order to "pull out" *amerta* water, some participants argued that the existing water in Sumberawan has already had*amerta* properties that could be used for various purposes without special rituals. They claimed that *stupa* (temple) was a washing or a transforming building, so that water coming out of the source has already had*amerta* properties. According to them, the *stupa*was located on the largest source of water that was flowed and accommodated to two ponds (*Amerta*Pond and *Sumber Penguripan* Pond) and a large pond outside the temple fence. Yet, there were also participants who said that the water from the source was sacred, so there was no need for 'tools' to transform the water. The water came from a sacred mountain (Mount Arjuno), so the water automatically had sacred properties.

According to the caretaker of Sumberawan Temple, *amerta* water was actually a symbol of sustainability of life. The gods and giants in *amerta* water mythology were symbols of love and passion in the procession of union of men and women. *Amerta*waterwas the symbol of the semen that gave birth to a new (human) life. Therefore, marriage was a sacred procession that began the birth of a new life. Human needs were symbolized by creatures and objects thrown out during the process of removing *amerta*water from the bottom of the Mantana Ocean. These needs were demanded by those who prayed at Sumberawan Site.*Amerta water* is a symbol of human kind sustainability.

Water source was the heart of life, especially for the agrarian community. They called it as *Sumber Penguripan* (a source of livelihood). *Tirta Amerta Sari* Cultural Carnival was a traditional ritual held to express a gratitude for the blessing of the water source while at the same time hoping that the water source will continue to be abundant. The ceremony was performed together with the *Bersih Desa*(Village Cleaning) ceremony in the month of Suro (the first month of the Javanese calendar). *Kamituwo*, the leader of Toyomerto village, said that this ceremony must be done in order not to be *kualat* (cursed/damned; Javanese: got a problem because of mistakes made against the elderly), resulting in the drying of the water source. Dry water sources meant that their lives were 'dry' or 'dead'. Not only villagers, water user groups from these sources also did*slametan sumber*(ceremony for water source).

Belief about amerta water was closely related to the naming of Sumberawan Site as *Kasurangganan* (Angel's Garden of Paradise). Most of the informants felt the calm and the peace in this place. This place was a good place for meditation as done by the ancestors. Some sources believed that the kings who have descended and were about to enter the world of saint meditated in Sumberawan/Kasurangganan. Some of them experienced*moksya*, the process of reunification of (spirit) of man and God. Meditation was the process of 'reuniting oneself'. This procession was done by bathing as a symbol of self-cleansing (physically and spiritually). Written on the walls of *Amerta*Pond: *mulat sarira angrasa wani*, which the free translation was: self introspection, freeing from all worldly attachment.

> *".... Some people say God is above, but God is in our souls. We can penetrate into the realm of God through nature. Beings must be united with nature so that there is something that is "diunggahno" (raised/lifted), 'means that there can be something that can enter into our soul ... ... In the process of purification, there must be something to be offered. Not only fasting, but there are also some rituals like bathing at night with the intention of 'metu soko lali', i.e. we must get out of the mistake, and 'jumeneng' i.e. on behalf of A Substance (God) Who Regulate Human .... "*

Some sources described Sumberawan Site as a place with a magical atmosphere and a great aura. Those who need "energy" were like filling themselves through rituals and meditation. A chaotic mind and atmosphere required a calm situation by removing the burden of worldly thought. The positive power would be derived by means of meditation, from A Substance (God) who regulate human, by connecting with ancestors' (properties) and uniting with nature. In addition to the place, the right time also determined the amount of energy. Sumberawan Site location was about 700 meters from the village (residential area) and surrounded by forest, which made it guarded in the silence. Before the greater use for tourism activities, access to the Sumberawan Site was quite difficult to reach. It was in the form of a path on the edge of rice fields.

The shape of the temple was in the form of a stupa, but it did not mean that Sumberawan Site belonged only to Buddha. The privilege of *kasurangganan* have existed prior to the construction of the temple. The caretaker of *Patirtan Watu Gede* stated that the stupa was a reminder to humans that life in this world was only 'half' from the real life. The caretaker Sumberawan also said that the temple of Sumberawan was*pepunden*, which was something that was "*pundi-pundi*"(being praised or glorified). The naming was not directed to the physical form of the temple, but to the 'something' it symbolized (ancestors, testament, meaning).

Participants of Buddhist leaders interpreted Sumberawan stupa as a symbol of Buddhist sanctity. Every year, Buddhists in Malang commemorate Vesak in Sumberawan. The Buddha was a man who attained perfect enlightenment, that was, united with God. Worship was done to God. Respect for the Buddha was to practice his teachings. Stupa became a reminder and appreciation of the Buddha and his teachings in an effort to 'reach' God.

3.2. Discussion

The universal system is shaped by a religious conception and cosmological image[16,17], namely: a sacred space that decides the homogeneity of space[6]. This space allows the connecting of two cosmic regions (the middle world - the upper world, the middle world - the underworld. It is around this point that our world lies (human). Thus, *axis-mundi* is located in the middle or 'navel point'[6,7]. In the settlement of pre-modern or early culture [3,35,36] which laid the foundations of urban spatial arrangement and architecture in cosmology, often the middle position is marked by a sacred symbol. The space around it becomes sacred, although there is also a nature that varies depending on the context of time and function[14,17,37].

Some sacred symbols indicate vertical elements such as pyramids, temples, totem poles, mountains, towers, tallest buildings, or they may be stone, cave, under floor, open space, whirlpool, tree, fire or smoke [6–8,11,13,14,16,19,22,24,38]. Sacred space or building as a bridge between the cosmos is an *axis-mundi*found in the most pre-modern/traditional settlements and houses[6,37]. Mountain or its representations become axis-mundi and *qibla*for most traditional settlements [15–17,28,36,39,40].

The existence of axis-mundi in pre-modern settlements shows its high hierarchy, also as a symbol of the human need to always connect and communicate with God, ancestors, or supernatural powers that support and become the important part of their life [3,8,10,20,22,41]. It gave birth to the concepts of social relations, political order, laws, norms, and policy in order to create stability, harmony, and safety. Previous studies revealed social relationships, assured power legitimacy, urban development and architecture, and resource management policies from the *axis-mundi* concept [8,11,12,14,24,25,29]. The meaning of a space will greatly affect actions, behavior, and policies on it [4,7]. Ritual procession at the specified time is made to 'open' the *axis-mundi* portal.

The concept of '*unggah*' as conveyed by participants indicated that God was 'domiciled' on the 'top'. Participants also used the term "penetrate" in the effort of meeting the union of man and God. Penetrating showed the existence of two different layers (cosmos). In some ways humans can *munggah* (go up), penetrate, and 'fused' with God. Ancestors were the nature, the example of nobility and effort of a man to be able to 'meet' and 'fused' with God. The idea of 'bridge between the cosmos' appeared in themes 3, 4, 5, and 6.

The representation of the idea of a bridge between the cosmos and sacred space was shown by five shrines, namely: wooden buildings, shrines in the *Amerta Pond*, 'altars' in front of the stupa, *Sumber Urip* Pool, and field. The sacred space was tentative, i.e., when used for religious ritual functions. However, sacred symbols confirmed that this space can not be used for any profane activity. It can not be used for washing clothes or household items at water source, destroying and taking something from the temple environment, cutting down trees, dating, and other bad taboos done in this place. Some participants did not dare to do it because they thought this place was guarded by *danhyang* (ancestral spirit). This belief resulted in the preservation of water resources and the environment.This also happened to the water sources in Java and other places outside, as Titisari and Siswadi [41,42].

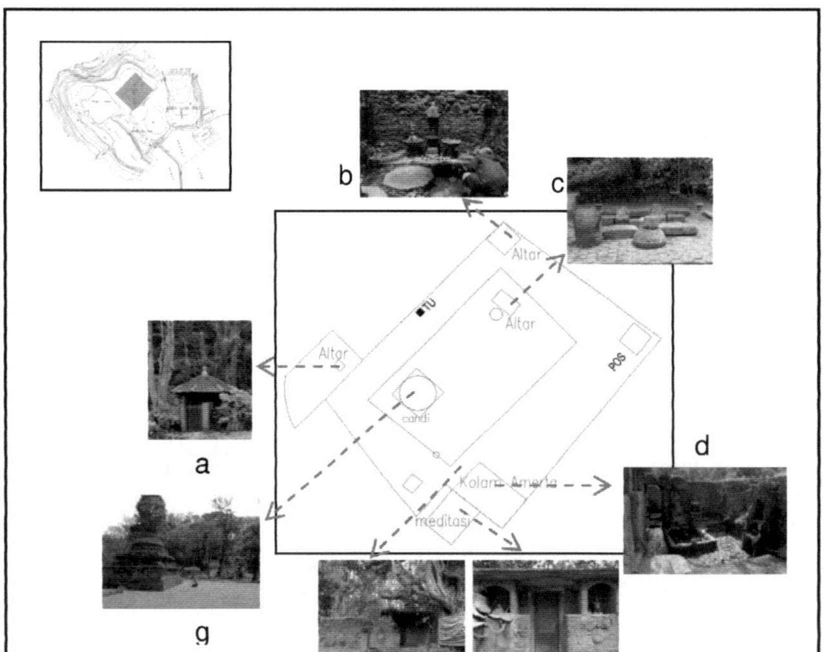

Figure 1. Lay out of Sumberawan Site, area of temple and water source. (a). Wooden buildings for meditation; (b). *Sumber Panguripan* Pond; (c). Worship area; (d). *Amerta* Pond; (e). The entrance to the meditation room insideamerta pond; (f). The front of Amerta pond; (g). Temple / stupa

The wooden buildings and shrines/worship area in the *Amerta* pond were closed (visual and accessibility), small, and the floors were elevated. This became a marker of the private character of its activities (Fig 1). The sacred atmosphere was felt from the closure and size of the room, the elevation of space, the Semar statue, the goddesses statue, the dragon head statue, the incense, *pelinggih*, the flowers, and the decoration of the pool wall resembling a cave wall (Fig 2). *Amerta Pond*was a closed pool. It was a square-shaped pond with a statue of *Dewi Dhanwantari* carrying a jug and head of Dragon of *Basuki*. There was a turtle statue (incarnation of Lord *Vishnu*) at the edge of *Sumber Panguripan* Pond.

Figure 2. Figures and decorations on *Amerta Pond* and *Sumber Panguripan* Pond

Another worship space was an open space shaded by trees and paving stone in the north of the stupa (Fig 1). There was an arrangement of ruins of top rocks of stupa nearby. Worship was oriented to stupa. The scents of incense, sacred artifacts, shrines, and large trees were wrapped in a plaid cloth of honor against *danhyang* who lived in a large tree. These fixed and semi-fixed elements served as markers of place sacrality.

Figure 3. *Tirta Amerta Sari*

Figure 4. Vesak

Worship that was oriented to stupa was performed at the Vesak (Fig.4) ceremony. Buddhists sat together in the field. Buddha statues, colored candles, incense, evening flowers, and offerings were added near the stupa. The people sat on the field and faced the Buddha statue as well as the stupa. *Pradaksina*was performed around stupa and field.

Ceremony of *Sedekah Bumi/Bersih Desa* (Alms of Earth/Village Cleaning)were done together with *Slametan Banyu*(water ceremony) of *Tirta Amerta Sari*. Villagers paraded parcels, foodstuffs, and tree seedlings from villages to water sources (Fig.3). The people gathered in the field to give thanks and beg for the sustainability of the water source for the sake of their agricultural life. After the ceremony was over, they ate the food they brought from the villagetogether. The vertical-transcendental axis was in the form of "cosmos bridge" of thanksgiving and prayer which were coupled with the intertwining of the horizontal-social axis in mutual cooperation shown in the ceremony of*Tirta Amerta Sari*. As Jones[43] stated that every activity is always connected with past experience and oriented to the future. Mills and Walker [44] dan Van Dyke and Alcock [45] have discussed ritual activity in its relation to the production of social memory, and we acknowledge the important role commemoration plays in the activities. The ritual has strengthened the attachment to this place from generation to generation. In addition, the implementation of the ritual in mutual cooperation becomes a means to maintain social relations [3,35,41,42]

Because the mountain was sacred, everything that came out of it was sacred [5,23,41].Sumberawan is located at the foot of Mount Arjuno. According to Marsudi [5,20] Arjuno mountain is a holy and sacred mountain, as evidenced by many sites of worship on its slopes. Water coming outof it isautomatically haveholy properties. Waterbecame something that was rewarded as a sanctifier or closer to the Most Holy (God/*Gusti*). The sources were well preserved through mythology, folklore[41,46],stories, rituals, prohibitions and invites were granted[3,35,36], location, and layout that affected the behavior in it[4]. This idea appeared in themes 1 and 2.

Punden (the tomb of the founder of the village) was commonly found near the water source. Both were representations of the starting point idea (navel point). Ancestral graves of the founder of the village (*punden*) weresymbols of past and present continuity, as well as the future[10]. The water source 'connected' the settlement with the mountain, as a symbol of the continuity of the two cosmos. *Tirta Amerta Sari, Vesak*, and rituals performed for both personal and collective interest were efforts to maintain the equilibrium of the cosmos through cosmogony. This effort was pursued through two things: the continuity of the vertical-transcendental relationship and the harmoniousness of social relationships. Water sources precisely 'kept away' from residential areas so that the inhabitants of the two cosmos were not disturbed [3]and so that the purpose of purification can be achieved. Although it was literally outside the village, but the essence of the water source was central. Themes 2, 5, and 6 led to the concept of the central point and the starting point.

Figure 5. Location of Sumberawan Site to residential area

The title of *Kasuranggganan* showed the strength of beliefs about the sacrality of this starting point. Belief in the sacredness of place and *amerta*water, gave birth to the name of the village of Toyomerto (*toyo/toya* meant water; *merta* was derived from the the word *amerta*, *toyomerto* meant water of life). Water became an intermediary for the formation of the cosmic bridge because of its nature and its 'energy'. Water was valued as the source and origin of life. The bathing ritual was the imagery of being sacred and clean of physical and spiritual form before facing the Holy One. There was the inscription which said "*Mulat sarira angrasa wani*" (fig 2) in Amerta Pond, which meant to clean the mind from the defilements of mind and feelings.

In contrast to Kassabaum and Nelson's research [8] which shows that water is the cosmic bridge of the middle and lower world, Sumberawan water source is the*axis-mundi*that connecting middle and upper world. In addition, in the Kassabaum and Nelson studies, the water is not a water source but a whirlpool in rivers or ponds. What does it means? For agricultural societies, water source is the most important resource for their life. Their dependence on water has spawned an integrated water resource conservation strategy in their cosmological philosophy.

4. Conclusion

The character of axis mundi as the cosmic bridge, the original/starting point, the central point, and the sacrality of space could be found in Sumberawan Site. The concept of axis mundi was attached to the water source as a source of life, which affected the sustainability of water resources. This process was done vertical-transcendentally (with respect to its axis mundinature) and horizontal-socially (one of which was represented in Tirta Amerta Sari). Although it was not located in the middle of the village as howthe common axis-mundi was, but the sacredness was actually maintained in a stronger manner. This model of environmental preservation as well as disaster prevention due to the scarcity of these resources combined the equilibrium of vertical-transcendental and horizontal-social relationships. Thus, this model needs to be studied further in subsequent studies.

5. References

[1] Abrams, N.E P JR. Cosmology and 21st-century culture. Science 2935536. 2001;1769-1770.

[2] Tuan Y-F. Space and Place: Humanistic Perspective. Dalam Springer Netherlands; 1979 [dikutip 26 Juni 2018]. hlm. 387–427. Tersedia pada: http://www.springerlink.com/index/10.1007/978-94-009-9394-5_19

[3] Lelono, T. H. Upacara Korban dalam Tradisi Mayu Desa, Tradisi Megalitik Tengger (Studi Etnparkeologi) Vol.34(2), 199-214. Berkala Arkeologi. 2014;34(2):199–214.

[4] Mazumdar S, Mazumdar S. Religion and place attachment: A study of sacred places. Journal of Environmental Psychology. September 2004;24(3):385–97.

[5] Marsudi. Bangkitnya Tradisi Neo-Megalithik di Gunung Arjuno. Sejarah dan Budaya. Juni 2015;9(1).

[6] Elliade, M. The Sacred and The Profane, The Nature of Religion. New York: Harcourt, Brace &World Inc.; 1987.

[7] Tuan, Y.F. Topophilia: A Study of Environmental Perception, Attitudes, and Values. New Jersey: Prentice Hall; 1974.

[8] Kassabaum MC, Nelson ES. Expanding social networks through ritual deposition: a case study from the Lower Mississippi Valley. Archaeological Review from Cambridge. 2014;29(1):103–28.

[9] Robin C. Outside of houses. Journal of Social Archaeology. 11 Agustus 2016;2:245–68.

[10] Rukmi, W. I., Djunaedi, A., Sastrosasmito, S., & Ahimsa-Putra, H. S. Situs Majapahit Trowulan: Menuju Tersambungnya Ruang Absolut. Jurnal Studi Sosial. 2014;6(1):59–64.

[11] Akkerman A. The North, Axis mundi and Gender Myths in the Rise of Civic Space. Dalam: Phenomenology of the Winter-City [Internet]. Cham: Springer International Publishing; 2016 [dikutip 9 Juli 2018]. hlm. 33–49. Tersedia pada: http://link.springer.com/10.1007/978-3-319-26701-2_3

[12] Siti Rukayah R, Puguh DR, Susilo ES, Indraswara MS. Local Wisdom of The Native Settlement as A Main Gate in The Northern Axis of Javanese City Center In Semarang. IOP Conference Series: Earth and Environmental Science. Mei 2018;152:012032.

[13] Crane JK. The Jefferson Courthouse: An Axis Exsecrabilis Mundi. Twentieth Century Literature. April 1969;15(1):19.

[14] Santosa, J. Arsitektur-Kota Jawa: Kosmos, Kultur, dan Kuasa. Jakarta: Centropolis Universitas Taruma Negara; 2008.

[15] Widodo J. Typological and Morphological Adaptations of Hakka Diaspora's Settlements in Cosmopolitan Southeast Asia: West Kalimantan case1. 2009;

[16] Singh, Rana. Cosmic Layout of the Hindu Sacred City, Varanasi (Benares). Arch & Company. 1993;9(2):239–50.

[17] Xu P. The Mandala Cultural Model can play a significant role in sustainable tourism: Interweaving Tibetan Buddhist landscape with architecture, nature, religious meanings and man's movements. 2006;16.

[18] Farrington IS. Ritual geography, settlement patterns and the characterization of the provinces of the Inka heartland. World Archaeology. Februari 1992;23(3):368–85.

[19] Mashuri, M. Perwujudan Konsep dan Nilai-Nilai Kosmologi Pada Bangunan Rumah Tradisional Toraja. Jurnal Jurnal Ruang. Maret 2012;2(1):1–8.

[20] Cahyono, M. D. Vulkano-Historis Kelud: Dinamika Hubungan Manusia–Gunung Api. Kalpataru. 2012;21(2):85–102.

[21] Pangarsa, G.W. Merah Putih Arsitektur Nusantara. Yogyakarta: Andi Offset; 2006.

[22] Brady, James E., and Wendy Ashmor. "Mountains, caves, water: ideational landscapes of the ancient Maya." Archaeologies of landscape: contemporary perspectives (1999): 124-145.

[23] della Dora V. Engaging Sacred Space: Experiments in the Field. Journal of Geography in Higher Education. Mei 2011;35(2):163–84.

[24] Christiane Barth. “In illo tempore, at the Center of the World”: Mircea Eliade and Religious Studies’ Concepts of Sacred Time and Space. Historical Social Research / Historische Sozialforschung [Internet]. 38(3). Tersedia pada: http://www.jstor.org/stable/23644525
[25] Lane BC. Giving Voice to Place: Three Models for Understanding American Sacred Space. Religion and American Culture: A Journal of Interpretation. Januari 2001;11(1):53–81.
[26] Reilly III FK. People of earth, people of sky: Visualizing the sacred in Native American art of the Mississippian period. Hero, hawk, and open hand: American Indian art of the ancient Midwest and South. 2004;125–37.
[27] Sardjono AB, Harani AR. Sacred Space in Community settlement of Kudus Kulon, Central Java, Indonesia. IOP Conference Series: Earth and Environmental Science. Desember 2017;99:012023.
[28] Sprajc, I. Astronomical And Cosmological Aspects Of Maya Architecture And Urbanis. Dalam: J.A. Rubino-Martin, J.A. Belmonte, F. Prada & A. Alberdi, editor. Cosmology Across Cultures; ASP Conference Series. 2009.
[29] Nelson, B. N., et al. Shaping Space: Built Space, Landscape, and Cosmology in Four Regions 2010. Dalam: Santa Fe Institute Workshop on" Cosmology & Society in the Ancient Amerindian World [Internet]. Santa Fe Institute; 2010. Tersedia pada: https://pdfs.semanticscholar.org/0070/d4bef9000d86c6a032890a48ed4d7e1dd014.pdf
[30] Dossey L. Prayer and Medical Science: A Commentary on the Prayer Study by Harris et al and a Response to Critics. Archives of Internal Medicine. 26 Juni 2000;160(12):1735.
[31] Creswell, J.W. Penelitian Kualitatif dan Desain Riset, Memilih di Antara Lima Pendekatan. 3 ed. Yogyakarta: Pustaka Pelajar; 2015.
[32] Muhadjir, N. Metodologi Penelitian Kualitatif, Telaah Positivistik, Rasionalistik, Phenomenologik, Realisme Metaphisik. Yogyakarta: Rake Sarasin; 1992.
[33] Rapoport, A. The Meaning of the Built Environment, a Nonverbal Communication Approach. Beverly Hills: Sage Publications; 1983.
[34] Groat, L, Wang, D. Architectural Research and Methods. Canada: John Wiley & Sons; 2002.
[35] Wiganingrum, Anditya And S, Leo Agung And Wahyuni, Sri. Nilai Kearifan Upacara Tradisional Susuk Wangan Sebagai Bentuk Solidaritas Sosial dan Pelestarian Lingkungan di Desa Setren Kecamatan Slogohimo Kabupaten Wonogiri. Jurnal Candi - Jurnal Pendidikan Dan Penelitian Sejarah. 2013;5(1).
[36] Morton, S.G. Ritual procession and the creation of Civitas among the ancient Maya: a case study from Naachtun, Guatemala. Canadian Journal of Archaeology. 2012;36(1):141–65.
[37] Rappenglück MA. The Housing of the World: The Significance of Cosmographic Concepts for Habitation. Nexus Network Journal. 19 September 2013;15:387–422.
[38] Barber JB. Óðinn and His Cosmic Cross. Sacrifice and Inception of a God at the Axis of the Universe. 2017;
[39] Pugh, T. 2001). Flood Reptiles, Serpent Temples, and the Quadripartite Universe: The Imago Mundi of Late Postclassic Mayapan. Ancient Mesoamerica. 12(2):247–58.
[40] Xu P. Feng-Shui Models Structured Traditional Beijing Courtyard Houses. Journal of Architectural and Planning Research. 1998;12.
[41] Titisari EY, Antariksa, Lisa Dwi Wulandari, Surjono. Sumber Air dalam Ruang Budaya Masyarakat Desa Toyomerto Singosari, Malang: Menggali Kearifan Lokal Mengenai Water Resources Sustainability. Jurnal Ruang. April 2018;5(1):77–90.
[42] Siswadi S, Taruna T, Purnaweni H. Kearifan Lokal dalam Melestarikan Mata Air (Studi Kasus di Desa Purwogondo, Kecamatan Boja, Kabupaten Kendal). Jurnal Ilmu Lingkungan. 28 Oktober 2012;9(2):63.
[43] Jones A. Memory and material culture. Cambridge University Press Cambridge; 2007.
[44] Mills BJ. Remembering while forgetting: depositional practices and social memory at Chaco. Memory work: Archaeologies of material practices. 2008;81–108.
[45] Van Dyke RM, Alcock SE. Archaeologies of memory. John Wiley & Sons; 2008.
[46] Wurianto, A.B. Aspek Budaya pada Upaya Konservasi Air dalam Situs Kepurbakalaan dan Mitologi Masyarakat Malang vol. 4 (2), pp. 80-88, Maret 2009. Humanity. Maret 2009;4(2):80–8.

Causal Factors in Resilience of Old Tambak Lorok Fishing Port in Semarang

Satriya Wahyu Firmandhani, Masyiana Arifah Alfia Riza
Department of Architecture, Faculty of Engineering, Universitas Diponegoro, Semarang, Indonesia.

Corresponding e-mail: satriya_firmandhani@yahoo.com

Abstract. Fishermen settlement of Tambak Lorok Semarang provides various facilities, including fishing port to support marine capture fisheries. An interesting phenomenon occurred in the fishing port of Tambak Lorok. Despite its poor condition with semi-permanent docks, fishermen still used it to auction off their catch down to date. In contrast, the new fishing port, which was inaugurated in 2007,was abandoned and collapsed right now. Looking at the phenomenon, it is necessary to conduct research to investigate causal factors in resilience of Tambak Lorok Semarang. The research employed descriptive qualitative method by analysing the existing condition in the field and interviewing local people and stakeholders. Results revealed two factors causing resilience of old fishing port: the community perception and the role of the fishing portas center of activity in fishing settlement.

Keywords: Resilience, Fishing Port, Tambak Lorok Settlement

1. Introduction

Tambak Lorok Semarang is a fisherman settlement that provides marine capture fisheries in the area of Semarang City. In order to support fishermen's activities in the settlement, there are special facilities, including fishing port of Tambak Lorok. In a fisherman's settlement, fishing port with its supporting facilities is crucial for carrying out auction of sea catches. Fishermen who want to auction their catch must dock at the fishing port, and then weigh their catch for auction at the fish auction place. An interesting phenomenon that occurs in the fishing port Tambak Lorok is the poor condition of the fishing port, with semi-permanent docks, but fishermen still continue to use it to auction off their catch. In comparison, the new fishing port was inaugurated in 2007, but only used 3 times after the inauguration [1]. Although the new fishing port was built with a complete and permanent facility located east of Tambak Lorok, it is abandoned and the building is collapsed right now.

Looking at the phenomenon, it is necessary to conduct continuous research to investigate the condition of fisherman settlement of Tambak Lorok so that it can provide guidance for its future development. In previous research,the usefulness of new fishing port Tambak Lorok was investigated, and results showed that the new fishing portwas not efficient because of various reasons, including unsuitable location, large waves, the absence of post-construction infrastructure maintenance, as well as the orientation of the building that was directly facing the sea[1].

Figure 1. The condition of old fishing port

Figure 2. The new fishing port building was collapsed

2. Methods

The purpose of this study is to determine the factors causing the new fishing port's resilience in Tambak Lorok Semarang. Based on that objective, this research employed descriptive qualitative paradigm that is by qualitatively analysing existing condition in field from existing variables and taking primary data result from interview with informants. Interviewees included stakeholders at auction and residents in Tambak Lorok.

In formulating the research problem, a research question emerged [2]. The research problem is represented in the following question: "What are factors that causeold fishing porttobe resilient and functioning?" That research question leads to the purpose and method that will be used in this research.

In collecting data regarding the resilience of a building, there are several contributing factors, including synergy, loyalty, activity, policy, and location[3]. In another opinion, resilience is an ability and capacity of society to withstand stress, survive, so people take action of adaptation, and rise again from crisis or disaster. Also some characteristics of resilience can be known if the identity persists in facing external and internal factors[4]. External factors here can be interpreted as pull factors, such as the environment, social life, and nature. On the other hand, the internal factors can be interpreted as driving factors, such as facilities and functions owned by a work of architecture[5]. This is consistent with the theory that a settlement will be strongly influenced by several factors, including: human, community, shelter, environment, and network[6]. This is also supported by the theory that the natural environment is the main capital in the development of infrastructure systems, which in turn the existence of infrastructure will sustain the economic system, and the economic system will sustain the social system[7]. Therefore, environmental conditions may affect the resilience of a building.

3. Discussion

In this section, the data of variables that may affect the old PPI resistance will be discussed. The discussion begins with the presentation of data that will be directly analyzed.

3.1 Research Locus

Tambak Lorok settlement is located in Tanjung Mas Village, North Semarang Subdistrict. Tambak Lorok has been known as fisherman's settlement since 1950 because of its location adjacent to the sea and the culture is passed down from generation to generation to date. The old fishing port Tambak Lorok that is located in the middle of the settlement can be accessed through the neighborhood and also through the river. The location of the old fishing port Tambak Lorok reviewed from the city of Semarang can be seen in Figure 3. The position of the fishing port is in the middle of the road in the fishermen neighborhood Tambak Lorok. In its surrounding, there is a traditional market and fish market. Here are the borders:

North: Houses of Residents
South: Houses of Residents
East : Pier
West: Road in the Neighbourhood

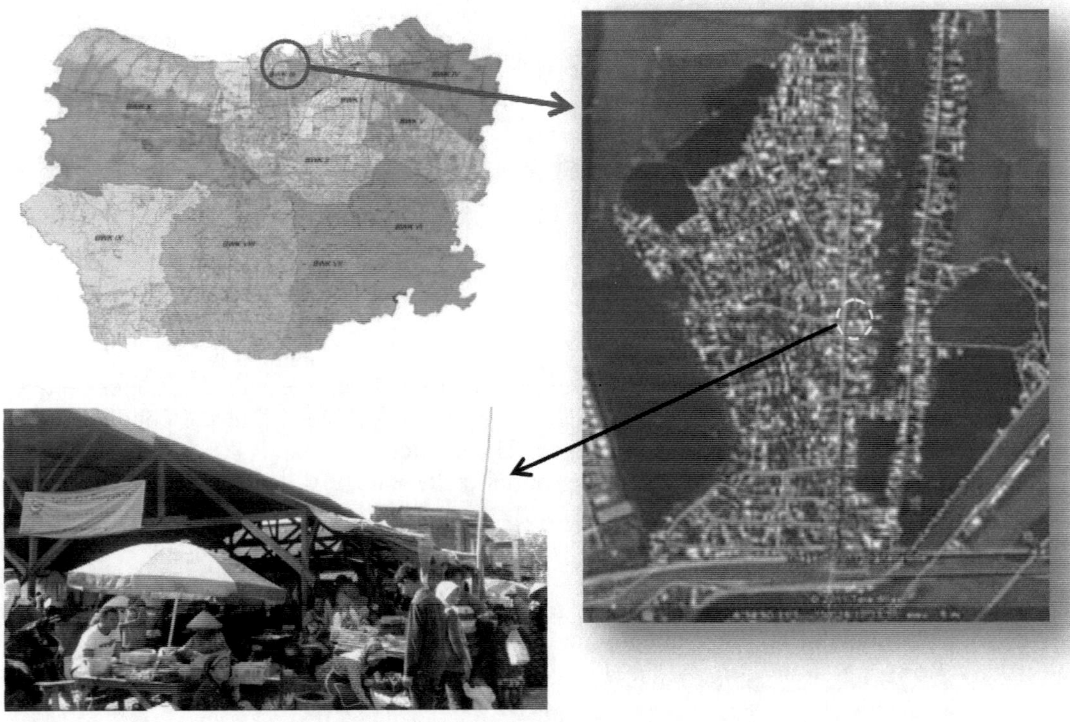

Figure 3. Location of Fishing Port

3.2 Fishing Port Building Condition

The physical condition of the PPI building is made of semi-permanent material that has been operating for decades. These materials include:

Column / Pillar	: Wood
Partition	: High brick wall ± 70 cm
Floor	: Rebates of concrete
Roof Frame	: Wood
Roof Cover	: Asbestos.

The physical condition of fishing port building can be seen in figure 4. Based on the observation and interview, the semi-permanent condition of the building at the fishing port Tambak Lorok is not an obstacle in conducting the auction. It is revealed that people remained enthusiastic waiting, followed the auction process and willingly stood or sit on the walls of the partition. However, some auctioneers wanted the fishing port to be expanded and built with permanent and good quality. This opinion could not be fully referenced, because the new PPI was built in Tambak Lorok with good building specification, but it was still not used. Based on this analysis, it can be seen that the current physical condition of fishing port building is not the cause of its resilience. Based on the analysis above, the culture and attitude of the user could be expressed as the local identity and social space of the settlement[8,9].

Figure 4. Physical condition of fishing port

3.3 Infrastructure

a. Access/Road

Access/road to the old fishing port can be seen in figure 5. From the picture, it appears that the access is less accessible, with pavement in the form of paving block, but the condition is often worsen by coastal flooding, which exacerbates road condition.

Infrastructure towards and around the fishing portis inadequate, especially on road access to port environments that is not fully finished. The road surface condition is also uneven because the road is often submerged in coastal flooding and passed by transporting vehicles. In addition to the quality of the road, the width of the road is also inadequate, with a width of ± 5 meters plus fish market activities on the sidewalk, so that it is not possible for vehicles to pass the road.

In addition to road access, the infrastructure that has a role in fish auction is fishing port, electricity flow, and water supply. In this case, the flow of electricity and water supply are adequate. However,the fishing port has a poor condition, because only in the semi-permanent material made of bamboo and wood. The condition of the fishing port can be seen in Figure 6.

Figure 5. Road access

Figure 6. Dock Condition

b. Electricity network

The electricity network in Tambak Lorok has been sufficient. All houses and buildings have been powered by electricity. The installation of electricity network is not only in every house or building, butin street lighting. However, street lighting has little role in fishing port's resilience, because the auction is in the morning and evening.

c. Water Installation

Tambak Lorok is at the outermost zone in North Semarang area, so it is directly adjacent to the sea. Therefore, the installation of clean water is from the state water company because ground water of Tambak Lorok has a less good quality coupled with the land subsidence that reaches >8 cm per year.

3.4 Accessibility

The location of fishing portis in the middle of Tambak Lorok fishermen settlement. This location facilitates the journey of the fishermen from their home and from the fishing boat docks. Despite the inadequate road conditions, the location of fishing port is considered strategic as it allows people to trade fish directly around the fishing port. The general market around the fishing port also supports this activity. The location of fishing port is at the center of activity that has a market, fish market, and fishing port itself, so that the facilities are mutually supportive of each other. The fishing port as the center of activity can be seen in figure 7.

Figure 7. Centre of activity in the fishing port

3.5. *Perception of local people or users*

Participation is defined as a tool that is interpreted as an active community involvements in the whole process of activities as a growth cohesiveness [10]. Regarding the much participation of the users that the fishing port could be still functioned, the perception of the local people/users is a dominant factor in determining whether space can function properly or not In qualitative research, the data in the form of direct interviews with informants in the field as the primary data that must gathered. In this study, the interviewees were the parties directly related to the activities in the fishing port, among others:

a. Fishermen
b. Traders/whole salers
c. Market traders
d. Local people

Some open questions were used to gather information from various sources. Questions directly lead to built-in background knowledge.

The functioning of a space depend on two factors, namely the pull factor is a factor of the physical environment that is happening and attracting people to conduct activities there. Another factor is the driving factor, which is the factor that pushes from within human to conduct activity in a certain space. The driving factor is commonly called the perception/perspective of a person or society in assessing environment and activities.

In the case of old fishing port in Tambak Lorok, it is known that it has inadequate physical condition and infrastructure, but the auction activity is still conducted there and even it is chosen over the new port, which is now abandoned despite its better building and infrastructure. Therefore, the data on the perception of society needs to be analyzed based on the results of interviews that have been gathered.

Based on the results of the interview, it is revealed that:

a. Public perception of the condition of the fishing port
   Based on public opinion, the fishing port condition cannot be developed anymore. This can be indicated from the interview answer:
   "*This has happened since long time ago*."
   Therefore, people actually realized that the condition of fishing port was not adequate and not ideal, but it could not be developed for certain reasons. The old fishing port cannot be developed due to illegal land tenure [1]. This forced people to conduct activities with the existing facilities in order to continue running the fisherman's settlement system in Tambak Lorok.

b. Public perceptions of the abandonment of the new fishing port and the resilience of old fishing port.
   The new fishing port was built and inaugurated by the Central Java Provincial Government in 2007, but it operated in just two days and then it did not operate again. This situation certainly raises a question on why it could happen. In terms of physical architecture, it is revealed that the location factors, such as its direct proximity to the sea, the waves, and shallow mooring contribute to the non-functioning of the new fishing port.

c. Public perception of the supporting infrastructure of the fishing port
   Based on the results of interviews, the public perception of supporting infrastructure of the old fishing port is that supporting infrastructure and facilities, such as road, lighting, drainage, and clean water are less adequate and less ideal. This is known because the residents and users of old fishing port deplore the bumpy road conditions that have to be elevated periodically. This shows the poor quality of infrastructure, especially the road as the main access to the fishing port.
   Despite such inadequate conditions, residents and stakeholders of auctions in the fishing port remain interested and participate in auctions routinely. In fact, there are some people who were not bothered by the road condition that was muddy and bumpy by walking downthe road on foot.

4. Conclusion

4.1. Conclusion

The analysis showed the dominant causal factors of resilience of old fishing port in Tambak Lorok, among others:

a. Public perception

Public perception showed that the community of auctioneers (fishermen, traders, middlemen, managers) were accustomed to facing the situation that occurred in the old fishing port in Tambak Lorok. Programs and systems that had been established in the fishing port had worked for decades and involved various parties that continue to exist every day. This perception can be influenced by several factors, such as revenue (financial gain) and education level. This is a strong cause of the longstanding function of the old fishing port in Tambak Lorok.

b. The fishing port as the center of activity in fisherman's settlement

The location of old fishing port is a contributing factor to its resilience. This is because the old fishing port is a powerful magnet that has been formed since the beginning of Tambak Lorok fishermen settlement. This is proven by the fishing port, fish merchants surrounding, and also growing market that provides basic needs for people around the old fishing port.

The fishing port is also accessible from the homes of the majority of fishermen, making it easier for them to rest when landing after auctioning their catch. One of the fishing port facilities that support its survival is the quiet dock that is far from large waves, thus the fishing boats do not crash into each other when anchored. This is important because the dock in the new fishing port becomes one of the reasons why the fishermen do not want to dock there as the waves were too large.

4.2 Recommendation

Based on the results of the study, the author proposes some recommendations, among others:

a. In designing the fish auction place of Tambak Lorok, it is necessary to take into account the public perception and the role of the fishing port as center of activity.
b. Further research is necessary in order to investigate influencing factors of public perceptions in Tambak Lorok with quantitative method to support previous research.

5. Acknowledgment

The author would like to thank to the parties who supported the completion of this paper, among others: residents and stakeholders in fishermen settlement of Tambak Lorok who provided primary and secondary data in the completion of this paper, and also other parties who supported the writing of this paper until its publication.

6. References

[1] Hardiman G, Firmandhani S and Marlina A 2015 The Usability of the New Fish Auction Anchorage (PPI) in Tambak Lorok Semarang from Building Conditions and Environment Perspective *Ssrn* 1–8

[2] Aliyah I 2017 *Gede Market: Resilience of Traditional Market as Component of Surakarta City Structure* (Universitas Diponegoro)

[3] Djunaedi A 2000 *Research Methodology* (Yogyakarta)

[4] G. S, Cumming; G, Barnes; S, Perz; M, Schmink; K, E. Sieving; J, Southworth; M, Binford; R. D, Holt; C, Stickler; T V H 2005 An Exploratory Framework for the Empirical Measurement of Resilience *Ecosystems* **8** 975–87

[5] Firmandhani S W, Setioko B and Setyowati E 2013 Faktor Pembentuk Persepsi Ruang Komunal Di Pemukiman Nelayan (Studi Kasus: Pemukiman Nelayan Tambak Mulyo Semarang) *Teknik* **34** 95–102

[6] Doxiad\=es K A and Dix G 1977 *Ecology and Ekistics* (Elek)

[7] Grigg N S 1988 *Infrastructure engineering and management* (John Wiley & Sons Australia,

Limited)
[8] Suprapti A and Iswanto D 2018 Residing Tradition of Muslim Community in Java Northern Coastal *J. Archit. Des. Urban.* **1** 1
[9] Pandelaki E E, Suprapti A and Firmandhani S W 2018 Typology of social space in Kauman Kampong Semarang *IOP Conf. Ser. Earth Environ. Sci.* **99**
[10] SUPRAPTI A, KIM S, PANDELAKI E E and FIRMANDHANI S W 2018 a Spatial Dialogue of Heritage Village Between Kauman in Semarang and Seochon in Seoul Toward Preservation Development *J. Archit. Urban.* **42** 16–23

Scenarios of village alliances promoting local development

Listen Prima[1,2]
[1]Leibniz Universität Hannover, Germany,
[2]Universitas Sriwijaya, Palembang, Indonesia

Corresponding e-mail: listenprima@gmail.com

Abstract. Rapidly expanding urban agglomerations create various industrial and developmental challenges, and trigger debates about globalization in many fields especially in relation to basic need, self-resilience, and eco-development. As a consequence, it is urgently necessary to promote the perspectives of peripheral villages in regards to the resilience concept based on sustainable local development. This research focuses on the peripheral area of Bangke Village, located in the peripheral hilly area of Kota Agung Subdistrict, Lahat Regency, South Sumatera Province, Indonesia. It has distinctive cultural heritage richness both material and immaterial, specific landscapes and natural resources, as well as social and historical values. However, the conditions are vulnerable, which not only threatened its existence of the local and national characteristics but also the subject of the local development. The objective of this research is providing a focused analysis of local potential toward the explorative scenarios as an initial step to promote sustainable local development within the framework of village alliance. The village alliance is an innovative and applicative analysis framework for the context of peripheral development area, especially in Indonesia. It connects Bangke Village as an axis with other eleven villages in the district as an alliance. Regarding the local potential, this research outlines five explorative scenarios such as food network, tourism, wood production, new village, and centre of learning. Hence, the methodological framework used the case study that consists of multiple tools of field survey, mapping analysis, and SWOT analysis.

Keywords: sustainable development, resilience, local potential, explorative scenario, village alliance

1. Introduction

The concept of sustainable development increases awareness of local development demands. The awareness relates with complex problems of social, territorial, economic, political and environmental frameworks due to globalisation. Hence, sustainability development framework is starting to redefine its significant underlying concepts, renewable energy solutions, environmental issues, green design, local government programs, corporate programs, tracking results, and future trends [1]. Regarding complex problems of environment, urban governance, and planning, implementation of the sustainable development concept would increase awareness of resilience of the area as well. Recently, topic of resilience is raised since area improvements are urgently needed to face the challenges of natural and urban transformation issues. Natural ecosystem services have to be integrated with the local development components, infrastructures, buildings, etc., and resilience projects must be able to increase the abilities of an area and minimise possible risks in future [2]. Thus, resilience improves abilities of an area to respond to local needs and priorities of its communities [3]. Moreover, to examine local community needs, an integrated action is needed to build the local capacity for development [4]. Local potential analysis is proposed as a tool of local development towards scenario planning as a strategic and communicative way to anticipate future transformation. This research outlines scenario alternatives as an explorative way to promote the local development, especially in the framework of villages integrated in an alliance within an area of Bangke Village in South

Sumatera Province, Indonesia. Hence, the scenarios alternative is outlined based on five dominant potential strengths of the village alliance: food and wood production, tourism capacity, local knowledge resources and human resource capability.
The idea of scenario alternative comes from the significant gap developing between urban area/city centre and peripheral villages in Lahat Regency. The regional regulation about space and regional planning in Lahat Regency does not address the local needs of the villages. Therefore, the implementation of the local development and planning in the villages is not effective, especially in terms of sustainable development concept. Moreover, the local development of villages through national agenda in Indonesia has begun and is managed by each individual village. Thus, there is no integrated program that coordinates villages to build strong connections as the way to promote sustainable local development. This research aims to create a framework for sustainable development by integrating nearby villages into an alliance. The framework can support the locals in expanding economic opportunities, preserving critical social and cultural heritage, as well as enhancing proper infrastructure and public facilities. This integration would contribute to a holistic perspective of local development planning to strengthen local potential of all villages and push local development towards adaptive and innovative programs in regards to future dynamics. The framework of village alliance would provide opportunities to analyse local potential in an alliance consisting complex components of heritage richness (the material heritage such as architecture buildings, and immaterial heritage such as local culture, traditions, rituals, and ceremonies), spatial and local development aspects (social, political, economic and ecological), and natural resources (scenic landscape as well as potential in some prospective sectors, such as agriculture, fisheries, and forestry). Regarding those potential strengths, this research proposes scenario planning as an explorative way to build on the valuable local potential through some specific objectives. The planning objective is outlined to fulfil the local needs in the future and reflected in more interesting programs, both for local people as the main actors and guests/people from outside as users. Therefore, this research would outline five scenarios alternatives: food network, sustainable tourism, wood production, new villages, and learning centres. Each of those scenarios has a principal aim based on main potential strengths of local development:

- The intensification of food production
- The advancement of heritage richness as resources of tourism industry
- The cultivation of the local forest as the source of wood production
- The development of responsive villages with the specific groups as the dominant inhabitants
- The valorisation of local knowledge as study centre.

2. Methods

This research used case study as research method by integrating all relevant components of heritage richness, spatial development, and natural resources in the framework of the village alliance as an integrated area. This approach accommodates the actual challenge related to the limited data, especially regarding GIS data at the level of the village alliance region. Moreover, the case study method used multiple tools, such as field work, mapping analyses, and SWOT analysis. The field work analysis was conducted in the case study area in terms of local capacity within the village alliance. Hence, mapping analysis could be described as the context analysis of the case study area and provided information on existing features as a solution for the condition of unlimited or undefined data. Then, result of field survey and mapping analyses formed the basis for the next steps, the qualitative analysis and SWOT analyses. SWOT analysis were constructed based on components of sustainable local development, such as spatial, social, economic, cultural, and government plans as the basic guidance to outline criteria for village alliance as well as alternative scenarios for the village alliance.

3. Discussion

Massive urban development contributes to environmental degradation that poses a significant concern in the context of sustainable development. Hence, it triggers to change a paradigm of development [5]. On the other hand, mass migration to big cities creates an opportunity for tourism back to villages as city-dwellers look for ways to remain connected to older traditions. It creates a chance to promote local development of some rural areas or villages, especially in South Sumatera Province.

Furthermore, this local development of those rural areas/villages will also be a creative economy strategy that is mainly dominated by low-income families. Local development based on local economy strategies would be more environmentally sustainable than any other large industrial development [6]. The proposed Bangke village alliance would balance the significance of its territory with components of spatial development as an important connection of local development. This connection is linear since the goal of spatial development is to set up an adequate advantage of certain territory [7]. Besides, the spatial development is beyond political-administrative, nation-state, or former geographic concepts of region [8] as unseparated components of local development in village alliance. The spatial development is also engaged with multi-sector and multi-stakeholder activities [9]. Moreover, "a robust urban agenda is expected to consider factors that improve the flexibility, adaptive capacities and sustainability of the urban system" [10]. Meanwhile, the concept of sustainability is challenged to create resilience capacity of a city [11]. Resilience should integrate all components of urban system, such as socio-technical and socio-ecological networks at multiple spatial level [12]. Peripheral areas are proposed as potential territories to build resilience in cities by a creative perspective to reconsider and replace processes of urban transformation based on initiatives arising from communities, local stakeholders, local administration and activation of more structured strategies [13] as a wider context of significance beyond the immediate area. Regarding the future development, "many cities (and local governments) are beginning to imagine alternative futures for themselves that go beyond the tradition of only providing for roads, rates and rubbish (and safety)" [14]. The alternative future will be addressed by scenario planning that respects the local history as well as its future [15] as a comprehensive perspective of development needs and alternative programs to address it [16]. Hence, 'scenario planning offers planners a way to bring together technical approaches and participatory planning in a systematic way to think creatively about the future' [17].

Furthermore, this step of the research outlines scenarios alternative as explorative way of planning. 'The explorative scenarios more often take their starting point in the future, compared to what-if scenarios, which are usually developed from the present situation' [16]. As reference, explorative scenario-building methodology has been tested in region Steinburg, Hamburg, Germany with three themes [18]; 'The Regional Village' that re-purposes local agricultural lands as productive multi-functional spaces for living, working, leisure, and tourism; 'Rural Towns' which reconnects small urban centre with the landscape; and 'Connecting Lines' that strengthens the relationship between Steinburg and the metropolis as connected by territorial development. Additionally, considering new complex metabolism of urban/rural settlement, several European countries, such as France, Germany, Poland and Italy revised territorial and rural models based on the current view of territorial relations [19] as an alliance. National Strategy 'Strategia Nazionale per le Aree Interne' (SNAI) Italy has been trying to boost development of Sicilian Inner Area 'Terre Sicane' which was formed by twelve towns of rural and villages in the hillside and mountains characterised by cultural, agricultural and landscape to become a connective and relational space [20]. SNAI represents an evolving government approach that addresses official authorities as well as informal governance groups to overcome the classic opposition between urban-rural [21]. Meanwhile, in Indonesia, planning policies come from the national level to regional as top-down practice. However, regarding peripheral area development, since 2015 Indonesian government has set up a program of village fund for every village and rural areas that focuses on village development to foster local economy [22]. In the three years since its implementation, this program has brought significant means to local development of villages and rural area in Indonesia. However, beyond infrastructures, the effectiveness of villages fund in the field implementation is still debatable.

3. 1. The village alliance

The Bangke Village alliance as the case study area is located in isolated hilly areas and consisting of eleven villages (see figure 1). There are three components as the main consideration in integrating the villages as an alliance defined based on the sustainable development concept: space, social-economy, and spatial planning and governance.

IOP Conf. Series: Earth and Environmental Science **213** (2018) 012007 doi:10.1088/1755-1315/213/1/012007

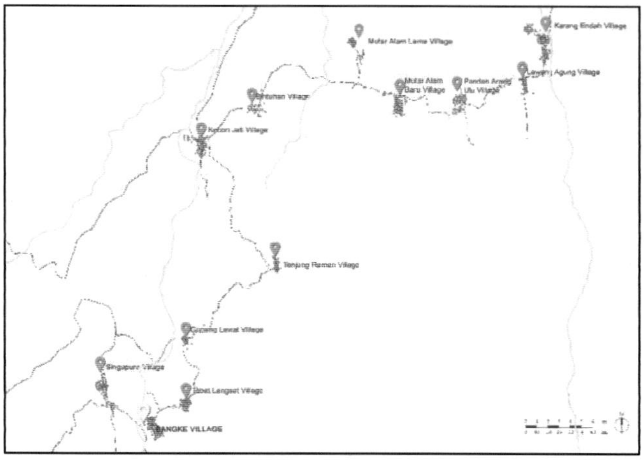

Figure 1. Map of Bangke Village and its alliance. Graphic. Listen Prima, 2018

The field survey was conducted to describe spatial context of the case study areas such as daily activities, traditions, culture, ceremonies. The mapping analysis supports the explanation about condition in several contexts. The results showed that the hilly area influences the creation of some local potential strengths such as agricultural farms in the upper area and at the border of the settlement. The result also indicated that the local people build the settlement clusters along with the watery area from the waterfall, which is located in the isolated forest among the villages. Moreover, the settlement cluster also influences the connecting line between each village, such as for access and transportation, buildings and facilities, and other facilities.

SWOT analysis constructs local potential of the case study area based on both quantitative and qualitative data from the field survey. The result of SWOT analysis defined the potential and challenges regarding space, social and potential values, as well as spatial planning and governance in the alliance (refer to table 1).

Table 1. Strength, Weaknesses, Opportunities, Threats Analysis Result

	Opportunities	Threats
Strength	• Vast capacity of heritage, natural resources as the primary sector of the development • Possibilities of the infrastructures network and alternative access • Transparent government practice with local governance instruments as the representatives in each village • Management of some potential areas to support local economic development, such as registered objects for protection, allocating fund for village to develop equally, and other activities.	• High frequency of the land use transition • Extension of the settlement area due to population increase • Modern perspective as a threat to traditional social/cultural values • Overlapping local regulations implementation and the dominance of some people/group
Weaknesses	• Loose control and management of environment • Unbalanced demographical composition • Economic over-reliance on the agriculture sector • Unequal focus of development of the villages • Limited potential of human resources	• Poor economic condition • Disaster may happen in many spots • Neglected local history • Ineffective work of local instruments

3. 2. Local potential

Based on the qualitative analysis of the case study, the local potential of village alliance could be defined in terms of the five scenarios outlined: food production, tourism objects, wood production, socio-demography, and local knowledge.

The villages have varied scales of agricultural/fishery activities (see figure 2,3) but they can be maximized to develop a more sustainable local economy. In Lahat Regency, total agriculture area is about 83% of the total region with the main food production of rice, maize/corn, sweet potato, peanuts and soybeans [23].

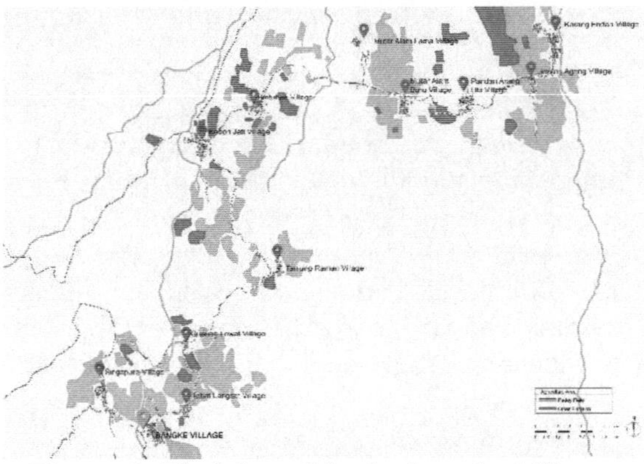

Figure 2. The Map of Agriculture in Alliance. Graphic. Listen Prima, 2018

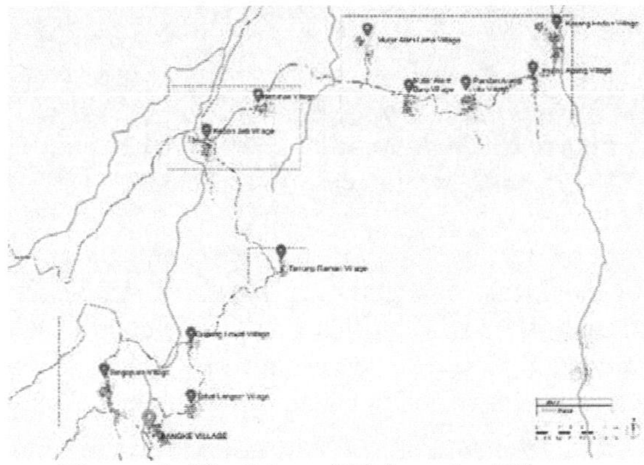

Figure 3. The Map of Fishery in Alliance. Graphic. Listen Prima, 2018

The village alliance has huge local potential for the tourism sector which is divided into distinctive groups: heritage architecture buildings and sites, natural resources, and the local values (see figure 4,5). The local potential of wood production in alliance has not developed yet. The total private wood production by individuals in Bangke Village alliance is 1316 Ha [23] and the locations of the production are spread in each village (see figure 6).

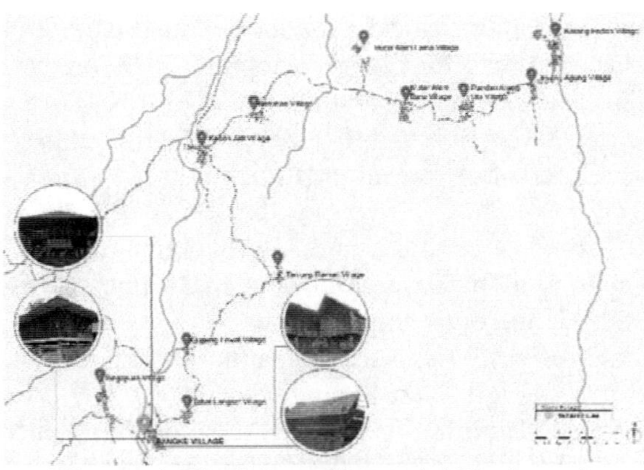

Figure 4. The Map of Material Heritage in Alliance. Graphic. Listen Prima, 2018

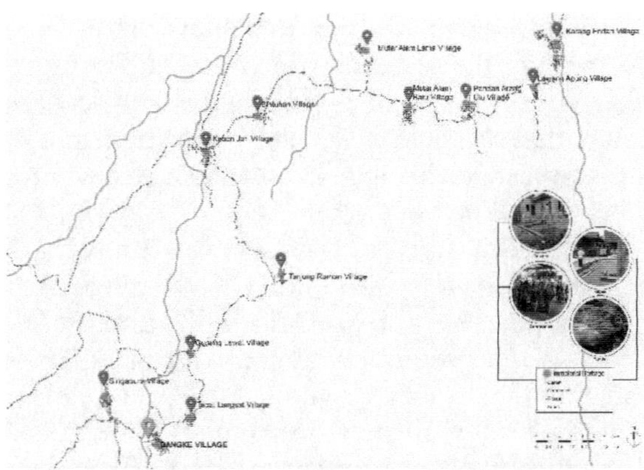

Figure 5. The Map of Immaterial Heritage in Alliance. Graphic. Listen Prima, 2018

The local potential of the wood production will support implementation of local development based on the sustainable development concept and the local identity and provide economic opportunities.

Recently, the demographic composition in the case study area is dominated by the elderly and children. Based on BPS-Statistics of Lahat Regency [23], the percentage of elderly was approximately 14% (over 60 years old), and children were approximately 30% of the total population, whereas many young people leaving the rural areas due to the job opportunities, education, and other reasons.

Regarding social demographic, three components have been defined: population number, public facilities, and organization (see figure 7).

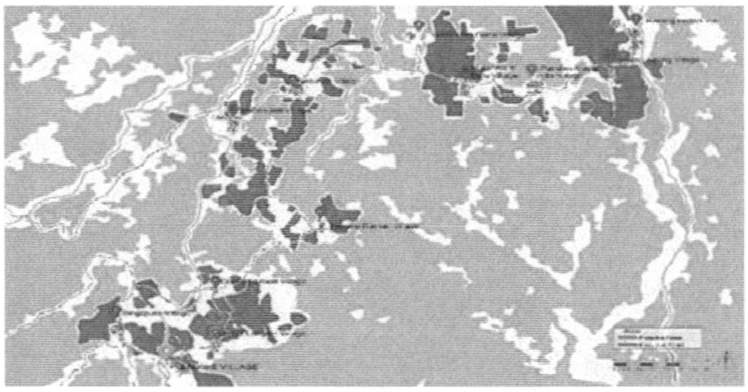

Figure 6. The Map of Forestry in Alliance. Graphic. Listen Prima, 2018

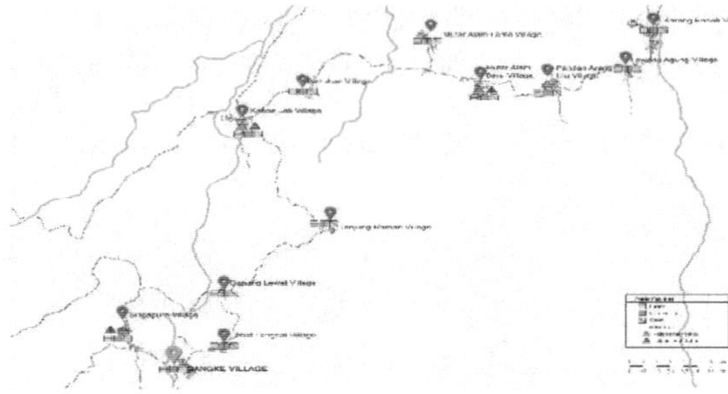

Figure 7. The Map of Demographic Composition and Public Facilities in Alliance. Graphic. Listen Prima, 2018

Beyond local characteristics, the local richness, such as cultural and heritage values in the village alliance offers possibilities as a source of information, cultural knowledge, experiences, as well as an interesting subject to learn both in formal and informal education programs. In the village alliance, schools are provided in some villages (see figure 7).

3.3. Scenarios alternative

3.3.1. Food network

The food network aims to build more sustainable agricultural practices to feed the villages more effectively with local agricultural products. It will also take advantage of the trend of consuming local and healthy food.

The food network scenario will outline three sub-scenarios: local market, regional market, and global market. The local market sub-scenario aims to increase agriculture production by focusing on the food plantation to fulfil the local daily need of the village alliance such as rice, fish, vegetables, and other product. It will build autonomous food supply within village alliance and anticipate food crisis in village alliance. Regional market sub-scenario promotes local product characteristics by increasing the varieties which may differ to any other products from other suppliers in Sumatera Island. The focus of global market sub-scenario is increasing export of Indonesian goods. These activities will intensify food production in agriculture, fisheries and other potential products by using recent technology and method, as well involving experts and related institution.

3.3.2. Tourism

The tourism scenario outlines two sub-scenarios of domestic tourism and international tourism. Domestic tourism sets the village alliance as a new tourism destination for local tourists in Indonesia. This sub-scenario will utilize local potential not only for activities such as cultural heritage tourism, tracking and river activities, agricultural activities such as paddy plantation but also for the local community to take active role under supervision of the local government.

International tourism targets tourists from around the world. By offering complex local potential and activities, this sub-scenario will set village alliance as the main tourist destination in South East Asia. Heritage richness will be reproduced in many activities not only as object of tourism but also as a subject to develop further model of tourism; for example, remodelling detail and constructing traditional houses. The natural landscape will also provide for many activities such as, agricultural tourism, nature sports, and other activities. Activities with active engagement with local communities will be held as the part of daily activities of local people, such as cooking, handcrafting, performance and arts, and other activities. In order to accommodate tourists, this sub-scenario will aim for facilities development, such as accommodation, commercials, and other facilities. Additionally, coordination and involvement with related experts and parties will be needed in this sub-scenario as the part of national tourism strategic planning.

IOP Conf. Series: Earth and Environmental Science **213** (2018) 012007 doi:10.1088/1755-1315/213/1/012007

3.3.3. Wood Production

Wood production scenario is the idea of maximizing capacity of local forest as a new source of wood production in region. Thus, this will strengthen local culture where wood is dominant material for local buildings, especially traditional houses as well as the most ecological material. Based on the coverage of wood distribution, scenario of wood stock outlines three sub-scenarios: local supplier, national supplier and global supplier. The local supplier sub-scenario aims to cultivate local forests as an agricultural side activities for local people. It will bring benefit not only a new economic source for local people, but also a way to protect the existence of productive forest in the alliance, and in the same time maintaining local price for wood production. Meanwhile, the national supplier sub-scenario will set the alliance as the main supplier in Indonesia region. This is scenario is significant since many regions need wood supply especially where traditional houses exist. The international supplier sub-scenario will bring back position of Indonesia as one of the biggest wood suppliers in international market. The forest cultivation aims to produce the maximum wood production. As the part of marketing strategy, village alliance will open for direct activities from market actors, both for investment and further manufacture, such as furniture, handcraft, and other adapted products.

3.3.4. New villages

The scenario of new villages responds to anticipate future needs of the local demographic within the village alliance, which is currently dominated by elderly people (over 60) and needs to retain with productive age (17-35). This concept will consistently group people with certain ages in the same preferences, such as capabilities, knowledge, hobbies, and other preferences as local characteristic of the environment and maximise those characteristics to bring benefit for the communities and development.

The old people sub-scenario sets all components or specific designs of the village structure to be elderly-friendly, such as ramps in common facilities, various groups of social activities, better health and psychological facilities, and other facilities. This sub-scenario maximises the capabilities of elderly as the resource of local values, such as culture, traditions, ceremonies, and other local values to support the local development by teaching and sharing them to young people and visitors. Moreover, this scenario will keep the local characteristics of warm and friendly people stigmatised by old people in the village alliance by showcasing the hospitality of the elderly people as part of village culture. Additionally, this scenario also brings benefit for old people to create the most valuable moment when they can spend rest of their life time in the place where they were born while building the character of a dependent society able to fulfil their own daily needs. Since in local society, old people typically bring additional responsibilities for their young family, this sub-scenario will increase the quality of life for both the elderly and their families by increasing the self-sufficient community as well as increasing their daily productivity based on the possible capabilities as the local cultural expert.

The young people sub-scenario aims to sustain demographic composition of young generation in the village alliance as the productive group of age. The society is designed to be more dynamic and contribute to vast local development by focusing on the amusement functions, such as cinemas and education facilities, workshop and training centre, public library, and other functions. Hence, development of mix sectors activities or local industries will support the productivity of young people as the most targeted users.

3.3.5. Centre of learning

This scenario aims to promote local knowledge beyond preservation. It brings forward local knowledge as a component to local development through academic and non-academic programs.

This scenario offers two types of learning activities as sub-scenarios: research aims and vocational skills. The sub-scenario of research aims set the village alliance as the centre of research or deep learning about specific cases/topics related to the local knowledge. The activities are more formal by involving academics and education institutions relevant to the subject. The number of participants will be limited only for the specific cases. The vocational skills sub-scenario offers the village alliance as the place for preparing and training specific vocations related to historic preservation, traditional construction, tourism/hospitality, agriculture and forestry, and other interests. The activities in this sub-scenario are more practical, such as workshop, training, and courses involving the local government, tourism board, vocational school, and all related parties. The target number for participants will be big groups with scheduled activities.

Future transformation due to the pressure of complex environment triggers each territory to increase effective local development. In response to these challenges and debates of globalization, interests in sustainable development has grown. The significance of sustainability development framework is started to apply in many programs of local development. This kind of awareness is related to the local development demands in urban and territorial planning.
The implementation of sustainability concept will be initiated by outlining the concept of village alliance within peripheral area. It improves local development by increasing transferability and effectiveness based on the dominance condition of peripheral villages in Indonesia. Hence, the context of peripheral area is significant, not only in transforming underdeveloped areas, but also in supporting the process to build resilience of local territories.
Analysis of local potential is proposed to identify future development opportunities. The analysis result will identify the potential components of local development as well as provide wide range of opportunities to strategically plan and promote local development; which is beyond conventional planning, with a more balanced environmental impact for the future.
The analysis results provide data for the next step of scenario planning, a creative approach which defines the significance of five local scenarios in village alliance: food, wood, tourism objects, socio-demographic condition and local knowledge. This scenario planning is an explorative way to construct alternative of development planning in village alliance before stating a prospective local planning development. This exploration guides a proper perspective of local development possibilities beyond the data and documents of the case study area. Furthermore, and it also offers an outlook of possibilities for future needs regarding collaborations with planning instruments and institutions, involvement of relevant groups and participants, as well as engagement within communities of village alliance.

4. Conclusion

A creative strategy that offers an outlook of future transformation by engaging components of politics, society, economy and culture in future development is urgently needed. This strategy must be based on the consideration of complex urban or rural environment as the initiatives to build resilience within sustainability development framework. The traditional planning method based on planning instruments creates development of gap between urban and rural area.
This research offers innovative scenario planning as a way to promote local development by exploring development within the context of village alliance as a comprehensive framework for sustainability. Hence, scenario planning alternatively provides more visionary planning perspective. This perspective gives more consideration to local potency in outlining prospective development approaches by involving local stakeholders and related institutions to form a more holistic perspective. Thus, it can be concluded that scenario alternative planning method provides a wide range of possibilities for future transformation and integrates all relevant components of local development, such as socio-demography, economy, culture, and politics.
Furthermore, this research is not limited to only one alternative scenario. Instead, the integration of scenarios are proposed to promote the local development. This method will promote each strong characteristic of those five explorative scenarios, namely tourism potential, food and wood resources, demographic composition, and knowledge sources as a comprehensive framework of scenario building. Meanwhile cultural heritage tourism will be used as the main goal, and this integrated scenario constructs all those main local strengths as powerful sources for local development.

5. Acknowledgment

This research was supervised by Prof. Dipl.-Ing. Jörg Schröder, Faculty of Architecture and Landscape Science in Leibniz Universität Hannover, Germany, and funded by BUDI-LN Scholarship of Kemristikdikti-LPDP (Indonesian Endowment Fund for Education). The author is also grateful for comments from Zoe Vercelli, MA., Fachsprachenzentrum in Leibniz Universität Hannover and for the supports from academics/staffs of Department of Architecture, Faculty of Engineering, Universitas Sriwijaya, Indonesia.

6. References

[1] Rossa, Stephen A. 2008 Sustainable Development Handbook. United State: The Fairmont Press, Inc. 44-45

[2] Jha, Abhas K., Minner, Todd W., Stanton-Geddes, Z. eds. 2013 Building Urban Resilience Principles, Tools, and Practice. Washington: The World Bank 2, 9

[3] Sattherthwaite, D., Dodman, D. 2013 Toward resilience and transformation for cities within a finite planet. *Environment and Urbanization: International institute for environment and development* **25**(02), 291

[4] European Commission 2015 Cultural Heritage Counts for Europe Report, and Krakow: CHCfE Consortium 9

[5] Steffen, W., Crutzen, Paul J., McNeill, John R. 2007 The Anthropocene: Are Human Now Overwhelming the Great Forces of Natures? *Royal Swedish Academy of Science* **36** (8) 614-621

[6] Fleming, Rachel C. 2009 Creative Economy Development, Sustainability, and Exclusion in Rural Areas *The Geographical Review* **99** (1) 61-80

[7] UN-Habitat III 2016 Urbanization and Development: Emerging Futures World Cities Report 2016 Nairobi, UN-Habitat

[8] Schroder, J. 2017 Towards an Architecture of Territories in Schroder, J et al. eds. 2017 Territories,
Rural-Urban Strategies, Berlin: Jovis 14-35

[9] Odendaal, N., McCann, A. 2016 Spatial Planning in the Global South: Reflections on the Cape Town Spatial Development Framework **38** (4) 405-423.

[10] Delgado-Ramos, Gian C., Guibrunet, L. 2017 Assessing the Ecological Dimension of Urban Resilience and Sustainability *International Journal of Urban Sustainable Development* **9** (2) 152

[11] Ahern, J. 2014 Urban landscape sustainability and resilience: the promise and challenges of integrating ecology with urban planning and design *Landscape ecol* **28** 1203

[12] Meerow, S., Newell, Joshua P., Stults, M. 2015 Defining urban resilience: A review *Landscape and Urban Planning* **147** 39

[13] Olivia, Jessica S. 2018 Resilience in City Core Peripheries: Flexible, Creative and Dynamic Urban Regeneration Process in Schroder, J., Carta, M., Ferreti, M., Lino, B., eds. Dynamics of Periphery. Berlin: Jovis 219,223

[14] Inayatullah, S. 2011 City futures in transformation: Emerging issues and case studies *Future* **43** 654

[15] Van Notten, P., Rotmans, 2001 The future of scenarios, Scenario and Strategy Planning **1** (3) 4-8

[16] Börjeson, L., Höjer, M., Dreborg, Karl-H.,Ekval, T., Finnveden, G. 2006 Scenario Types and Techniques: Towards a User's Guide *Future* **38** 723-739

[17] Chakraborty, A., McMillan, A. 2015 Scenario Planning for Urban Planners; Towards a Practitioner's Guide *Journal of the American Planning Association* **81** (1) 18-29

[18] Ferreti, M. 2017 Scenariobuilding for Regiobranding in Schroder, J et al. eds. 2017 Territories, Rural-Urban Strategies. Berlin: Jovis 88-95

[19] Carta, M. 2017 Planning for the rur-urban Anthropocene in Schroder, J., Carta, M., Ferreti, M., Lino, B. eds.Territories, Rural-Urban Strategies. Berlin: Jovis 40

[20] Contanto, A. 2018 The-periphery-city of the inner areas: local development in the Sicani territory in Schroder, J., Carta, M., Ferreti, M., Lino, B.eds. Dynamics of Periphery. Berlin: Jovis 195-197

[21] Lino, B. 2017A New Rur-Urban Utopia? Social innovation and the case of the Sicani area in Siciily in Schroder, J., Carta, M., Ferreti, M., Lino, B. eds. Territories, Rural-Urban Strategies. Berlin: Jovis 111-112

[22] Ministry of Finance Republic of Indonesia 2017 Buku Pintar Dana Desa. Jakarta: Ministry of Finance 11-16. Accessed at 23/05/2018
[23] BPS-Statistics of Lahat Regency 2017 Lahat Regency in Figure 2017, BPS- Statistics of Lahat Regency 32, 111. Accessed at 19/03/2018

Cultural Significance of the Place: the Key Factor in Urban Heritage Programme

Yohannes FIRZAL[1)], Irham TEMAS[2)]
[1,2)] Universitas Riau, Pekanbaru - Indonesia

Corresponding e-mail: yfirzal@eng.unri.ac.id

Abstract. Siak Sri Indrapura city on the east coast of Sumatera was the centre of Malay Sultanate in 18th-20th century. The development of the city combined local and Europe architecture. In one side, this creates a unique urban character on how this architectural sense meets with Malay culture, which has maintained as the way of life. However, despite recognised as the regency capital in the reformation period today, the city struggles to maintain this architectural sense, in particular for continual urban development that is dealing with more socio-cultural challenges. Therefore, this paper investigates how Siak Sri Indrapura city deals with this situation, and attempts to adopt the urban heritage programme (*Program Kota Pusaka*) in last two years. The programme is assumed not only giving an opportunity to redevelop and to maintain the architectural sense, but also embracing Malay socio-cultural life in the same time. This paper used qualitative research approach through a systematic process that consists of the preparation stages, field visit, data analysis and interpretation, community empowerment, and the findings. This paper affirms that the urban heritage programme at Siak Sri Indrapura city has mainly determined by a cultural significance factor. The factor has not only used to maintain an exclusive set of collective local identities of the city, but also to become the main key factor in today's urban heritage programme in Indonesia.

Keywords: Cultural Significance, Urban Heritage Programme, Architecture, Malay

1. Introduction: Implementing Urban Heritage Programme

Siak Sri Indrapura city was found by the Malay Sultanate in18th century. As built in centuries, character of this city has influenced by several developing scenes, and most of them still can be seen today. Traditional houses, old market, China town, Malay tradition, colonial buildings, and new contemporary buildings are some urban attributes that can be easily found. This city became a new mixture frontage of modern urban Malay development. In particular today, along with Indonesia's reformation period, it is giving more flexibility in utilizing urban development approach to local governments in order to build their city based on privilege needs such as local character, identity, and Malay culture [1], [2].

However, altering parts of the city is not only giving some advantages. The local government has trying to develop the city in sense of local tradition; by preserving old buildings such as the palace, the sultan mosque, the *Balairung*, and is also embracing Malay culture for people in their daily life tradition. In this matter, the rapid changing can potentially re-characterizing the whole character of the city into another direction that is probably not aligned to local agenda. Therefore, the local government has finding alternatives approach by identifying significance factors in rebuilding the city.

The significance factors can be seen in several views. Some of them can be rooted in the values attributed to the heritage [3]. In this view, significance is not only found on historic district, but is also involving inhabitants and their way of life in the historic district. This leads to legitimacy of the cultural values of the district [4] that are possible in the forms of tangible and intangible dimensions [5]. Therefore, cultural significance of the place is represented in the place itself, its fabric, setting, use, associations, meanings, records, related places and related objects.

In light of the above, every place is unique, and may has range of values that makes the place to become different one to another. The cultural significance in this paper is defined as the sum of the qualities or values of the place. Therefore, this paper investigates the importance of the cultural significance as the main key factor in urban heritage program at Siak Sri Indrapura city.

2. Methods: From Research to Practice

Investigating, surveying and analyzing process are aimed to ensure that all decisions and actions that are proposed and documented in the urban heritage program can show the qualities or values of the city. However, there are two important aspects in collecting and analyzing proses. First aspect is related to first hand field data [6]. This means the data have to obtain by directly engage to the place [7], and become as primary data. Secondly, local people involvement is important whether it is categorized as empowerment or participatory [8]. This involvement is seen as reliable ways to understand what and how local people want their city for following years. In short, it is important to listen what the local expectations.

These first hand field data and information gathered from the local people need to be validated through front desk research [9]. This validation can be obtained from previous research made, old maps and corresponding letters that had been made in sultanate and colonial time. This validation process is also to become a secondary data that support whole analyzing process [10], [11]. Gathering information from key informants also can contribute in validating process. In this ways, key informants have two roles; helping to identifying and correcting field data, and putting as resource of information [12], [13]. Moreover, consultation to the officers of local government is seen as significance role as well. It not only served as ways of confirmation, but also found what data and information unavailable openly to public.

3. Discussion: Embracing Heritage and Cultural Context

In Indonesia, urban heritage has well acknowledged as a programme for many years [14], and is popularised in early 2000s [15]. Today, this urban heritage programme is also known as P3KP - *Program Penataan dan Pelestarian Kota Pusaka* in term of protection of Indonesia's urban heritage that is sponsored by the government and has closed working together with BPPI – *Badan Pelestaroan Pusaka Indonesia*, and JKPI - *Jaringan Kota Pusaka Indonesia*. According to its guidance [16], the programme consists stages within range five to six years to its completion. Early stage is known as the proposal and action planned stages. The local leader signing commitment to conduct following stages encloses both stages. This stage becomes crucial as the city has to able to show and prove its distinction. In other word, the most significant part of the city is the primary reason why the city is qualified enough to obtain the programme.

In planning, urban heritage programme can be understood in various ways [17]. First, the program is seen as a process of managing change. In this way, the program can help to preserve layers of urban life such as memory, history, and diversity. Secondly, the program is seen as a process of protection. This protection is aiming to take care and to guide any planned intervention on historical, social and cultural significance. Another way, the program can be found as the recognition and planned retention of the built environment as heritage. By well known all means above, the urban heritage program can be smoothly drawn in practice [17].

In mid 2016, the local government of Siak Sri Indrapura City has been decided to joint to urban heritage programme (P3KP). A proposal submitted, and accepted in the end of year (fig. 1). As highlighted heritage and cultural context, the proposal consists of a list of tangible and intangible

IOP Conf. Series: Earth and Environmental Science **213** (2018) 012008 doi:10.1088/1755-1315/213/1/012008

attributes of Siak Sri Indrapura city that are all assumed relate to local culture, and are part of the city such as historical information, old maps, urban fabric, traditions, and also all administrative paper requirements.

Figure 1. The urban heritage proposal, 2016.

Figure 2. The proposal of the urban heritage action planned, 2017.

In 2017, the local government continued to submit another document named the urban heritage action planned (RAKP) (fig. 2). This document is not only used to explain about their planning for following years, but is also required a statement of significance. Statement of significance is understood as a concise and distilled summary of the cultural significance of the place [19]. The statement, in fact, needs to be justified by depth analysis of every attributes or criteria [20] that have related to the heritage and cultural context of Siak Sri Indrapura city. This process has found as the hardest process as require more sources and evidence in justifying and summarising all attributes or criteria that have proposed such as urban fabric, pattern of the city, setting, zoning, significant buildings.

The first attribute is about the Siak Sri Indrapura city itself. In brief, the city was built close to the *Jantan* River by the Malay sultanate in 18th century. The river was not only served as main transportation line heading to the strait of Malacca (fig. 3), but also used as income generated for the sultanate by taking trading tax along the river for hundreds of years, and the sultan's palace also built facing to the river in 1889. Architectural of the palace was found as combination style of European, Malay and Arab architecture (fig. 4), and was designed by Germany architect. From this perspective, the river has important role in not only constructing the city but also connecting the city to other parts of world that was contributing in shaping local tradition as well. This shows that the city has been widely influenced beyond traditional values through riverine life since the beginning.

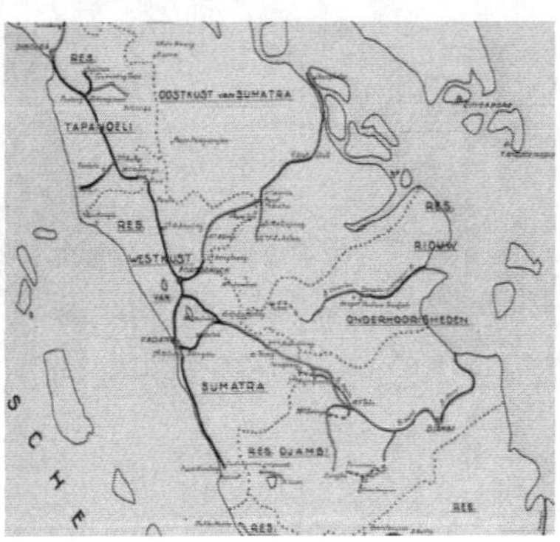

Figure 3. The city is connected by river to Malacca strait.

Figure 4. The Palace became the centre of the city.

IOP Conf. Series: Earth and Environmental Science **213** (2018) 012008 doi:10.1088/1755-1315/213/1/012008

Historical urban fabric and pattern are another important attributes in understanding how the city grew at the beginning [21]. One important aspect of the urban fabric is to know where the first center of city in order to locate the oldest part of the city. This area might be potentially valuable to become heritage district (fig. 5). By locating the area on today's map, it can be used to figure out of how the city has growing and where its direction and trends to be as well [22]. In the case of Siak Sri Indapura city, the oldest part of the city is still stay at riverbanks, and new development is growing to dry-land side (fig. 6). From this point, urban fabric and pattern can be used to highlight where urban heritage district will be focused; at riverbanks area. This shows how the river has still important to the city.

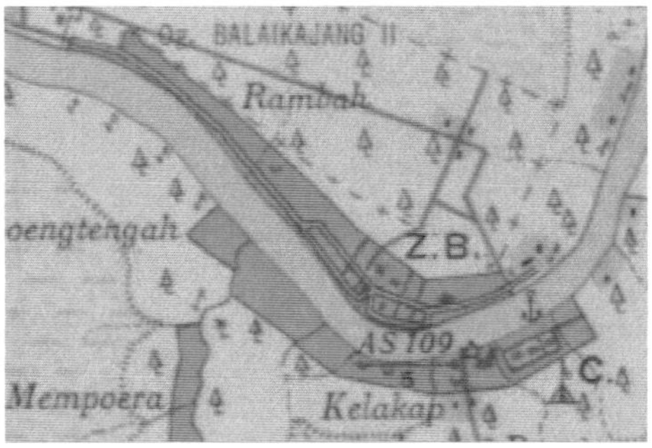

Figure 5. The earliest urban fabric formed in 1898.

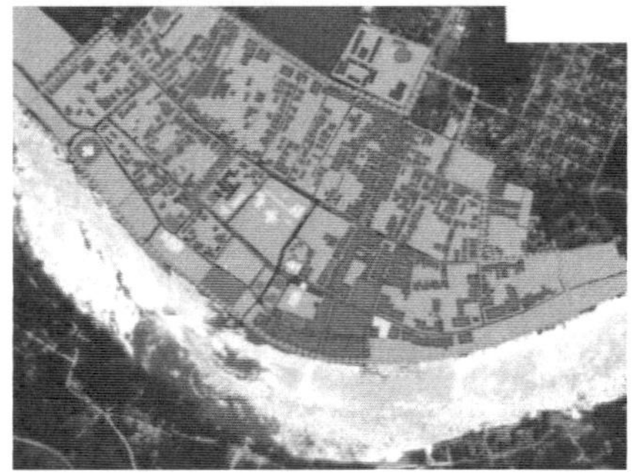

Figure 6. Today's zoning of the city, 2017.

Another attribute of the heritage and cultural context is by reviewing on built forms of the city. Despite new contemporary buildings growing up, remaining old buildings can be seen in this city. They have significance values not only in architectural perspective, but are also from historical and tradition values apart of how they were built [23]. Some of the buildings have related to the sultanate such as the *Balirung* (Court Hall) (fig. 7), and some of them had colonial experiences (fig. 8) that was definitely not a bright side of the city's history. However, from this built forms attribute, significance buildings are playing siginificant role for incoming heritage district [24]. They had setting and charactering on city's built form through their unique architectural style, and have continuing in influencing on today's buildings style at Siak Sri Indrapura.

Figure 7. The *Balairung* as representation of local architectural style.

Figure 8. Remaining of the Dutch Military building.

Thus, the urban heritage action planned document is enclosed with statement of significance. In order to highlight distinction of the city, the local government has proposing the statement as *Terwujudnya Kota Siak Sri Indrapura Sebagai Kota Warisan Budaya Melayu Berbasis Ruang Sungai Dan Memiliki*

Peran di Asia Tenggara which literally means to becomes a city of Malay cultural heritage based on riverine tradition in South East Asian region. From the statement, it shows that heritage and cultural context become basis of the plan, and have used as main factor in their urban heritage program proposed.

The proposal has written through several documentation approaches. For built environment attribute, all-important old buildings whether concluded as heritage object or historical and landscape matters are identified and mapped (fig. 9). This means to figure out where significance built forms of the city are scattered. It was continued to mark and to delineate them by zoning and cell technic. From this approach, the attributes are grouped into three main areas (fig. 10). Then more detail object analysing was conducted to validate each significance values. From these approaches, delineation of heritage district has obtained by combining three main areas and the river (fig. 11). The district has also recognised and approved by the central government thru the Minister of Education and Culture, and by national heritage expert team –TACBN in mid 2018 [25].

Figure 9. Significance buildings are scattered both side of the city

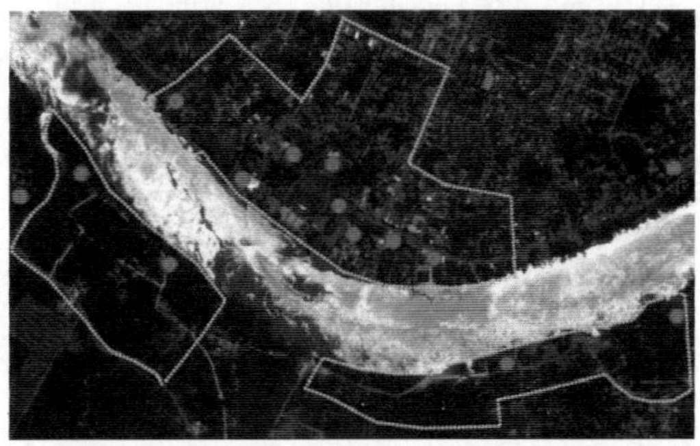

Figure 10. Three main areas of significant buildings

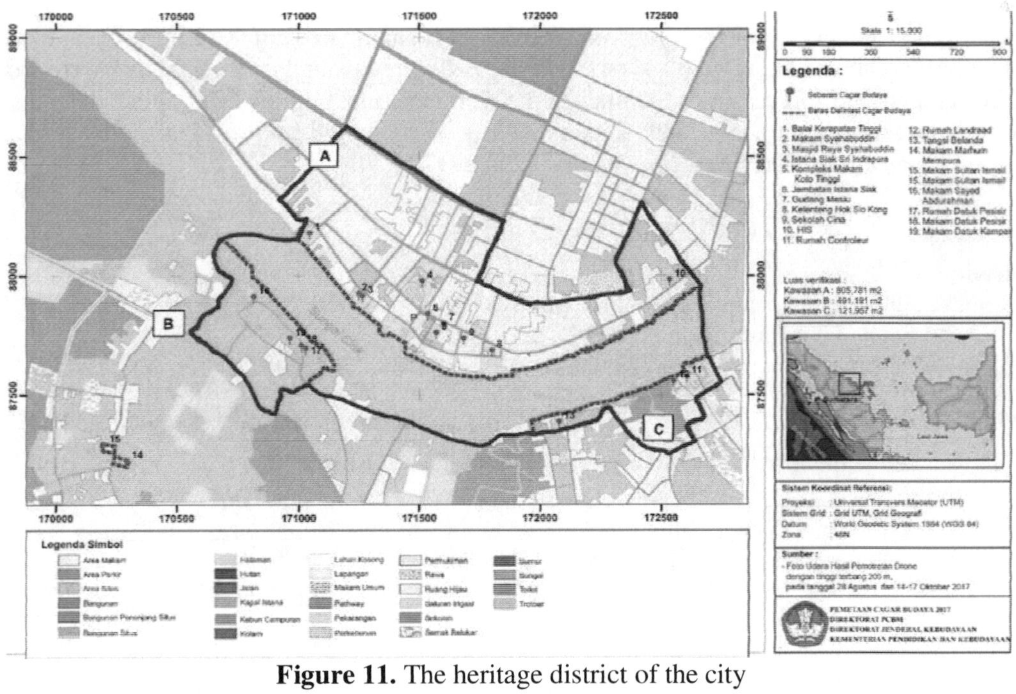

Figure 11. The heritage district of the city

Conducting the urban heritage program has brought different development experience to the Siak Sri Indrapura city. Despite its complexity, doing whole process absolutely need serious commitment not only from the local government, but is thru involvement of more local people, NGOs, and stakeholders [26]. The program is not only used to bring back historical cultural background for today's context [27], but has also officially acknowledged urban space for all local people, apart of what their ethnic background such as Malay, Chinese, Arab, and other incomer people. Re-arrangement of the space is under way by proposing urban heritage district design guidelines (RTBL-P) in 2018. This will provide a clear heritage district highlighted to be developed according to based on its cultural significance [28].

Moreover, the heritage planning is going to provide a planning document [29] how to develop the district for next few years, and how local people and NGOs can actively and sustainably involve in the programme [30]. The last, despite initiated by the local government, the urban heritage program is basically coming from and will be giving back to local people [31]. In this respect, the program has prepared for local people how to conserve and pass along for next generation without loss of heritage and cultural values [32]. This is why implementing urban heritage programme at Siak Sri Indrapura is found as translation of theories to the ground - from research to practice that have conducted by many stakeholders.

4. Conclusion

In the case of Siak Sri Indrapura city, heritage and cultural context had significant roles, and became the basis to construct conception of the cultural significance in the urban heritage programme. This cultural significance is defined as the sum of the qualities or values represented and as representation of the city itself whether through interpretation, association, relatedness, and perception of the attributes and criteria. By doing this, the cultural significance becomes the main key factor in the urban heritage program. It is also useful for maintaining an exclusive set of collective local identities of the city as well. Further research can focus on monitoring tool on how the urban design programme runs and its co-relation to unpredicted future impacts on the city.

5. Acknowledgments

This paper is a summary from the proposal (P3KP) and the urban heritage action plan documents (RAKP) submitted to the urban heritage programme. All materials and pictures above were from the reliable documents and sources. As a member of the working team, authors can publish this part of the documents for academic purposes only. This paper is also supported by Riau University through the research grant, *Penelitian Unggulan Universitas*, from 2017 to 2018.

6. References

[1] Reid, A., 2001. Understanding Melayu (Malay) as a Source of Diverse Modern Identities, Journal of Southeast Asian Studies, 32(3), pp. 295-313
[2] Long, N.J., 2013. Being Malay in Indonesia. Copenhagen: NIAS Press.
[3] Hoi An Protocols., 2003. [Online]. Available at: http://unesdoc.unesco.org/images/0018/001826/182617e.pdf (Accessed: 2 June 2018).
[4] The Nara Document on Authencity., 1994. [Online]. Available at: https://www.icomos.org/charters/nara-e.pdf (Accessed: 2 June 2018).
[5] The Burra Charter., 1979. [Online]. Available at: https://australia.icomos.org/wp-content/uploads/Burra-Charter_1979.pdf (Accessed: 2 June 2018).
[6] Patton, M.Q., 2002. Qualitative Research & Evaluation Methods. 3rd ed edn. London: Sage
[7] Peattie, L.A., 1992. Aesthetic Politics: Shanty Town od New Vernacular? T D S R, 111(11), pp. 23-32.
[8] Kellett, P., 2000. Voice from the Barrio, Oral Testimony and Informal Housing Processes, Third World Planning Review, 22(2), pp

[9] Lincoln, Y.S. and Denzin, N.K., 2000. Handbook of Qualitative Research. 2nd edn. London: Sage Publications.
[10] Creswell, J.W., 2003. Research Design: Qualitative, Quantitative, and Mixed Method Approaches. 2nd edn. Thousand Oaks, California: Sage Publications.
[11] Richards, L., 2009. Handling Qualitative Data: A Practical Guide. 2nd edn. London: SAGE.
[12] Burgess, R.G., 1991. Field Research: A Source Book and Field Manual. London: Routledge.
[13] Tremblay, M.A., 2011. The Key Informant Technique: A Non□Ethnographic Application, in Burgess, R.G. (ed.) Field Research: A Sourcebook and Field Manual. London: Routledge.
[14] Eryudhawan, B., 2017. Urban Conservation in Jakarta since 1968, SPAFA 1(1), pp. 1-20 [Online]. Available at: http://www.spafajournal.org/index.php/spafajournal/article/view/174 DOI: http://dx.doi.org/10.26721/spafajournal.v1i0.174 (Accessed: 1 June 2018).
[15] JKPI., 2009. Brief History of Jaringan Kota Pusaka Indonesia (JKPI). [Online]. Available at: http://www.indonesia-heritage.net (Accessed: 2 June 2018).
[16] P3KP., 2015. Program Penataan dan Pelestarian Kota Pusaka. [Online]. Available at: http://sim.ciptakarya.pu.go.id/kotapusaka/ (Accessed: 2 June 2018).
[17] URA – Urban Redevelopment Authority & ICOMOS Singapore., 2017. Conservation Technical Handbook Volume 1 Introduction. Singapore: URA Press.
[18] Sveinung K. Berg., 2017. Conditional values of urban heritage: context and scale, Journal of Cultural Heritage Management and Sustainable Development, https://doi.org/10.1108/ JCHMSD-01-2016-0008
[19] The Burra Charter, 2013. [Online]. Available at: https://australia.icomos.org/wp-content/uploads/The-Burra-Charter-2013-Adopted-31.10.2013.pdf (Accessed: 3 June 2018).
[20] Sobhani Sanjbod, H., Hermans, L. M. C., Reijnders, D. A. H., and Veldpaus, L., 2015. Captain, where can we find the attributes? In S. Vigneron, & R. D. Pickard (Eds.), Conference Paper. International Conference on the Implementation of the 1972 UNESCO Convention Concerning the Protection of the World Cultural and Natural Heritage Legal and Practical Evaluation of the Management of Cultural World Heritage Sites, 14-17 January 2015, Kent, United Kingdom. pp. 1-13.
[21] Kuora. Hisako KUORA., 2016. Spatial Transformation and Sustainability of the Block Structure of Historic Urban Centres in Japan. Journal of Comparative Cultural Studies in Architecture, 09(2016), pp. 4-12.
[22] Dijk. Terry Van, And Weitkamp. Gerd., 2017. Place from the Past in New Tows: Hongkong's Old Villages, METU JFA Advance Online, pp. 1-24. DOI: 104305/METU.JFA.2017.2.51
[23] Ginting. Nurlisa. and Rahman, Vinky., 2016. Preserve Urban Heritage District based on Place Identity. Asian Journal of Environment-Behaviour Studies, ajE-Bs, 1(1), Maiden, December 2016 (p.67-77)
[24] Ginting, Nurlisa. and Wahid. Julaihi., 2016. Recalling the Past: Maintaining continuity in urban heritage tourism. Journal of ASIAN Behavioural Studies 1(1), pp.81-89
[25] Kemdikbud., 2018. Keputusan Meteri Pendidikan dan Kebudayaan Republik Indonesia Nomor 164/M/2018 tentang Penetapan Kawasan Cagar Budaya Pusat Pemerintahan Kesultanan Siak Sri Indrapura sebagai KawasanCagar Budaya Tingkat Nasional. Biro Hukum dan Organisasi Kemdikbud.
[26] Ripp. Matthias, and Rodwell. Dennis., 2017. Governance in UNESCO World Heritage Sites: Reframing the Role of Management Plans as a Tool to Improve Community Engagement, in Makuvaza. Simon. (ed) Aspects of Management Planning for Cultural World Heritage Sites: Principles, Approaches and Practices, New York; Springer. DOI: 10.1007/978-3-319-69856-4_18
[27] Neve. Mario., 2018. Would Urban Culture Heritage be Smart? Culture as a Land Factor and Italian Cities' Smartness, Journal of Communication and Languages – Cities of the Future, 48

[28] Adishakti. Laretna T., 2009. Tantangan Kota Pusaka Indonesia. [Online]. Available at: https://www.researchgate.net/publication/320498373
[29] Guzmán, P.C., Pereira Roders, A.R., Colenbrander, B.J.F., 2014. Bridging the gap between urban development and cultural heritage protection. Conference Paper. 34th Annual Conference of the International Association for Impact Assessment. DOI: 10.13140/2.1.4633.7923
[30] Guzman. Paloma., Ana R. Pereira Roders., and Colenbrander. Bernard., 2018. Impacts of Common Urban Development Factors on Cultural Conservation in World Heritage Cities: An Indicators-Based Analysis. Journal Sustainability, 10(853), DOI:10.3390/su10030853
[31] Cinà. Giusepp, Demiröz. Merve, and Mu. Qi., 2017. Participation and conflict between local community and Institutions in the conservation and regeneration processes The Case of Novara Old Town. [Online]. Available at: https://www.researchgate.net/publication/321213377
[32] Piagam Jakarta, 2003. [Online]. Available at: https://www.icomos.org/charters/indonesia-charter.pdf (Accessed: 3 June 2018).

Spatial Conflict in Urban Kampong Development A Case Study in Kampong Pendrikan, Semarang

Bangun IR Harsritanto[1*], Wulani Enggar Sari[2], Rizqi Jamaluddin[3], Ratih Widiastuti[4] , Rona Fika Jamila[5]
[1] Universitas Diponegoro, Architecture Department, Semarang, Indonesia
[2]Parahyangan Catholic University, Architecture Department, Bandung, Indonesia
[3]CV Luhur Mubarok, Demak, Indonesia
[4]Universitas Diponegoro, Vocational School, Semarang, Indonesia
[5]Universitas Mercu Buana, Architecture Department, Jakarta, Indonesia

Corresponding e-mail: bangunirh@arsitektur.undip.ac.id

Abstract. Urban development triggers transformation in kampongs. The concept of a kampong as a residential area that has transformed into mix used area creates spatial conflict. This study aims to identify and give brief explanations about spatial conflicts in urban kampongs. The research was conducted in qualitative research paradigm which operated through site observation, questionnaire and literature. The result showed multiple inhabitant perceptions and area development lead spatial conflicts in urban kampong, such as discussion about: vacant spaces, houses' transition area and houses' front yards. The evidence of spatial conflicts are clearly shown by signage, barriers, and even buildings installations.

Keywords: urban kampongs, spatial conflicts, perception, human settlement

1. Introduction

No man's land phenomena often appear in many residential spaces because public spaces in urban are considered as the secondary space and are owned by the city or individuals. Designers of contemporary neighborhoods, new towns and cities are dealt with challenges in creating open spaces and outdoor environments which are collective and unifying many frameworks for betterment. Urban betterment advocates buildings as separated objects in the landscape, considering the bigger fabric of lines and open spaces. The contribution of the urban designer reaches at best after the advocacy for the betterment [1]

The space and the temporary neighborhood open spaces system gives the materials for cognitive schemes, the social effective responses, and the cultural factors as the perceptual inputs, [2-4]. Those conditions include the meaning of neighborhood and communication, the nature of social relationships and connection, and the design of activity places. Neighborhood spaces have meaning for people, invite access for all, and encourage use and participation. The good neighborhood spaces are the ones well-maintained by their users. Therefore in planning process, users' perceptions and preferences are important factors. The perception and preference have meaningful sources of information for environmental planning [5-7]. Those factors are the important body of research however they are not frequently included into the process of neighborhood planning.

Urban planners are not enough to make new projects of residential and neighborhood environments. Therefore urban planners must learn how the working system of environment runs well [8-10].

IOP Conf. Series: Earth and Environmental Science **213** (2018) 012009 doi:10.1088/1755-1315/213/1/012009

Designers and researchers are demanded to realize that people may utilize their environments in predictable ways. However, in some conditions, people are unable to enjoy the planned environments and many available opportunities in the surrounding areas since designers and researchers are unable to predict them. Therefore, if planners and researchers want to learn how to improve the designs, they will see how people use space as the important thing.

The study of spatial conflict and conflict resolution various levels cacn show fundamental concepts of modern geography and social significance [11-12]. This concept has relevance in contemporary society and has generated large and developing areas of multidiscipline researches.

In spatial conflicts, the disadvantages may be caused by multiple factors, especially economic status [13]. Nowadays, several groups with less developed skills have severely limited access to the political infrastructures, and they also become less obidient; hence difficulty in simulating those factors. The advantages are usually related to money, mobility, and majority.

The main similarity between the simulation and the real conflicts is the arguments of social issues, especially the social issues related to selfish character or neighborhood objectives [14-15]. Members of the neighborhood frequently argue about the occupation of area such as trespassing of the landfill in their area which will have negative impacts and will escalate greater conflict. Residents of the neighborhood state that the landfill area will have an effect on the nearby commercial district affecting the local economic and social environments. Everybody uses the commercial area for shopping, dining, or entertainment, so this point is usually quite effective in winning the support at least from those living in the middle class neighborhood like urban kampong.

The scope of this paper is limited to the study of the 'spatial conflicts' in public spaces at the Pendrikan community and how people show them. Particular emphasis will be placed upon the spaces that are integrated with the physical design of the neighborhood. There are areas in Pendrikan that surrounding residential buildings that are poorly designed, and these areas deserve attemtion. There is a need to examine how such neighborhood spaces can be an appropriate and meaningful territory to serve people. Suggestions are based on site observations and literatures. Activities were observed, users were interviewed, and their patterns of outdoor space use were recorded. Useful neighborhood spaces were defined, and what forms of design that did and did not work were analyzed.

2. Methods

The methods of this paper were literature studies, site observations and questionnaires to identify the occuring or the future space conflicts. In next phase, compilations of the conflicts were organized and explained using literature reviews per case studies. The paradigm is the qualitative-rationalistic research which was explained in inductive ways using several evidences.

3. Discussion

3.1. Defining the spatial conflicts

According to the first-ranked scenarios of land use change, all coalitions will assign almost the total of the case study area to public land uses. The public land use spaces in combination with high priorities expressed for the Kampong environments are the major factors leading into the formation of a commonly scenario; that proposes the study area into vacant and transitional spaces. The issue of spatial dependence recognizes that neighborhoods are not isolated islands. Social interaction and access to community resources transcend neighborhood boundaries [16].

Through the previous literature reviews, the spatial conflict is defined as the neighborhood, as community or personals, having objectives on accessing the public space around their residence. The objectives of people in urban kampong are related to many factors hence variations of results. The different objectives in accessing public spaces around the residence are the paper limitation regarding spatial conflict term.

3.2. The spatial conflicts areas

Regarding to the previous study in Pendrikan, the main spatial conflict trigger the land use transformation [17]. The kampong's open spaces which are identified as spatial conflicts are vacant space (no man lands), space in front of houses, and the houses' front yards. The vacant spaces or no man lands are the part of streets having no house across from them (It can be the house backs or the house sides). This area usually is undeveloped and unauthorized by anyone until the local authority from RT (the chief of negihborhood unit) to city councils established a legalized plan on it. However the plan itself will never be realized if the the neighborhood community never considers the land as their region.

The second area is the space in front of the house (the house connector). This area is also the part of the street development program made by the authority. However the development of Pendrikan urban kampong is usualy grassroots proposal instead of top down regulation. In addition, Kampong Pendrikan is one of many kampongs in Semarang designed as indigenous settlement areas since Dutch Colonial era.

The third is the house front yards. In kampong transformation, the yards can be public spaces since they are the transition space and perhaps the only access between houses and streets in the front. The second and third areas are connected and related as spatial conflict area.

The three areas were defined as spatial conflict areas by people in urban kampong as result of different objectives and the 'as mine-your-our space' idiom. The parameters which determine the mine-your-our-space idiom in this paper are accessing the space (access), controlling the space (control), securing the space (secure).

3.3. The evidences of spatial conflicts

Vacant space or no man lands are located in many parts in Pendrikan. The location of vacant spaces can be identified on figure 1 as void areas between the street line and solid building. Some parts of this area are occupied by food stalls, personal/communal parking shelters, temporal house extensions (but in further development it becomes permanent without legalization), and unused areas that become garbage disposal.

The house connectors are noticed by the area between solid house and street line. It might be small and shallow depending on the house width, but the expansions to the street body depend on the social values and street width itself. These areas are occupied by food stalls, parking areas, and house extensions. The purposes are similar to the vacant lands, but the detailed evidences are different.

The house yards are actually private space, but in some cases, they are transformed into shops and houses and then defined as public space. The food stalls, stores and parking areas have transformed these private areas into public spaces.

Figure 1. Kampong Pendrikan from solid and void mapping

IOP Conf. Series: Earth and Environmental Science **213** (2018) 012009 doi:10.1088/1755-1315/213/1/012009

The house yard space conflict (showed in figure 2) occurs as the private access and public access collided in the same space. Some houses divided their access by making more than one (usually two) gates and doors. The house owners wished that if they use this area their activities can be controlled and secured and the separate access still keeps owner of houses in-out access.

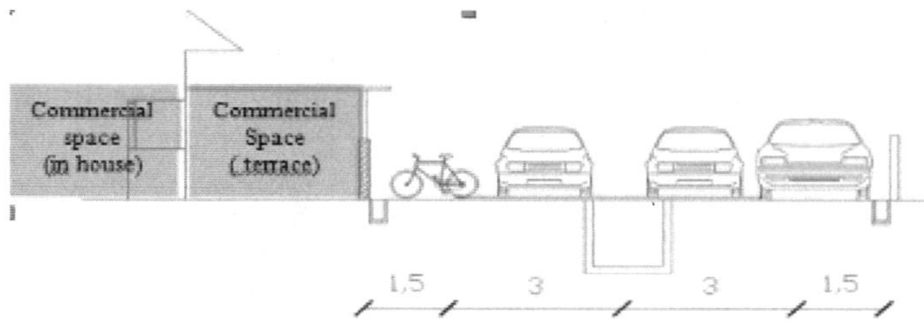

Figure 2. The house yard conflict

However those wished only permitted and designed by the house owners but not their next house neighbor. Several spots showed signage and fences in connection with the conflict area (see figure 3) made by the owners to assert their areas [18]. The conflicts in this area are clear and justified toward this area in the view of the house guests (permitted or without permitted) and not for the house inhabitants. The house owners show their control-access-secure area and so do their neighbors, and the trespassing activities may trigger social conflicts among them.

Figure 3. The spatial conflict signage

The transition areas also have the similar evidences. The signage of authority is found in this area as the street furniture. Further reactions of conflict are also showed by the instalation of landscape elements such plants in the soil or pots, small parks, flag pole foundations, surface material differences, and common traffic signs use (see figure 4). It was found that in order to show the spatial conflict in soft reaction there are landscape elements installation instead of hard text signage.

The vacant areas have slightly different evidences since these areas might not related to any inhabitants in short terms. No sign of spatial conflict responses in several spots. In deeper interview to the food stalls/shops' owners, most of them said that they must pay tax to local authorities for their illegal activities. The tax prices are varied and disclosed. The spatial conflicts occur when their activity happens on street parking and when there are more serious of garbage problems. The occupied spaces cause the street jams and environmental pollutions (see figure 5).

The selfish perception about space among inhabitants in the urban kampong and people visiting in short time can be considered as the main factors of spatial conflicts. The conflict itself needs time and

IOP Conf. Series: Earth and Environmental Science **213** (2018) 012009 doi:10.1088/1755-1315/213/1/012009

space to show up. The catalysts of the space conflicts are the bigger size, the more frequent occupation time, and the inhabitants' demands for space expansion.

Figure 4. The landscape spatial conflict evidences

Figure 5. Vacant space conflicts

3.4. Proposal of conflict resolutions

Three areas with three catalysts of spatial conflicts and a main factor can be studied to offer the problem solving strategies. In this chapter, we avoid comparing the disadvantages and advantages of spatial conflict in relation with eliminating the problem or keeping the problem with options. The basic strategy regarding the areas, catalysts and main factor are strengthening relationship among community members in order to discuss every conflict happening in their area, starting from the smallest and rare case. The weaker the relation among the community members is, the bigger and the more frequent spatial conflicts may come.

4. Conclusion

There are three areas in urban kampong residences potential for spatial conflicts in the neighborhood, such as: vacant space, house transition area and front yards. The evidences of spatial conflicts are shown in the form of signage, fences, gates, buildings, and landscape elements installation, parking slots and localized commercial area localization. Design element did influence on the spatial conflicts. The strengthening of neighborhood may bring solution to these conflicts. The longer and more frequent the conflicts may occur, the more difficult efforts all parties must create.

5. Acknowledgments
Authors would like to express their gratitude to colleagues, mentors and family for their encouragement, DAFT UNDIP research funding and department managerial system for the supporting, and, last but not least, the community members of Pendrikan for the data and their assistance in this research.

6. References

[1] Tawfiq M. Abu-Ghazeh 1996 Reclaiming Public Space : The Ecology of Urban Open Space in Town of Abu Nusseir Jordan *Landscape and Urban Planing* **36**.197-216
[2] Robert R. Churchill & Ronald D. Liebowitz (1990) Spatial conflict and conflict resolution, a classroom simulation, Journal of Geography in Higher Education, **14:2**, 151-156
[3] Rapoport, A., 1976. Socio-cultural aspects of man-environment studies. In: A. Rapoport (Editor), The Mutual Interaction of People and Their Built Environment. Mouton, The Hague.
[4] Boyden, S.V., 1979. An integrative ecological approach to the study of human settlements. MAB Technical Note 12, UNESCO, Paris.
[5] Michelson, W., 1977. Environmental Choice, Human Behaviourand Residential Satisfaction. Oxford University Press, London.
[6] Rapoport, A., 1977a. Human Aspects of Urban Form: Towards a Mat-Environment Approach to Urban Form and Design. Pergamon Press, New York.
[7] Zube, E.H., 1984. Environmental perceptions and behavior. University of Chicago Reporter, **209**: 60-83
[8] Rapoport, A., 1977b. Socio-cultural aspects of mat-environment studies. In: A. Rapoport (Editor), The Mutual Interaction of People and the Built Environment: a Cross-Cultural Perspective. Mouton, The Hague.
[9] Corbet, M., 1981. A Better Place to Live: New Designs forTomorrow's Communities, Rodal Press, Emmaus, PA.
[10] Lang, J., 1987. Creating Architectural Theory: The Role of Behavioral Sciences in Environmental Design. Van Nostrand Reinhold, New York.
[11] Elhance, A. (1983) Location of social facilities: an analytical approach, Indian Journal of Regional Science, **15**, pp. 53-78.
[12] Barton, S.E. (1984) Conflict resolution as necessity, practice, and ideal, Journal of Planning Education and Research, **4**, pp. 96-102.
[13] Teitz, M. (1968) Toward a theory of urban public facility location, Papers and Proceedings of the Regional Science Association, **21**, pp. 35-52
[14] Caldweix, L.K. (1984) Garrison diversion: constraints on conflict resolution, Natural Resources Journal, **24**, pp. 839-863
[15] Merriam, D. & Stein, M. (1985) Caught in the takings muddle, Planning, **51**, pp. 23-39.
[16] Santorinou, A et al (2007) Spatial conflict management in urban planing, University of the Aegean, Department of Environment
[17] Harsritanto, Bangun I.R. (2008) Hubungan Persepsi Penghuni dengan Pola Ruang Terbuka Publik terkait Kegiatan Komersial, University Diponegoro
[18] Harsritanto, Bangun I.R et al (2017) Spatial Transformation Pattern Due to Commercial Activity in Kampong House, Proceeding of 15th international conference of Quality in Research 2017 – I dwell section-, pp. 724-731

The community role in green area sustainability as a model of energy-efficient buildings in the humid tropical region

Sri Yuliani, Wiwik Setyaningsih
Universitas Sebelas Maret, Surakarta, Indonesia

Corresponding e-mail: sriyuliani71@staff.uns.ac.id

Abstract. Green design models have the potential to improve the environment's microclimate and to decrease energy consumption in a building. The successful implementation of green design model is determined by government policy, the concern of architects and the role of the community. The role of the community is important because it has the potential to be the part of the green sustainability model; that is, to provide green spaces and follow up on the sustainability of green governance elements. The aims of this research are to describe the role of community in the understanding and implementation of green elements in buildings. In addition, it also aims to relate the understanding of the community with sustainability of the green design model in energy-efficient buildings. This research used descriptive methodology by purposive random sampling in three categories of society: low income, middle, income and high income categories. The results of this research have brought contribution to theories of development in term of ecological architecture concept, and this concept can be done through the community involvement in the implementation of the green design models that are adaptable to humid tropical climate.

Keywords: community roles, green area sustainability, energy efficient buildings, humid tropical regions

1. Introduction

Each architectural work should consider environmental impacts. The implementation of green architecture design gives positive influence to the microclimate environment, among others in the decrease of building temperature. Decreasing the temperature of the building will directly contribute to the temperature decrease of the area and energy efficiency of the building. This positive contribution creates an artificial ecosystem environment that harmonizes with the natural environment, as it is the concept of ecological architecture. On a broader scale, Samir Younis [1] declares that development in the world should use ecologically constructive practices in building construction, agriculture, economic cooperation, and energy consumption; affecting the earth in the future. For architects, sustainable buildings should be the primary goal of sustainable urban development. For all humans, environmental sustainability is an existential goal that has implications for aspects of life in order for the earth to survive and sustain. Sustainability of the concept application of ecological architecture that emphasizes green design is determined by three contributors, namely government its policies, the architects with their design works, and the general public as the user. To that end, the background of this study considers the rationale of the research findings related to the role of society as one of the success determinants of sustainable development and green design, as part of the embodiment of the concept of ecological architecture that takes into account the buildings energy efficiency.

According to Maria Ignatieva and Karin Ahrnè [2] the biodiversity approach to landscape architecture is based on the best achievement of innovative landscape design solutions, which will incorporate the flora and fauna components. The research and its implementation require interdisciplinary collaboration, including stakeholders, policy makers, and horticultural managers. Studying urban grass, green roofs and green walls should combine different perspectives; that is, social and ecological in planning, design, and management of sustainable city. Alternative potential sustainable solutions are unlikely to be found without understanding the social motives behind the strong attachment, from modern society to disguise unsustainable elements, such as grass from urban green areas or attractions of having green roofs and green walls. The research shows the role of community in the successful implementation of landscape architectural design.

On the other hand, Malik et al. (2017) conducted a study of urban space management approaches related to ecological functions, explaining how far the role of stakeholders in urban areas, which will further strengthen the importance of green open spaces. Specifically in the city of Depok, there are so many problems related to the provision and utilization of open space which need certain approaches. These approaches were initially applied to the private sectors through the provision of contributing applications, thus firstl used by the governments in managing public assets. The study used descriptive methods, at the beginning of the study that explained the existence of the city's green open space realm as part of the urban space from the theoretical perspectives of space, functions, and the role of various problems that occur within it. The results showed that there are problems in the management of green open space in Depok. In asset management, it has provided important findings to accommodate the participation of existing stakeholders in the management of green open spaces. The asset management approach is a participatory approach by involving all parties. This approach also provides space for people who previously had difficulty in expressing their criticism and in accessing information [3]. The most important thing in this research is; it turns out that the optimization of open green space planning can be through asset management that accommodates stakeholders in development. The role of society can be gained by providing space for their opinions either individually or collectively within a community.

Luca Tricarico [4] studied of communities in development and found that the concept of community has maintained a constant and growing interest in urban studies and many related fields. The origins of this sustained interest seem to stem from the importance of community concepts in various forms of language and political interpretation in different planning practices. Contributions through analysis of various ethical and planning theories provide a framework for renewal of community action. Based on the purpose of this study, the argument will continue through literature reviews on four ethical theories and three key aspects related to spatial planning, as well as matching these theoretical analyses with practice examples. The ultimate goal is to provide genuine analysis of the movers and outcomes of various forms of community, raising common issues referring to spatial planning, social organization, and regulation [4]. This study further described how to use the role of communities in a society that contribute to development planning, especially in urban spatial planning. The research generally concludes that the community has a strategic role in the sustainable development.

While in terms of building energy management, a study on the evaluation of building energy by [5] describing that methods for environmental assessment and 'green' certification of buildings are increasingly being adopted by local governments and other entities as a means to meet environmental objectives, related to energy. The literature of building and environmental assessment has examined how these tools are interpreted by stakeholders in design and construction, but it still receives little interest for the interpretation and use as policy objects although this position has a controlling role with enormous impact. The analysis shows that for policy makers, the importance of green building does not lie in terms of the assessment of a set of environmental benchmarks but as a standard that derives commercial value from policy objects. This research also conducts open discussion required on the logical thinking that encourages the use of green building tools in public policy and its goals.

The research focused on the performance of policy makers, and discussions are needed for encouraging the role of communities in energy management for green design. Therefore, this discussion was certainly not limited for policy makers alone, but it is also for public.

The research [6] concluded that energy efficiency is important, if carbon emissions are reduced significantly, then it is essential to consider the meanings and levels of service and the types of consumption and demand that continuous supportts in policies related to efficiency. Energy efficiency can be started from the households although these area ae not the largest part of the world's energy consumer. Energy efficiency in households is determined by the role of society in managing the energy, and it can be done by providing green spaces that physically help to condition the environment. Therefore, in connection with the findings of Elizabeth Shove, this study aims to identify the level of public knowledge of the green elements benefits at home, and the community awareness in managing green elements to be more productive.

Elizabeth L. Hewitt et al [7] suggested several frameworks incorporating social and psychological elements of behaviour that are environmentally significant, with most considered cognitive and deliberate decision making. The behaviour of household energy consumption however, reaches the spectrum from being grounded and deliberate to unplanned and automatic. The purpose of the paper was to advance the knowledge of reasonable and unplanned behaviour in the context of pro-environmental action. Using the survey results provided to residents of a green residential urban building, this study explored five household consumption behaviours, and tested the hypothesis that unplanned behaviour would be poorly predicted by a behavioural framework based on values and reasons. The research findings indicated that the badly-planned behaviour will have an inefficient impact on building energy. This finding was used as a reference in building energy management for residential scale and necessary to educate the public to know the benefits of green space in the building. A deeper understanding of how unplanned, automated, or inefficient behaviour in society, can help policy makers and building designers more wisely to respond occupants' behaviour and to consider the design performance.

The building sector is a major energy consumer, and empirical research reveals that carbon dioxide emissions from the building sector outweigh industrial and transportation sectors, as presented in [8]. Contextual alignment has generally been applied in designing new buildings in historical districts. In the field of design, a philosophy-based design approach requires objective measurement. When new glass buildings are considered, the question of greenhouse gas emissions becomes significant. To minimize energy consumption and to reduce solar glare, brise-soleil and horizontal shading fixed devices have been used as building façade elements. Currently, algae as part of constructing façade systems have been used in creating a healthier indoor environment. The use of secondary leather buildings with the role of plants, contributes to the decreasing of building temperatures, and it can also be significant contribution to both design approaches and associations with choice of building materials and provision of potential energy alternatives. Ultimately, the evidence can be used as a justification for eco-friendly principles for architectural design. The findings of green design applications are very important to be known to the public at large. The development of the research findings may be considered for the occupancy sector. When the green design innovation prototype is ready for implementation, it is necessary to communicate this design to the community. Conversely, when the green design is not communicated to the community properly, the community will not understand it certainly and impact on green design cannot be sustainable.

In terms of communication on green design, this research [9] stated that the development of green design can be realized through social learning. Social learning is an approach for some development actors. For successful development with green design concepts, the support from the policy makers is urgently needed. In this research, the policy makers are the important actors of the development. The approach was then implemented more extensively with the same method which resulted hence creating social learning to the community as green design users.

From the background of some research that has been done, it turns out that the role of society is very strategic in development, specifically for the sustainability of the green design as a model of energy efficient buildings. Therefore this study aims to obtain a relationship between the roles of society and the sustainability of green design in buildings. This research is useful to know how the correlation between the role of society and sustainability of green design model can be held optimally. The sustainability in the implementation of green architecture is important to contribute to the energy efficiency of buildings in humid tropical climates.

2. Methods

This research used quantitative descriptive method to show the relationship pattern between the successful implementation of green model in residential building and the community role. Data were collected through questionnaires by purposive random sampling. Questionnaires focused on three economic levels: low-income communities, middle-income communities, and high-income communities. The research location was in Indonesian humid tropical area of Central Java, particularly Surakarta City. The location was chosen because the city could present the urban residential characters in general.

The questionnaire model included questions about the status of residential ownership having the opportunity to provide green spaces, knowledge of the benefits of green space at home, and the community's concern about the importance of green space in the dwelling. This research was also conducted with field observation on the phenomenon of green elements in residential area. Observations included the conditions and green layout model of urban residential buildings, and the green living conditions of residential areas at the questionnaire distribution sites. According to [10], research variables are determined by the nature or value of research objects having certain variation to be studied and examined, as a reference to determine the conclusions of the research undertaken. This research put dependent variables as the research object, namely the sustainability of green design as the energy building efficiency model that can have different values because influences from other variables. The independent variable was the role of society since the society is one of the determinants in the success of the development. Meanwhile, the moderator variable is the economic level of the community, the potential welfare of the community that can affect their active role the in development.

Data processing was done through inferential statistical analysis by disclosing the problem descriptively into the data level of public knowledge about the benefits of green space and community awareness to participate in the optimization of green space. The results of data analysis were used to verify the associative hypothesis on the formulation of associative problems in the form of causal relationships, namely the role of society and the success of the green building model.

3. Discussion

The role of society in development is a strategic part that can determine the success of the concept of development. No matter how good the policies issued by the government or how beautiful the building design is, when the inhabitants or building users do not play the expected role as specified by the government and the intend design by architectural then certainly the development is not as optimal as expected. Development can be sustainable with the support of policy makers, policy controllers, and development executors. Furthermore, the community, as one of the determinants of the sustainable development, should be able to support the success of development, both as the executor and the controller of development policy. Empirically, the development can be constrained due to the lack of community support and even resistance.

This study discusses the role of the community in the sustainability of green design in Indonesia, particularly in the city of Surakarta in which the location can be seen in Figure 2. Surakarta is located in the lowlands at an altitude of 105m above sea level and in the city centre 95m above sea level, with an area of 44.1 km^2 (0.14% of Central Java). The city of Surakarta lies between 110 45` 15 "- 110 45` 35" East Longitude and 70`36 "- 70` 56" South Latitude, and borders with; the Karanganyar Regency and Boyolali Regency in the north, the Karanganyar Regency and Sukoharjo Regency in the east and west, and the Sukoharjo Regency in the south. Surakarta is divided into five sub-districts, each headed by a head of sub-district (*camat*), and 51 villages each headed by a head of vilage (*lurah*). The five sub-districts in Surakarta are [11]: the sub-district of Pasar Kliwon (postal code 57110) consisting of nine villages; the sub-district of Jebres (postal code 57120) consisting of villages; the sub-district of Banjarsari (postal code 57130) consisting of thirteen villages; the sub-district of Laweyan (postal code 57140) consisting of eleven villages; and the sub-district of Serengan (postal code 57150) consisting of seven sub-districts.

Figure 1. Location of research: Kota Surakarta, Jawa Tengah, Indonesia

To obtain the community role, the research was conducted by collecting data through polling in questionnaires distributed to five sub-districts in Surakarta. The study identified targets in the kampong area based on the *kampong* potential, explicitly stated in [12], namely that the alleys within the kampong that can be used as a public space by applying adaptive design principles. The use of bamboo in the design is the right solution based on its construction and material properties. The problem of limited public space in the kampong can be overcome by applying adaptive architectural principles in the alleys or lanes. Public spaces can be customized to meet the conditions in this area. One good way is to provide an adaptive design alternative [12]. Therefore, the potential of kampong, which is the smallest ecosystem in the community, can be a benchmark of the quality of society's role related to the sustainable green design. Empirical land use data in 2017 shows that each district in Surakarta has similar pattern, mostly for housing as depicted in Figure 2.

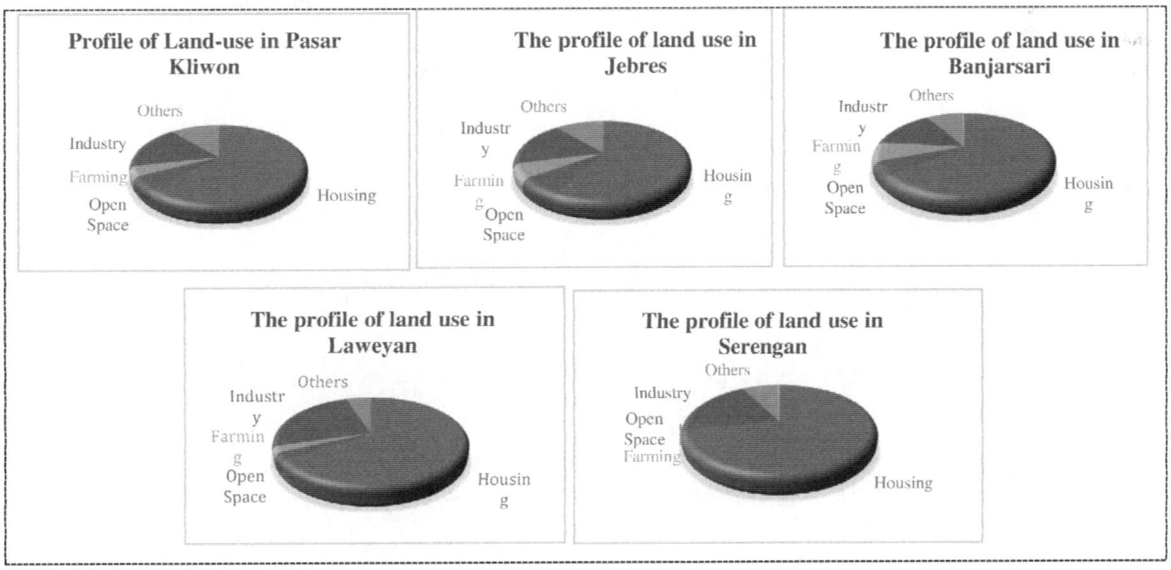

Figure 2. The profile of land use in Surakarta based on Statistic of Surakarta Municipality [11]

The land use profiles in each district indicate that housing has an important role as the sector that should be targeted for the implementation of the green planning in Surakarta. The average percentage rate above 50% is already very strong indicaction for the success. It is important that the housings need layout, and the community as users is educated about the potential role in the sustainability of green design on the buildings, hence natural potential in tropical climates that can achieve building energy efficiency. These findings have been strengthened by the research of [6] which identified energy efficiency in well-planned-based home activities. Added by [5] the government needs to play an optimal role as the policy maker and policy control tool in the implementation of green design. When energy management at the occupancy level can be controlled, the energy efficiency of buildings can be macro-suppressed. This goal can be achieved because the profile of the urban land use in humid tropics such as Indonesia is dominated by land use for the housing sectors. Energy consumption determinants for the housing sectors controlled within the policy determinant corridor can be implemented through consideration of the criteria of activities undertaken at home.

In different areas, Malcolm Miles [13], conducted a research in London about participation in housing where this housing tends to be seen as a sub-sector of the construction industry. In other areas, the opposite is true; housing is still considered to be an activity to meet basic needs for shelter. Thus, this process is often done by the users themselves in the informal settlements that surround most of the city. Although these settlements were once considered a threat to the urban order (or urbanization), there is now increasing recognition that self-sustaining and self-managed housing can meet the needs of urban development in a way that is usually more sustainable and also lower in cost than the standard housing scheme (both in the public or private sector). This is not seen as an example, but only two cases can be compared and contrasted in the field of the new approaches to building the city for the future [13]. This example clearly shows how the community plays a role in providing colour in development. Therefore, the housing sectors have enormous potential in the sustainability of green design.

In a more detailed survey, the study determined one of the districts viewed in depth research, namely in the Jebres sub-district. The survey showed the opportunities of people who played roles in the provision of green space reached 100% as seen in Figure 3. The data survey analysis showed that the higher the economic level of society, the greater the potential role of society in green design sustainability. High-income communities have a 100% chance of providing green space for buildings, but these opportunities need to be assessed more directly by considering the willingness or the motivation of the community. Another point, the availability of plants owned by the community, shown in Figure 4, has the same profile, which is directly proportional to the economic level of the community, the higher the economic level of the community, the greater the chance of providing plants in the surrounding environment. The availability of plants is certainly an opportunity of the role of society. Both figures 3 and 4, describe the role of society that can be seen from the economic level; and therefore, the green design planning can be implemented optimally with cross subsidy model.

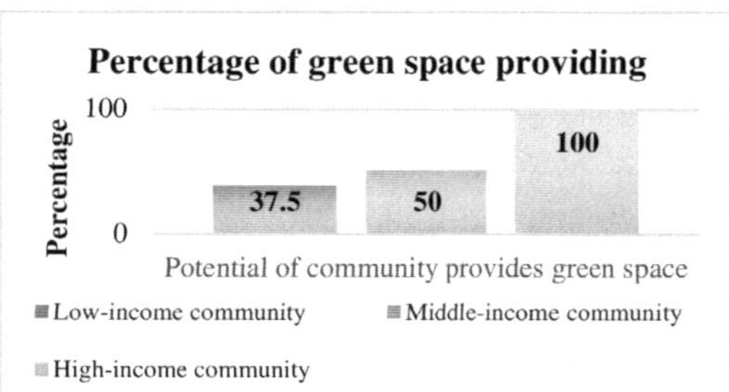

Figure 3. Community potential provides green space based on community income.

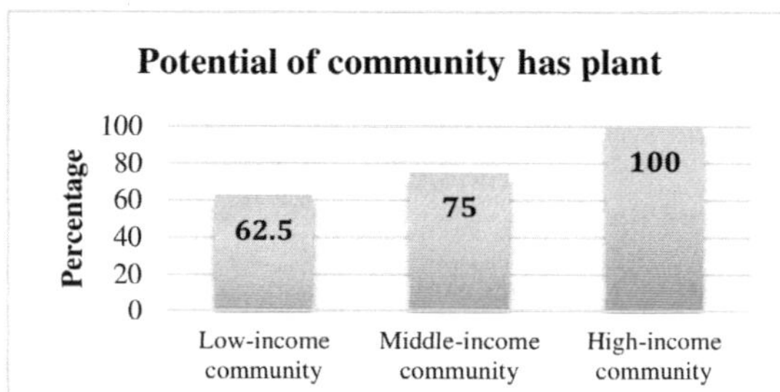

Figure 4. Community potential provides plants based on the community income.

In line with [2,3,13] and et al [7] about community participation in development, the analysis of the research data in Surakarta, shown in figure 3 and figure 4, describes the existence of the dependence correlation forms; for instance, the provision of green space and the supply of plants to the community depending on the economic level of society. Therefore, it can be concluded that the role of society becomes important for the sustainability of space and green elements in buildings.

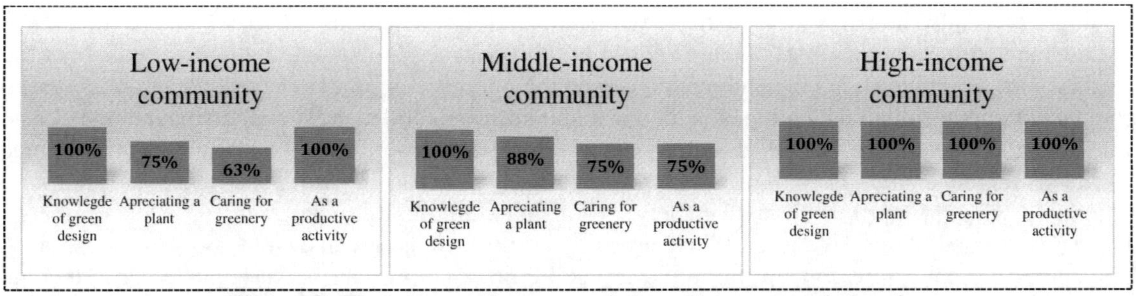

Figure 5. The community responds based on economic level.

In terms of the further role of the community, research results show that 100% of the people have already known the benefits of green design for environmental sustainability. Several parameters to express community responses to green design sustainability showing that low, middle, and high-income people share the same view. The differences in community response are at the level of appreciation, maintenance, and assessment of green design productivity as seen in Figure 5. In low-income communities, the knowledge of green design and benefits is embedded, but the appreciation in the practice of daily life needs to be improved. Among the middle segment of society, the consideration about the values of green design productivity is lower, but their appreciation and willingness to maintain is higher. And for high-income people, green design has gained great attention and support from them.

With the pattern of responses obtained from the research on the role of society in green design sustainability, the result obtained is that there is a very significant relationship between the appreciations from the community on green design with environmental conditions around the dwelling; hence the higher the communitiy's appreciation, the better condition of green area environment. For urban tropical areas such as Surakarta, efforts to optimize sustainability of green design are likely in the housing sectors. While linked to the economic level of the community, sustainability of green design requires cross-subsidy management methods among communities that can be managed by them [4] with mentoring through social learning [9]. Finding that is useful in green design innovations, such as the finding of algae as a façade from Widjaja Martokusumo [8] is an example of a green design application that needs to be socialized to society by means of social learning. The middle class is the society that needs education about green design productivity that gives effect to the energy efficiency of the building. While low-income communities have the opportunity to make green design as a productive area, as well as providing knowledge about building energy efficiency through green design. For high-income communities, they are important asset in achieving green design sustainability in the humid tropics. The role of society needs to be managed and optimized through a community in order to have good relationship among determinants of successful development [3] [4].

In relation to building energy efficiency, low-income communities share the same view with high-income communities. As shown in Figure 5 lower income people consider green design as productive activity, hence a productive space. In more in-depth research, low-income communities perceive that the productive values of plants give benefits whereas for high-income communities, they consider green designs provide comfort space, fresh air that contributes to health and environmental improvement. Although the views are different, it is good to see these findings because they are rich with information.

From this discussion, it is clear that the community has a role in green design sustainability as a model of energy building in the humid and tropical regions.

4. Conclusion

The results showed the importance of involving the community in implementing the green plan model as a space that reduces the buldings' heat in order to achieve building energy efficiency in humid tropical regions. The community's role contributes to the success of green sustainability provided in buildings, especially in urban areas with humid tropical climates such as Indonesia where the largest land uses are for residential areas. What needs to be improved is the community knowledge about the correlation of green building elements with building the energy efficiency, so that sustainability green design can target residential development. Community knowledge is directly proportional to the public awareness for the sustainability of the green plans in residential buildings. Construction of buildings and infrastructures, especially housings implementing green design, will be more optimal and sustainable with the optimization of society's role. In addition, it is necessary to educate energy management on buildings, so energy efficiency in buildings through green design, as the application of the ecological architecture concept in humid tropical climate areas, can be sustainable. The findings recommend the importance of further research related to optimizing and managing the potential power of communities in the tropics in the implementing green building model for energy efficiency of buildings in humid and tropical regions.

5. Acknowledgments

Authors hereby express their gratitude to PNBP UNS 2018 which has provided their research with financial support through the Competitive Grant selection proposed in 2018 and realized in Contract: 543/UN27.21/PP/2018.

6. References

[1] Younes S 2014 Nature and city: a co-evolutionary project, *Jornal of Architecture and Urbanism.* **38** 3 198

[2] Ignatieva M and Ahrne K 2013 Biodiverse green infrastructure for the 21st century: from "green desert" of lawns to biophilic cities, *Journal of Architecture and Urbanism.* **37** 1 6

[3] Malik A A M 2017 The Role of Stackeholders Related to the Management of Ecological Function of Urban Green Open Space, Case Study: City of Depok, Indonesia, in *International Conference of Sustainability in Architectural Design and Urbanism*, Semarang.

[4] Tricarico L 2017 Community action: value or instrument? An ethics and planning critical review, *Journal of Architecture and Urbanism.* **41** 3 230

[5] Goulden S, Erell E, Garb Y and Pearlmutter D 2015 Green building standards as sosio-technical actors in municipal environmental policy, *Building Research & Information,* 1

[6] Shove E 2017 What is wrong with energy efficiency?, *Building Research & Information,* 1

[7] Hewit E L, Anderws C J, Senick J A, Wener R. E, Krogmann U and Allacci M S 2015 Distinguishing between green building occupants' reasoned and unplanned behaviours, *Building Research & Information.* 2

[8] Martokusumo W, Koerniawan M D, Poerbo H W, Ardiani N A and Krisanti S H 2017 Algae and building facade revisited, a study of facade system for infill design, *Journal of Architecture and Urbanism.* **41** 4 296

[9] Hojem T S M, Serensen K H and Lagesen V A 2014 Designing a 'green' building: expanding ambitions through sosial learning, *Building Research & Information.* **42** 5 596

[10] Sugiono 2017 Metode Penelitian Kuantitatif, Kualitatif dan R&D, Bandung: Penerbit Alfabeta

[11] Statistic of Surakarta Municipality 2017 Surakarta Municipality in Figures 2017, BPS-Statistics of Surakarta City, Surakarta

[12] Widharto S D, Muchlis N, Laksono S H and Setijanti P 2015 Design concept for green public space in kampong, *Journal of Architecture and Urbanism.* **39** 2 116

[13] Miles M 2013 Participation: Housing and Urban Vilability, *Journal of Architecture and Urbanism.* **37** 3 224

The Sundanese and Javanese House Local Wisdom to Respond to The Climate Changes

Handajani Asriningpuri, Estuti Rochimah
Institute Technology of Indonesia, South Tangerang, Indonesia

Corresponding e-mail: hasriningpuri19@gmail.com

Abstract. Climate change affects the ongoing seasons, including the season displacement, the out-indoor thermal comfort, building usage, healthy space and the confusion in the building – infrastructure execution schedule. However, in a place like West Java or Middle of Java that condition do not work out because some local customs, tradition and nature condition give difference and specific characteristics (uniqueness) called Local Wisdom. Climate change also influences three elements in architectural discipline. There are nature as a place that influences the building, building in nature as human product to fulfill human necessity, and human as designer and user or occupant of the nature. Therefore, it is better not to oppose the climate change but to obey and fill the pursuit because those elements are linked to each other. Those reasons made author interested to do the research especially in Sundanese and Javanese traditional houses. The research objective is to proof the hypothesis that local wisdom influences the climate change in certain zone condition. The discussion referred to some theory like Lippsmeier's and Triharso's theories. The descriptive, qualitative, casuistic methodology used in this research. Results showed that adaptation endurance to climate change was possible through maintenance-preservation and conservation the traditional house as local wisdom and teaching the occupant to maintain together. The result become an architectural product and will be continued as new hypothesis in the next research.

Keywords: Climate Changes, Local Wisdom, Traditional House

1. Introduction

1.1.Background

Indonesia has many tribes, so that many traditional houses will be found, like vernacular stilt house in Sundanese represented by Kampong Naga housing complex and vernacular landed house as Javanese house in Middle of Java represented by Kampong Bade. The differences of vernacular typology house include the background of nature influence culture, local customs-traditions respond, building materials, a uniqueness and colors of architecture. That diversity is called local wisdom or genius loci.

In addition, the ongoing climate change also influences the three elements of architectural discipline, like nature as a place that influences the building, building in nature as human product to fulfill human necessity, and human as user or occupant or designer of the nature. Therefore, it is better to obey and fill the pursuit then oppose the climate change, because those elements are linked each other. The house commonly has saddle rooftop model and the expansion, with a specific characteristic of detail, materials, and meaning. It depends on the location. The traditional house does not have an impact on the climate change. The Local wisdom defends and protects it, so that the background is to seek and solve the problem on how can local wisdom endure the climate change.

The object of the study was in Kampong Naga in West Java located in highway province between Garut - Tasikmalaya in Neglasari Village sub district Salawu approximately 25 km from Garut.
The other object was Kampong Bade located in highway between Karanggede to Gemolong approximately 27 km from Gemolong- Karanggede district - Boyolali regency-Mid Java.

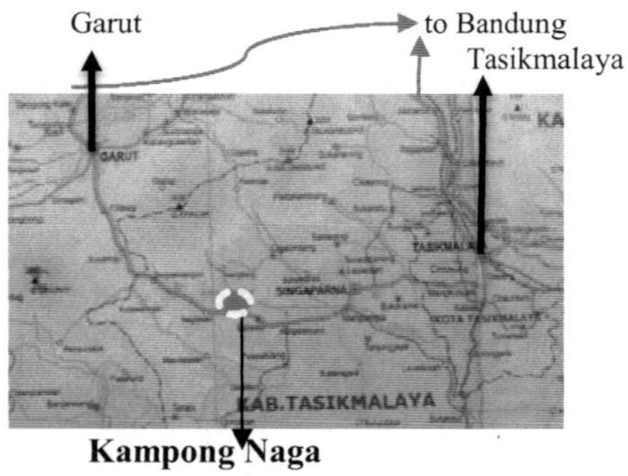

NORTH

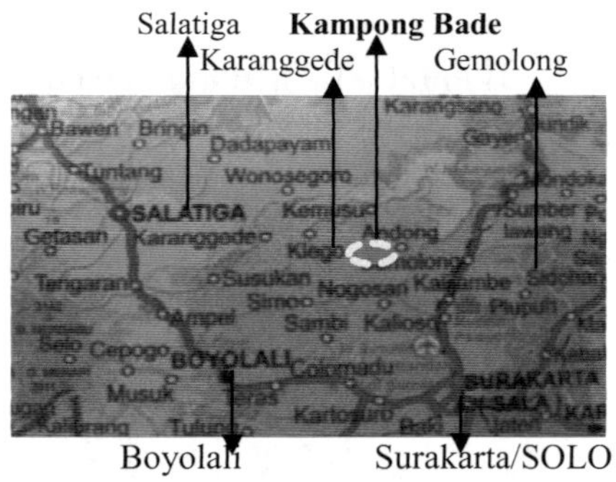

Figure. 1. West Java Map
Source : Tourism Map.CV INDO BUWANA

Figure. 2. Center of Java Map
Source : Tourism Map.CV INDO BUWANA

Each kampong has different uniqueness responding to the surrounding. Related to the weather climate and the geographical location, those objects were analyzed based on the aspects of micro climate, like the weather, nature, ventilation system, day lighting system, temperature and humidity. The table shown Kampong Naga had lower degree of temperature then Kampong Bade.

Table 1. Micro Climate and Weather
Source: Private Document (Observation product)

No	Aspect over view	Kampong Naga	Kampong Bade
1	Climate	Tropical Humidity	Tropical Humidity
2	Seasons	Dry and Rainy	Dry and Rainy
3	Geographical location of the object	7° 56' 49"- 7° 56' 49" SL & 107° 25'8" EL	7° 50' 27"- 7° 56' 49" SL & 108° 18'32" EL
4	Temperatures degree of the object	21,5°C - 24,1°C (peak of dry seasons) in June 25th - 29th 2018	30,5°C - 32,8°C (peak of dry seasons) in April 25th – May 3th 2018
5	Humidity	70% - 80%	40% - 60%
6	Temperature of comfort	21,5°C - 24,1°C	30,5°C - 32,8°C
7	Sun Light	Clear and bright	Clear and bright
8	Day Light	Clear & sometimes become overcast/cloudy	Bright, temporary blinded
9	Air movement velocity	Breezy, temporary cool in the night & very cool in the early morning	Hot, sometimes warm & cool in the early morning

1.2. Site Condition

Kampong Naga lies between the Singaparna hills and Karan Mount hills. Its slope is surrounded by steepness topography. In East side "Ciwulan" river flows along the village, used as sewage system, two springs are located in up of the two hills as drinking water resources and many rice fields and fishponds are used as livelihood inhabitant. The occupied houses, the tomb of grandparents Karuhun/ancestral

Singaparna and the livelihood inhabitant are composed by an ordering system with the mosque and an *"alun-alun"* (little piazza) as a center.

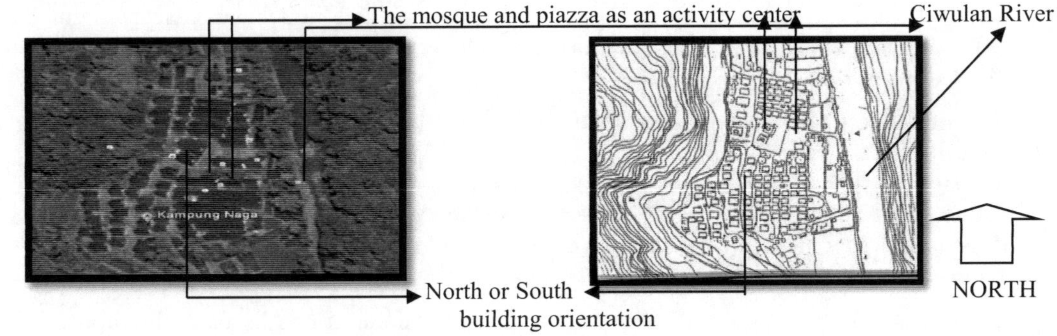

Figure 3 & 4. Site Plan of Kampung Naga
Source : https://maps.google.com/and Journal of Kampung Naga by I. Indartoro (1987, pp 9)

Kampong Bade is also in slope slightly level land, in the beginning of the 'Kendengs' mountain range, separated into two areas: South side which faces the reservoir "*Waduk Bade"* and the North Side which faces a river "*Kali Serang*". The site condition review aspects are topography and surrounding situation, house orientation, typology and house dimension, materials, construction elements etc. as shown below:

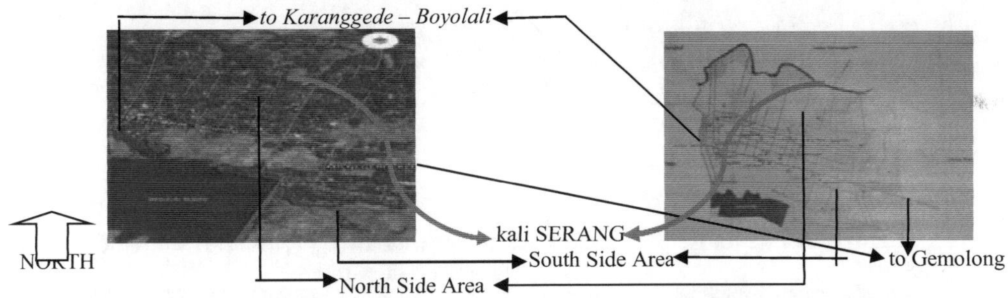

Figure 5 & 6. Site Plan of Kampong Bade
Sources : https://maps.google.com &Bade Village head office

Table 2. Site Condition
Sources : Private document

No	Aspect over view	Kampong Naga	Kampong Bade
1	Topography	8° - ±45° slope slightly level land	25° slope slightly level land
2	Surround situation	"Ciwulan" river in The East side	"Kali Serang" river in The West and North. "Waduk Bade" in Southwest part
3	House Orientation	Side by side in orderly to South or North	To North or South not in orderly, is dependent on the site form.
4	Typology and House Dimension	Stilt House, Square edge 6.00m x 6.00m or Rectangular 6.00 m x 9.00 m (6.00 mx10.00m).	Landed House. Rectangular edge 9.00 m x 12.00 m or 9.00m x 15.00 m
5	Construction element, Materials	Foot step foundation made by stone The Body made by wood Manglid TYPE and bamboo. Roof made by Palm & Tepus Leaf.	Cube step foundation made by brick. The Body made by Teak Wood tree type (tectona grandis sp-teak wood/Jati).The Roof made by Corrugated Tile

1.3. Object Study

1.3.1. Sundanese House

This Sundanese house was represented by housing typology in Kampong Naga. This study selected a sub village named Nagawir in Neglasari Village – Salawu District-Garut Region; West Java Province. It was easy access to reach the location, approximately 800 m from the slope downward.

The houses are stilt houses where usually the stilt houses have an optimal crossing air flow capacity because the distance between indoor floor surface and the ground can pierced by the crossing air flow. Those houses are assembled, facing to north or south, and not eroded by the influence of culture. The materials are bamboo & wood called "Manglid". The roof shape is "Suhunan Jolopong " (Gable) and covered with Palm and Tepus leaves, made the cooling air temperature of the room above possible. The shape and edges of the house are shown below.

The occupants always precede every activity with ritual and faithfully obey the rules of custom. The tradition is done enthusiastically, especially when they prepare materials for repairing the house, like praying to god, preparing food and beverage as tribute for the nature guardian.

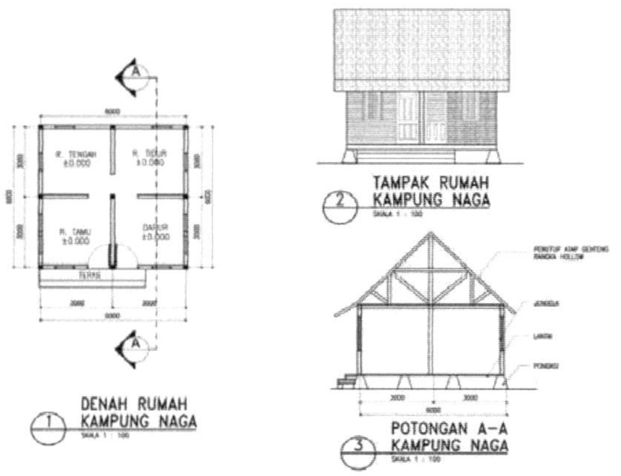

Figure. 7. The house plan and elevation of Sundanese house in Kampong Naga
Source: private document

Figure 8 & 9. Sundanese house shape and edges
Source: private document

1.3.2. Javanese House

Kampong Bade housing represented the Javanese house in a sub village (*dukuh*) that is located at the high way from Karanggede to Gemolong. It was easy to reach the location. Its highest level of the side is along the province highways that is $\pm$ 186 meter up to sea surface. Slope slightly appears on the land surface from the highway to north. The lowest site is in the South where slope slightly comes to the basin. There are many Jati (*tectona grandis sp*-teak wood) forests along the street. Climatic influences affect the landscape, buildings, air system and lighting system. The temperature was recorded on the average of 30, 5°C to 32, 8°C, and 26, 3°C to 28, 5°C under the densely vegetation area.

The houses are landed house, dispersed in North & South area, facing the north or south. The floor surface is topsoil, but several houses are covered with mortar. The roof is made of woods covered with corrugated tile. There are two types, namely:" ***Sinom*** and "***Dorogepak***". The optimal crossing airflow capacity happened because the crossing airflow pierced the room in upper part and made the cooling air temperature above the room.

Everyone did the activity in the house or during the cultivate season. People always precede their activities with ritual's conviction. They faithfully obey the rules of custom – tradition although the influence of culture alteration eroded several houses, new technology and materials, etc.

"*Sinom"* (the wide type) The house of Nyoto Widodo -Ds Pelang.

Figure 10 & 11. The shape and edges of house Type I "*sinom*" (along become wide)
Source : Private Document

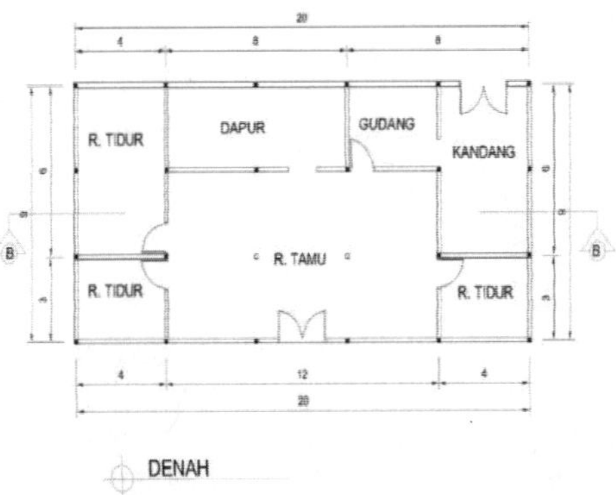

Figure 12. Plan of type I *"Sinom"*
Source: private document

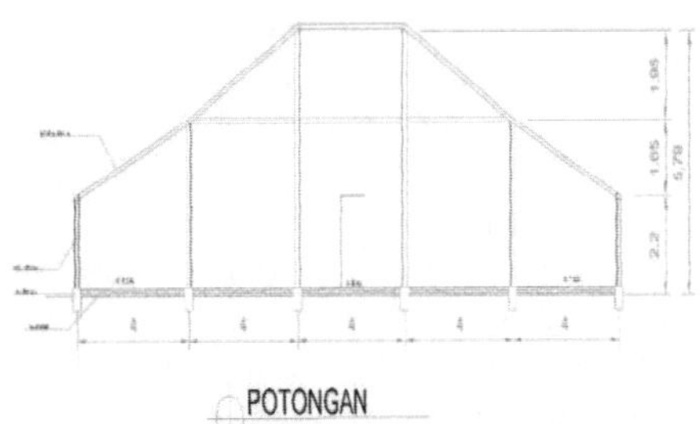

B - B The section of type I

Figure 13. Section of house type I *"Sinom"*
Source: private document

Dorogepak (Stretch along a Length) type" The house of Mrs Sulastri's - Ds. Bade

Figure 14 & 15. Shape and edges of house type II "*Dorogepak* (Stretch along a Length)"
Source : Private Document

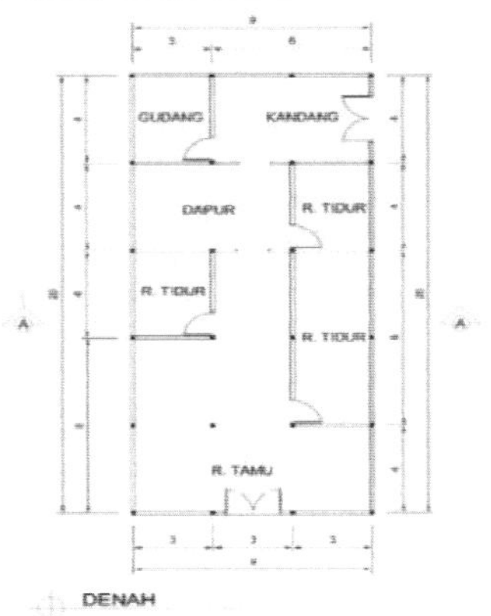

Figure 16. The Plan of house type II *"Dorongepak"*
Source: private document

A –A The section of Type II.

Figure 17. Section of house type II *"Dorongepak"*
Source: private document

2. Method

The research began with GOALS INVENTIONS as first hypothesis, found by study of the FACT and NEEDS (from Field Survey), worked in **Survey Stage** refer to the IDEAL CONCEPT (from Literature Survey). Then it used the theory (ideal concept), and matched with field survey in **Analysis Stage**. (Problem Seeking Step) Furthermore problem statement was examined to intervene the problem solving. As a result of discussion, many alternatives were nominated (**Synthesis Stage**). The last stage was to find the alternative problem solving, as an output of the research and then **Conclusion** or new hypothesis of the next research.

IOP Conf. Series: Earth and Environmental Science **213** (2018) 012011 doi:10.1088/1755-1315/213/1/012011

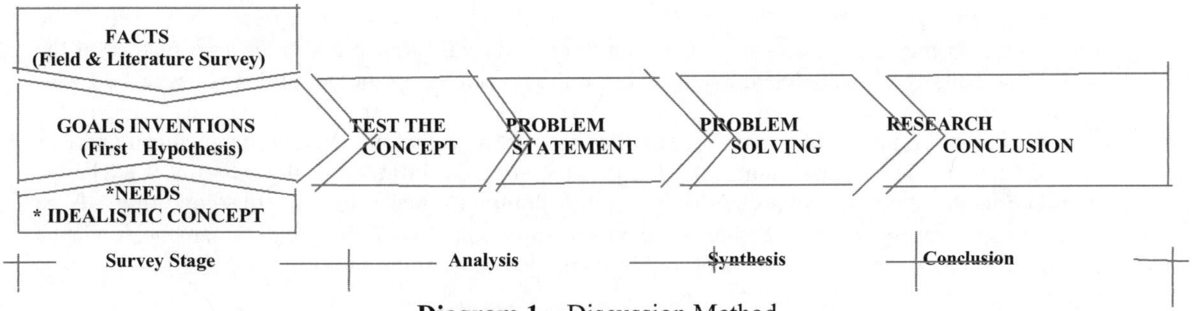

Diagram 1. Discussion Method
Source : Private Document

3. Discussion

3.1 Philosophy

Philosophy is a group of belief and the people way of behaving about the truth of everything, such as various reasons, rights and the basic way of thinking theory or logical activity, which contains logical metaphysics, epistemology and aesthetic. Kampong Naga and Kampong Bade occupants have their own uniqueness philosophy.

3.2 Local Wisdom

Local wisdom is defined as a basis of tradition and expressed in human creation that is culturally passed over generations of the community or tribes. Local wisdom is a physical and color of social cultural custom inhabitant condition that has a specialized local characteristic (Yudohusodo S, 1990). Using a new modern materials and technology is like to delineate a defiant attitude local wisdom by modernism (Haryanto B, 2012). Therefore, a local wisdom regarded as backwardness causes slow action. The Green Building Council of Indonesia (GBCI) stated that the green aspect is one of local wisdom requirements.

3.3 Village

It is a collection of residential buildings, as a communal-homogeneous with the physical condition of the hilly or flat topography and lay out of buildings. There are three types of villages:
First, indigenous villages has traditional viscous nature, very strong relationship family culture, the ability to accept outside influences in low terms of maintaining authentic tradition and strong enough ecosystems relationship. Second, outskirts village (rurality) is the character as a heterogeneous settlement where some places are homogeneous, traditionally characterized by a sense of strong mutual help, located on the shores of the city and the buildings clustered solid enough. Third, village city (urbanity) is a community settled by the heterogeneous nature of the inhabitant, with very strong individualistic, high sense of competition where the nature of mutual assistance is less and the layout of the buildings is clustered very close no distance buildings.

3. 4. The Traditional House

Traditional house is same as vernacular house. According to Bernard Rudofsky (1964) a vernacular house is an architecture form produced by local people to respect local ways of life and culture. It applies simply of technology as well as stabilizes local environmental resources by using local materials. Therefore, the name is the vernacular architecture product because of concerns to the local characteristic.

3.5. Climate and Seasons

Climate is the composition of the atmosphere and weather in the period in a particular area in certain places. It is divided into two climatic conditions, macro and micro. Macro climate is the climate that has the basic properties depending on the geographically location, sea level, coastal conditions and wind direction. Micro climate is the climate in the air layer near the surface of the earth. Air movement is smaller or larger than temperature difference and the rough surface of the earth. The country climate depends on the natural environment principal and it is associated with an average temperature, humidity and season that creates characteristic. Indonesia located in the humid tropical climates have two seasons that are wet and dry. It has many tropical rain forests and savannas, rain with high humidity, high temperatures, a little wind, the small heat exchange with moderate to strong solar radiation. In the humid tropical climates the condition of the mountains generally have moderate temperatures and solar radiation greater than the lowlands where the temperature difference is quite high; cold in the morning, hot in noon, and heat frosty in night.

3. 6. Climate and humidity

According to George Lippsmeier (1997) regarding the limit of comfort on the equator, temperatures range from 22, 5°-30° C with a relative humidity of 20-50% during dry season and during wet seasons is around 22,5 ° C - 29 ° C with 80-100% humidity. The weaker wind blows when the higher humidity comes and the sun shines 12 hours per day (Sangkertadi, 2013). The characteristic of temperature in tropical humid climate is 27° - 30°C with humidity below 60%. The rainfall intensity should be low in dry seasons, then temperature is 23°-26°C humidity 70%-80% and rainfall should be high intensity. The temperature, humidity and rainfall should be controlled by using ideal ventilation system, wide aperture, gable/hipped roof top model, a wide roof terrace, as a backstop between exterior - interior spaces.

3.7. Healthy House

A proper house should have a good proportion of floor and site ratio, compatibility to the occupant number, comfortable space and compliance with the utility, including ventilation, daylight, outdoor-indoor temperature, vegetation, drink water resources, sewage drain and energy. It should include the renewable material and energy. The vegetation density and the building capacity should be arranged to get an ideal freshness living into the house.

3. 8. Ecosystem

Ecosystem is reciprocal relationship of mutual influence between mortal invitation organism's environment and abiotic environment. Because each requires maintenance life then there is a balance between living beings. Natural ecosystems do not require maintenance or energy subsidies because each component can meet their needs and can be in balance. Therefore, as the natural inhabitants, the natural balance of the ecosystem associated with the environment in which we live in needs to be maintained.

3.9. System for the air flow

Natural ventilation is the intentional movement of air from outside to the inside of the building. Natural ventilation system is system for the air flow, the air changes naturally without the tools, or it is natural cooling the buildings by using the physical properties of the air to remove heat or cool. There are three types mechanism of natural ventilation: due to wind gusts, due to differences of the air and the effects of the hot air flow (Noor Cholis Idham, 2016). Using a system of cross ventilation is the best way by putting the openings positioned cross each other. The type, size and position of the window holes can also affect the natural ventilation mechanisms (Georg Lippsmeier, 1997). The human role is to solve the problem of solar heat entering from the outside of the building to be dominant in the humid tropical climates. Therefore, it should be noted that a high roof in a large of rooftop space volume with the space as ventilation holes on the roof can help reduce heating underneath space of the roof (Karyono Tri Harso, 2016). The tropical region with high humidity is part of tropical climate. One of the indicators is thermal comfort, thus humidity in the house should be controlled by natural ventilation. Aperture dimensions will affect the airflow into the house. To receive natural ventilation the amount for living, dining room, kitchen is 20%, for bedroom 25 % and for toilet minimum 15 % floor area.

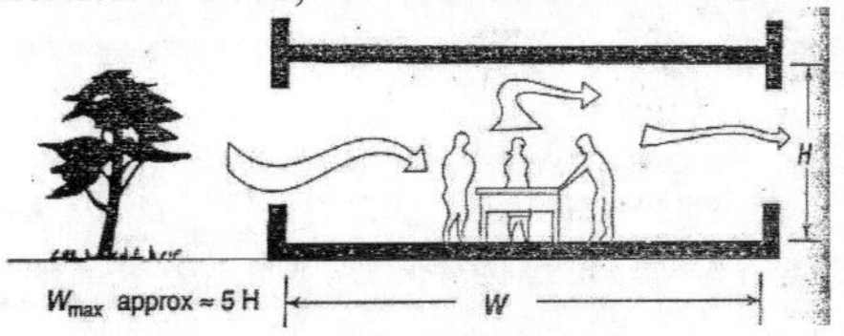

Figure 18. Cross Ventilation
Source : Susan Clare Roaf, (2001.pp. 156)

Figure 19. The air flow in rooftop and space beneath the floor
Source: Tri Harso Karyono, (2016 p. 71).

3.10 Building Orientation

Structuring the mass of the building and building orientation produces a variety of effects on the movement patterns of air flow and air velocity (Tecky, 2014). Building orientation should be facing toward the North and South in order to receive a little more heat (Karyono T.H, 2016). The purpose of planning the orientation of the building is getting proper airflow to the room as well as the controlling. The arrangement of buildings on the site will also dictate the flow of the surrounding air, and the air around the site will be entered into every building.

3.11 Day lighting

Day lighting is the light comes from the sun (nature) that can be used to support day-to-day human activities. A room is not a room without natural lighting (Louis Khan in P. Manurung, 2012). The direct use of day lighting for houses is the major factor in a healthy and ideal living, Location factor becomes one of the effects to optimize the incoming sunlight (P.Manurung, 2012). The utilization of lighting during the day can be used with day light entering through the openings of windows, doors, openings or gaps that exist in the wall (wall board, woven bamboo walls, and others). Because the shelter does not require the level of illumination lighting in the room, generally it only requires 250 Lux (± 250 Lux in space can be used to write and read, Wahyudi Agung, 2012)

3.12 Discussion and Result

3.12.1. Discussion

Traditional house is usually single layer floor, the floor up from the ground. For example stilt model from Sundanese house, the uniqueness characteristic is a distance between floor and the ground, named "*kolong*" (hollow space) where the distance depends on the topography soil condition. Related to the climate, hollow space makes the air flow in the alleys as a cooler of the space above. The bedroom (*enggon*) is a private area that needs a fresh air so it should be located in back side. A living room (*tepas*) is a public area that should be in front part. A kitchen (*pawon*) makes a dirty and hot air, thus the living room and the kitchen should be in front. The space between living room and the outdoor is called "*golodo*g", made from bamboo, used for sitting, smoking and chatting.

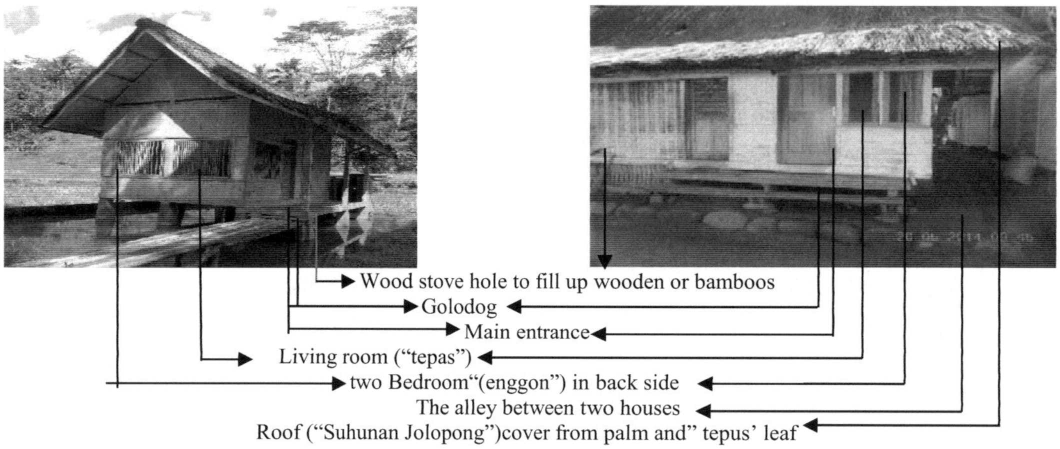

Figure 20 & 21. Illustration of distribution space in Sundanese house
Sources: Private document

Kampong Naga is a native village, which has genuine local knowledge. The inhabitants still maintain the ritual ceremonies, the natural resources and living etiquette, like the regulation of housing orientation. Local wisdom in Kampong Naga is a custom and tradition from their belief or conviction in living regulation. People are very ritualistic, so when they intend to do something such as to repair, fall trees, to start trading, etc. they do the ritual if they have collided the regulation. The observation selected 12(twelve) houses in Kampong Naga that represented the Sundanese.

Figure 22 & 23. People in group prepared the ritual to build a house in Kampong Naga.
Source: Private Document

The houses floor In Kampong Bade usually landed into the ground without cover. They depended on the environment, nature, land typology, soil condition, local customs and tradition. People also have genuine local knowledge as native people. The inhabitants still maintained ritual ceremonies, like the house regulation. The settlements spread into many group *(dusun),* without assembling order. This study selected 15 houses in Kampong Bade to represent the Javanese house.

Tabel 3. Traditional Village Character
Source: Private Document

No	Village's Name	Place	Accesibility	Character
1	Kampong Naga	The genuine name is Kampong Nagawir (the popular name is kampong Naga) in Neglasari village Salawu district Garut regency;	The location is easy to reach, between Garut-Tasikmalaya.	The permanent houses are not eroded by the culture, new technology and materials .etc influence. The location is accessible.
2	Kampong Bade	In the early decade people called Kampong Wates (cause there were two "*wates*" name in Karang gede & Simo district) then the government change Wates in Karanggede into *"BADE"* which means *"kehendak"* in Indonesian it means that people in this kampong have many "wish" to do something"	The location is easy to reach, in highway Karanggede-Gemolong	Starting form ± 1995 (reformation era) many people went out/moved to big city to develop their knowledge, perspective, salary, so that after the people brought the new perspective, many of traditional houses are eroded by the influence of new perspective.

Table 4. The object study requirement list
Sources: Private document

No	Item for examine	Name	Material	Form
1	Typology of house	Stilt or Landed	Wooden or bamboo	Square edges or rectangular edges
2	Foundation	Pedestal/ Foot step	Stone or Brick	Cube or flat
3	Floor	Top Soil or ground / plat form	Covered or uncovered	Flat and square edges
4	Colom	Six or four main pillars	Wooden or bamboo	Pile or post
5	Walls	Dividing Wall	Wooden or bamboo	Wooden plank or bamboo plaited
6	Doors,	Orifice edges	Wooden or bamboo	Rectangular plank or plaited
7	Window	Shuttle or bar	Wooden or bamboo	Plank or lattice
8	Ceiling	0n / off	Wooden or bamboo	Rectangular /plaited
9	Roof	Saddle or pyramid roof	Wooden or bamboo	Square
10	Roof cover	Organic or un organic	Leaves or Tile	Plaited of leaves or tile

The two traditional houses were examined by the list above. The positive and negative values of the houses were examined.

Kampong Naga had 32 positive points showing that the Sundanese house was held out from climate change. They had sturdy viscosity into the local wisdom although they were open to the newest information. Therefore, the local wisdom should be preserved as our cultural heritage. In the meantime kampong Bade should be preserved to eliminate the eroded swell.

3.12.2. Green Building aspect assessment

Green Building Council of Indonesia (GBCI) requires the GREEN ASPECT ISSUE to proof that a locus has an original uniqueness as local wisdom. Therefore, the object study reviewed as below is a breakdown of green aspect assessment.

Table 7 shows that total value of Sundanese house in Kampong Naga was more obedient to the regulation of custom tradition of green building aspect. The analysis from number

1,2,4,5,6 assessment item showed that the Sundanese house in kampong Naga scored higher then Javanese house in kampong Bade.

Table 5. Green Assessment for Tradition Village & Houses

Source: Private analysis

No	GREEN ASSESSMENT	Kampong Naga	Kampong Bade
1	Appropriate Site Development (ASD CATEGORIES)	25	24
	Energy Efficiency & Conservation (EEC CATEGORIES)	15	11
3	Water Conservation (WAC CATEGORIES)	9	9
4	Material Resources and Cycle (MRC CATEGORIES	28	25
5	Indoor Health and Comfort (IHC CATEGORIES)	19	18
6	Building Environment MANAGEMENT (BEM CATEGORIES).	23	18
	TOTAL VALUE	**119**	**105**

3.12.3. Structural element

The main structure used dowel and hole to connect the construction element in unrigidly frame. It is the best system to respond the building from earthquake and support the building load constantly.

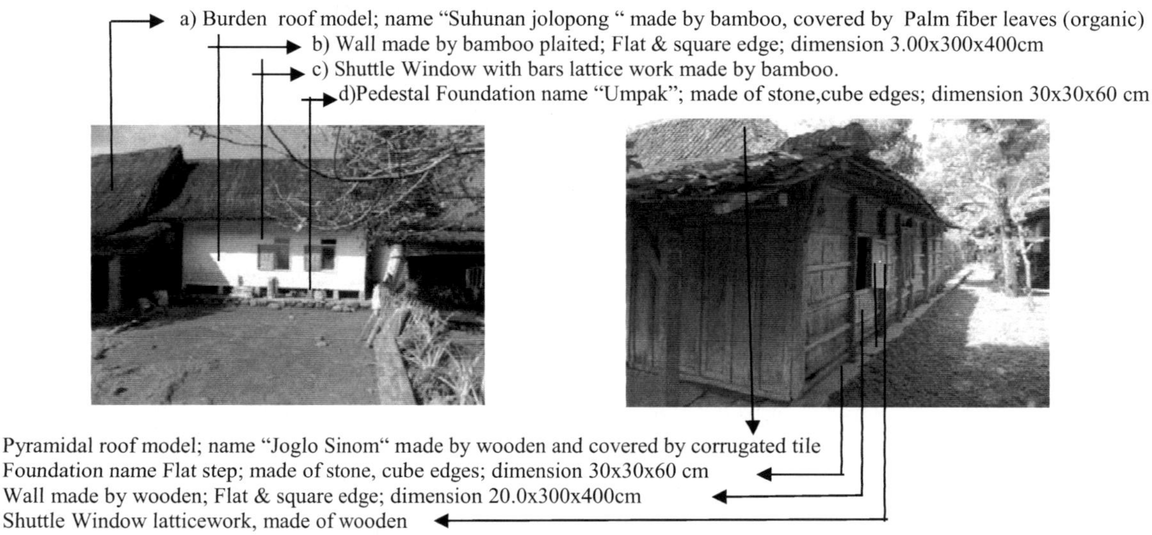

Figure 24 & 25. Structure Elements of Sundanese and Javanese house.
Source: Private Document.

4. Conclusion

Apparently the traditional houses that followed the local wisdom can hold out the climate change influence, proven by:

Table 6. The Conclusion (product of problem solving)
Source: Private analysis

No	Examine Item	Theory	Sundanese House	Javanese House
1.	Philosophy	Base on the meaning of philosophy:	The occupants have their own unique philosophy.	The occupants have their own unique philosophy.
2.	Village	Village characteristic theory :	Kampong Naga belongs to	Kampong Bade as Outskirts

			Indigenous villages	village
3.	Traditional House	Bernard Rudofsky theory: And Agung Wahyudi and C Widi Pratiwi	Stilt house type, Square and Rectangular edges 	Landed house type, Rectangular edges 
No	**Examine Item**	**Theory**	**Sundanese House**	**Javanese House**
4.	Healthy House.	Karyono T.H's theory:	Sundanese house: Have a good ventilation, daylight, outdoor-indoor temperature, vegetation, ideal freshness living into the house, energy efficiency, renewable material	Javanese House Have two types of shapes and edges of rooftop" ***Sinom*** " and "***Dorogepak*** " make the crossing airflow pierce the room in upper part.
5.	Ecosystem	Heinz Frick dkk theory : Natural ecosystems do not require maintenance or energy subsidies because each component can meet their needs and can be in balance.		
6.	Climate&Seasons.	Georg Lippsmeier's theory: Climate is the composition of the atmosphere and weather in the period, in a particular area, in certain places. Air movement is smaller or larger than temperature differences & the rough of earth surface.	The object located in kampong Naga in the humid tropical climates has wet & dry seasons, between two mountains. Refer to table 1. Micro Climate & Weather no 9. **The temperature is chilly in the morning, warm in noon frosty at night.** Humidity 70% - 80%	The object located in kampong Bade in the humid tropical climate with two seasons, wet & dry, a little wind, rain with low humidity, high temperature The heat exchange with moderate to strong solar radiation. The temperature is quite high; cold in the morning, hot in noon and night. Humidity 40% - 60%
7.	Building Orientation	Karyono T.H's theory: 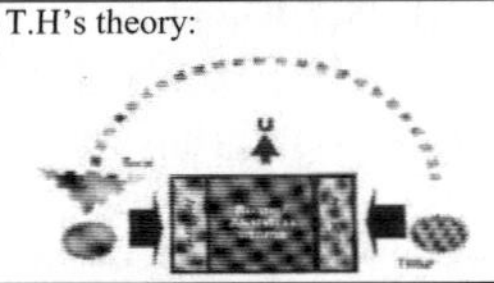	Related to table 2. Site Condition no 3. **House Orientation: South or North, side by side in orderly .**	Related to table 2. Site Condition no 3 **Face to North or South,** depend on the site form, not in orderly
8.	The air flow system: * Ventilation	Georg Lippsmeier's theory: a system of cross ventilation is the best way 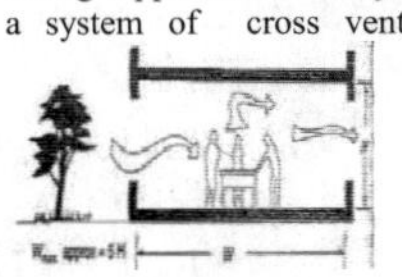	Refer to figure 22.  Shuttle or bar, lattice form Aperture dimensions, amount minimum 15 % to-25% floorarea.	Refer to figure 24  Shuttle, wooden plank Aperture dimensions, amount minimum25 %-40 % floor area
9.	Day lighting.	Louis Khan's theory in P Manurung : P.Manurung' theory : Wahyudi Agung theory :	Refer to table 1 no 7and no 8. Aperture dimensions: amount (wide) 60 cm x (high) 120 cm for window and (wide) 70 cm x (high) 210 cm for door.	Refer to table 1 no 7 and no 8. Aperture dimensions: amount 2 x 60 cm (wide) x 20 cm (high) for window and 80cm (wide) 210 cmx (high for door
10.	Local Wisdom	Siswono Yudohusodo Theory: The Green Building Council of Indonesia (GBCI)	The Sundanese house showed how deep fidelity to custom and tradition	Javanese house consistently implementation & maintain the living regulation community

5. Acknowledment

The authors would like to thank the Management of Research and Public Services Directorate. Reinforcing Research and Development Directorate General, Ministry of Research, Technology and High Education. Referring to the Contract No 044/KM/PNT/2018. March 6th 2018 for PDP Hibah Programme.

The authors would also like to thank The Management of Indonesian Instiute of Technology Serpong. Specially to DR. Tris Dewi Indraswati, DR Ratnawati, DR Joelianingsih, DR Iyus Handrawan and DR Riana Herlina for their support and help in all aspects.

Many thanks to the surveyor team for the hard work to complete this research; to Ms Fajar Kurniawati & Ms Milla Andina Fajriah for the help of technical aspects to finish this research.

6. References

[1] Indartoro,I *KAMPUNG NAGA – a study of the relationship between beliefs, tradition and the pattern of a traditional village.*-1987. Yogyakarta-Laboratory of the History and Development of Architecture p 9-13.

[2] Yudohusodo, Siswono. dkk *–Rumah Untuk Seluruh Rakyat-* 1991. Jakarta, UP Bharakerta.

[3] Maria, Siti, dkk -*Sistem Keyakinan Pada Masyarakat Kampung Naga dalam Mengelola Lingkungan Hidup (Studi Tentang Pantangan dan Larangan)*-1995.Departemen Pendidikan dan Kebudayaan RI. pp 11.

[4] Lippsmeier, Georg. -*Tropical Building (terjemahan)*-1997. Jakarta, Erlangga.

[5] Frick, Heinz, -*Dasar-dasar eko-arsitektur*-1998. Yogyakarta, Kanisius.

[6] Salura Purnama.-*Menelusuri Arsitektur Masyarakat Sunda*.- 2007. Bandung, Ciptasastrasalura Publisher. p 31-61.

[7] Frick,Heinz and Tri Hesti M -*Ecological Architecture*- 2006. Yogyakarta, Kanisius.

[8] Sangkertadi-*Kenyamanan Termis di Ruang Luar Beriklim Tropis Lembab*-2012.Bandung, Alfabeta.

[9] Anwar, Hendi.-*Ethnic House of Sundanese*- 2013, Depok, Griya Creations.

[10] Idham,Noor Cholis.-*Arsitektur Kenyamanan Termal*-2016,Yogyakarta,AndiPublisher.

[11] Karyono, Tri Harso -*Arsitektur Tropis*- 2016, Jakarta, Erlangga.

[12] Handajani A,Galih Pambudi, Fajar Kurniawati - *The Ancestral Heritage Green Technology in Parahyangan Land West Java, Indonesia.*-2014-. Proceedings of UNIKOM National Conferences: GREEN TECHNOLOGY, TO BUILD THE PRESENT PERIODE, AND MAINTAIN THE FURTHER PERIODE School of Architecture, Faculty of Technology and Computer Sciences Indonesia Computer University, Bandung , October 23st . pp. IV 11-19

[13] Handajani Asriningpuri, Fajar Kurniawati, Galih Pambudi,- *The Ecological Architecture existence in Sundanese Land.*-2015-. Proceedings of TAU International Conferences: MITIGATING AND ADAPTING BUILT ENVIRONTMENTS FOR CLIMATE CHANGE IN THE TROPICS. School of Architecture, Tanri Abeng University,Jakarta,Indonesia March 30th – 31st. pp 420.

[14] Handajani Asriningpuri, Ratih Budiarti, Harlisa – *Study of Bioclimatic Application to The Spatial Habitation Along The River Bank CILIWUNG.*-2015- Proceedings of *2th* International Conferrence, on Engineering of Tarumanagara University, URBAN ENGINEERING for FUTURE GENERATION. Jakarta,Indonesia October 22nd – 23th pp. AE-12/1-8.

[15] Handajani.A, Aliviana Demami, Fajar Kurniawati -*Tropical Minimalist Housing Style Inspired by Local Wisdom.*- 2016 -. Proceedings of *2th* LIVABLE SPACE International Seminar, Trisakti University, APPLYING LOCAL KNOWLEDGE FOR LIVABLE SPACE. Jakarta,Indonesia 1st December pp. 146-152.

[16] Handajani Asriningpuri, Fajar Kurniawati, Keiko Sugianto - *Local Wisdom Fidelity as an Evidence of Traditional House Preservation* -. Proceedings of *3th* International Conferrence, on Engineering Tarumanagara University, "SMART ENGINEERING for FUTURE CITIES". Jakarta, Indonesia October 4th – 5th. pp AE-15.

[17] Susan Clare Roaf,-*Natural Ventilation of Buildings in India*-2001- Journal of Climate Responsive Architecture. ISBN 0-07-463218-3 Tata McGraw-Hill Publishing Company Limited. pp. 156.
[18] Amin Sumadyo. H - *Peranan Bahan dan jarak bangunan terhadap Kondisi Termal pada Rumah Tradisional Duku Pancot Tawangmangu* – 2004 - KOMPOSISI Journal of Archhitecture, October. ISSN 1411-6618. Architecture Programme - Faculty of Engineering- Atmajaya University Yogyakarta Indonesia pp 131
[19] Agung Wahyudi.dan C Widi Pratiwi – *Efforts Sunda Traditional Building Design Approach for Local Wisdom, Green And Save* (*Upaya Perancangan Bangunan Tradisional Sunda Sebagai Pendekatan Kearifan Lokal, Ramah Lingkungan Dan Hemat Energi)* – 2012- Journal of National Simposium RAPI XI 4 Mart ISSN 1412-9612 Engineering Faculty - Muhhamadyah University Surakarta Indonesia. pp A30
[20] Gerarda Orbita Ida Cahyandari.-*Tata Ruang dan Elemen Arsitektur pada Rumah Jawa di Yogyakarta sebagai Wujud Kategori pola Aktivitas dalam Rumah Tinggal* - 2012 - KOMPOSISI Journal of Archhitecture. October ISSN 1411-6618.Architecture Programme-Faculty of Engineering-Atmajaya University Jakarta Indonesia pp 103.
[21] Manlian Ronald Simanjuntak.dan Muhammad Agung Wibowo -*Identifikasi Factor Konstruksi dan Pengaruhnya dalam Fungsi Bangunan* - 2012- ARSITEKTUR Journal of Archhitecture November.ISSN 2086-5740.Architecture Departement – Engineering Faculty -Tarumanagara University Jakarta Indonesia pp 1.
[22] Tecky, Andrian, Yani dan Dita, *Telaah Penghawaan Udara Alami pada Ruang Dalam Rumah Kuncen di Kampung Pulo*, Online Jourrnal Institut Teknologi Nasional – Juni 2014, pp. 3.
[23] Dwi Wahjoeni Soesilo Wati, *Matrilineal Architectural Values On The Construction of The Minangkabau House*.- 2013 - KOMPOSISI Journal of Archhitecture. April. ISSN 1411-6618.Architecture Programme-Faculty of Engineering-Atmajaya University.Yogyakarta pp 195.
[24] Setiawan Hardono *The Matrilineal Architectural Values On The Construction of The Minangkabau House*.- 2014 - DIMENSI On line journal of Archhitecture and Built Environment. July ISSN 2338-7858. At E-Journal http://dimensi.petra.ac.id indexed by Google Scholar Petra Christian University. Surabaya.Indonesia pp 51.
[25] HandajaniA, Aliviana Demami, Fajar Kurniawati, *The Influence of Micro Climate Aspect to Minimalist Landed Houses and Vernacular Stilt House* -2017- Livase on line journal of Livable Space April 26th ISSN 2548-7515 Trisakti University Jakarta. at E-Journal@trisakti.ac.id pp 45
[26] www.google.maps
[27] https://m.tempo.co/read/news/2012/01/11/204376575/menikmati-keindahan-arsitektur-kampung-naga, by Aditya Herlambang Putra
[28] https://maps.google.com/ and Jurnal of Kampung Naga by I. Indartoro (1987, pp 9).
[29] http://arsitekturdanlingkungan.wg.ugm.ac.id/2015/11/20/pengaturan-penghawaan-dan-pencahayaan-pada-bangunan/) by Dwita Hadi Rahmi access at January 20 2017 9:00 a.
[30] http://arsitekturdanlingkungan.wg.ugm.ac.id/2015/11/20/pengaturan-penghawaan-dan-pencahayaan-pada-bangunan/ Kontributor: Dwita Hadi Rahmi Posted on November 20, 2015 by arsitekturlingkungan

Textile Tourism Image as an Identity of Cigondewah in Bandung City

Karto Wijaya[1], Asep Yudi Permana[2],
[1] Universitas Kebangsaan, Bandung, Indonesia,
[2] Universitas Pendidikan Indonesia, Bandung, Indonesia

Corresponding e-mail: kartowijaya@universitaskebangsaan.ac.id

Abstract. Cigondewah is one of the priority areas of textile industry development in Bandung and a center of fabric shopping tourism area for local and national tourists. The majority of people inhabiting Cigondewah are traders and entrepreneurs who are mostly residents or those living surrounding the area who become the main economic actors in Cigondewah. The growing and developing trading activities can affect buildings along the right and left corridors at Cigondewah Street. The existence of the textile industry area results in its own impact and phenomenon on the change of built environment function on Cigondewah Street that can improve the society's economy. Cigondewah is declared as one of the creative economy areas in Bandung. It is closely related to the built environment which provides an area for fabric shopping tourism in Bandung. This research described the existing phenomena in the area by examining the shape, activities, characteristics, changes, relationships, similarities and differences of phenomena of the built environmental along Cigondewah Street.
Image and identity of Cigondewah as a fabric shopping tourism have a special meaning for the community of Bandung and surrounding areas as an alternative to fabric shopping tourism. The location of Cigondewah that is located around the fabric industry center in Bandung and Cimahi areas makes Cigondewah very close to the textile industry. This research intends to acknowledge activities done in the residential area, existing image and identity of the study area, built environment changes, and its layout.

Keyword: Built Environment Function, Identity of the region, Image of the region, Textile Industry

1. Introduction

Indonesia's creative industry is in a significant period of growth. Based on the results of the study of Indonesian Creative Industries of 2009 conducted by the Ministry of Trade, the comparison of the development of the national Gross Domestic Product (GDP) shows a clear contribution from the creative industry sector. Based on the sequence, the creative industry has contributed as much as 7.8% or around Rp. 235,633 Billion and it is placed in rank 6. The position was higher than the Finance, Real Estate, and Corporate Services, as well as transport and communication sectors [1]. This significant development should be a better starting point by incorporating it into the curriculum of creative industry development in universities.

The development of creative industries requires a strong synergy between academia, business, and government. First, academia is used to deliver creative and original ideas. Second, business is needed to improve the abilities of economic value or marketing ideas in trading. Third, the government is useful for keeping the political will related to the development of the creative industry. The potential of Indonesia is significant in developing the creative industry. Since the global financial crisis hit the world, it has forced state managers, economists and businesses to think critically about the fragile global economic system. The global financial crisis has finally opened the eyes of the world that there are still other sectors that can provide large incentives in addition to the financial sector that has been honored. One of which is the creative industry sector based on creativity and intellectuality.

The selection of democracy as a model in state management also contributes to the development of the creative economy in this creative industry. The government system chosen by the Indonesian people will have implications for the development of cultural, social, legal, economic, and other systems. The democratic model chosen by the Indonesian people provides wider opportunities for creativity [2]. One of the most important models is decentralization process that is normatively expected to create independence and economic power based on the wealth of local economic values. Respect for individual rights provides a vast opportunity for the development of creativity based on original ideas, which will have an effect on the growth of the creative industry.

Creative industries were born along with the development and openness of communication technology. Wide information does not suspend the development of human life. Now, the development of lifestyle information no longer has obstacles, because the information technology has been through the barriers of nation and state. In line with the opinion from Marshal McLuhan, world is finally no more than a global village. The model of this democratic country also allows the creation of economic sectors based on a community based economics where the economic system is a part that can not be separated from the activities of Indonesian people independently. Independence and creativity become the cornerstones to run a democratic economic system, not monopoly or oligopoly system. In the community based economics system, creativity becomes an important starting point. The next development is the emergence of the term Creative Industry. The sector has not been much considered by the people of Indonesia. The system of regional autonomy and cultural diversity becomes a strong capital to develop creative industries throughout Indonesia. For comparison, for example, the Indian State has more or less the same conditions as Indonesia that has diverse ethnic and culture. The diversity underlies India now to be one of the centres of the technology, intellectual, and entertainment industries.

Bandung as one of the creative cities in Indonesia has enormous potential. One of the potential areas to be a creative area is Cigondewah textile trading area. Architecture does not merely perform physical development but also can take part in supporting a region to be an area with more value. One of the ways is by arranging the land. Setting the land is important to determine the best option for allocating certain functions so that in general it can give an overall picture of how the area in a region should function [3]. In line with the opinion, [4] stated the ability of a type of activity to place itself in a strategic location depends on the level of productivity owned by the activity. Activities with high levels of productivity are commercial and industrial activities. Both activities have a great potential ability to conserve activities in a particular field. As the actors of economic activity, urban communities tend to choose a strategic area.

A large number of land use that has been converted into commercial buildings and warehouses in Cigondewah street has caused by industrial activities around the area. The residential function to be economic and trade activities along the street as a circulation room has two direct effects on environmental quality. Environmental quality has a strong influence on the continuity of commercial activity and strong regional visual character of the physical structure as well as the form of the city [5]. The irregularities of spatial pattern in the development of the Cigondewah area has caused almost all the existing buildings in the area are not well organized and they seem shabby to be inhabited and for the business area. Commercial and industrial activities as high productivity activities are potential for conservation of a land that can develop and change in function [6].

The development of land conversion occurred in the area of Cigondewah resulted in the image of the impressed area into a business area that grows so rapidly but it ignores the factor of visitors' convenience and also the quality of the environment.

The existence of industrial activities in Cigondewah converts the residential function to be a rapid business area and generate some buildings to be unorganized. The building appearances along the Cigondewah street corridor seem inappropriate, as the building function as residential has switched into warehouses and commercial buildings. In addition, the activities of commercial activities and processing of textile industry products that do not care about the surrounding environment have formed the area to have a special and unique identity.

1.1 Environmental or Town Functions

Mirsa, (2012) stated that understanding a city according to its physical function is a built-up area that is denser than the surrounding area. Meanwhile, according to the economy, a city is an area where there are very diverse business activities with the dominance of non-agricultural sectors, such as trade, industry, services, offices, transportation, [7]. Areas having trade and industrial activities create a built environment that switches many functions to support the sector of environmental functions and develop the economy of urban areas. In addition, [8] an urban area basically accommodates human activities that require similar steps in all activities of its citizens. The ability of the services of an environment can be seen from its primary function for example as a place of face-to-face or indirect human communication and as a place of economic activities. Various efforts were conducted by humans in forming a physical environment. The ultimate goal is to build togetherness of various functions manifested in different shapes and characters. In addition, the ability of an environment can be seen from how many second and tertiary functions possessed by the environment. There are also some environmental elements that have a function as a place, in which the physical element of the environment acts as a means of binding the social, economic and cultural life of society. According to Poerbo (1999) the modern sector with its affluent lifestyle is served with an adequate standard of infrastructure and buildings. Meanwhile, high-density suburbs where most workers of informal sectors reside live with lack of infrastructures and social facilities, chaotic spatial planning, poor sanitation, and drainage. Based on the statement, city or environment will always experience changes such as the occurrence of social facilities. This will result in changes that occur in the physical, social, and also the function of the environment or city.

1.2 Land Use

A key element in the urban design is the determination of land use that can create a circulatory relationship to regulate the density of activities in urban areas. Land use is the use of land to create the best option to allocate certain functions, so it can provide an explanation for an area experiencing changes in physics and function [3]. Along with that, Land Use Planning is a concept that is contrary to the fact stating that all social activities are still on the ground. However, it relates to the use of land for a specific purpose [9]. At first, this kind of spatial planning movement focused on the problem of urban planning. Various methods have been developed to analyse or forecast future developments (socially and economically) and translate them into the need for land and its spreading in space. This generated the problem of land-use planning drafts in the past that lack of understanding of the diversity of regional land-use.

1.3 Building form and massing

The form and massing of a building are created by the height or magnitude of the building, and appearance and configuration of the building massing. The massing of the region includes buildings, the surface of the land, the objects that form the urban space and the patterns to define the building form and massing with the principles and thoughts behind the physical form of the city. Spereigen (1965) made a synthesis of a building form and massing, which has a scale relating to views, circulation, and size of adjacent buildings.

The basic urban design principles and techniques conveyed by [10] bring many of the issues pertaining to the shape of the building into a unity, among them:

1. The scale of its relationship with human vision, circulation, adjacent buildings and the surrounding environment.
2. Urban space as the main element of urban design, as well as the rise of an urban form, scale,

and type of urban space.

3. Building massing, ground surfaces and other objects capable of forming urban space and patterns of activity, both large and small.

1.4. Growth and Development of the Area

The most dominant force determining an environmental growth is an economic power, although other aspects also have effects on the changes. The development of the built area provides a wide potential especially in its development as an area that has excellent products or specialization in the scope of tourism development plan of the city [11]. Therefore, this economic aspect is a prominent factor in influencing the changing physical environment. The residential area that seems to change is basically the need for the development of the area towards the commercial area. Changes inbuilt functions indicate the success of an area from settlements into commercial areas [12]. The need for economic improvement is greatly felt in the previously unopened areas to change land functions in various ways such as changing the function of buildings or selling the existing land to get an additional value elsewhere.

1.5. Regional Image & Identity

Image of an area can be built by the presence of a location link to space and meaning. The location link between objects in urban areas is an important reference that allows humans to carefully recognize the various objects and locations because they are aware of their position on the environment and compare the interests of various clues in the form of guidance of street directions in the exploration of the environment of different areas. [13] To add the image of an area, there are three points affecting the mental image of a person in an area. They are: (1) Identity, in which people can understand the picture of the region including objects identification, objects differences, and knowable subject; (2) Structure, in which people can see the pattern of an area including the relation between objects, relation between subject and object, and visible patterns; (3) Meaning, in which people can understand the space of the region including the meaning of objects, the meaning of the subject and object, and the feeling that can be experienced. Regions that have specificities on the identity, structure, and meaning provide an image of the function of the area.

Identity is the feature of the area, while the image is what is perceived by the community about the area experiencing spontaneous development[14]. Identity is usually sent simultaneously along with information sources received by humans through communication media in the form of buildings and regions or cities. Urban identity is basically acquired through various social media networks that are territorially attached. The form of the identity is not only shown at the level of settlements, but it can also be seen in the development of places such as streets, districts, cities, and even wider until the state[15]. The developing area always makes its own identity, which promotes the area to be more known by the society and the surrounding environment.

2. Method

The analytical approach used in this study is the descriptive approach. Descriptive approach is a method of analysis performed by looking at the state of research object through the description, understanding or explanation of the analysis of the location being observed. According to Bungin [16], qualitative descriptive research aims to reveal, describe, and summarize the various conditions, situations, or various phenomena[17][18]. The approach analyzes social realities existing in the community that became the object of research and attempts to bring the reality to the surface as a characteristic, model, sign, or description of a particular condition, situation or phenomenon[17]. Based on the above statement, this study intends to determine the condition of residential area activities, existing image and identity of the study area, and changes of built environment and spatial area.

The purpose of this qualitative descriptive research is to describe the phenomena that exist in the area by examining the shape, activity, characteristics, change, relationship, similarity and difference of built environment phenomenon that occur in Cigondewah street. Image and Identity of Cigondewah as the area of fabric shopping tourism have a special meaning for the community of Bandung and surrounding areas as an alternative fabric shopping. This study describes various conditions of the situation or various phenomena of changes in the function of residential buildings that exist along the Cigondewah area as the center of the creative industry so that it is depicted as reality or a built environment condition that develops and grows as it is.

3. Discussion

Bandung is one of the economic growth centres of West Java whose development is very dynamic. The role and function of Bandung itself as an industrial centre, trade centre, educational centre, tourism and cultural centre, as well as the central government of West Java Province, underlay the condition. With higher economic growth rate than the average economic growth of other cities in West Java, Bandung has become a city attracting migration of residents from other regions to perform various activities that potentially make the city of Bandung as one of the densest city in Indonesia.

Cigondewah is a residential area that was transformed into a fabric sales centre. Generally, Cigondewah area is used for trading by traders and entrepreneurs who are mostly residents as the main economic actors. Hundreds of shops lining the streets of Cigondewah sell fabrics. The fabric shopping area is located in Cigondewah Rahayu sub-district, Bojongloa Kaler district. However, the textile area itself covers the area belonging to three different villages, namely Cigondewah Rahayu, Cigondewah Kaler and Cigondewah Kidul located in Bandung Kulon Sub-district. This area in the document of Bandung City Government Planning is planned to be developed as a fabric shopping area in Bandung. The area is near Kopo Kencana Secondary Center which is designated as the center of the Core City to serve areas within the Tegalega Development Area. This area in *RT / RW* of Bandung is an industrial area having environmental orientation and a creative industrial area in Bandung [19].

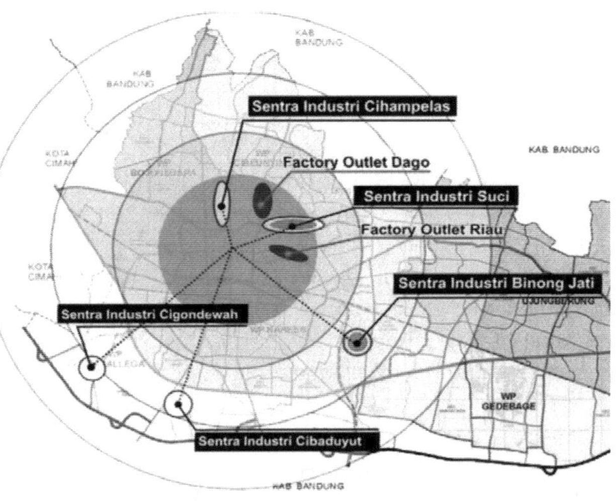

Figure 1. Creative Industrial Area of Bandung in the period of 2011-2030 (Source image: Bapeda Kota Bandung)

Structuring of this area provides an opportunity to utilize, to a greater extent, the function of commercial areas and industrial areas aiming to improve more the economy of the surrounding community. Cigondewah in the map plan of *Bapeda* Bandung belongs to the category of creative industrial centers that make Cigondewah as a fabric shopping tourism.

3.1 Building Massing of Cigondewah Area

According to the *RTBL* of Bandung City, Cigondewah is promoted to be a creative tourist area supporting the economy of Bandung, so that the residential buildings of Cigondewah experienced many changes in their function into commercial buildings. Buildings along the Cigondewah street corridor are almost all seen selling fabrics and yarn and the rest of the fabric itself.

The condition of ground figure (solid void) corridor of Cigondewah street shows the existence of void square (inside the building) and void street (street), whereas its solid building is quite varied both the massing and the wide of building tread. It can be seen in more details in picture 4.17 which shows that solid void corridor of Cigondewah street seemingly is not structured and designed well by the government of Bandung. Cigondewah has a pattern being able to bring up the forms that will illustrate suitability between the organization of physical space and social space. The area of Cigondewah has a composition of building blocks massing that tends to be homogeneous. This is indicated by the similarity of patterns in the region.

Most of the buildings located along the main streets are blocks that serve as the edge of the area dominated by residential buildings functioning as commercial buildings or warehouses that coincide with residential buildings. Meanwhile, the residential buildings of Cigondewah community are in the form of single blocks adjoining each other. The buildings have developed following the existing street network. Cigondewah community has built their buildings according to the availability of land they own and many buildings tend not to have their Building Permit (IMB) as legality. Therefore, they often ignore the orientation of buildings and the distance between buildings. As the result, the area of Cigondewah is classified as a densely populated area forming the building massing that is adjacent to each other.

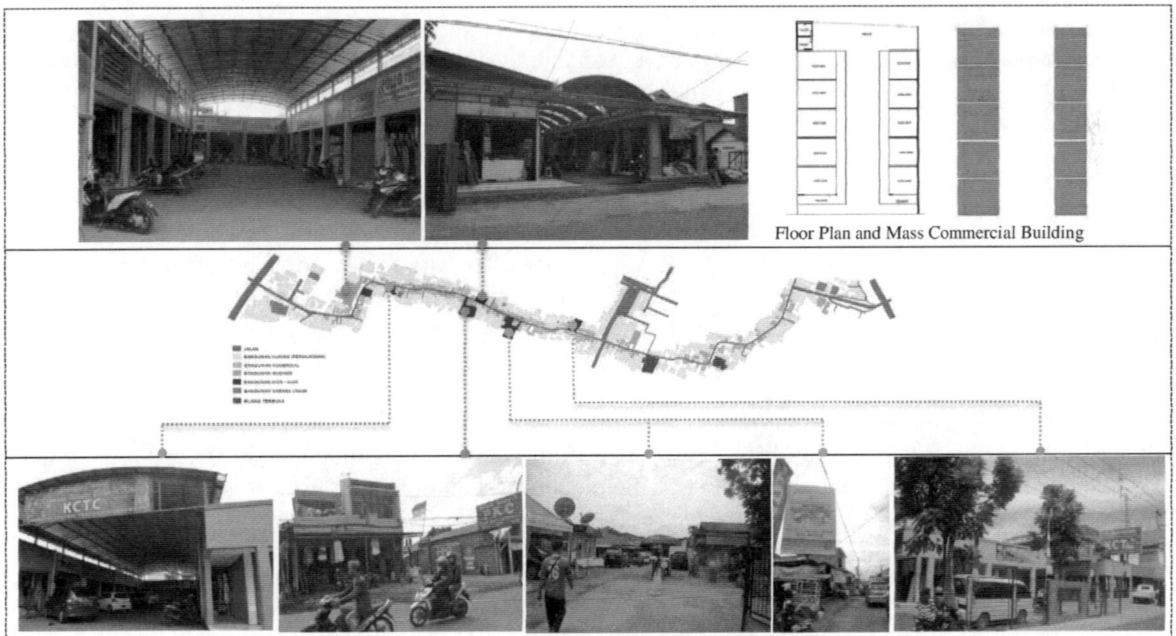

Figure 2. Shelter that turns into stalls form clusters Jl. Cigondewah Kota Bandung
(Source image: Processed from Peta Bandung 2017)

The above picture is the result of the identification of the Cigondewah street showing the building blocks located along the Cigondewah street corridor. When it is viewed from the perspective of the land area, the building blocks do not experience much change in the form of the building massing. The change occurred only during the development period in accordance with the needs of its users. The

Cigondewah area texture pattern has a stack of building massing that accumulates to form an agglomeration massing resulting in a very high density.

Based on the condition of the area relating to solid and void elements and from the land area point of view, there is not much change in the building blocks form. The changes occurred only at some building massing that experience the development in accordance with the needs of its users. Changes related to regional logging in transition to commercial buildings and warehouses do not significantly affect changes in extent. However, there is a change in the composition of the constructed land or building massing towards open space. It is clear that the density level of Cigondewah street corridor belongs to a very high-density level marked by the reduction of open space to be commercial buildings.

Cigondewah area of Bandung City is dominated by blocks of building massing with unplanned formations, which follow a gradual growing building pattern. Therefore, functionally a relationship does not focus on one of the existing blocks of the area. The direction of the changing pattern of the built environment on the Cigondewah street corridor follows the path of the Cigondewah street circulation. There is a relationship of collective massing form, mass relationship and open space around Cigondewah Street. As the result, the buildings along the Cigondewah Street have changed the function of the built environment from a residential area to be commercial buildings.

3.2 Identity and Image of Built Environment Function in the Textile Tourism Area of Cigondewah Bandung

Along with the development of the region, then the identity of the region will emerge as a feature of the area that is received by the community. In line with the above statement, Permana and Wijaya [15] explained that identity is usually sent simultaneously together with sources of information received by humans through communication media in the form of buildings and regions or cities. Urban identity is basically acquired through various social media networks that are attached territorially where the implantation does not only show the level of the residential area but the development of places such as streets, districts, cities [14]. The theory reinforces that the identity of an area experiencing changes is shown by the changes in residential function and activities that have high productivity level, which is the commercial and industrial activities. Both activities have a great potential ability to conserve the activities in a particular field.

The use of residential building functions as a trading area and land use activity on street corridors as a space for movement (circulation) caused two direct effects on the quality of the built environment, namely the continuity of commercial activity and strong visual quality of the physical structure and physical form of the city. Commercial activities will strengthen the public spaces as complementary to each other that gives effect to the identity and image of the area. Changes in the built environment function that give the impact of unity to the environment do not emerge in the Cigondewah Street Corridor. Based on the observation result, activities and transfer of occupancy function generate an identity of this area as textile tourism area. Given in the corridor Cigondewah street changes in the built environment function that looks dense distance between buildings and impressed chaotic to give the image of its own area so that characterize the unique area. Changes in environmental functions built in the Cigondewah street corridor affect the identity of the region. In this case, the community shows less attention to harmony and design of the function of the building and the physical form of the building. Hence, it will affect the identity of Cigondewah area as a textile tourism area in Bandung.

3.3 The distance between buildings and height of buildings in the area of Cigondewah

This study explains that the distance between buildings on the Cigondewah street corridor gives the impression to the unique regional identity of Cigondewah image with the massive density of the building massing. According to the density of the distance between buildings, the identity of Cigondewah was formed as a densely populated residential area. It changed the residential area into a trading area that allows the Cigondewah community to widen the area of their buildings. In addition, the height of the existing buildings in Cigondewah provides its own identity in the author's observation that the average building has a building altitude of about 1-2 floors. The height of the building existing in Cigondewah gives the special meaning that the existing buildings in Cigondewah were built naturally and not designed as the author and respondent thought during this research.

Commercial buildings located in the area of Cigondewah were built irregularly showing chaos, slum, and inconvenience in the area. Many buildings in the area of Cigondewah were formerly residential and have been renovated to be a group of commercial buildings lined up so that the distance between the existing buildings in Cigondewah is very close.

3.4 Building Orientation and Building Area in Cigondewah area

The orientation of the buildings in Cigondewah has an effect on the structure of the area. The orientation of buildings along Cigondewah Street is facing Cigondewah street corridor so that the structure of the area is very clearly seen in the street corridor itself. Based on observations, Cigondewah street corridor provides direction related linking structure between the city of Bandung, Cimahi City, and Bandung regency. Hence, the area Cigondewah is easily recognized and understood by the people of Bandung and surrounding areas.

The very strategic area of Cigondewah is located in the structure of Bandung which is located in the southwest part of the Bandung center. In comparison with other areas, Cigondewah has advantages of being adjacent to the textile industry. So the location of the area has its own meaning that the distance is still in the textile industry in the city of Bandung. This condition provides a distinct advantage for the Cigondewah community to establish the site as a trading and textile warehouse.

The width of the existing building bases in Cigondewah has its own influence on the meaning of the region as a collection of building massing in a very dense residential environment so that the area seems to be narrow. People visiting Cigondewah will feel a distinctive atmosphere that other regions do not have.

3.5 The characteristics of changes of built environmental function in Cigondewah

The residences located on the left side of Cigondewah Street Corridor are residential dwellings that have another function in economic activities to be shops, warehouses, and home industry. Public facilities in the form of places of worship and schools are also available on the street. The dwelling that functions as a residence is also still present in the area, but the longer the number is reduced because the community is more concerned with the use of buildings for economic activities. Cigondewah community itself prefer to use their house for trading than only for the living.

The shift and change of building function evolve along with the change of this area from partly as a residential area to commercial area or trade. This is supported by the Bandung government program that includes Cigondewah as one of the 6 priority areas of development of Bandung to be one of the creative cities in Indonesia. A description of the changes in the built environment function that exists in the Cigondewah street corridor provides an illustration that the built environment function and the building base area has a significant effect on the change of built environment function on Cigondewah street corridor such as (1) one of four of a residence has changed into a commercial building; (2) one of two of a residence has changed into a commercial building; (3) three of four of a residence has changed into a commercial building; (4) a residence has fully transformed into a commercial building. Changes in the built environment function in Cigondewah area dominated by the number of residential houses that are converted into commercial buildings are influenced by economic factors that play an important role in the Cigondewah community.

Physically, Cigondewah street corridor was formed by the ranks of buildings whose orientation are towards the street. Mostly, the facade of the buildings along Cigondewah street corridor has changed in appearance and a small part still retains the existing buildings. Cigondewah after being set by the government of Bandung as a creative economic region has a significant role along the corridor of Cigondewah street so that people maintain the existing buildings to be commercial buildings. The price of land along the Cigondewah street corridor is now quite expensive after the region is set to become a creative economy area. The buildings that were once residences on the side of the Cigondewah Street have turned into growing trade buildings and they are increasingly crowded.

Based on the discussion, the change in built environmental functions in the corridor of Cigondewah street, according to respondents in this study, describe that the image of Cigondewah regions shows a very strong identity, especially in the distance between existing buildings. Cigondewah giving the image to the community as a collection of trade buildings fabric promotes the identity of this region as a textile tourism area in Bandung. This area is formed by the activities of the people who sell fabrics so that the existing buildings along the corridor of the Cigondewah street experienced a major functional shifting as residences so that many people who have landed on the streetside of Cigondewah will use to change and add the function of the building.

4. Conclusion

The identity and image of Cigondewah, from the observation of researchers, was a set of residential buildings that changed the function as the result of industry around this area. Thus, the form and function of the buildings did not reflect the textile trading area. Cigondewah community was not concerned with the factors of building facades form and building function, but the area was more concerned with other factors, such as the quality of economic improvement and the quality of its own community to further develop this area spontaneously. Therefore, the improvement of the quality of existing buildings along the street corridor of Cigondewah was not maximal.

Plans of structuring and developing in the Cigondewah area of Bandung City should also consider and adjust the condition of the area and the surrounding environment, and the needs of the community to change their residential buildings into commercial buildings. Furthermore, the government needs to note the change of *KDB* and *KLB* of buildings so that they should follow the rules. This is due to the increasing number of changes in the existing building function in Cigondewah. As the result, the density of the building is increasingly uncontrolled so that the area is less well-structured.

5. References

[1] M. H. Sutanto, "Gelombang Ekonomi Ke Empat, Gelombang Ide dan Gagasan," *KOMUNIKATOR*, vol. 6, no. 5, pp. 27–34, 2014.
[2] A. Y. Permana, N. Sumarna, and K. Wijaya, "Membangun Kampung Kreatif melalui Kolaborasi Mahasiswa dengan Masyarakat: Kasus Kawasan Balubur-Tamansari Kota Bandung," 2017, pp. 51–58.
[3] E. Darmawan, *Ruang publik dalam arsitektur kota*. Semarang: Badan Penerbit Universitas Diponegoro, 2009.
[4] D. Sujarto, *Beberapa Pengertian Tentang Perencanaan FIsik*. Jakarta: Bhratara karya Aksara, 1985.
[5] G. Cullen, *Consept of Visual*. London: McGraw-Hill Publishing Company, 1961.
[6] K. Wijaya, B. Setioko, and T. W. Murtini, "Pengaruh Perubahan Fungsi Lingkungan Binaan Terhadap Citra Kawasan Wisata Tekstil Cigondewah," *KOMPOSISI*, vol. 11, no. 2, pp. 67–76, 2015.
[7] R. Mirsa, *Elemen Tata Ruang Kota*. Yogyakarta: Graha Ilmu, 2012.
[8] H. Shirvani, *The Urban Design Process*. New York: Van Nostrand Reinhold, 1985.
[9] H. Poerbo, *Lingkungan Binaan Untuk Rakyat*. Bandung: Yayasan AKATIGA, 1999.
[10] P. D. Spereigen, *The Architecture of Towns and Cities*. New York: McGraw Hill, 1965.
[11] E. Hasanah, Iswatun; Setioko, Bambang; Setyowati, "Pengaruh Activity Support Terhadap Kualitas Visual Koridor Jalan K.H. Agus Salim Semarang," *Teknik*, vol. 35, no. 2, pp. 61–67, 2014.
[12] K. Wijaya, "Perubahan Fungsi Lingkungan Terbangun Di Koridor Jalan Cigondewah Kota Bandung," *ARCADE*, vol. 1, no. 1, pp. 8–13, 2017.
[13] M. Zanhd, *Perancangan Kota Secara Terpadu*. Yogyakarta: Kanisius -Soegijapranata University Press, 1999.
[14] A. Y. Permana and K. Wijaya, "Kota Bandung: Kota Pendidikan antara Citra dan Identitas

Kota," in *Reinterpretasi Identitas Arsitektur Nusantara*, 2012, pp. 27–34.
[15] A. Y. Permana and K. Wijaya, "Education City As Identity of Bandung City," in *International Conference on Urban Heritage and Sustainable Infrastrcture Development (UHSID)*, 2013, pp. 15–19.
[16] M. B. Bungin, *PENELITIAN KUALITATIF: Komunikasi, Ekonomi, Kebijakan Publikdan Ilmu Sosial lainnya*. Jakarta: Kencana Prenada Media Group, 2008.
[17] J. W. Creswell, *Qualitative Inquiry and Research Design*. London: Sage Publications, 1998.
[18] Junaidi, "Mass Media's Perspective on the Challenges of Information and Communication Technology (ICT) in Indonesia," in *International Conference on Environment and Technology (IC-Tech)*, 2017.
[19] Bapeda, "Dokumen Peraturan Daerah Rencana Tata Ruang Dan Wilayah Kota Bandung 2011-2030," Bandung, 2017.

Residential space transformation as the legitimacy space A case study: Magersari *Ndalem* Sasanamulya Baluwarti Sunanate Palace of Surakarta

Avi Marlina
Universitas Sebelas Maret, Surakarta, Indonesia

Corresponding e-mail: avimarlina@staff.uns.ac.id

Abstract. Magersari is courtiers' residence and their descendants who inhabit *cêpuri* of Surakarta Sunanate Palace under the orders of the King or the Prince. The transformation has been taking place from long time ago and along with hereditary dwellings. The transformation usually occurs in the main residence, but at the Magersari residence, there has been a different pattern of transformation of residential spaces due to overall changes of pattern at the Magersari residence. This article focuses on the act of transformation as a step made by the palace to gain the space legitimacy. How is the legitimacy is manifested through space transformation? This article aims to find how legitimacy is produced in the space transformation. The objective of this study is to find the pattern of residential space transformation as the manifestation of the space legitimacy practice that cause changes in the pattern of the Magersari residence and to find its cause. The research used descriptive qualitative with the case study approach. It began with observing transformational forms at the residential space, and then continued with searching for transformation patterns and conducting in-depth interviews to key informants to find the cause of the space legitimacy practice in the transformation. The results of this study indicate that the legitimacy of space is found in every practice of transformation that is always followed by the restrictions imposed by the rules from the palace in the form of *dhawuh* (orders). Submission to *dhawuh* is the way to maintain the authority of the palace continuously. The phenomenon of transformation as the practice of space legitimacy can provide a discourse to the policy maker of the palace to establish the transformation rules of residential space.

Keywords: transformation of Magersari residential space, restriction of space transformation, legitimacy space, *dhawuh*(orders), authority of the palace

1. Introduction

Magersari Baluwarti is the residence of the courtiers (*abdi dalem*) and their descendants at *cêpuri* in the Baluwarti of Surakarta Sunanate Palace. This place is located between the center of the walls of the palace (*kedhaton*) and the walls of Baluwarti [1,2]. The spatial pattern of Magersari residence refers to the hierarchical concept of palace's layout situated at the center as the representation of the concept of macrocosm-microcosm [3-5]. In the beginning, Magersari embraced a mutualistic symbiotic pattern based on the relationship between the aristocrats or nobles and *abdi dalem*, on the one hand the courtiers worked to serve the nobles in return for a residence in the noble courtyard where the residence is around the noble dwelling; on the other hand, the nobles receive their manpower [6]. The Magersari residence in Baluwarti is divided into two groups, namely the courtiers who live in the courtyards of the nobles (*ngindung*) and the courtiers who live outside the nobles courtyards [7]. This article will discuss the case of *ngindung* residency in Magersari because it can represent the case that aims to find the answers in the objectives of the study.

Magersari has no land certificate, but it has a permit called *palilah* (a permit to occupy the buildings in Magersari) based on the decree of the *Pengageng Parentah Keraton* in the form of *Palilah Griya Pasiten*. The land and the palace are belong to the Surakarta Sunanate Palace while the buildings at the Magersari land belong to Magersari [8]. In the present, when the palace has changed its status to become the preserver of the cultural tradition of *keratuan* or kingdom, and *Sinuhun* remaines as the king in the sense of the palace's cultural tradition and no longer as the real authority holder of the Sunanate [9]. Therefore, there is a change in the system which brings an impact on residential system in Magersari. The Magersari houses have undergone many changes along with the lifestyle changes of the nobles and types of the residence. Magersari is mostly inhabited by the third or fourth descendants of the courtiers who used to serve kings or nobles [6,10].

The change in the palace system triggers conflicts of land as a result of the enactment of the Basic Agrarian Act (UUPA) of 1960 stating that the residency more than 20 years have the property rights[10]. However, Magersari residents occupy the palace's land which is tied to the palace's land status according to the Presidential Decree No. 23 of 1988 [8]. Various steps have been done by Magersari residents to still stay in the palace's land, and one of their ways is to transform the occupancy in order to occupy the land as an effort to strengthen the status of the land. Meanwhile, the nobles, who place themselves as lord of the Magersari's inhabitants, legitimize the residential spaces in Magersari to maintain their existence.

How space transformation in the Magersari residence can be regarded as the practice of space legitimacy that can be reviewed first from the discussion about transformation. Transformation itself is interpreted as a gradual adjustment process done by the user [11-13]. This approach occurs in most of the space transformation process. What makes this article different is that transformation happens not only in physical changes but also in its practice to maintain the authority. The practice of transformation can also be done in order to protect the socio-ecological system of a community [14]. The difference is that the community system has no practice of exercising authority. The case of Magersari is different because it emphasizes the practice of legitimacy, so the apropriate approach able to reveal the linkage between space transformation and legitimacy practice is needed.

There are various ways of transformation. Such as cutting the mass and unifying the mass [15], finding the transformation pattern of the initial occupying design [11], rotation [16], by performing spatial integration [17,18] and doing spatial practice to create a comprehensive and functional form on the limited land [16, 19]. Meanwhile, transformation is interpreted as socio-spatial and socio-economic action [20]. The difference is that such an approach is only physically focused action, so it can not be used as an approach to see the transformation in Magersari which involve the ruler of the palace.

Harrison and Todes have different opinion about the transformation. They state that the transformation is more to a spatial control action [21,22]. Marlina, who support their opinion [13]. reveals that each transformation is controlled by spatial characteristics which then create differences in each type of the transformation. The transformation approach as the spatial control can be used as the starting point in this research in ordert to find the type of spatial transformation as a practice of legitimizing power.

The transformation support as the spatial control corresponding to its characteristics is a cooperative spatial management system [23]. The explanations of spatial transformation management are the steps to improve the relationship between physical space and organizational aspects such as culture and artifacts which become the priority in changing space [24]. The act of spatial control is different from Chi Sing's opinion who states that transformation can occur in all parts of both the center and the periphery [25]. This approach has its drawbacks because it does not recognize the characteristics of transformation and spatial characteristics, so I refuse to use the Chi Sing approach. I rather offer the theory of transformation as spatial control action.

The discourse on transformation was also revealed by Moglia that transformation depends on an understanding of the values and rules of decision makers [26,27]. I agree with Moglia that transformation is related to the values and decision makers' rules. Transformation actions related to values and rules will result in creative acts. The concept of transfomation can be interpreted as a creative act in utilizing space that can be categorized as an intelligent transformation [28]. I will use two transformation approaches in writing this article, the approach of the concept of transformation as the spatial control and the concept of transformation associated with the values and rules of the ruler.

Regarding creativity, it can be implemented in changes of the spaces necessary for business and economic development of occupants [29-31]. This approach only focuses on the act of physical transformation itself to develop its economy. Thus, this business and economic-related transformation approach can not be implemented in the Magersari *ngindung* land since the transformation of space in the Magersari residence is not economic-oriented. My opininon is that the spatial transformation of Magersari residence is related to the spatial control in the palace land associated with the royal authorities, and it is not merely physical changes; but it focuses more on non-physical issues like status, authority, and legitimacy. Therefore, I use Max Weber's theory and Anthony Giddens's theory of authority to explain the phenomenon of the space transformation that occurr in Magersari.

According to Weber [33], the use of authority as an aspect of social relations lies in the legitimacy order that applies to communities who have belief in the truth of the system. Power has control over the behavior of its community [33]. The approach of authority and control is closely related to the traditional society having rulers and subordinates pattern [34]. Authority and control still take place in the Magersari residence because it is located at the palace's land that lives and thrives with strong traditions and beliefs where the patrimonial system is still in progress even though the palace does not have real power anymore.

Beliefs in a legitimacy system also blend with social stratification encompassing status differences and existing in traditional community with strong tradition [33]. Magersari residence settlement can represent Javanese tradition and has a clear division of social strata of noblemen and courtiers [1,6,7] so the social stratification system in Magersari palace can answer the case of spatial transformation as the means of reproducing legitimacy. I agree with Giddens's opinion stating that domination as a social system can be expanded by reproducing dominant structures through power that can be manifested in the organization of space and time in spatial and territorial development [35]. This article focuses on an architectural transformation related to residential space in the Magersari residence that will be seen as a practice of legitimacy.

The objective of this study is to find the forms or patterns of residential space transformation in the Magersari residence as the manifestation of the practice of space legitimacy causing changes in the Magersari residential pattern and to find its causes.

2. Methods

This research used the descriptive qualitative method with case study approach which was done through a unique case exploration analysis to answer the question of 'how' and 'why' [36] in the case of residential transformation pattern in Magersari. The case study was employed to examine a community related to processes and activities limited by place and time [37]. Therfore, this method is appropriate to examine the Magersari residence that still exists. The first step was data collection through field observation which was followed by in-depth interviews to see the phenomenon of residential transformation as a practice of space legitimacy that caused changes of residential patterns in Magersari. Qualitative data collection was done through several stages namely: 1) observing the transformation pattern of the residential space with purposive sampling technique in Magersari Sasanamulya where 33 people dwell there. This technique is in accordance with what was described by Creswell [38]. The next step was 2) conducting in-depth interviews with 41 key informants to collect accurate information with snow ball techniques. These key informants, moreover, introduced other informants hence more in-depth information. The next step 3) was matching the patterns to relate some cases similar to some theories [36]. For data validation, it was done by triangulation [39].

3. Discussion

3.1. The location and the characters of Magersari residence

The location of Magersari Baluwarti residence in Sunanate Palace of Surakarta Central Java is the part of Baluwarti Village, Pasar Kliwon Subdistrict. Baluwarti region borders the Kauman and Kedunglumbu Subdistricts in the north, the Pasar Kliwon and Gajahan Subdistricts in the south, Gajahan Village in the west, Semanggi and Pasar Kliwon Villages in the east (figure 1). The research locus is located in the north of the main palace (kedhaton) which is Magersari Ndalem Sasanamulya located in cêpuri Baluwarti (figure 2).

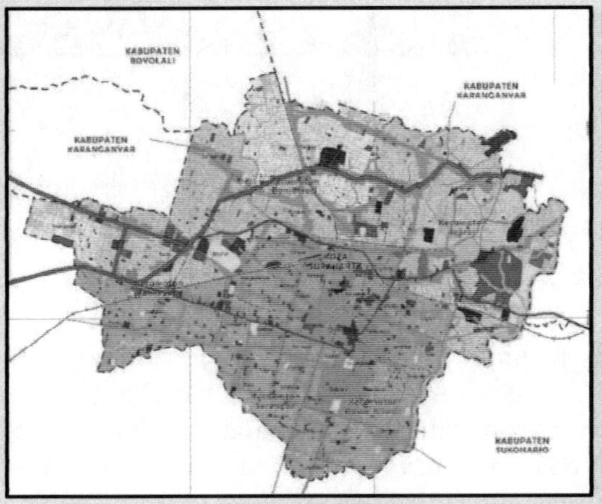

Figure 1. Location of Magersari Baluwarti Palace of Surakarta
Source : gis.surakarta.go.id (2016)

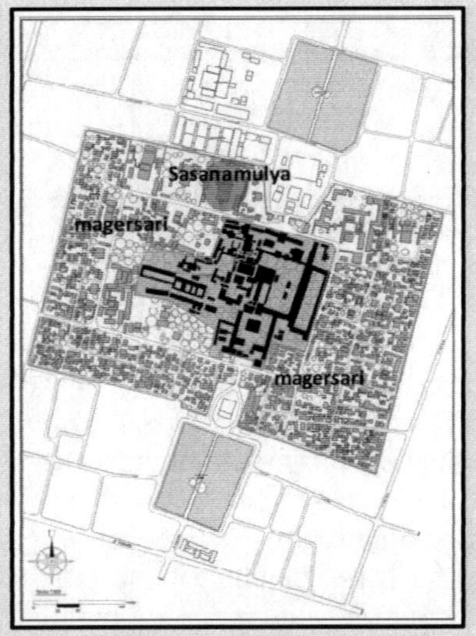

Figure 2. Research Locus in Magersari *Ndalem* Sasanamulya

Ndalem Sasanamulya was built during the reign of *Paku Buwono* IV. This place is characaterized by the surrounding 6 m walls called *cêpuri* Sasanamulya. In this residence, there is a clear division based on the social strata of the kingdom, in the middle is the nobles' residence called *ndalem* (figure 3), while the residence around the *ndalem* is the place for courtiers and his descendants, the noble assistants who work for the nobles, as well as the noblemen descendants which is located on the left-right-back of *ndalem*. This place is called *ngindung* (shelter), and it is separated by *cêpuri* wall as the space barrier (figure 4, figure 5 and figure 6).

This study includes 33 Magersari residences with a total population of 128 people divided into 41 families, and located around *ndalem*. Public facilities located in this area are public toilets, shared wells, drying areas, and open spaces. Some residences already have their own toilets and wells. The use of shared facilities in a traditional residential has advantages in land efficiency in this settlement, and it also strengthens the familial system in social, economic and cultural relations [40]. The people at Magersari settlements share the facilities and live with their relatives, and this tradition has been done since their ancestors. One place of living can be inhabited by some descendants such as parents, children, son-in-law, grand children, nephews, and cousins. This tradition is called *sak brayat* [41].

Figure 3. *Ndalem* Sasanamulya as the center of the residence located in the center of the settlement and and dominant

Figure 4. *Cêpuri* wall Magersari courtiers surrounding *Ndalem* Sasanamulya.

The status of Magersari land is the palace's property having the right to occupy the building based on the permit called *Palilah Griya Pasiten*. Thus, Magersari residents do not have the certificate of land[8]. Magersari has the obligation to pay double taxes namely the Land and Building Tax paid to the municipality and the *duduk lumpur* money paid to the palace. *Palilah* or permit is valid for 3 years and must be renewed in *pasiten* palace. *Palilah Griya Pasiten* also has arranged all the regulations in Magersari such as obeying all the orders from the palace, forbidding renting or selling, obligating them to repair or build parts of buildings in which they must ask permission to the palace, and obligating them to return if the buildings provided the users pass awayd to the palace[8]. These hereditary rules have become the habit since the first era of the Sunanate Palace.

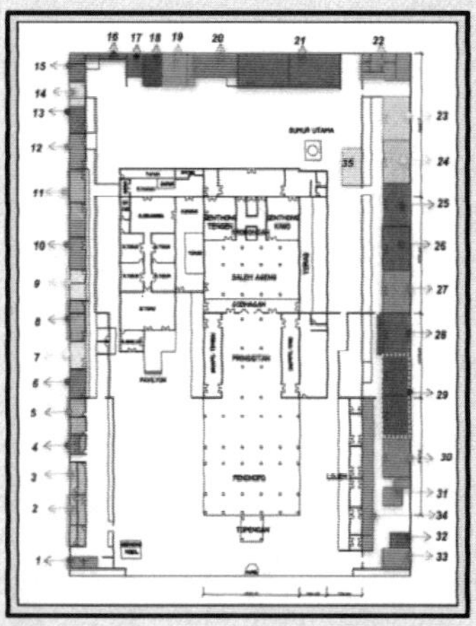

Figure 5. Magersari dwelling location surrounds *ndalem* (core occupancy) which limited by *cêpuri* wall to separate or distinct between *ndalem* noble with Magersari courtiers

Figure 6. *Cêpuri* wall as the arbiter between *ndalem* noble with Magersari settlement

This system has been built down through the generations by the palace to maintain public trust in its cultural systems and traditions, such as the concept of authority conveyed by Weber [33] stating that traditional authority is based on beliefs that will legitimize traditional societies. The right of Magersari people to live in the palace by accepting all the rules from the Sunanate Palace is the form of the Magersari's acknowledgement of the Sunanate Palace's power, and such action is actually the strengthening of legitimacy. I interpret this relationship as the strengthening of power from the palace by extending the traditional mutualism symbiosis. This kind of symbiosis, like in Magersari, is manifested in modern residence by building relationships between landowners and the tenants [40,42]. The difference is that this system takes place in public settlements while this case takes place on the area of the palace which has strong tradition.

The concept of traditional authority is also manifested in the pattern of mass management of buildings and indirect interpersonal relations between the nobles' residence and Magersari residence. These residences are separated by the courtyard and the cêpuri walls surrounding the nobles' home where the dominance of space lies at the center of the residence. The ndalem layout at the center as the most dominant building is a way of presenting power. This concept is in line with the Giddens Structure theory [35] which reveals that authority resources can be made through the organization of space as a means of extending the nature of power within a community.

The Magersari settlement has its uniqueness in occupying and dividing its residential space. In occupying and dividing the space is a tradition passed down continuously from generation to generation based on the lineage set by the kings or nobles based on orders or *dhawuh*. *Dhawuh* has an absolute value of obedience as *Sabdo Pandito Ratu* (royal words) so that the Magersari residents always obey the *dhawuh*. In the spatial context, the agreement with the nobility is the power to divide division the space which is supported by the *Palilah Griya Pasiten*, the letter or permit to occupy from the palace. Obedience to *dhawuh* is based on the belief in the power of a king in the form of *Sabdo Pandito Ratu*, and these king's words is believed to have the power of *walat* (sanctions). Therefore, the belief in this myth supports the process of legitimacy, and the traditional authorities rely on the myth [43], and legitimacy will be easy to grow in religious communities [33]. The difference is that the myth in the king's palace represents and protects the people as a *Gusti-kawula* [44] and it will shape obedient behavior to the king all his orders as *Sabdo Pandito Ratu*.

3.2. The transformation pattern of residential space and legitimacy in Magersari

The transformation pattern of Magersari residential space is the extension of the space toward the front of the building approximately less than 3 meters in which the characteristics of the original building and area is unchanged (figure 7), the addition of the space is done by making the partition with plywood or cabinet (non-permanent material), addition of rooms' types with mezanin wooden without changing the original building (figure 9), addition of space's types in the form of the building in front of the residence located separately from the original building having fixed shape building (figure 7, figure 8, figure 9, and figure 10). The building is transformed into a residential function, and the new *façade* as a cover of some the original *façades* is still able to make the original *façades* visible (unchanged) by pasting the new *façade* in front of the original *façade* so that the original *façades* is seen as the border in internal space (figure 7).

The transformation pattern findings in this resarch are the extension, the addition of the space, the addition of the space type, the change of space function, and the change of the façade in all Magersari residence where its spatial transformation characteristics occur at outside of the original building which still retains its original form. The transformation of the residence with all the restrictions set by the palace occur in Magersari causing changes in the residential pattern at Magersari (figure 11).

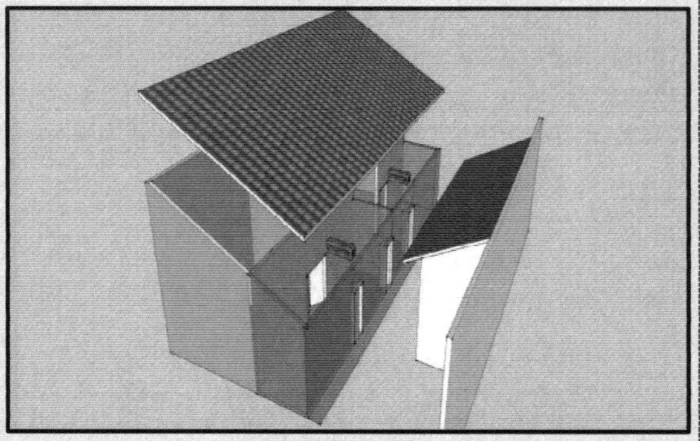

Figure 7. The space expansion without changing the original building. The new façade adheres to the original façade and builds the kitchen, bathroom, washing room, drying space located outside at the main residence as the separated building and is adjacent to *cêpuri* wall without changing the original building. The original façade can be seen on the wall inside the building

Figure 8. Kitchen, bathroom, washing room, drying space is separated and have similar paint on the façade wall to show the family descendants.

The uniqueness of Magersari transformation space is that it does not alter the original building, so it is clearly visible to see the original building façade, the doors and windows and the order of teak wood (the original building element) on the inner wall of the space extension (figure 7). The other uniqueness is building a new type of space in front of the residence located separately from the main residence in the area adjacent to the *cêpuri* Magersari courtiers wall (figure 7, figure 8, figure 9, figure 10). The similarity of transformation at the Magersari settlement is building a kitchen, bathroom, laundry, drying place located outside the main residence as a separate building adjacent to the *cêpuri* Magersari wall without altering the original building (figure 8 and figure 10).

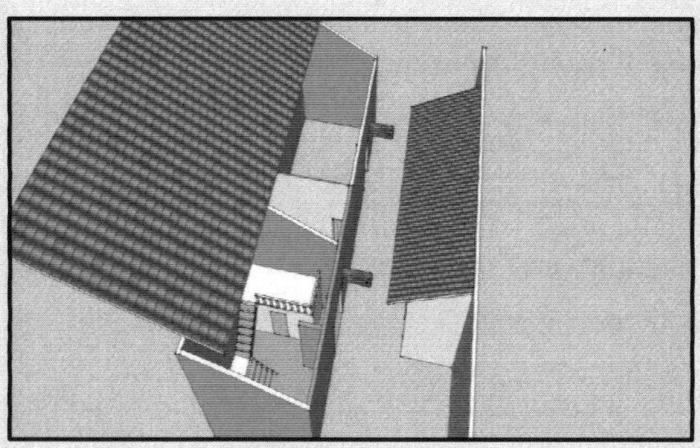

Figure 9. Building a new room in front of the residence separated from the main occupancy adjacent to *cêpuri* wall and the use of non permanent wood for *mezaninne* in the residence.

Figure 10. The original building is still maintained, the new building located in front of the residence, is shown by similar paint colour to describe their descendant.

Total changes in space function occur in the residence without damaging or dismantling the original building. The types of buildings that has changed the function are the *gedhong mobil*, the *gedhong kereta*, the *gedhogan* jaran (stables), the wood warehouse, the *prantunan* (kitchen), the concubine residence, the empty yard that switched its function to be the residence courtiers and their descendants. Space function changes did not happen in *ndalem* (central *cêpuri* the residence of *Ndalem Sasanamulya*).

The findings of the spatial transformation characteristics in the Magersari are as follows: 1) the original building retains its authenticity; 2) the expansion related to ways, directions, and distances; 3) the increasing amount of space associated with the manner and the direction of the addition; 4) the addition of the space type associated with the location, manner, and the type; 5) changes in spatial functions related to the limitation of new space function types and restriction of the spaces that can change their function; 6) altering of façades associated with the restrictions on the changeable façade and how to alter the façade. Characteristic of the transformation at Magersari space is always related to the act of restriction in changing spaces. Restriction of spatial transformation as a social-spatial organizing action to extend the authority and Magersari community as a vessel for saving authority, as expressed by Giddens that space and time relate to power theory [36]. Every practice of spatial transformation in the Magersari residence is an act of changing accompanied with all restrictions from the royal court in the form of orders (*dhawuh*). All the space transformations at the *cêpuri* palace are based on *dhawuh* from kings or nobles who have higher positions. Like the changing of supporting structures such as *prantunan*, *gedhogan jaran*, *gedhong mobil*, *gedhong kereta*, the wood warehoused, and kitchen is *dhawuh* from th kings or nobles. The *dhawuh* from the palace is believed to be *Sabdo Pandito Ratu* that must be and can not be rejected or denied where no one dares to argue *dhawuh* because it is believed to contain the mystical element associated with *walat*, as described by Coleman that the belief in this myth supports the process of legitimacy and the traditional authorities rely on the myth [44]. *Walat* is believed by Magersari community as the sanction as a result of violating or disobeying the *dhawuh*.

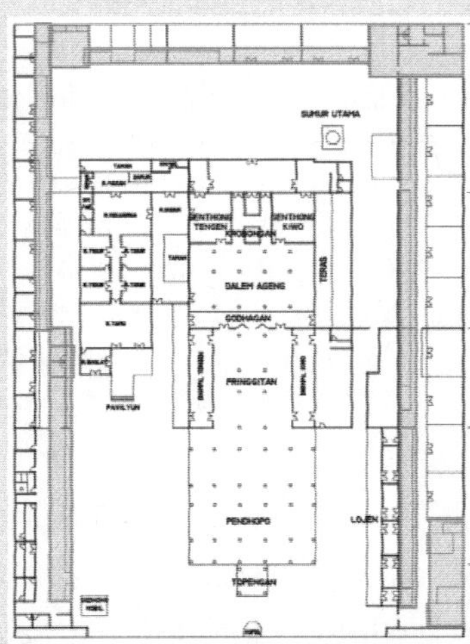

Figure 11. Changes in spatial settlement patterns of Magersari due to the transformation of residential space (limitation of transformation lies between the walls boundary *cêpuri* Magersari)

Thus, *dhawuh* is believed by Magersari residents because it has the power that governs behavior towards space. The belief of Magersari residents in *dhawuh* is in line with the beliefs in the cultural system. They believe that all spaces or buildings in the palace have a transcendent power which is built upon the authority of *dhawuh*. In its narrative, it is explained that authority is a 'orders' from generation to generation and does not require written rules in its exercise. This command is in the form of order from the higher social structure to gain legitimacy [33]. Traditional authority as a order (*dhawuh*) was built by the palace from generation to generation to maintain the system through tradition.

Magersari inhabitants occupy the palace's lands and transform the residential space based on *dhawuh* (order) and the order of restrictions the use and the transformation of spaces. This act is supported by the obligation to obtain permit from the palace. The act of asking permission to the palace is actually a practice to create the obedience among Magersari residents who have been granted a right to occupy the palace's land, and this kind of practice has been formed continuously in the *dhawuh* tradition. This authority refers to legitimized authority which means that a ruler has the right to rule and is expected to be obeyed. In addition, it also means that there are rulers, ruled communities, the ruler's will in the form of order, the obedience, and the evidence that the ruled community must obey ruler's orders [33]. This legitimacy system occurs from traditional actions based on customary rules of the palace. Legitimacy is obtained through the regulation of residences and the space transformation of the residences.

In conducting the transformation of space like improving the building, increasing the space, increasing the space width, adding the type of space, changing the function of space, and changing the façade of the residence, it is essential to ask permission to the king or nobles. In addition, there are some courtiers appointed by the palace or nobles who go always around to control the residences that are changed by residents and report it to the palace. Usually after receiving the reports, the palace immediately summon the Magersari residents who do not have permission to repair or build the residences and remind them about the permit. The control over the transformation of residential space is an action to control the behavior of Magersari residents towards their space, in which the palace places itself as the ruling party and expects obidience from the residents so that the recognition of the palace authority can still be maintained in the Magersari tradition system. Such action is the way of the palace to gain legitimacy from Magersari residents. This control is supported by the assertion that there is power that controls the facilities and infrastructure [33].

The submission of Magersari residents to the restrictions in connection with the placing of space and changing of space is an act of space legitimacy. This submission is expected by the palace in order to preserve its authority which merged with traditions from generation to generation. As in Weber's narration that obidience is an aspect to maintain the continuity of authority [33]. *Dhawuh* is the practice to strengthen the legitimacy of space functioning as the order in determining which space is allowed to transform and which place is not allowed to transform and determining how to conduct this transformation with all the restrictions. *Dhawuh* also serves as the main controller in making spatial changes or transformation as well as in restricting the space use and the space change.

4. Conclusion

Magersari *Ndalem* Sasanamulya Sunanate Palace of Surakarta is known as its Magersari which becomes the residence of *abdi dalem* (courtiers) who experience the spatial transformation of space. By analyzing the pattern of the spatial transformation, it is found that the practices of reproducing the palace's power is an act of obtaining the legitimacy from Magersari residents. This act is in the form of authority recognition manifested in obedience to *dhawuh* (order) as the proof. The legitimacy of space is found in every transformation practice that is always followed by the restriction from the royal court in the form of *dhawuh* (order). The *dhawuh* (order) in connection with the transformation is manfested in the restriction on changing the form of original building in order to maintain its authenticity; restriction on making extension to ways, directions, and distances; restriction in adding the amount of space associated with the manner and direction; restriction on adding the types of space associated with the location, manner, and the type of space; restriction on changing the spatial functions related to the limitation of new space function types and on the location having changeable functions; restriction on changing the facade which is related to limiting the façade that can change and how to change the façade. These traditional authorities are endowed through generations in orders to expand their authority and acquire the legitimacy of space from Magersari. In the practice of expanding the legitimacy of space, it requires rules in making spatial changes to create a Magersari settlement in accordance with the concept of spatial Java, for it requires further research related to the concept of spatial Magersari in Java.

5. Acknowledgments

The author would like to thank: (1) Sunanate Palace of Surakarta, especially *Pengageng Ndalem* Sasanamulya who have granted permission to make the place the research locus, (2) Baluwarti village as the regional office of Baluwarti Keraton Kasunanan of Surakarta.

6. References

[1] Soeratman, D 2000 *World of the Palace of Surakarta* 1830-1939; Foundation for Indonesia: Yogyakarta, pp **105-112.**

[2] Adrisijanti, I 2000 *Islam Mataram Urban Archeology*; Publisher *Jendela*: Yogyakarta, p **184**.

[3] Geldern, R. H 1982 *Conception of the State & The position of Kings in Southeast Asia*; Rajawali: Jakarta, p **2-6.**

[4] Tjahjono, G 1989 *Cosmos, Center, and Duality in Javanese Architectural Tradition : The Symbolic Dimensions of House Shapes In Kota Gede and Surroundings*; University of California: Berkeley, p **197**.

[5] Behrend, T. E 1982 *Kraton And Cosmos in Traditional Java*; University of Wisconsin-Madison: Madison, pp **160-183**.

[6] Priatmodjo, D 2004 *The Palace of Surakarta Nowadays: An Anthropological Study about Reposition of Traditional Kingdom* ; Dissertation; University of Indonesia: Depok,p **227- 230**

[7] Widayati, N 2015 *Baluwerti toward "Independence Village" Residential Study of Abdi Dalem and Sentana Dalem in Surakarta Palace*; Dissertation; Technique of Architecture: Jakarta, p **19-20**

[8] Isbandiyah, K 2008 *Surakarta Palace Policy of Hadiningrat on Building and Land Management by President decree Number 23 Year 1988 about Status and Palace Management of Surakarta in Baluwarti Village, Surakarta City;* Thesis; Diponegoro University: Semarang, p **79,93,108**

[9] Setiadi, B.; Hadi, Q.; Trihandayani, D 2001 *Kings in Republic Era of Surakarta Palace and Paku Buwono* VII; Bina Rena Pariwara: Jakarta; p **116.**

[10] Farkhan, A 2002 *Transformation of formation and structure of the settlement environment in Baluwarti Surakarta*; Thesis; Diponegoro University: Semarang, p **69-125.**

[11] Aryani, S. M.; Mulyadi; Wahyuningsih, I. E. S 2015 *The house design transformation; the preferences and the patterns*. Procedia Environmenta Sciences, 28, **723.**

[12] Permana, A. Y.; Wijaya, K 2017 *Spatial Change Transformation of Educational Areas in Bandung*. International Conference on Sustainability in Architectural Design and Urbanism (ICSADU), Semarang; p **11**.

[13] Marlina, A.; Hardiman, G.; Setioko, B.; Sari, S. R 2015 *Spatial Transformation Pattern of Dwellings of Javanese Nobles's Houses at the Palace of Surakarta Sunanate:Case study of Ndalem Purwodiningratan and Ndalem Suryohamijayan*. Advance in Environmental Biology, 9 (27), **374.**

[14] Ge, D.; Long, H.; Ma, L.; Zhang, Y.; TU, S 2017 *Analysis Framework of China's Grain Production System : A Spatial Resilience Perspective*. Sustainability, 9, **16.**

[15] Cahyono, U. J.; Setioko, B.; Murtini, T. W 2017 *Transformation of form in the growth of modern Javanese house in Laweyan Surakarta.* Journal of Architecture and Urbanism, 41, **294.**
[16] Munzer, S.; Fehringer, B. C. O. F.; Kuhl, T 2018 *Specificity of mental transformations involved in understanding spatial structures.* Learning and Individual Differences , 61, **49.**
[17] Liu, J.; Liu, Y.; Yan, M 2016 *Spatial and temporal change in urban-rural land use transformation at village scale-a case study of Xuanhua district, North China.* Journal of Rural Studies, 47, **434.**
[18] Magoni, M 2017 *Metabolism in Spatial Palnning:which possible integrations.* City, Territory and Architecture, 4, **7.**
[19] Wang, F.; Liu, X.; Zhang, Y 2016 *Spatial landscape transformation of Beijing compounds under residents' willingness.* Habitat International, 55, **178.**
[20] Chen, Y.; Wen, Y.; Li, Z 2017 *From blueprint to action : The Transformation of the planning paradigm for desakota in China.* Cities, **464.**
[21] Harrison, P.; Todes, A 2015 *Spatial transformation in a "loosening state"; South Africa in comparative perspective.* Geoforum, 61, **160.**
[22] Quinn, B.; Wilks, L 2017 *Festival heterotopias:Spasial and temporal transformation in two small-Scale settlements.* Journal of Rural Studies, 53, **41-43.**
[23] Shakya, L.; Takada, M.; Kanki, K 2018 *Spatial structure of a courtyard-style settlement originating from a Buddhist monastery: A Study of a cooperative space.* Japan Architectural Review, **12-13.**
[24] Hansen, M. A. C. S. a. G. K 2017 *Change your space, change your culture:exloring spatial change management strategies.* Journal of Corporate Real Estate, **104-105.**
[25] Chi-sing, C 2011 *Typological transformation:study of traditional urban dwelling and urban fabric of Guangzhou*; The University of Hong Kong: Hong Kong.
[26] Moglia, M.; Cork, S. J.; Boschetti, F.; Cook, S.; Bohensky, E.; Muster, T; Page, D. 2018 *Urban transformation stories for the 21st century: Insights from strategic conversations.* Global Environmental Change, **231.**
[27] Mellano, P 2017 *Regaining the culture of cities.* City. Territory adn Architecture, **4-5.**
[28] Kumar, H.; Singh, M. K.; Gupta, M.; Madaan, J 2018 *Moving towards smart cities: Solution that lead to the Smart City Transformation Framework.* Technological Forecasting & Social Change, **10-13.**
[29] Correa-Garcia, E.; Velez-Correa, J.; Zapata-Caldas, E.; Velez-Torres, I.; Figueroa-Casas, A 2018 *Territorial transformations produced by the sugarcane agroindustry in the ethnic communities of Lopez Adentro and El Tiple, Colombia.* Land Use Policy, **8-13.**
[30] Remali, A. M.; Salama, A. M.; Wiedmann, F.; Ibrahim, H. G 2016 *A Chronological Explorationof the evolution of housing typologies in Gulf Cities.* City, Territory and Architecture, 3, **13.**
[31] Mirsa, R.; Soetomo, S.; Mussadun; Asnawi 2016 *Spatial Settlement Pattern in Laweyan as Supportive Batik Production Activities.* Indonesian Journal of Geography, 48, **87-89.**
[32] Weber, M 1978 *Economy and Society an outline of interpretive sociology*; University of California Press: London; Vol. 1, pp **212-400.**
[33] Sell, C. E 2017 *The Two Concepts of Patrimonialism in Max Weber:From The Domestic Model To The Organizational Model.* Federal University of Santa Catarina (UFSC) Florianopolis,SC,Brazil: Brazil; Vol. 07, pp **315-340.**
[34] Moedjanto, G 1987 *Legitimacy concept of Java : Applied by Mataram Kings*; Publishert Kanisius: Yogyakarta; p **102.**
[35] Giddens, A 2010 *Theory of Structuralism of the Basic Foundation of the Formation of Social Structure of Society*; Pustaka Pelajar: Yogyakarta Indonesia; Vol. 1, pp **401-410.**
[36] Yin, R. K 1989 *Case Study Research Design and Methods*; SAGE Publications Inc: London;Vol. 5, pp **1-4.**
[37] Stake, R. E 2010 *Qualitative Research How Things Work.* In Qualitative Research Studying How Things Work; The Guilford Press: London; p **23.**

[38] Creswell, J. W 1998 *Qualitative Inquiry and Research Design Choosing Amanong Five Traditions*; SAGE Publications Inc: California; p **118.**

[39] Benz, I. N. a. C. R 1998 *Qualitative Research Methodology Quantitative Eksploring the Interactive Continuum*; Southern Illinois University Press: America; p **52.**

[40] Budiarto, L 2005 Magersari : *The spatial-culture of kampung settlements as an urban strategy in Indonesia cities and urban housing*. XXXIII IAHS World Congress on Housing Transforming Housing Environments through Design, Pretoria South Africa ; pp **1-15.**

[41] Marlina, A 2017 *Residential Space Transformation of Java Ndalem Noble and House of Abdi Dalem Magersari in Baluwart i the Palace of Surakarta*; Diponegoro University: Semarang Indonesia; p **261.**

[42] Firmansyah, A. Y 2016 *Architecture metabolism approach which integrates the concept Magersari in supporting balanced development with green agricultural land in suburbs.* Procedia - Social and Behavioral Sciences, **609-616.**

[43] Coleman, J 1997 Control in Modern Society. *In Authority,Power,Leadership:Sociological Understandings*; Octagon Books: New York; pp **31-44.**

[44] Larson, G. D 1990 *The period leading up to the Palace Revolution dan Political Life in Surakarta,1912-1942*; Gadjah Mada University Press: Yogyakarta Indonesia, Vol. 1; p**174.**

The morphological changes factors in Kotagede fringe areas

Bambang Setioko, Bangun IR Harsritanto, Mustika K Wardhani
Universitas Diponegoro, Semarang, Indonesia

Corresponding e-mail: keliek2000@yahoo.com

Abstract. The growth and development of the city takes place through a continuous process of urbanization. This process creates inverted urban growth where urbanization initially took place in the urban core, but in the last decades it has spreade to the fringe and border areas. The settlement growth phenomenon to the fringe areas also occurred in historic cities, and it changes the characteristics of the cities. Historic cities should maintain their ancient configuration although it has undergone changes due to the process of modernization. City configuration is a reflection of the way of life, the society and the network of its inhabitants [6]. The urban development influenced by various factors such as economic, social, cultural, technological, and natural conditions. A city can be seen as a collection of artifacts built from various eras [6]. Parts of a city which are the most easily detected when there are changes are fringe areas. Nowadays the growth of Yogyakarta City overflowed to the fringe areas namely Kecamatan (Sub-district) Kotagede, a sub-district that borders Bantul Regency. As a conservation area, Kotagede undergoes the rapid growth of Yogyakarta City. The morphology of its residential areas, which was originally strong characterized by organic and traditional settlements, has transformed into mixed use. It is feared that the uniqueness of morphological cultural heritage area will be disappearing. This study aims to explore the factors of morphological changes in Kotagede fringe areas based on a field survey using quantitative approach with rationalistic descriptive method.

Keywords: fringe, border, cultural, heritage.

1. Introduction

In recent decade, the growth of urban areas in Indonesia tends to follow the global urban trend characterized by peripheral urbanization. This phenomenon has the potential to encourage the growth of the urban area to be very broad and unlimited [1]. Morphology of cities is changed dramatically, from inside to outside, causing the role of border areas increases, while the urban core's role decreases [2].

In the beginning, this process takes place in the city cores, but in recent decades it is even more prominent in urban fringe area especially in big cities. The problem caused by city growth is the shifting tendency of city functions to urban fringe, commonly known as urban physical appearance spreading process to the outside [4]. Cities are said to represent the greatest achievement of human civilization, the place where culture and the great traditions are created and preserved [5]. This phenomenon is clearly visible in the Kecamatan (sub-district) Kotagede, the fringe area of Yogyakarta. This historic city should still leave its ancient configuration although it has undergone changes due to the process of modernization. City configuration is a reflection of the way of life, the society and the network of its inhabitants [6].

This research was conducted Yogyakarta city; the capital of the Yogyakarta Special Region. This region is located in the South of Java Island. The population data indicated that there was a population shift from the city center to the fringe and border areas. This tendency is predicted to continue since the gross density level of population and buildings in fringe areas is indeed low.

The increasing interest of people to live in fringe areas is motivated by several factors, including the booming of silver home industry. Sub-district Kotagede is the locus of this research located in the bordering area of Yogyakarta municipal area and Bantul Regency. The Morphology of Sub-district Kotagede has three different zoning characteristics consisting of: (1) traditional housing areas, (2) silver home industry and arcades, and (3) mixed used.

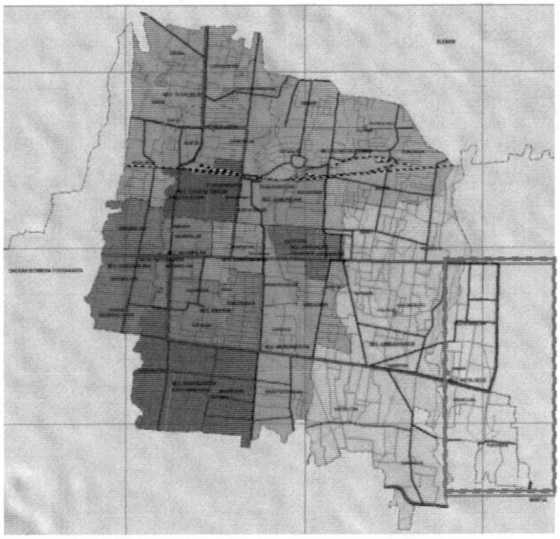

Figure 1. Maps of Yogyakarta Municipal Area

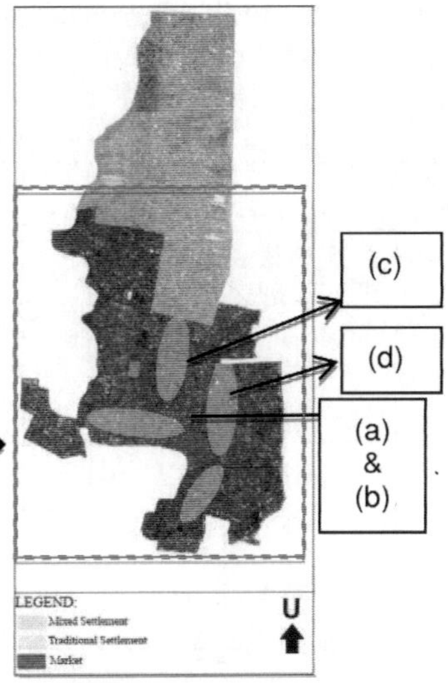

Figure 2. Sub-district Kotagede Maps

Figure 3. Building Characteristics in Kotagede*'Sopingen'* Cluster in Mondorakan street

Figure 4. Limasan roof

Figure 5. Changes near *'Legi Kotagede'* market

Figure 6. Silver Industries in Jalan (Street)Kemasan street

2. Methods

This paper aims to explain the factors of morphological changes in Kotagede as the fringe area of Yogyakarta. The research process began by conducting factors analysis using SPSS 16.0 for Windows software. Random sampling was done by distributing 100 questionnaires in the locus of research. This research employed the quantitative approach within rationalistic paradigm. Statistical analysis was applied to minimized bulk of many factors becoming several dominant factors. The purpose of using statistical analysis was to reduce variables of morphological changes.

Table 1. Identity of variables

Name of Variables	Sub Variables	Code	Number of Questionnaire	Name of Variables	Sub Variables	Code	Number of Questionnaire
X1	Structural	(X1.1)	1	X6	Resilient	(X6.1)	16
	Functional	(X1.2)	2		Constancy	(X6.2)	17
	Visual	(X1.3.1)	3		Advantage	(X6.3)	18
		(X1.3.2)	4	X7	Public facilities	(X7.1)	19
X2	Form	(X2.1)	5	X8	values and norms	(X8.1)	20
	Space	(X2.2)	6		Cosmology	(X8.2)	21
X3	Physical	(X3.1)	7	X9	Planned Settlement	(X9.1)	22
					Unplanned Settlement	(X9.2)	23
	Non-physical	(X3.2.1)	8	X10	Ecology	(X10.1)	24
		(X3.2.2)	9		Technology	(X10.2)	25
X4	Street Pattern	(X4.1)	10		Government Program	(X10.3)	26
	Building Blocks	(X4.2)	11				
	Settlement pattern	(X4.3)	12				
X5	Addition	(X5.1)	13				
	Elimination	(X5.2)	14				
	Movement	(X5.3)	15				

Source: Analysis Result, 2018

3. Discussion

The number of respondents was 100 (level 95% confidence interval). The process of reliability test was done by using Pearson correlation to determine variable reliability, which can be seen from the generated variables having significance value > 0.60. Furthermore, the reliability level was checked through Corrected-Correlation Item-corrected with SPSS software (Table 2).

Table 2. Reliability Statistics

Cronbach's Alpha	Cronbach's Alpha Based on Standardized Items	N of Items
.759	.771	26

The results of reliability test showed all the variables of morphological changes factors reached 0.759 on Cronbach's Alpha, Therefore the variables were included in 0,60 to 0,80 category. The variable proofed to reliable predicate, based on those category. Not all factors are analyzed using statistical analysis. The KMO and Bartlett test was used to eliminate the variable x in order to achieve the most reliable results. Element forms variable has mean almost equal to factor analysis on scale 1,2,3,4,5 with affirmative answer was the smallest and the negative was the biggest.

Table 3. KMO and Bartlett Test before Extracting Factor

Kaiser-Meyer-Olkin Measure of Sampling Adequacy.		.585
Bartlett's Test of Sphericity	Approx. Chi-Square	939.306
	Df	325
	Sig.	.000

Table 4. KMO and Bartlett Test after Extracting Factor

Kaiser-Meyer-Olkin Measure of Sampling Adequacy.		.712
Bartlett's Test of Sphericity	Approx. Chi-Square	541.817
	Df	153
	Sig.	.000

From this test, there were several variables that examined MSA< 0.5. There were variables X1.2 (functional), X2.1(Form), X3.2 (non-physical), X4.2 (Building Blocks), X5.1(Addition), X6.1 (Resilient), X10.1(Ecology), X10.3 (Technology). In accordance with the determination of the MSA value <0.5, then the factor was the smallest value omitted. After elimination of variables with MSA value <0.5, the KMO value at the end of the analysis could be determined. The final KMO and Bartlett Test reached 0.712 (table 4 – right side).

After identifying the correlation between variables and KMO values, the next step was to assess the components of the formed factors. There are 18 factors components from the statistical analysis in which only (six) factors that obtain eigenvalues > 1.

Table 5. Total Variance Explained table

Component	Initial Eigenvalues			Extraction Sums of Squared Loadings		
	Total	% of Variance	Cumulative %	Total	% of Variance	Cumulative %
1	4.117	22.872	22.872	4.117	22.872	22.872
2	2.315	12.861	35.733	2.315	12.861	35.733
3	1.957	10.871	46.603	1.957	10.871	46.603
4	1.518	8.432	55.035	1.518	8.432	55.035
5	1.204	6.688	61.724	1.204	6.688	61.724
6	1.095	6.084	67.807	1.095	6.084	67.807

Extraction Method: Principal Component Analysis.

Table 5 shows that total 6 (six) factors with Eigenvalues >1 were 67.808 % while 32. 192 % were influenced by other factors beyond model of the research. In the chart below, there was a tendency to stop to form a horizontal line, the result of the scree plot:

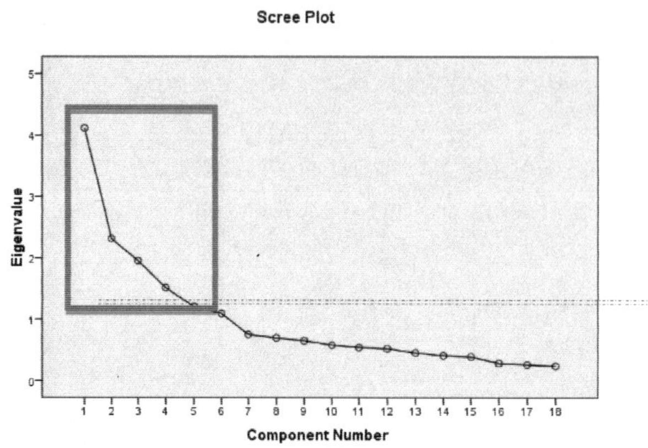

Figure 7. Scree Plots Graph
Source: The Analysis Result

On the graphic above (figure 9) the consistently declined pattern formed to lead optimal amount of component use. The starting point was at 6, so there should be 6 components taken. Each loading factor should be identified and examined to establish an interpretive factor that reveals the basis (Hair, et.al, 2009). Factors that have a value greater than ± 0.50 are the main factors. Table 6 is the result of orthogonal rotation (Varimax with Kaiser Normalization method).

Table 6. Component Matrix[a]

	Component					
	1	2	3	4	5	6
X1.1	.711	.272	-.209	-.035	.020	-.230
X1.3.1	.537	.290	-.500	.025	-.031	-.092
X1.3.2	-.130	.676	-.277	.312	.061	-.207
X2.2	.678	.374	-.293	-.066	-.111	.017
X3.1	.307	.050	.127	-.161	.801	-.058
X3.2	.657	-.050	.356	.038	-.147	-.076
X4.1	.151	-.185	.611	.431	.273	-.099
X4.3	-.110	.551	-.108	.477	.174	.378
X5.2	.180	.287	.637	-.039	-.372	.043
X5.3	.400	.486	.370	.101	-.135	-.026
X6.2	.619	-.090	-.009	-.284	-.275	-.162
X6.3	.510	.059	-.143	-.057	.118	.723
X7.1	.689	.046	-.028	-.306	.050	.067
X8.1	.594	-.564	.051	-.022	.106	.003
X8.2	.576	-.475	-.173	.199	.170	.002
X9.1	.254	-.427	-.083	.525	-.339	.357
X9.2	.292	-.227	-.265	.665	-.031	-.381
X1.0.2	.375	.327	.546	.166	.120	.009

Extraction Method: Principal Component Analysis.

a. 6 components extracted.

According to table 6 the Factors were labelled as

1. Component 1 = Linkage
2. Component 2 = Form and space
3. Component 3 = Physical

4. Component 4 = Building configuration
5. Component 5 = Transformation
6. Component 6 = Regulation

The summary stage after performing statistical test:
Stage 1: Specifying the origin variable. There were 26 selected based on literature studies.
Stage 2: Conducting the reability test. Variables had Cronbach Alpha 0.759, it defined in reliable Category. 100 respondents answered questionnaires.
Stage 3: Conducting KMO and Bartlett Tests. The Measure of sampling increased from 0.585 to 0.712 by eliminating 8 variables which have the lowest correlations. This number means all the variables were ready to analyze. The Extraction process was then used to derive the number of factor based on the Principal Component Analysis by Scree plot Criterion.
Stage 4: Rotate the factor using the orthogonal rotational method to form a group of factors consisting of fine-grained subdivisions. The values of loading factor changed based on the group of factors namely (1) Linkage, (2), Form and Space (3) Physical, (4) Building Configuration, (5) Transformation, (6) Regulation
Stage 5: Interpreting the factors, hence 6 factor, the variables viewed in Table 6 having loading factor greater than 0.5.

4. Conclusion
A series of quantitative analysis done successfully concluded 6 morphological changes factors in Kotagede. The linkage is a most dominant factor in the morphological change. Although Kotagede has undergone more varied changes and economic activities, but the community network activities still have traditional character. At the empirical level, the long exsisted traditional markets still survive as a centre of community socio-economic activities. Meanwhile, the form and space are the second dominant factor that can be interpreted that the urban tissues of the Kotagede do not undergo many changes. From the visual point of view, the building mass in some parts has changed, but the form and space still have a strong traditional Javanese character.
Changing factors in physical and regulatory aspects are interrelated due to the traditions of society in managing buildings, and these traditions; moreover, are more adhered than the design guideline set by municipal government. Building configuration implying the configuration of residential groups has not changed much. Changes mostly occur on the building façade, but the room configuration still has Javanese traditional pattern. In Kotagede, there is no significant transformation of building mass and forms. Street and pathways mostly have narrow dimension and have organic patterns, and building block and settlement are characterized by Javanese traditional patterns.
The policy of Yogyakarta Municipality in managing the growth of the Kotagede shall accommodate the local values. This study highly recommends not using the "ideal city model", which is blindly adopted idea from western countries. Some evidences indicate that using "ideal city" model to solve problems in region having historical heritage is not fully compatible with the Indonesian socio-economic condition.

5. Acknowledgments
The authors would like to express their gratitude to the Faculty of Engineering of Diponegoro University for the financial support of this research project through the strategic research grant 2018. The authors also would like thank to the community members and the official staffs of Sub-distric Kotagede, Yogyakarta who were very supportive in providing the data for this research.

6. References
[1] Rossi, Aldo. (1982).The Architecture of the city, The Graham Foundation: New York.
[2] Gillham, O. (2002). *The Limitless City, A Primer on the Urban Sprawl Debate.* USA: Island Press.
[3] Graham, S. a. (2001). Splinter Urbanism. London: Routledge.
[4] Setioko, Bambang; Olivia, Deasy. (2015). *Shaping Factors of Urban Fringe Characteristics,*

with Case Study Semarang City, Indonesia. AENSI Journal.
[5] Hutchison, Ray. (2010). *Encyclopedia of Urban Studies.* SAGE Publications.
[6] Trancik, roger. (1980*). Finding Lost Space, Theories of Urban Design*, New York: Oxford University Press.
[7] Hair, J.F. 2009. Multivariate Data Analysis. Seventh Edition, Internasional Edition. Prentice Hall.

Investigation of Urban Places in Seoul Digital Industrial Complex (G-Valley)

Suha Ryu, Saehoon Kim
Department of Landscape Architecture, Urban Design Concentration, Graduate School of Environmental Studies, Seoul National University, Seoul, Korea

Corresponding e-mail: wwwkac@seoul.ac.kr

Abstract. The National Industrial Complex was designated by the Korean government to strategically nurture the country's key industries. It is larger than a normal industrial quarter, playing a major role in boosting national economic development. In Korea, industrial complexes began to be created since the post-war period. During the period, the country experienced the worst economic condition with many problems. Facing the problems, factories for light industry started to be established, then the focus has shifted towards heavy chemical industry [1]. Seoul Digital Industrial Complex (G-Valley) was established through the development of small to large manufacturing factories such as clothing workshops. However, because of the rapidly industrializing areas out of Seoul, a number of old factories moved to other locations in the 1980s and 90s, and G-valley was replaced by venture groups and IT related companies. Since its establishment in the 1960s, physical infrastructure has been aged and some social problems have been raised. In this paper, we will investigate the problems of G-valley and seek the future solutions by analyzing the current situation based on its urban spaces. It is significant to look at the common problems and future visions not only of G-valley but also of the industrial complexes in other areas of Seoul and its metropolitan region

Keywords: deteriorated industrial complex, national industrial complex, industrial structure acceleration

1. Introduction

G-valley is an industrial complex that has been developed in the 1960s and changed due to the changing industrial structure and aging environment. As a solution to these problems, G-valley is attempting to build a multi-purpose facility (called Knowledge Industrial Center) for improving the working environment of high-tech industries. It is a decent solution to tear down the old factory and rebuild buildings to accommodate more businesses and people. However, in order to solve problems and continuously activate industrial complexes, it is necessary to think about urban space in relation to current issues.

The G-valley consists of a total of three complexes over Gasan-dong and Guro-dong. The authors decided to focus on the 3rd complex because the oldest buildings remain with heterogeneous industrial structure. Compared to that, the 1st and 2nd complexes are smaller than the 3rd complex, and most of the existing factories have been replaced by new knowledge industrial centers.

This paper starts with the question: what can be urban space solutions suitable for the current G-valley's state? Therefore, the purpose of paper is to propose a new perspective on G-valley by analyzing the most pressing issues.

2. Methods
First, the authors described the problems through the literature review and analysis of urban space in the site. After that, the authors presented a potential solution. The site analysis involved the location, the industrial structure, policies and issues.

3. Discussion

3.1 Location

Seoul Digital Complex is located in the southwest part of Seoul. The size of the complex is about 1,922,261 m². There are 9,670 companies and 151,813 employees [2]. Among them, the 3rd complex is the largest part with the size of 1,131,000 m² and is located to the left of the complex, facing the western highway and Anyang Stream on the left side. The western highway is the main road that mitigates and transports traffic in the south-western region of Seoul. It also has a very busy traffic volume of 110,600 cars per day on average. Anyang Stream is a tributary of the Han River. Bicycle roads are well developed and its traffic is high. The opposite side of Anyang Stream is the city of Gwangmyeong and is mostly a residential area.

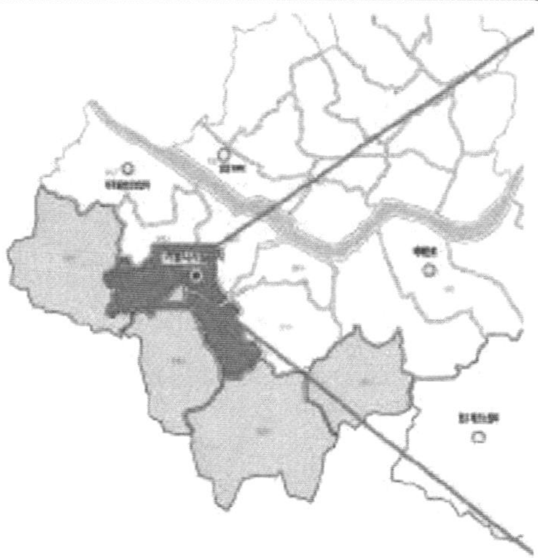

Figure 1. Current state of Guro-gu and Geumcheon-gu

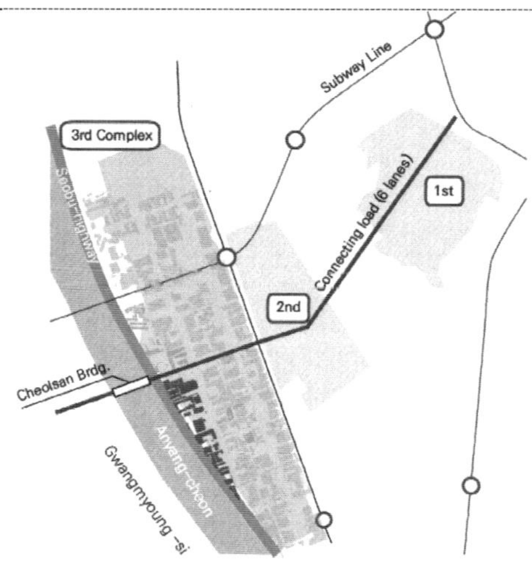

Figure 2. Current state of Seoul Digital Complex

3.2 Industrial Structure

Until the late 1990s, the complex was mostly occupied by textile apparel, simple machinery, and electrical and electronic companies. However, in December 2000, the name of the complex changed from Guro Industrial Complex to Seoul Digital Industrial Complex, and the transition from the labor-intensive manufacturing industry to the advanced industrial complex centered on venture, R & D, and advanced information and knowledge industry began. Therefore, the number of related industries has increased, and it has become an urban high-tech industrial complex with a focus on on knowledge-based service industries.

Table 1. Industrial structure of companies in G-Valley [2]

	Number of Industries in G-Valley (Percentage %)					
	1967	1980	1990	2000	2010	2017
Food	-	1(0.4)	2(0.8)	4(0.6)	29(0.3)	75(0.8)
Fiber	7(22.6)	71(31.7)	81(31.0)	90(13.8)	533(5.2)	638(6.6)
Paper	-	14(6.3)	24(9.2)	96(14.8)	229(2.2)	234(2.4)
Petro-chemistry	4(12.9)	15(6.7)	23(8.8)	9(1.4)	225(2.2)	339(3.5)
Non-metal	-	1(0.4)	1(0.4)	16(2.5)	15(0.1)	18(0.2)
Steel	-	3(1.3)	3(1.1)	2(0.3)	13(0.1)	10(0.1)
Machine				147(22.6)	611(5.9)	651(6.7)
Electric electronic	13(41.9)	100(44.7)	109(41.8)	213(32.8)	2389(23.1)	2,244(23.0)
Transportation equipment	-	-	-	15(2.3)	31(0.3)	43(0.4)
Etc.	7(22.6)	19(8.5)	18(6.9)	58(8.9)	150(1.4)	190(2.0)
Non-manufacture	-	-	-	-	6124(59.2)	5,282(54.3)
Total	31(100.0)	224(100.0)	261(100.0)	650(100.0)	10,349(100.0)	9,724(100.0)

3.3 Related Policy and Issues

3.3.1 Western Underground Road Tunnel Project

In 2015, Seoul city decided to launch a western underground road tunnel project. 10.33km of the western highway will be buried underground, and the upper part will be constructed with general roads and parks (figure3). The road is scheduled for opening in 2020[3]. The goal of the project is to reduce the traffic congestion by dispersing about 50,000 cars per day, and to improve the living environment by changing the upper part into an environmental-friendly space.

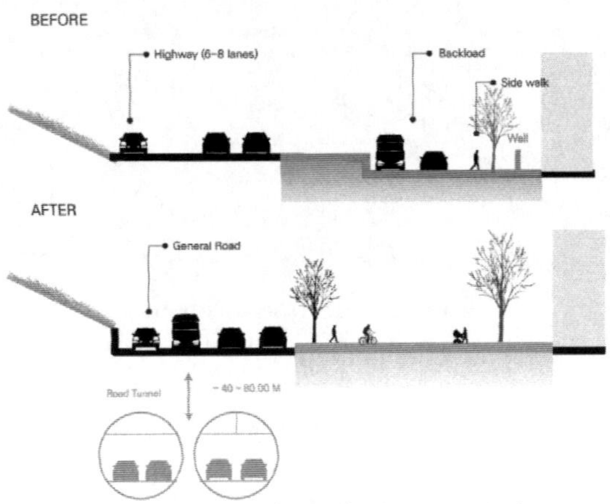

Figure 3. Before and After the completion of Western Underground Road Tunnel Project

IOP Conf. Series: Earth and Environmental Science **213** (2018) 012015 doi:10.1088/1755-1315/213/1/012015

3.3.2 Structural enhancement of G-valley

Structural enhancement project is to promote the attractiveness of enterprises and enhance the competitiveness of tenant companies by strengthening corporate support services, and maintaining and repairing public infrastructure of the industrial complex[4]. Examples of the project include public spaces, sports facilities and welfare facilities, and conducting cultural events for companies.

3.3.3 Government's support for start-up

Nowadays, an important trend in Korean government 's economic policies is about promoting start-up businesses. Therefore, the government initial budget for supporting start-ups amount to about 729 million US dollars this year. They support start-up funds, run start-up centers, and provide workspaces to run the company. And IT-oriented start-ups are being expanded.

3.4 Problems

I would like to summarize the problems of the complex based on the above points. Currently, G-Valley has complicated problems that cannot be solved within the limited land and financial capital. Two major problems are observed as follows.

3.4.1 Old or missing infrastructure

Most of the buildings on the site are 20-30 years old, and most of them are manufacturing factories (Figure 4). Compared to the working population, the infrastructure is not enough, showing the lack of supportive facilities such as green spaces, nurseries and public health centers (Figure 5). The amount of open space seems to be sufficient due to the existence of public space in front of the knowledge industrial centers, but areas with old factories have few open spaces.

The lack of transport infrastructure is also a problem (Figure 6). Large parcels that are partitioned to build factories in the past are now difficult to access by cars, and the traffic volume is not dispersed, causing a traffic congestion. The access road between the western highway and the site is well-known for severe traffic congestion. It is very serious that there are problems in the infrastructure, and improvement is inevitable.

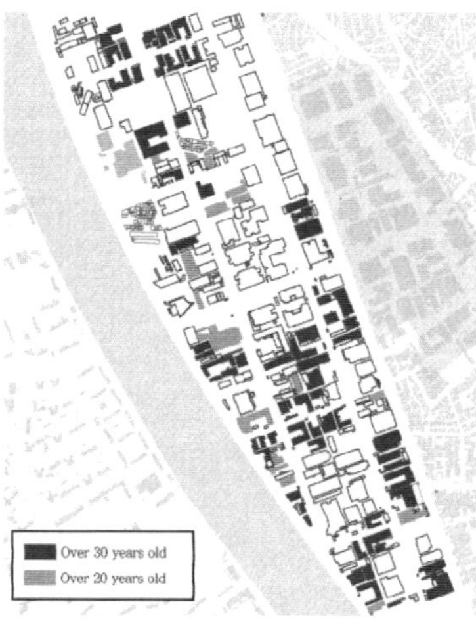

Figure 4. Deterioration of Buildings

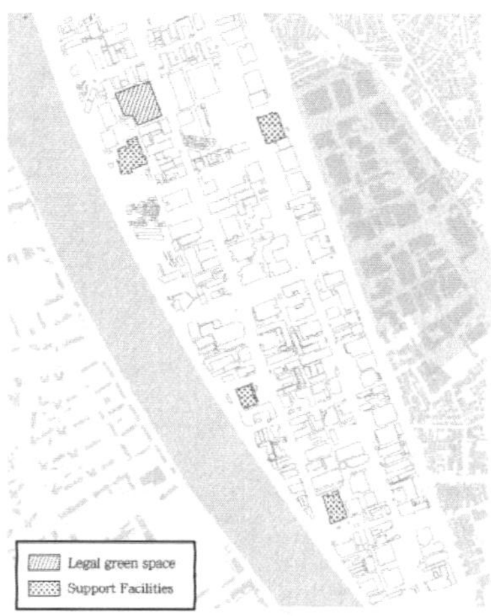

Figure 5. Location of green space and supporting facilities

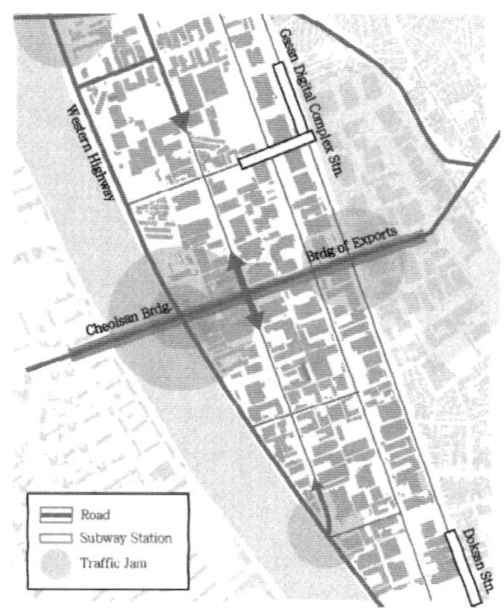

Figure 6. Traffic

3.4.2 Hollowing-out of workers

More than half of the consumer population of Gasan-dong and Guro-dong are G-valley commuters (Figure 7). In other words, G-valley is strongly influential on the vitality of a local commercial area. The real problem is that commuters get off after work and leave on holidays. This result in the hollowing out of the populations and serves as a big obstacle to local commercial economy. Therefore, G-Valley is expected to play a major role beyond the industrial use towards revitalizing the surrounding area.

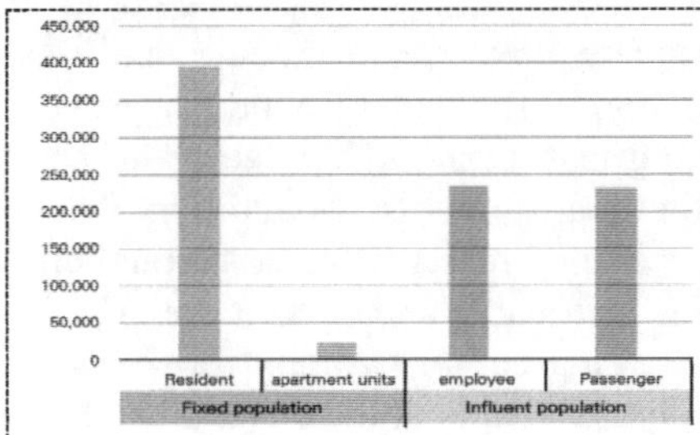

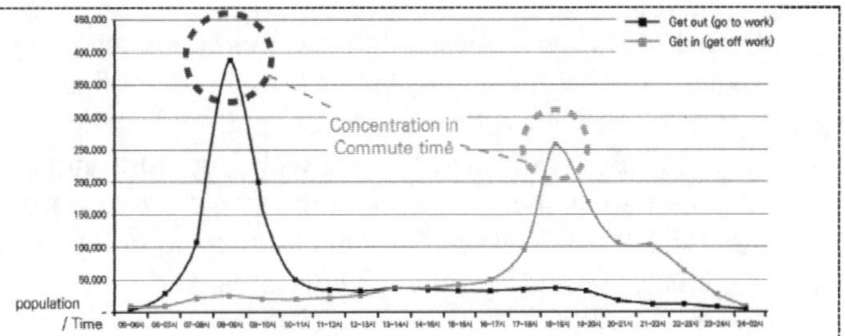

Figure 7. Consumer in Gasan, Guro dong **Figure 8.** Get in and out of Gasan Digital Complex Station [5]

4. Conclusion

A solution for the complex is proposed as follows.

- As the western highway will be buried underground, additional available/developable sites will be obtained.
- Supply deficient public infrastructure to the area and create a new program.
- Do road maintenance with the implementation of the underground project of western highway.

4.1 Traffic problem solving

The entry of the western highway has a tangled ramp. Because of the underground project, it will become unnecessary. The entry ramp will be removed and replaced by another road (figure9). In addition, 300-400 m large parcels will be divided and bypass roads will be created (figure 10).

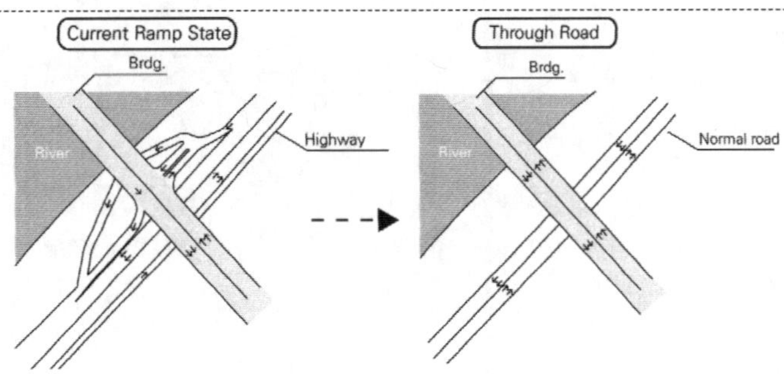

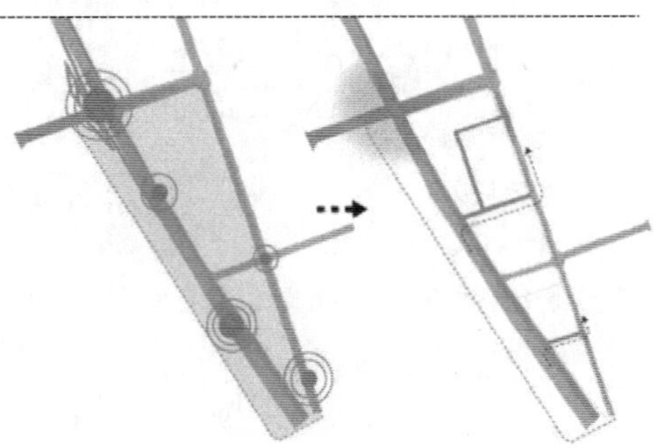

Figure 9. The entry ramp **Figure 10.** Dividing large parcels

4.2 Infrastructure supplement

Supportive facilities such as worker welfare centers and sports facilities will be inhabited in a newly obtained area. Also, the lack of green spaces, which was one of the major problems in G-valley, can be solved through building a connection with the Anyang Stream.

4.3 The Young Start-up incubation

A business center, a community place, and a dormitory, etc. that can induce young start-up entrepreneurs will be established in a newly obtained area.

The western highway is located on the boundary between the Anyang Stream and the site, and the upper space will become an important urban space after the completion of the underground project. In the site, the waterfront offers a very special experience for walking and sitting. The waterfront can play a role in improving the pleasant working environment and the quality of life. In addition, since the road is owned by the government, the upper space formulated after the project will also be publicly owned and used. This will enable the G-valley to do business that is affordable and benefit the public. Taken together, the underground project of wesetrn highway proposes a great opportunity for the new vision of the G-valley. The newly obtained area will be a 'public space' and is able to contain urban infrastructure and programs beyond the limits of the development.

G-valley, which is in a difficult situation to redevelop, also has tried hard to solve these problems. However, if G-valley has many opportunities to change in a short period of time like our suggestion (figure11), we anticipate that G-valley will be able to establish a new identity. We also think that the situation that G-valley is experiencing is a common problem that other urban areas are also experiencing, so the suggestions of this paper can provide useful information for other industrial complexes' restoration and maintenance in urban areas.

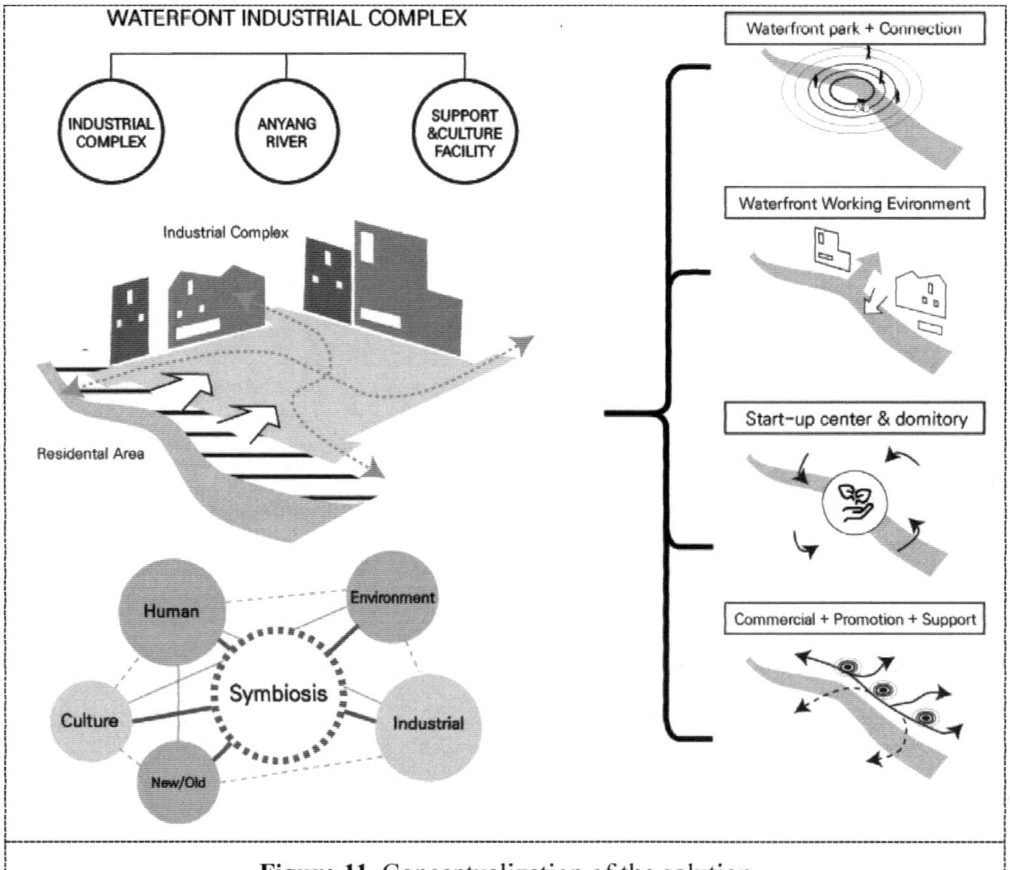

Figure 11. Conceptualization of the solution

IOP Conf. Series: Earth and Environmental Science **213** (2018) 012015 doi:10.1088/1755-1315/213/1/012015

5. References

[1] Korea Industrial Complex Corporation 2014 *Achievements and Development Tasks of 50 Years of Industrial Complex*
[2] Korea Industrial Complex Corporation 2017 *Major National Industrial Complex Industrial Trends*
[3] Seoul city 2011 *Policy Library*
[4] Seoul city 2015 2030 *Semi-industrial area comprehensive development plan*
[5] Statistics Korea Homepage (http://kostat.go.kr)

What is the sustainable method enough for our built environment?

A. Djoko Istiadji[1], Gagoek Hardiman[2], Prasasto Satwiko[1]
Universitas Atma Jaya, Yogyakarta, Indonesia[1]
Universitas Diponegoro, Semarang, Indonesia[2]

Corresponding e-mail: istiadji@yahoo.com

Abstract. The sustainable movement has grown and evolved rapidly in the 21st century. This movement is a response to the various problems of the relationship between humans and the earth that appear more disruptive and seemingly intractable problems. After the era of sustainable development emerges, many concepts and new methodologies such as biophilia, cradle-to-cradle, resilience planning, natural step, permaculture, living buildings, eco-districts, transition cities, and integrative and biomimetic design, etc. The superabundance of these theories indicates the notion of sustainability that is still not yet rigid and massive and rises curiosity to seek its end-line. This paper has done the extensive literature reviews on the field of Green Building Sustainable Development, Biophilic, Biomimetic, Resilience, Restorative and Regenerative for mapping it, created the constellation, and seed the relationship between these strategies and how they fit together. The mapping results with the matrix method of relevant research factors have succeeded in displaying the focus and assumptions built by each sustainability method. The results of this study have also provided a picture of a complex and complicated universe of how to achieve sustainability. One way or one method is not enough to gain sustainability Therefore, situational and contextual problems encountered must be handled with various approaches, and complex strategy must be elaborated in a systemic, organic, holistic and nonlinear way. The wave of sustainability ideas still seems continuously find the united and rigid concepts based on the existing theory and method.

Keywords: *Sustainable Development, Biophilic, Resilience, Regenerative, Rigid Sustainability Concepts*

1. Introduction

Environmental issues that are still and continuously faced by environmental experts and practitioners today are the climate change and its effect on human life. Population growth and industrial development are thought to be the causes of the increasing energy use. The consequence of this usage is the significant accumulation of CO_2 emissions causing global warming [1]. Climatic anomalies have been evaluated statistically and also have shown a significant change. Climate changes in three decades describe that the earth's surface is warmer in every decade since 1850. In the Northern Hemisphere, the period from 1983 to 2012 is the hottest 30-year-period within the last 1400 years. The combined global ground and sea surface data computed with linear trend show heating of 0.85 ° C (0.65 to 1.06) 2 during the period 1880 to 2012 [2]

Climate change makes broad impacts on living systems on the Earth [2] Some of the impacts can be large or small, have a positive or negative consequence, and have a direct or indirect effect. The impacts have various forms such as the occurrence of weather changes, polar ice melting, sea level rise, changes in the hydrological cycle, changes in wind cycles, and other phenomena. These impacts then trigger other problems such as more extensive and frequent disasters, changes in animal biological cycles, the emergence of new epidemics, increasing the death toll, and even the extinction of individual species. Climate change gives a very significant effect in all aspects of life. The World Health Organization estimated 600.000 people worldwide died from natural disasters due to weather conditions in the 1990s, and most of them lived in developing countries [3]. The Global Humanitarian Forum report presented that since 2009 there has been more than 300.000 deaths, and around 125 billion dollars have lost each year. Most of these; thus, because of worse floods and droughts happening in developing countries [4]. In addition to the previously mentioned terrible phenomena, the climate changes also affect the cost of

adaptation in which it is estimated that adaptation cost for the increase of the global average temperature (as many as 2.0 degrees Celsius) is between 70 and 100 billion dollars per year [5] .
Experts have responded to these environmental issues since the early middle ages. Environmentalists have come to understand that the Mechanistic-Reductionism paradigm that drives people to dominate and excessive nature exploration can adversely affect the "SUSTAINABILITY" of life on earth. Experts and practitioners have been cultivating the idea of sustainable development nearly three decades, and by reviewing at the development of environmental problems faced today, it appears that its approach and problem-solving strategies are still effective. The results of the research conducted by Pamela Mang and friends indicated that the problem lies in two of many aspects, namely [6]
The role of the man who forgets that he belongs and depends on a complex network of nature and the fact that he in the network to preserve the nature. Experts and practitioners more focused more on how they technically resolved environmental issues. They talked more about aspects of strategy and methods than to talk more about the essence of how this life system works. For nearly 20 years, practitioners and experts talked about a green building and its measuring tools that tend to be technological sustainability. In response to this indication, it appears that aspects of human and natural relationships are the key problem. The importance of creating togetherness of man and nature has been understood for a long time, but in practice, the balance of both is not maximally applied yet. This matter encourages the wave of thoughts on sustainability continues to grow and bring new concepts and approaches. Environmental-Architectural concepts such as Biophilic, Biomimetic, Resilience, Restorative, Permaculture and Regenerative have emerged to realize the harmony in life.
Based on the above facts, this study was conducted with the aim to examine how these new sustainability concepts accommodate human and natural relationships. Mapping the existence and constructing their constellations in order to make them more precise when applied in environmental activities. We hope that this research can find a new conceptual framework based on those various concepts and approaches.

2. Methods

This study did a critical review of the literature regarding the new sustainability concepts and approaches. The focus of the study is to understand, map and find the constellation among the approaches. At the initial stage, the study was oriented to find the basic theory, conceptual framework and empirical indicators of each concept. The theories' system of thought was also explored to understand the universe and widespread environmental issues. The results of the study were then compiled, tabulated and analyzed for their constellations. The final phase of the study would review its role and contribution to the achievement of sustainability. This stage aims to see the opportunities and ways in which existing approaches can be integrated into a new sustainability approach.

3. Discussion

3.1 Timeline and Constellation of Environmental Movement

Environmental movements have emerged since the early middle ages. They are social and moral movements. The concern on excessive nature exploited for industrial interests was the background of the movement. Preservation and conservation actions by building some national parks were proposed for accomplishing the actions to save nature and their habitats. Although the call for environmental existence is continuously voiced, technological developments for the industry were also rapidly progressing and even increasingly endangering the earth's safety. The advance of pesticide technology resulting in soil, water and air degradation, free usage of Chlorofluorocarbons (CFC) which caused ozone deflection, and the use of nuclear technology where its security was still not guaranteed. These issues have led to a stronger environmental movement which transforms itself into ecological movement since the 1960s. In the past, ecological movements were predominantly from Americans. After that, concepts of sustainability develop progressively and become popular internationally.

Furthermore, the conditions of livelihood got worse, and the energy crisis caused by the oil embargo happened encouraging environmentalists to do more to protect the environment. They proclaimed Earth Day and established environmental protection institutions such as Green Peace, Occupational Safety and Health Administration (OSHA), Environmental Protection Agency (EPA), the American Green Building Council. As time goes by, the awareness related the importance of preserving nature develops and and continues to grow for maintaining the life sustainability on earth.

The issue of sustainability is not only a mere ecological problem but also a systemic problem to other disciplines like economic and social. The impacts of green building actions should also be measured in the economy and social points of view in a balanced way to achieve an optimal sustainability; namely sustainable development [7]. This approach refers to nature carrying capacity measured by the maximum number of environmentally-supporting organisms in a particular habitat that is considered determinants for the development [8]. The grand idea of self-reliance, self-sufficiency and economic sovereignty through the exploitation of natural resources must consider the limit of carrying-capacity that the balance and sustainability between ecological, social, economic, cultural, existential lives and human beings are still well-maintained and sustainable [9]

The development of environmental movements in Japan looked different. They concerned more about the aspects of disasters which frequently happened with the larger impacts. Environmental issues related to the existence of nature were not the core of the problem. It is possible because eastern culture respects nature, so they always tried to preserve the nature well. From disasters side, some programs to overcome the impacts were created such as Hyogo framework program which was all about how to reduce impacts and build resilience. Then, Sendai framework was held to increase more impact in reduction programs yielding seven clear targets and four action priorities: (i) Understanding disaster risk; (ii) Strengthening disaster risk governance to manage disaster risk; (iii) Investing in disaster reduction for resilience and; (iv) Enhancing disaster preparedness for effective responses, and "Build Back Better" in recovery, rehabilitation, and reconstruction.

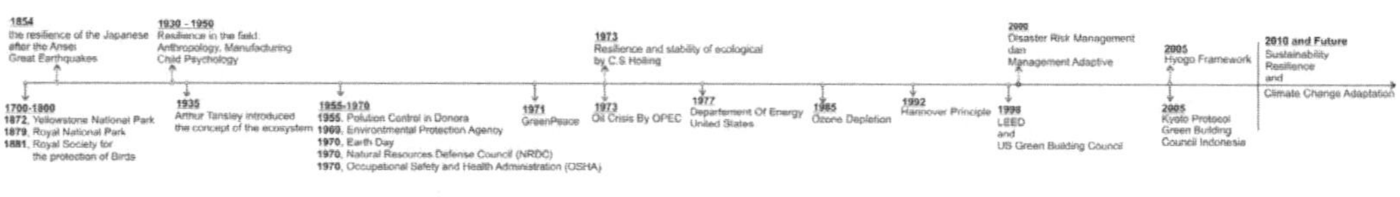

Figure 1 Environmental-Ecological Movements Timeline
Sources: (10),(11), (12), (13) (14)

Timeline of Environmental-Ecological Movements is shown in Figure 1. These actions did not just happen nationally in United States, Japan or some other countries, but then became a global action through some international conferences, started from Stockholm 1972, Rio de Janeiro 1992 and Johannesburg 2002. The event at Rio de Janeiro in 2012 discussed a lot about sustainable development. Conferences on climate changes were held in Kyoto Protocol 1992, Bali Road Map 2007 and Copenhagen 2009. Disaster aspects were studied in three conferences: Yokohama in 1994, Kobe in 2005 and Sendai in 2015 as the latest one. The above conferences are described in Figure 2. Some conferences involved some countries in the world indicating that environmental issues have become a critical agenda of the world community, and there were three focuses namely sustainability, climate change, and disaster risk management.

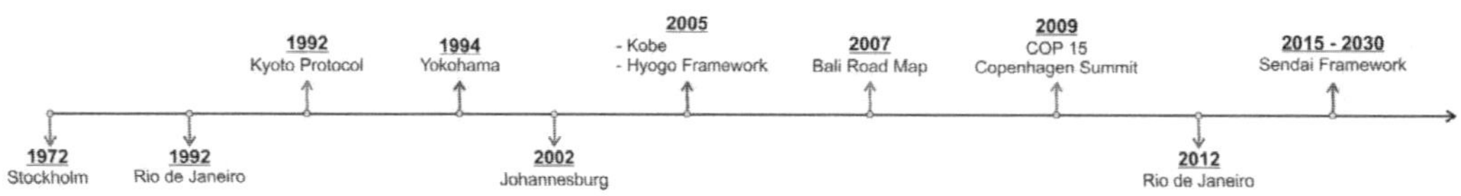

Figure 2 Environmental International Conferences Timeline
Sources: (10),(11),(12),(15)

Environmental aspects are always related to the building. During this time, built-environment is known to have negative impact toward nature, and it affects sustainability. Therefore, environmental agendas also become a concern to architects. They collaborate with environmental experts to create environmentally friendly architectural design concepts as shown in Figure 3. Since the early 20th century, world great architects have attempted to adopt environment concepts to finish their works. Bioclimatic Architecture (1906-19), Environmental Architecture (1963-1972), Energy Conscious Architecture (1972-1983), Sustainable Architecture (1980-1993), Green Architecture (1990-2006), Carbon Neutral Architecture (2006-2015), and the last, Regenerative Architecture is a series of concepts and approaches in connection with the environmentally friendly concept [15]. The worsening of

IOP Conf. Series: Earth and Environmental Science **213** (2018) 012016 doi:10.1088/1755-1315/213/1/012016

environmental conditions encourage the engineering of problem solving methods. Bioclimatic is a response to environmental and climate problems while Green Building and Sustainable Development tend to solve more complex and integrated technical-environmental problems. Considering the environmental conditions that are worsening indicated by the increasing number of disasters, the concept of Resilience arises. Natural conditions that are not conducive are responded by acknowledging the importance of the presence of natural elements. Humans and nature are interconnected. Bionic, Biophilic, and Biomimetic designs appear to encourage the passive use of natural elements for the humans. Knowing that the latter approach is still more inclined to human interests and there is no equality of interest between human and nature, the Restorative, Permaculture and Regenerative are proposed ones. These last concepts carry a living system approach. Humans are the part of nature and play an active role in co-evolution. Their activities must contribute positively to nature, besides for themselves. These approaches replace the old concept of 'giving less negative impact to nature'.

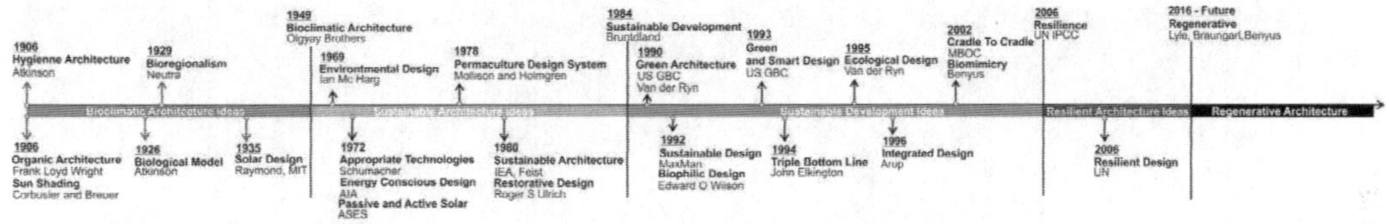

Figure 3. The Timeline of Environmental-Architectural Design Concept
Sources: (10),(11), (12), (13), (14), (15)

Observing and analyzing some environmental concepts and approaches since the early middle ages, it is clear that these are broad issues. Corresponding to the civilization, Environmentalists and Architects have attempted to understand and resolve environmental issues, following their interest. From merely social and moral movement efforts to engineering movement, and the last is the living system. There are three essential things that have been observed until today: (i) Environmental issues cover three significant aspects: sustainability, climate change, and disasters risk reduction. (ii) Environmental issues cannot be looked at just as environmental issues, but they are related to some other disciplines. Competitive modern lifestyle demands encourage exploitation of natural resources use and development of less environmentally friendly technologies. Furthermore, these various issues seem to be overlapping and more complicated; hence integrated problem-solving based on other disciplines. Social, economic, psychological, and cultural aspects must be considered. Besides expanding from the disciplinary side, it also enlarges to a wider scale. The scope of the issues has spread more extensive involving larger geographic areas. The conferences held showed the efforts to solve the problems and to overcome the complex network of issues. (iii) There are three focuses related to solving environmental issues i.e. environmental engineering (Bioclimatic, Sustainable & Green Architecture), Disaster issue (Resilience) and the human-nature relationship (Permaculture, Restorative & Regenerative Design). It can be seen from the timeline, that experts have discussed the three topics above since the early 18th century, but they emerge to be a popular environmental approach one by one. It shows the great effort of humans to understand and solve environmental problems that are complex and broad. This finding is also proof that at the first, the experts preferred to use technological approaches to achieve sustainability rather than to use ecological approaches. This approaches are useful but do not effectively solve the problem. These facts match with Pamela Mang's opinion [6] stating that the issues are not only about engineering but also about the human-nature relationship beside the engineering matter. Therefore, it means we need both approaches for solving the problem (technological and ecological sustainability). Beside to understand technically, we need to understand the way our universe works (ecological matter). We have to support our living system.

3.2 Environmental-Architectural Design Concepts and Approaches

In more details, the study on concepts and approaches of environmental-architectural design is shown in Table 1. As previously mentioned, there are 3 different approaches in Figure 4. The first group (G1) consists of Bioclimatic, Green Building, and Sustainable Development while the second group (G2) consists of Biophilic, Biomimetic, Restorative, Permaculture, Regenerative approaches. The last group (G3), then, is Resilience. In the first group, they attempt to make the design have high performance

Table 1. Concept and Approaches of Environmental-Architectural Design

Approaches	Background and Aim	Concept and Definitions	Elements
Green Building **Paradigm :** Mechanistic-Reductionist **System of Thinking:** Linear, One way-Open Loop and Partial.	Green building is an effort to change the way the built-environment designed, built and used. Green building is a building with an integrated approach started from early planning phase until the use of building, environment and entire community. Therefore, the green building is the building that is ready to incorporate climate adaptation to reduce the negative impacts of climate in the future (16)	The high performance green building is the building designed and built in which it can be operated by considering environmental conditions and having economic value from time to time (17)	Land Development, Energy Conservation and Efficiency, Water Conservation, Material Use and Waste Management, Environment Comfort and Health, Built Environment Management (18)
		Building designs give less bad impacts towards environment and consider the right use of technology (16)	Sustainable Site Design. Water Conservation and Quality, Energy and Environment, Indoor Environment Quality, Conservation of Materials and Resources (19)
	One way to measure the success of the implementation of green building is one of the benefits which can be shown in the lower cost compared to other conventional buildings. The reduced cost is also followed by the decline of energy consumption in green buildings, compared to conventional buildings (20)	Green building is able to accommodate health purposes, surrounding environment, built environment and design with an effective cost and measurable building performance (21)	• Can be operated and maintained by energy usage and resources efficiently and effectively • Give high performance and support users' productivity • Create a better and healthier built environment (22)
Sustainable Development **Paradigm :** Mechanistic-Reductionist **System of Thinking :** Linear, One way-Open Loop and Partial.	The sufficient availability of natural resources, preserved ecology process, social-cultural environment and suitable economy (22)	**Humanity has the ability to make development sustainable to ensure that it meets the needs of the present without compromising the ability of future generations to meet their own needs** (23)	• Social • Economy • Ecology
	The harmonic two-way relationship from economy and social factor will create an equitable relationship. The connection between economy and environment will maintain sustainability, both activities work with mutual and viable benefits. The connection between social and environment will create resilience (24)	Sustainability can also be defined as an urge to get more benefits in investing but still consider the nature without ignoring it. (17)	• Adaptation toward climate on site and adjustment toward condition changes. • Not producing waste • Promoting society's health and well-being as well as ecosystem health. • Supposedly consist of integrated energy saving system which maximize efficiency and comfort. • Improving health and local biodiversity ecosystem instead of degrading it (25) • Futurity: focusing not only toward initial or short term of time but toward long term in the future. • Equity: be able to give equity toward social life in present time (26)
	The main act of sustainable development is ecology conservation which aims to ensure that the action and decisions made in the present time will not inhibit chances of future generation to be able to make use of the same natural resources (24)	The key concept of sustainable architecture contains responsibility in minimizing negative impacts of building toward environment with efficiency of material and energy usage (24) Environmental value: giving substantial effect in enhancement environmental value (27)	
Biophilic **Paradigm :** Mechanistic-reductionist **System of Thinking:** Linear, One way-Open Loop and Partial.	The application of biophilic design is considered successful if the design can improve physical, mental and behavior in better condition toward humans in built environment (28)	Biophilic designs include nature elements inside human life to improve human life and health quality (28) The main idea of biophilic approach is believing that human being evolved 99% because of adaptive biological responses caused by nature and not by man-made.	• The direct experience of nature: referring to contacts between built environment with nature. • The indirect experience of nature: referring to contacts which represent contacts with nature and later are transferred to some forms and conditions inside. • The experience of space and place: referring to spatial characteristics in natural environment which benefit in health and human life quality (28)
	The biophilic-based design aims to maintain productivity, function and resilience owned by nature in long term of time (29)	Biophilic designs take into consideration more in balancing nature condition between human activity and life as a final result (28)	
Restorative **Paradigm :** Organic-Holistic **System Thinking:** Non-linear, Closed loop and Systemic.	Decreasing and un-prioritizing of environment quality make the environment gives bad impacts toward human health. Restorative design is needed to close the distance between man and nature (32)	Creating a healthy comfortable environment in which several sectors of nature elements being included, such as form, vegetation use, lighting, sound, materials and room smell to reach a good relationship rate between human and nature. (33)	• Biophilic Design • Responsibility of nature • Responsibility of well-being • Application of restorative approach often used on healthcare building and the city park. (33)
	Restorative is the beginning of environmental psychology development aims to reduce stress or treating mental health experienced by humans. This approach believes that humans	Restorative approach means maintaining and conserving nature or environment more importantly for humans to maintain sustainability. (33)	

IOP Conf. Series: Earth and Environmental Science **213** (2018) 012016 doi:10.1088/1755-1315/213/1/012016

	have relationship with nature much stronger than relationship between humans and the built environment (33). While a restored condition can evolve positively after the intervention, the success of the process is usually dependent on further human management (34)	Restorative approach focus on understanding the difference between natural and artificial or man-made condition. Maximizing the use of restorative based approach is believed to improve human life quality so human diseases can be prevented because nature give positive impacts for humans (35)	
Permaculture **Paradigm :** Organic-Holistic **System Thinking:** Non-linear, Closed loop and Systemic.	Permaculture is a design system aims to integrate fabrication system, nature, temporary, sosial and ethnics to be one pattern of form and benefits each other in the systems according their functions. Ancient society could manage natural resources only by using materials from nature. Permaculture conserves, supports and cooperates with nature, grow up with in the same time. This means humans learn from nature, not the other way by going against or competing with nature (36) (37)	Permaculture focuses on how connection of each components and how function of each systems instead of focuses on just one component system. This proves that sustainability aims to protect all systems of line in earth (37) The main focus of permaculture is value and ethics to be back to natural function. Permaculture improves the ecosystem life quality to fulfill human life need in the future. Permaculture attempts to fix earth ruined by humans (37)	• Care of the Earth: Provisions of all life on earth to stay continuously and multiply. • Care of people: Provisions on all humans to continue using available natural resources. • Fair share, or share the surplus: By adjusting what we need and design what earth has so that the resources can overflow for all human and other species on earth (37) • Permaculture principles: Observing and interacting, Capturing and restoring energy, Earning results, Self-respecting and receiving feedback, Using and respecting natural resources, Waste-free, Designing from patterns to details, Integrating better than separating, Using and respecting diversity, Edge-effect, Facing up every change more creatively (38)
Regenerative **Paradigm :** Organic-Holistic **System Thinking:** Non-linear, Closed loop and Systemic.	Regenerative approach aims to eliminate waste from industrial products or making the waste usable in other processes (39) Regenerative approach aims to create a healthier and more resilient living quality and equity of community. Regenerative aspects supposedly can be accepted by society and made use to change built environment faster. This change can make built environment more adaptable and resilient to face climate change situation (40) Regenerative development aims to restore or create the capacity of ecosystems and biogeochemical cycles to function optimally without constant human intervention.(34)	The main foundation of regenerative-based development is a method to integrate or involve regeneration in practical challenge of land use and community development (6) Environmental problems occur because of broken relationship between nature and human. The main problem lies on cultural and human psychological rather than technical or technological. It is a fully conscious awareness that the health of an ecosystem is dependent on human health and human health is dependent on the health of the whole ecology. It is co-evolutionary (6)	• Regenerates: adding potential factors related to the bigger systems • Improve: changing focus toward how to add capacity owned by humans and nature. • Permaculture approach and holistic management • Maintain: attempting to maintain impacts and effectivity of performance to deal with uncertain environment conditions. • Resilience, cradle to cradle and green movement • Operate: increasing energy and material use efficiency, eliminating toxic and reaching standard level of interdisciplinary capacity (6)
Resilience **Paradigm :** Mechanistic-Reductionist **System Thinking:** Linear, One way-Open Loop and Partial.	Disaster risk reduction is an objective set of rules in managing the reduction risk of disaster which also contributes in strengthen resilience and achieving sustainability level (41) Damage and loss level when disasters happen can be indicators of social and economy conditions failing in their sustainability: 70-85% of investment done by investors show that they did not have a backup plan in considering an abnormal condition (43)	The building can withstand physical damages, quickly recover, maintain main building function, not only when the disturbing incidents occurring, but also restoring operational target level faster after the incident had happened (42) Resilience as the network components ability aims to revive toward the desired performance condition after experiencing disturbance. Resilience as an aim of four interacting paradigms: reliability, vulnerability, survivability and recovery (44)	• Preparation (Prevention) • Response • Recovery

and less negative impact on nature, whereas the two other groups act to achieve a better quality of the environment; hence balancing the living system. This variation shows a different solving phase. The first attempts to make energy and resources efficient technologically (engineering approach) while the other ones try to support positively human-nature relationship and creating abundance (social-cultural approach)[6].

Furthermore, discussing the second group, Restorative, Permaculture, and Regenerative concept (G2b) is a bit different than other concepts in this group. They actively build a relationship between human and nature and give positive impacts on the environment aside to human. Restorative and Permaculture attempts to improve environmental quality by doing a thing to nature, whereas Regenerative acts more actively by making the environment stronger and healthier and supporting human-nature living process positively. It means the other concepts, besides Regenerative, are passive. They attempt to drag and turn back nature in the design (reconciliatory) for bringing the benefits to human life. It fits to the hypothesis that humans will be healthier and better in the natural environment.

Based on this above fact, (G1-G2a), it is founded that one of the weakness indicators of those various environment concepts is human domination to stay controlled. They utilize nature for their own benefits. Then, taking a closer look at the definition of sustainable development in Table 1, it could be seen clearly that the definition still takes on the human-interest side. There are no words that give an indication toward the sake of nature. This further confirms that sustainability issues lay on human's part in their relationship with nature. Humans should not only stand outside of nature, but they should also be inside and give mutual benefit, just as proposed in Regenerative. Human and nature should do a co-evolution.

Comparing the background of Resilience concept to the other concepts, the most prominent difference is their substance. Resilience arises driven by hazards issues, whereas the others by ecological issues. All this time, the ecological concepts attempt to prevent natural damage because of human's exploitation. They attempt to reduce or heal the environment's diseases. Whereas, Resilience concept act to reduce hazards impacts (the pain) that caused by natural system damages such as global warming and climate change. The earth is sick, and Resilience attempts to prevent the pain. It attempts to survive the shock and recovered fast. It could be seen that both concepts appear to solve the environmental issues faced from two different but interconnected perspectives. The diseases and the pain should be treated so the condition does not get worse.

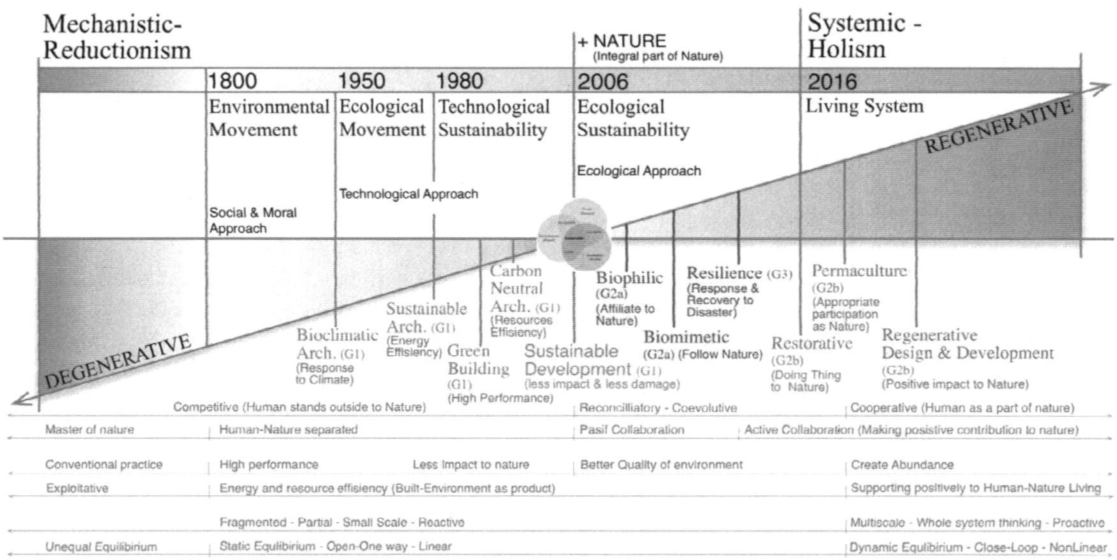

Figure 4 The Map of Environmental-Architectural Concept
Sources: (6),(10),(11), (12), (13), (14), (15) combined & modified

Maintaining natural balance is the core of the Sustainability Concept. The first group with concepts based on technological sustainability believes that earth and all living in it will just be fine if the natural balance can be maintained statically. This balance means if humans take from nature for say ten units, then humans should give back also ten units in order to keep the natural balance. System thinking should be made in such a way so that balance input and output can happen precisely. This paradigm is called steady-state equilibrium [6] which is believed to be the basic thinking of environmental movements in the early of 20th century (G1-G2a). Hence, to achieve it, attempts are made by preservation and conservation approaches and strategies. Natural conditions that should not be touched and exploited by humans will be maintained, while others that could be used for human life will be preserved. Conservation is an attempt to produce product optimally by using resources as less as possible. But this still gives an indication that one day all those resources will run out. But, to renew resources is not easy to do, especially following entropy which states energy transfer process takes place in one-way because of recovery time issues. Returning one liter of gasoline for a motorcycle takes millions of years. Responding to this, another idea comes up. This idea assumes that balance will always be changing. Balance is dynamic (dynamic equilibrium) just as Heraclitus said in Ancient Greek Era. The balance tends to be dynamic and moving forward, never stops at one point hence the change itself is the only permanent cause by the life activities on earth [45]. This incident happens constantly so the only thing left to do is surviving by changing and adapting following Homeostatis concept. Nature life has Autopoiesis Dissipative behavior. Universe has the ability to produce, regenerate and readjust itself in the form of dynamic, absorbing energy and materials process from environment and simultaneously produces product and residue which functions as energy and materials for the other life systems. It continuously happens, and through that process, every organisms and life systems regenerate, form and maintain themselves alone or together [9]. All component of life in the universe viewed as one large organism in unity and interrelated. They support and allow the universe to develop, regenerate and form a new balance [46]. Nature is a big and complex system. Accordingly, they do not discuss and think about balance again but think about the relation system for regenerating and adapting. This is shown in sustainability concepts in G2b. They discuss ideas on eco-literacy which encourages the use of ecological principles and human-nature living system as the base of development in modern society [9].

Solving the big and complex of environmental issues, according to Capra, lies on the main characteristics non-linear of the system. Complexity itself is a set of concepts and mathematical techniques related to other subsystems in a non-linear way, forming a complex network. They cannot be understood in the method that is characteristics fragmented. This perspective asserts that we could never sort out the universe to be smallest units standing independently. The universe is a relationship that related to each other as a whole system [46]. The old system of thinking with Mechanistic-Reductionism appears to be less conducive used. The Mechanistic-Reductionism paradigm has a linear and partial of thinking, a causality conception that runs one direction, simplifies the problem and fills the limitations with many assumptions. This approach will give a chance of failure significantly when be used for solve the ecosystem problems that are complication and complex.

Inspired by the successes of the information system in modeling a complex problem, right now, there is a theory that offers systemic, holistic and non-linear called complexity theory. The main feature of complexity theory is an anti-reductionist through the fundamental rethinking of the system of nature so that conceptualization interrelationship and dynamic-relationship can happen better simultaneously between each phenomenon randomly [47]. The "study of complexity" attempts to find common principles underlying the behavior of complex systems. They are large collections of components interact in nonlinear ways. This behavior implies that the system cannot be understood simply by understanding its individual components. They interactions cause the whole to be "more than the sum of its parts" [48]. The advantages are the modeling closes to the real complexity of the phenomena as well as the broader network that covers multi-disciplines, multi-scales, and multistage. However, up to this moment, there are only a few people using this complexity theory to study life network and doing research on this field [46]. Adopting this theory might help the various concepts mentioned above to pick opportunities to achieve better sustainability.

After reviewing environmental concepts above, the result shows that the concepts and frameworks of Sustainability Concept has been being developed, progressing toward one more rigid concept. This goes in line with Gou and Xie's saying that to answer all critics on sustainability, it is not by bringing more ambiguous and suggesting more complicated indicators, but by the influences, relationships and context [33]. It is more important to develop a holistic approach by expanding the indicators more comprehensive and the framework more flexible that to be adjusted with different contexts, directed to co-existence and co-evolution through technological sustainability approach as well as ecological sustainability. Human should be active giving positive contribution to nature, and supporting the human-nature living system. The universe is a complex system that needs a approach and method based on complex situation. Using Complexity Theory could be a change to achieve it. It is the new light of sustainability.

4. Conclusion

a) The emergence of various concepts and approaches discussed is a wave of ideas to obtain sustainability. Concepts and approaches have emerged, even though literally separate, both technological sustainability, ecological sustainability, and living system is co-related and all needed.
b) One of the sustainability issues that always need to be observed lies in human's role and position that conscious or unconscious, standing outside of environmental activities. Humans should understand their roles and become a part of the nature to do co-evolution. They must know the way our universe works and support our living system that is the 'new' sustainability approach.
c) Mechanistic-reductionism approach is not conducive for maintaining the sustainability, because basically nature is a complex system. The sustainability paradigm needs to be shifted toward Systemic-Holism approach that has dynamic networking and dynamic equilibrium.
d) Moreover, it should utilize a suitable characterized; systemic, dynamic, organic, holistic and non-linear approach. It could be accomplished by adopting the complexity theory. However, complexity theory should be researched in detail for its application.
e) The wave of sustainability thinking still seems to be moving forward to find the unity and rigid of the concept based on the existing theories and methods.

5. Acknowledgement

We would like to thank for the opportunity to use library's facilities available in laboratory of 'Struktur dan Teknologi Bahan Jurusan Arsitektur Fakultas Teknik', Central Library of Universitas Diponegoro and Central Library of Universitas Atma Jaya Yogyakarta. The easy access to get articles, scientific papers from journals and reference books with Sustainability and Environmental Concept topic accelerated and helped to accomplish this research.

References

[1.] Lam HL, Varbanov PS, Klemeš JJ, Yan J. Green Applied Energy for sustainable development. Elsevier. 2016;161:601–4.
[2.] IPCC. Summary for Policymakers. Climate Change 2014: Synthesis Report. Contribution of Working Groups I, II and III to the Fifth Assessment Report of the Intergovernmental Panel on Climate Change. 2014. 2-26 p.
[3.] Chan M. How climate variability affects health. 2012;4:1–4.
[4.] Global Humanitarian Forum. The Anatomy of a Silent Crisis. Hum Impact Rep Clim Chang. 2009;135.
[5.] Merchan M, Pdf E, Giz SS, Giz SS, Taylor H, Pdf E, et al. Energy and Climate Change Adaptation in Developing Countries. 2017;
[6.] Mang P, Haggard B. Sustainability, Regenerative Development and Design A framework for evolving. 2016. 1-274 p.
[7.] Baker S. Sustainable Development. 2006;264.
[8.] Lincoln B& C. A Dictionary of Ecology, Evolution and Systematics. 2nd ed. Oxford Paperback Reference; 1998.
[9.] Keraf DAS. Filsafat Lingkungan Hidup. Daerah Istimewa Yogyakarta : Sleman: Kanisius; 2014. 1-224 p.
[10.] Sonnebon L. The Environmental Movement. New York: chelsea house; 2008. 1-129 p.
[11.] Miranda Schreurs EP. Historical Dictionary Of The Green Movement : Second Edition. United States of America: Scarecrow Press; 2007. 1-391 p.
[12.] Baweja V. What ' s Next for Architectural History ? Sustainability and the Architectural History Survey University of Florida. 2014;452–7.
[13.] Vernon R. A Brief History Of Resilience : From Early Beginnings To Current Constructions. Community Plan to Foster Resil Child. 2004;13–26.
[14.] Mcaslan A. THE CONCEPT OF RESILIENCE Understanding its Origins, Meaning and Utility. Torrens

Resil Inst. 2010;(March):1–13.
[15.] Attia S. Modern History of Sustainable Architecture. Regen Posit Impact Archit Learn from Case Stud. 2018;(Porteous 2013):7–12.
[16.] Zahedinia SS. The role of Bionic architecture in achieving sustainability. 2016;7(11):211–7.
[17.] Sajjadian SM, Lewis J, Sharples S. Sustainability in Energy and Buildings. 2013;22:903–12.
[18.] GBCI. GREENSHIP RATING TOOLS GREENSHIP untuk Gedung Baru. 2012;2014(April).
[19.] Ragheb A, El-Shimy H, Ragheb G. Green Architecture: A Concept of Sustainability. Procedia - Soc Behav Sci [Internet]. 2016;216(October 2015):778–87.
[20.] Dwaikat LN, Ali KN. Green buildings cost premium: A review of empirical evidence. Energy Build [Internet]. 2016;110:396–403.
[21.] Green Building Council Indonesia. Greenship Rating Tools. Jakarta: Green Building Council Indonesia; 2016.
[22.] Larson L, Rajkovich N, Leighton C. Green Building and Climate Resilience: Understanding impacts and preparing for changing conditions. Univ Michigan [Internet]. 2011;260.
[23.] WCED. Report of the World Commision on Environement and Development: Our Common Future. Oxford Pap [Internet]. 1987;Report of:400.
[24.] Dao V, Langella I, Carbo J. From green to sustainability: Information Technology and an integrated sustainability framework. Elsevier [Internet]. 2011;20(1):63–79.
[25.] Neyestani B. A Review on Sustainable Building (Green Building). MPRA. 2017;
[26.] Langston CA, Ding GKC. Sustainable Practices in the Built Environment. Oxford: Reed Educational and Professional Publishing; 2001. 288 p.
[27.]Iyengar K. Sustainable Architectural Design : An Overview. New York: Routledge; 2015. 1-295 p.
[28.]Stephen R. Kellert EFC. The Practice Of Biophilic Design. 2001;
[29.]Isle N. Biophilic Design. 2003;(September):1–25.
[30.]Zari MP. Biomimetic design for climate change adaptation and mitigation. Archit Sci Rev. 2010;53(2):172–83.
[31.]Zari MP. Biomimetic Approaches To Architectural Design for Increased Sustainability. Design. 2006;(April):2006.
[32.] Yannick Joye AEVDB. Restorative Environments. 2012;57–66.
[33.] Gou Z, Xie X. Evolving green building: triple bottom line or regenerative design? J Clean Prod [Internet]. 2017;153:600–7.
[34.] Jenkim S, Zari MP. Rethinking our built environments: Towards a sustainable future [Internet]. 2009. 55 p.
[35.] Nunes LS. The handbook of salutogenesis. The Handbook of Salutogenesis. 2017. 415-421 p.
[36.] Wolfe S. Regenerative Permaculture Design. 2014. 64 p.
[37.]Wolfe S. Regenerative Permaculture Design: A beginner's guide to designing a truly liveable world. 2014. 64 p.
[38.] Design CA. 10 Parametric Plugins Every Architect Should Know ! 2017;1–15.
[39.] Gabel M. Regenerative Development: Going Beyond Sustainability. 2005;10.
[40.] Zari MP. Regenerative design for the future. Build 115 Sustain. 2009;(115):68–9.
[41.] UNISDR. Man-made and Technological Hazards. 2017;
[42.] Sanchez AX, Osmond P, Van Der Heijden J. Are Some Forms of Resilience More Sustainable than Others? Elsevier [Internet]. 2017;180:881–9.
[43.] UNISDR. Sendai Framework for Disaster Risk Reduction 2015 - 2030. Third World Conf Disaster Risk Reduction, Sendai, Japan, 14-18 March 2015. 2015;(March):1–25.
[44.] Castillo Borja F, Vázquez Román R, Quiroz Pérez E, Díaz-Ovalle C, Mannan S. A resilience index for process safety analysis. J Loss Prev Process Ind [Internet]. 2017;50:184–9.
[45.]LIM TL, KOW MY, MAHDZIR A, ABU BAKAR N. Fragments The Collected Of Heraclitus [Internet]. Proceedings International Conference on Innovation and Management. 2016. 61-77 p.
[46.] Latif M. ORIENTASI KE ARAH PEMAHAMAN FILSAFAT ILMU. Jakarta: PRENADAMEDIA GROUP; 2014. 340 p.
[47.] Walby S. COMPLEXITY THEORY, GLOBALISATION AND DIVERSITY. 2013;
[48.] Mitchell M. Complexity A Guided Tour. 2009.

Investigating problems of vehicle parking in a low-rise residential area in Nangok-dong, Seoul

Seungwoo Seo[1*], Saehoon Kim[2]
[1]Department of Landscape Architecture, Urban Design Concentration, Graduate School of Environmental Studies, Seoul National University, Seoul, Korea
[2]Associate Professor, Department of Landscape Architecture, Urban Design Concentration, Graduate School of Environmental Studies, Seoul National University

*Corresponding e-mail: s9228211@naver.com

Abstract. A number of low-rise residential areas in Seoul came to be filled with single –family houses in the 1970s. But as the housing pattern changed to multi-family housing and demand of cars explosively increased, vehicle parking in the area became problematic. Although the local government tightened regulations on parking standards and built a public parking zone to solve this problem, the residents' complaints worsened as parking many vehicles along a narrow alley caused community conflicts and pedestrian safety was threatened. In the study, Nangok-dong was investigated as most of the cars parked on a narrow road caused degeneration of pedestrian and residential environments. This study investigated the actual parking conditions in the study area and proposed a few ideas for solving the complex parking matter in a low-rise residential area.

Keywords: low-rise residential area, parking space, illegal parking problem

1. Introduction

Nangok-dong is one of the low-rise residential areas in Seoul and is designated as the city's urban regeneration area [1]. There are many reasons regarding why parking is problematic in the low-rise residential areas in Seoul. First, the areas were initially occupied by single–family houses through Land Compartmentalization and Rearrangement Projects in the 1970s. But some of the houses began to be changed to multi-family housing. Second, the demand for vehicles use explosively increased from the 1980s [2]. Accordingly, available parking space became deficient and many vehicles came to be parked along a narrow alley. This caused community conflicts and pedestrian safety problems.

The local government has implemented three policies to solve the parking problems. First, they constructed the public parking lots. Second, the resident priority parking area was designated on the road to manage the illegal parking and to add more parking space in the low-rise residential area. However, the vehicles occupied the alley of the residential area, and pedestrian safety was threatened. In addition, there is a limit to the parking supply fundamentally. Finally, 'Green parking project' was implemented to provide parking space by removing the fence of the houses. However, this project is effective only for some single-family houses [3].

Despite these efforts, the policies have not solved parking problems in Nangok-dong. Because constructing the public parking lot and designating the residents' priority parking area have limitation of space for supply [4]. Since only 7% of the total houses are single-family houses in Nangok-dong, the green parking project is not effective.

2. Method

At first, the data on low-rise residential parking problems are collected through the internet, literature surveys and other papers. The author confirms the number of cars parked illegally in the area and identifies problems in the area. Second, the author categorizes each problem, formulates a solution strategy for it, and proposes an action plan under the proposed strategy.

3. Results and Discussion

3.1. Present conditions of illegal parking

The author investigated the number of illegal parking vehicles and the locations through on-the-spot inspection. As shown in Figures 1, 2, and 3, most of the vehicles are illegally parked on the road. During the daytime, 377 cars were illegally parked and at night 547 cars were illegally parked. This is a residential area, so more cars were illegally parked at night. Also, 190 parking spaces were located on the roads to resident priority parking areas. Most of these areas are located on the roads having a width of more than 6 meters. Especially in the case of the Site A and B, the space was too narrow to walk along the road due to the illegally parked vehicles. (see Figure 1, 2)

Figure 1. Illegal parking on the narrow road at daytime in Site A.

Figure 2. Illegal parking on the narrow road at night in Site B.

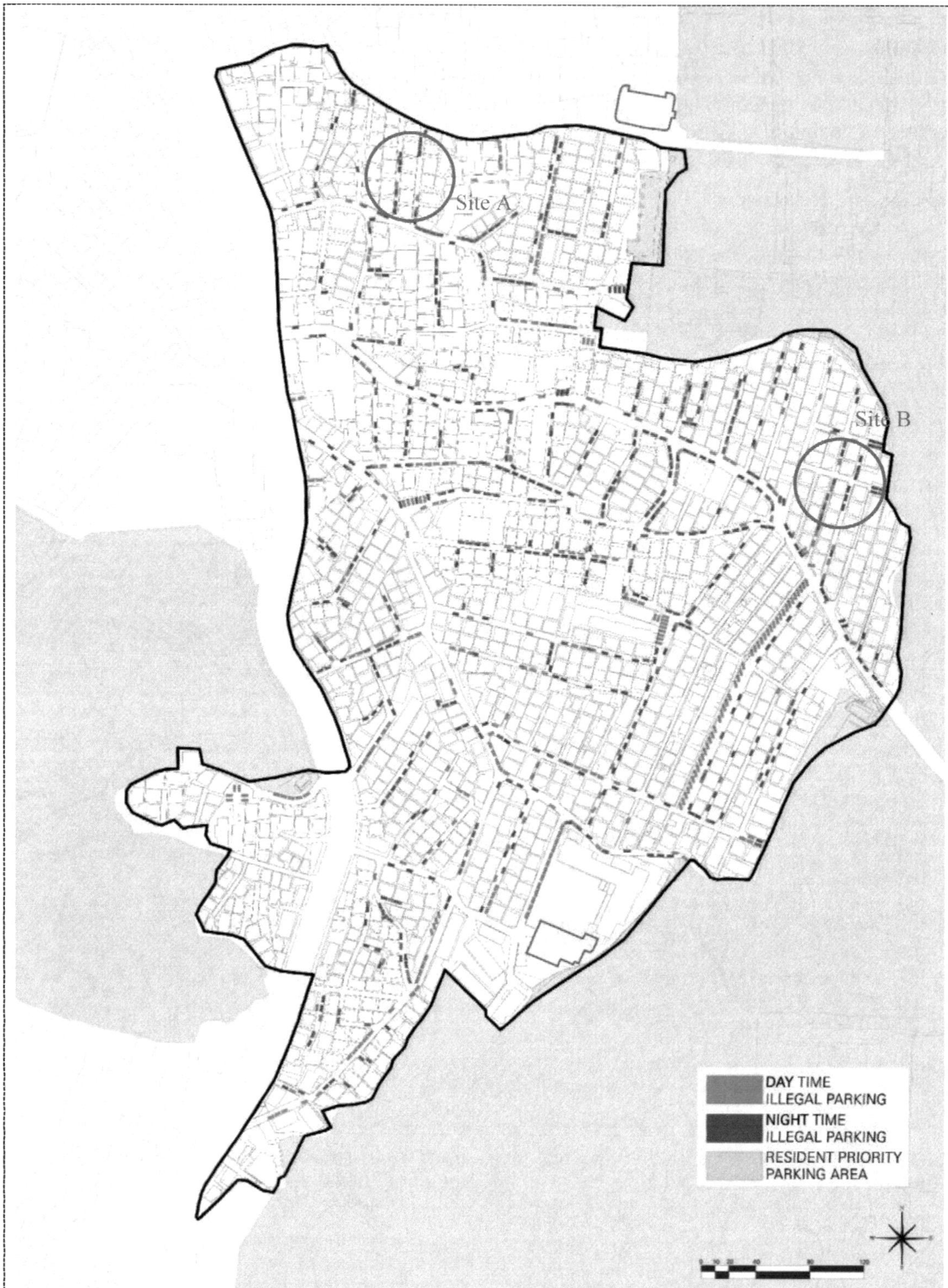

Figure 3. Illegal parking status on day/night time and resident priority parking area

3.2. Problems
Vehicles parked illegally on the road threaten the safety of pedestrians, interfere with vehicle traffic, and cause conflicts among residents. There are three reasons for the problems [5].

3.2.1 Quantitative shortage of parking lots
The author analysed the statistics of the area. As shown in Table 1, the number of vehicles registered per household is 0.52, and it appears that the number of parking lots is insufficient compared to the number of registered vehicles. So, there are no parking places for 1059 vehicles in Nangok-dong.

Table 1. Nangok-dong statistics in 2016

Statistics of Nangok-dong	*counts*
The number of households	12869
The number of vehicle registration	6689
The number of parking spaces	5630

3.2.2 Vehicle-dependent environment
There are two reasons why many residents in the area are dependent on vehicles. First, most public infrastructure is concentrated to the west of the neighborhood. On the west side, there is a wide road called 'Nangok road'. In this area, there are restaurants, cafes, convenience stores, and market, etc. Second, the altitude difference between the west and the east is about 60m. It is difficult for people to walk to their homes from the west to the east, because of the high slope. With this terrain and road infrastructure, residents have little choice but to rely on vehicles.

Figure 4. Concentrated facilities on the west in Nangok-dong

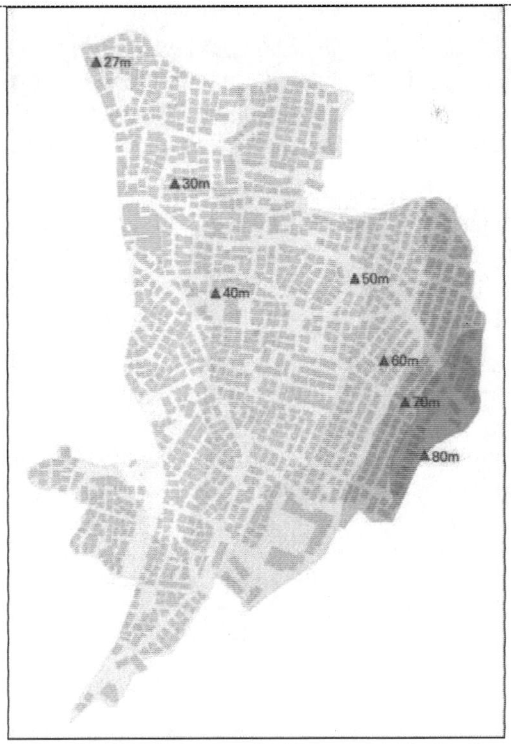

Figure 5. Altitude status in Nangok-dong

3.2.3 Inefficient use of parking space

During the daytime, the residents' priority parking area on the road is fully occupied, but the public parking lot is empty. This pattern can be seen as a result of the reduction in the available walking space and open space. (Figure 6) Another reason is the use of parking lot for other purposes. As shown in Figure 7, the parking lot is used as a warehouse, and the car is illegally parked on the road in front of it.

Figure 6. Full of resident priority parking area at the daytime

Figure 7. Use of the attached parking lot for other purposes

3.3. Potential solution

3.3.1 Supply of the public parking lots

Since there is absolute shortage in the amount of parking lot, more public parking should be supplied. This may help to provide about 50% of previously deficient parking spaces. The remaining 50% may be solved by different methods, such as efficient use of the existing parking space.

3.3.2 Reduction in vehicle demand

Reducing vehicle demand is about making it convenient for people to live and commute without vehicles. There are two ways to do this. One is to place alternative mobility infrastructure to the east side. The other is to connect the east and the west with a walkable route and public transportation.

3.3.3 Parking lot management system

The structure of the parking lot management system is shown in Figure 8. The supplier input available parking lot to the system, which provides the consumer (e.g., vehicle owner) with the information. Then, the consumer pays the parking fee to the system, and the system distributes charge to the supplier. The system flexibly accepts different parking demand by time of day, makes the street occupied by cars an open space.

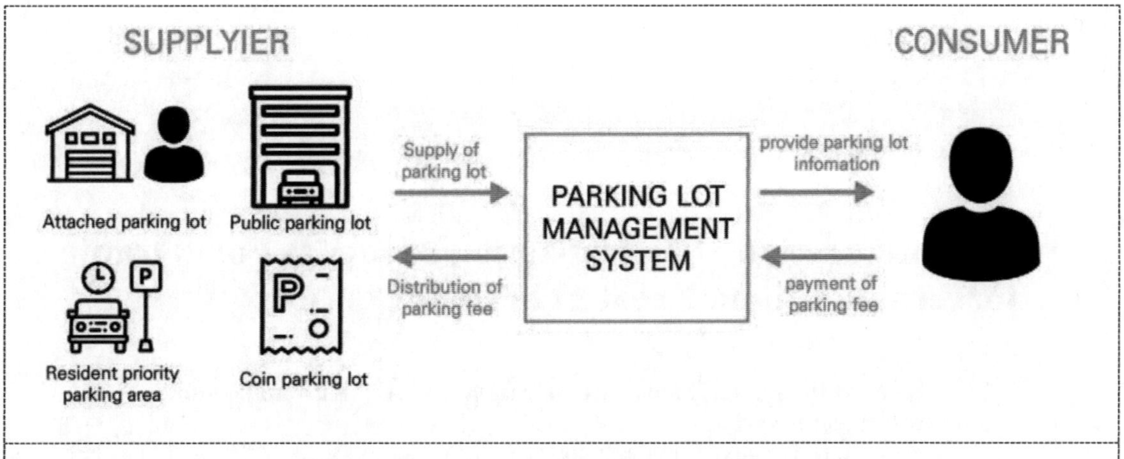

Figure 8. The structure of parking lot market system

4. Conclusion

In conclusion, we can categorize problems of parking as follows: 1) quantitative shortage of parking lots, 2) vehicle-dependent neighbourhood environments, and 3) inefficient use of parking space. The author would like to propose potential solutions to each problem in three ways: the supply of the public parking lots, reduction in vehicle demand, and comprehensive parking lot management system.

5. References

[1] Gwanak-gu Homepage(www.gwanak.go.kr)
[2] Statistics Korea Homepage (http://kostat.go.kr)
[3] Seoul real estate information plaza Homepage((http://land.seoul.go.kr/land/))
[4] The Seoul Institute 2013 *A Study of Efficient Use of Parking Space in Residential Areas* 135
[5] Traffic accident analysis system Homepage(http://taas.koroad.or.kr)

Spatial Configuration of Optimization Model at Fort Oranje Area in Ternate City to Create Thermal Comfort

Asri A. Muhammad[1], **Firdawaty Marasabessy**[2], **Arif Kusumawanto**[3], **Ardhya Nareswari**[4]
[1,2]Khairun University, Ternate, Indonesia
[3,4]Gadjah Mada University, Yogyakarta, Indonesia

E-mail: asriam@yahoo.co.id

Abstract. Ternate's geographical condition as a coastal city with a volcanic mountain and on the equator makes this city potential for the phenomenon of urban heat island. This condition causes the air temperature in Ternate to be quite hot around 26-34°C. Consequently people feel reluctant to conduct activities in the public space during the day. This situation occurs at Fort Oranje public space area, where people visiting this place are not so crowded. It occurs because thermal condition is inconvenient for people to conduct any activities. This research aims to optimize the configuration of both the building and vegetation space to obtain a convenience thermal condition at Fort Oranje. The method used empirical measurement and modelling/simulation (engineering of area) using Envi-MET software. This method was used to conduct a simulation of thermal optimization at Fort Oranje. The results indicated that the engineering of spatial configuration with the intervention on the building by utilizing the vent and the use of vegetation could improve thermal comfort at Fort Oranje compared to the existing condition.

Keywords: thermal comfort, space configuration, urban public space, Envi-MET.

1. Introduction

Thermal comfort at Fort Oranje public space in Ternate is quite impossible to be created with solid building mass configuration minimum vegetation. Solid building mass configuration causes uneven wind movement patterns, so that it increases air temperature and decrease humidity, therefore it cannot create thermal comfort [1]. Thermal comfort in outdoor area will affect the townspeople to conduct outdoor activities [2]. Urban microclimate causes specific changes such as slowing down the wind speed or storing solar radiation by the construction material used [3]. To control the rate of increase in temperature, people must pay attention to the duration and intensity of sun exposure as well as the sun angle [4]. In addition, air movement is also very helpful in controlling the existing temperature. This is a parameter caused by building and environmental interaction [5]. Considering these parameters, thermal comfort in the urban public space is possible to be created. Monitoring the thermal comfort is an indicator for identifying the environmental condition in the area [6]. This research aims to optimize the configuration of both the building space and vegetation to create thermal comfort in the Fort Oranje public space area.

2. Methods

The study focused on the optimization by engineering Fort Oranje area that was divided into several area models according to the study variables to see the effect on thermal comfort and Fort Oranje public space area sustainability. This study used experimental method with the simulation using computer application and field measurement. Simulation was a controlled study method in real context with the purpose of studying dynamic interaction in an arrangement [7]. There were three simulation processes, i.e.:

1. Empirical measurement, to show thermal condition and thermal analysis result using data that measured on site with specific measuring instruments.
2. Simulation used ENVI-met system software to determine the effect of existing space arrangement on thermal condition.
3. Optimization simulation with space configuration engineering using ENVI-met software to the better thermal comfort condition.

3. Discussion

The analysis process of Fort Oranje thermal comfort was conducted through some stages:

1. Making changes to some Fort Oranje areas according to the research variable (space variable), namely vegetation and building to achieve optimal thermal comfort in the area.
2. Making models (3 models) of the area according to the space variable (building and vegetation).
3. Conducting an analysis of thermal condition in each model using a micro-climate simulation model with ENVI-met software system with fixed climate data (measurement result) as in the simulation of the existing condition. Simulation data of each model was taken to represent the 3 different times morning, noon, and afternoon.
4. Analysing the Fort Oranje Area Model Optimization Simulation

3.1. Analysis of Fort Oranje Area Model Optimization Simulation

The analysis was performed on the elevation of 1.20 meters with the assumption that in this elevation the user will strongly experience the thermal comfort. The climate data used for the simulation was fixed meaning that there was no change in climate data in the existing model simulation or the three models created.

The optimization simulation was conducted using ENVI-met system software to determine the effect of spatial configuration change (which was divided into 3 models) on the thermal condition of Fort Oranje area and to obtain a better thermal comfort compared to the existing condition. Using the optimization simulation, it is discovered that there is a relationship between the spatial configuration and area microclimate. Spatial configuration at Fort Oranje area is various and it appears on the different existing thermal comfort.

Optimization simulation was conducted with the consideration that the area:

1. Has a high building density
2. Lack of vegetation in the area

The condition above allows the set of spatial based on the variables of the building and its vegetation so that there are changes in the thermal variables. The simulation was conducted on 4 created models using ENVI-met Eddi, namely:

1. Existing Model, this model represents the existing condition of Fort Oranje area without making any changing to its variables.
2. Model 1st (Vegetation), this model is Fort Oranje model with a setting/change on the one of the spatial configuration variables namely vegetation. The change in this model includes adding trees (see figure 1. number 1,2,3,4 & 6) and grass (see figure 1. number 5) in certain areas based on the result of the existing simulation.

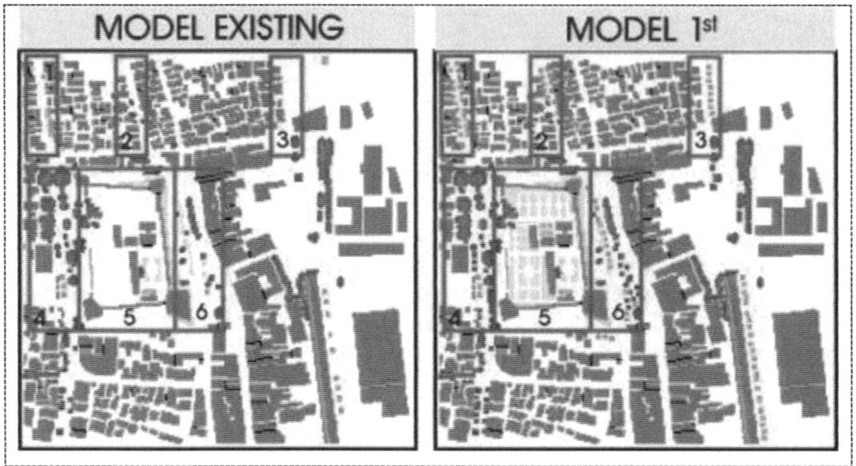

Figure 1. Model Existing and Model 1st of Fort Oranje Area

3. Model 2nd (Building), this model is Fort Oranje model with a setting/change on the one of the spatial configuration variables namely building. The change in this model includes making a wind tunnel (see figure 1 number 2) and set back the building that is close to the road (the building uses an arcade).

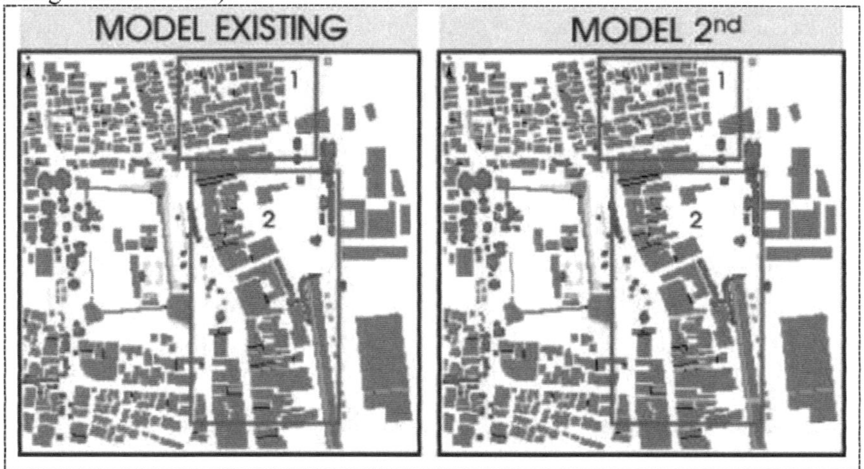

Figure 2. Model Existing and Model 2nd at Fort Oranje Area

4. Model 3th (Vegetation and Building), this model is a combination of model 1 and 2. The changes were performed to the two space variables, namely building and vegetation (see figure 3.). The building changes in this model are indicated in figure 3. numbers 3 and 8, while the changes in vegetation are indicated in figure 3. number 1,2,4, & 5 (trees) and figure 3. number 6 for grass.

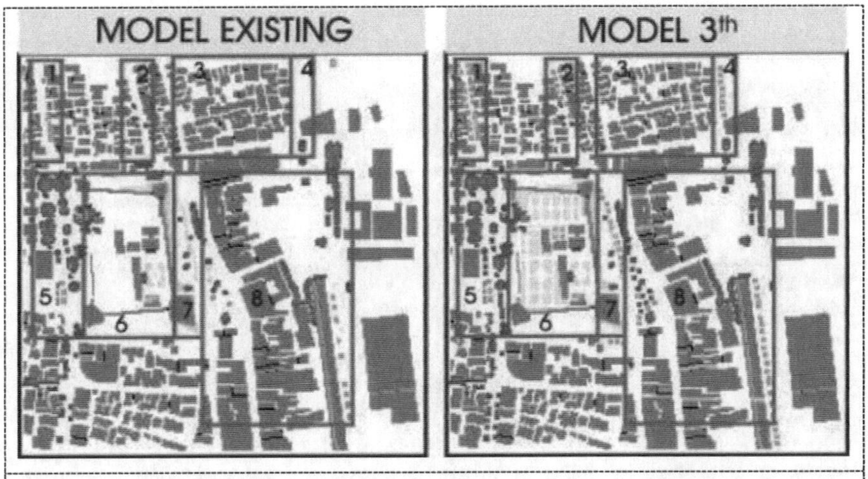

Figure 3. Model Existing and Model 3th at Fort Oranje Area

3.2. Thermal Comfort Analysis on Fort Oranje area Model 1st

The change performed in this model was by adding vegetation to the area as previously explained (see figure 1). After simulating the model 1 using the ENVI-met system, and there was a change in the temperature and humidity in the area changed (see figure 4). The temperature condition in the morning (at 9:00 a.m.) clearly indicated a decrease in the corridor area with trees, indicated by the colour changing in the visualized simulation result (see figure 4, temperature, figures 1,2,3,4 & 6). The area in Fort Oranje did not indicate any significant changes (see figure 4, temperature, number 5 at 9:00). The decrease in the temperature was also followed by humidity increase. Humidity increase is clearly indicated in the front area of Fort Oranje (see figure 4, moisture, number 6).

During the day (at 1:00 p.m.), the difference in temperature and humidity between the existing model and model 1 was clearly visible. The air temperature in the existing model looked high (see figure 4. Existing temperature, number 1,2,3,4,5 & 6) but different condition was indicated in model 1, where at the same time the model 1 air temperature was lower than the existing condition (see figure 4, temperature model 1, number 1,2,3,4,5 & 6). This condition also occurred in the humidity, where at the same time the humidity of model 1st was higher than the existing condition (see figure 4. humidity, number 5 & 6). The temperature and humidity condition in the afternoon (at 4 pm) were not much different from the morning and noon, in which the temperature condition was lower than the existing condition and the humidity was higher than the existing condition.

Wind movement in the existing area and model 1 was not much different. This appears in some areas intervened (see figure 5, number 1,2,3, & 4). However, different condition occured in the inner and front area of Fort Oranje (see figure 5, figure 5 & 6) in which the wind speed was lower than the existing condition. This situation occurred because the added vegetation reduced the wind movement passing through it (see figure 10, existing and model 1st).

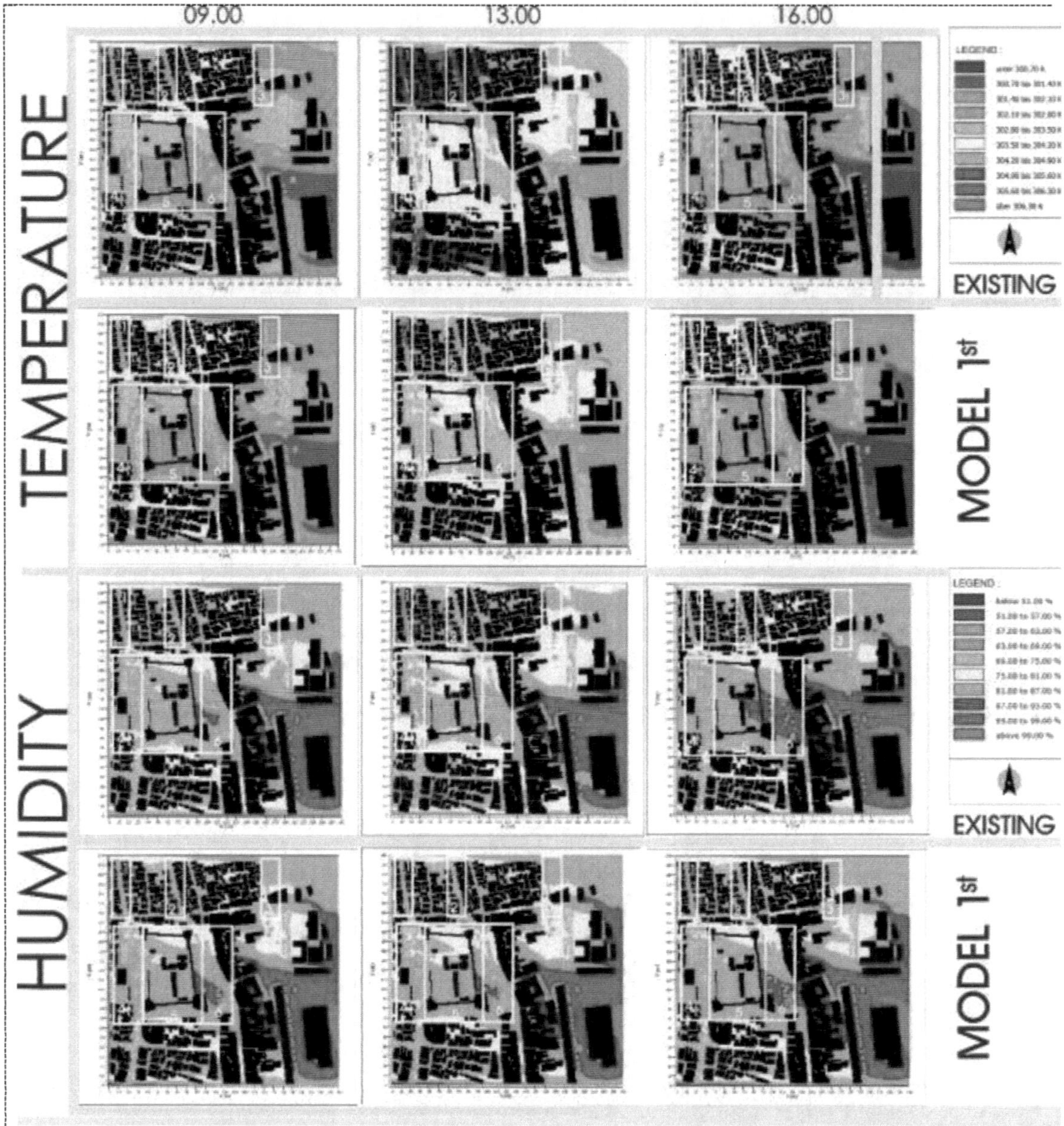

Figure 4. Result of Temperature and Humidity Existing and Model 1st

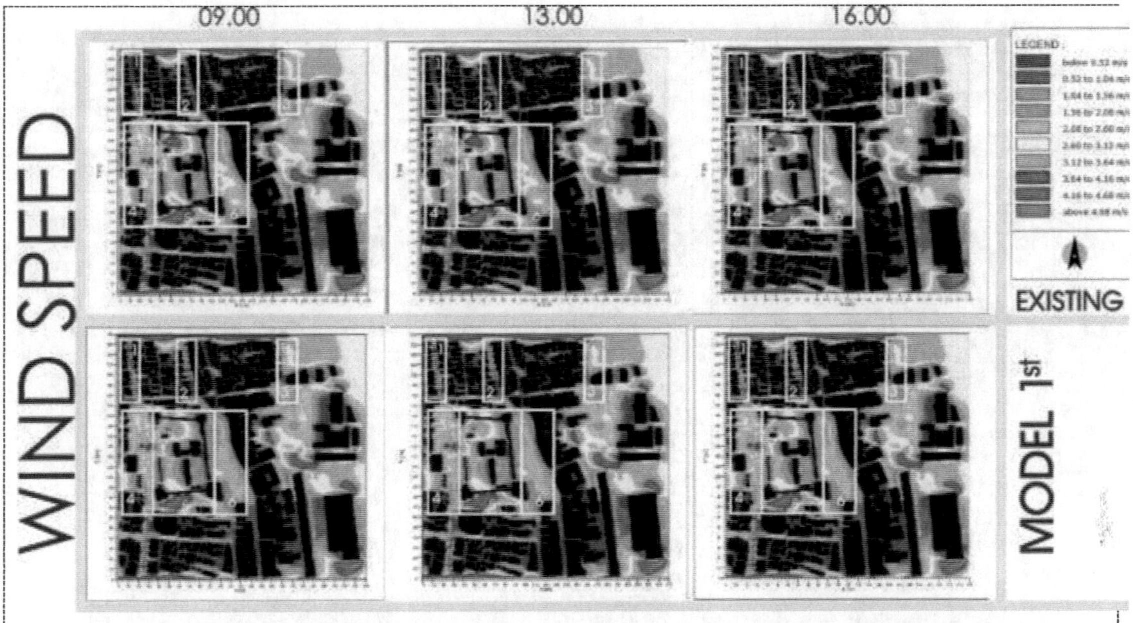

Figure 5. Result of Wind Speed Existing and Model 1st

3.3. *Thermal Comfort Analysis on Fort Oranje area Model 2*

In this model, the intervention was conducted only in 2 areas (as indicated in Figure 2). The intervened variable was the building whereas the vegetation was the same as the existing condition. Climate data used for the simulation was the same as the existing condition and model 1. Based on the simulation, in the morning (9 am) the temperature change occurred in the area (see figure 6, temperature, number 1 & 2) was not significant, but if it was compared with the existing condition, the air temperature in model 2 was lower than the existing condition. Temperature changed was evidenced during the day (at 1 pm) where the air temperature in the area with the intervened building mass was lower than the existing condition. In addition, the changes made in the area affected the temperature decrease around it (see figure 6, temperature, number 1 & 2). This condition was caused by the unobstructed wind movement as it passed through the changing area. As in the morning and noon, in the afternoon the condition of air temperature in the existing model was higher than the model 2 (see figure 10, existing & model 2).

The decrease in temperature was followed by increase in humidity. From the three times, only during the day where the condition of air humidity in model 2 was clearly seen as higher than existing condition (see figure 6, humidity, number 1 & 2). The wind movement area that experienced change in its building mass was more dynamic than the existing. Here, the moving wind could go through wide road corridors, and the existing condition did not allow the wind to move freely through the road corridor because the building was too close to the road. The buildings that were too close to each other also affected the wind movement through the area. (see figure 7, item 1 and picture 10 of existing & model 2). The existence of corridor that cut the direction where the wind came (see figure 7, number 2) greatly affected the wind movement to the public space of Fort Oranje because the wind that moved toward the area was blocked by the existence of the buildings on the side of the corridor. With the creation of wind tunnel in a building, the wind could move through the space without move up to pass the building (see figure 7, number 2 and figure 10, existing and model 2).

IOP Conf. Series: Earth and Environmental Science **213** (2018) 012018 doi:10.1088/1755-1315/213/1/012018

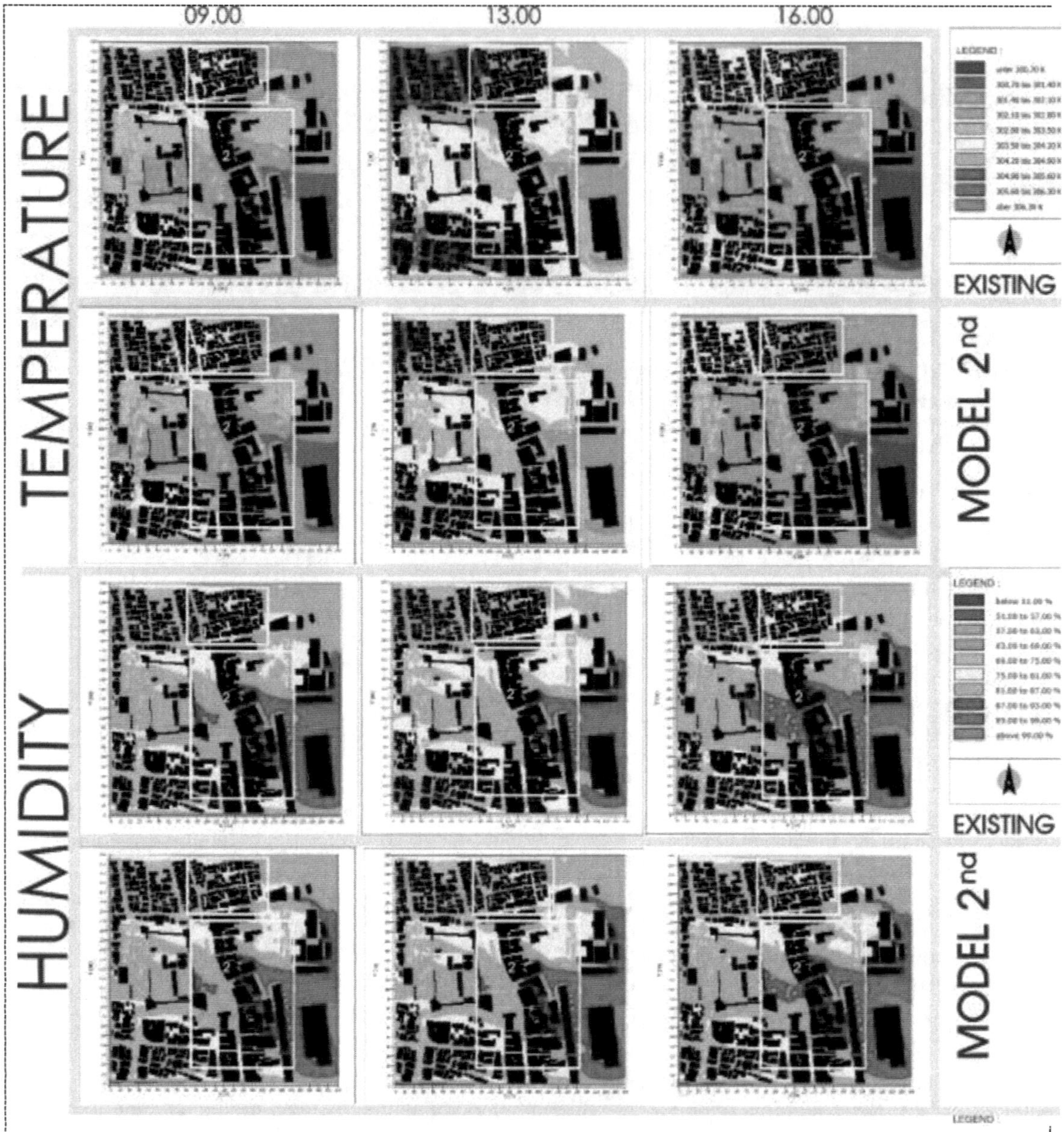

Figure 6. Result of Temperature and Humidity Simulation Existing and Model 2

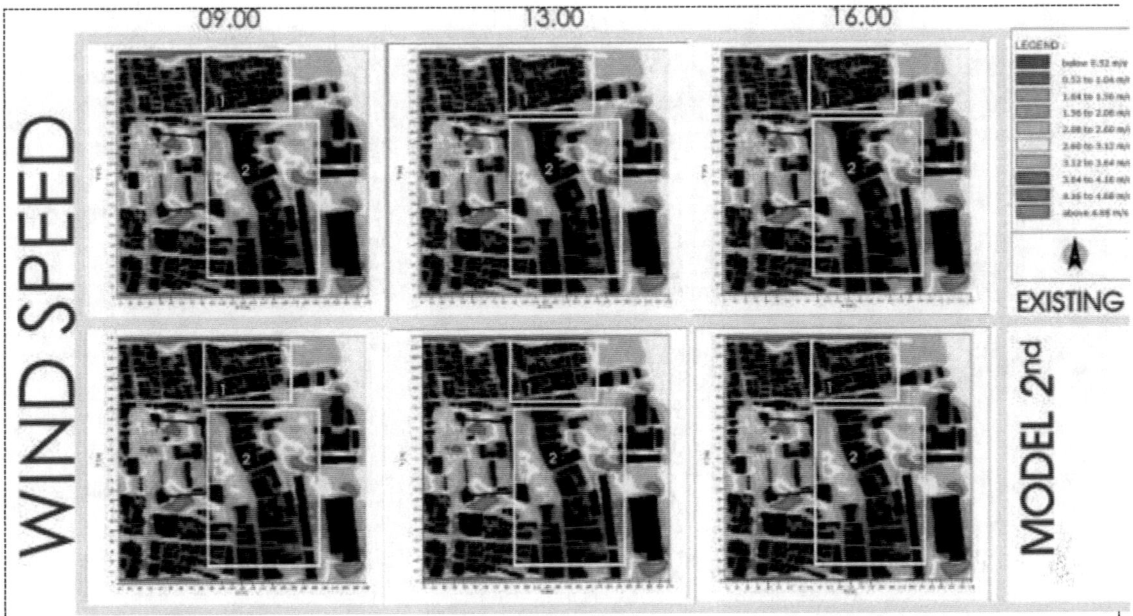

Figure 7. Result of Wind Speed Simulation Existing and Model 2nd

3.4. *Thermal comfort analysis on Fort Oranje area model 3*

This model is a combination of model 1 and model. Changes were made to both space variable namely building and vegetation (see figure 3). Change in spatial configuration affected the decrease in temperature and increase in humidity in the area. Temperature changes occured in all changed areas. This condition was clearly seen when compared to its existing. Significant air temperature changes occurred during the day (at 1 pm) (see figure 8, temperature, number 1,2,3,4,5,6,7 & 8). The addition of grass (see figure 1, model 1 number 5 and figure 3, model 3 number 6) at Fort Oranje that served to reduce the area exposed to direct sunlight could decrease air temperature at Fort Oranje (see figure 8, number 6). The decrease in air temperature was directly proportional to the increase in air humidity in the changing area (see figure 8, humidity, 1,2,3,4,5,6,7 & 8). As with air temperature, significant increase in humidity occurred during the day (13:00).

Wind movement occurred in this model also experienced change. In certain areas the wind speed increased, (as indicated in Figure 9, item 3). This increase was caused by the changes in building mass by creating arcades in the building. In addition, there was a decrease in the wind speed (see figure 9, number 7). This was due to the existence of trees added to the area that reduced the wind speed passing through it (see figure 10, model 1 and model 3).

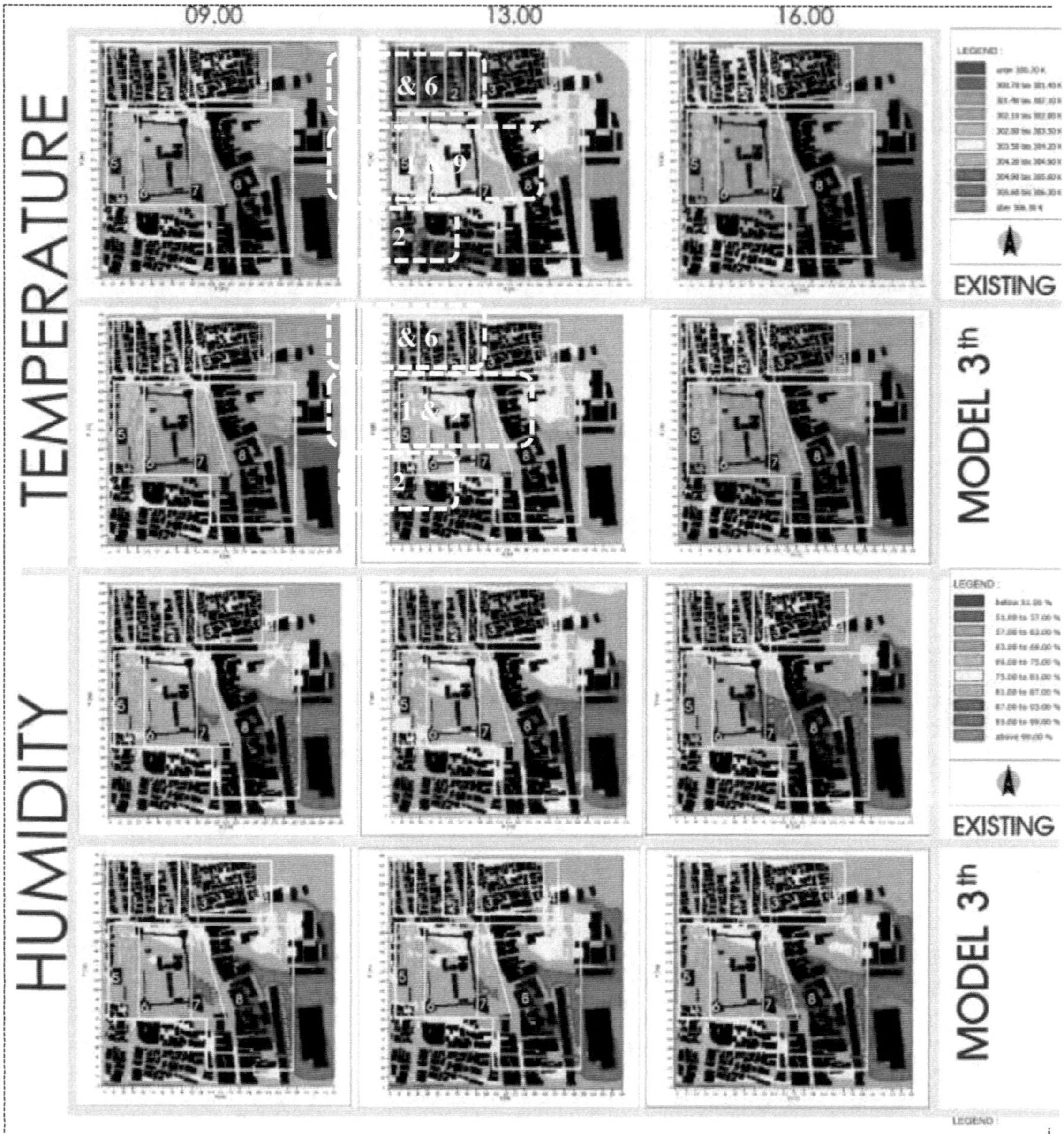

Figure 8. Result of Temperature and Humidity Simulation Existing and Model 3th

IOP Conf. Series: Earth and Environmental Science **213** (2018) 012018 doi:10.1088/1755-1315/213/1/012018

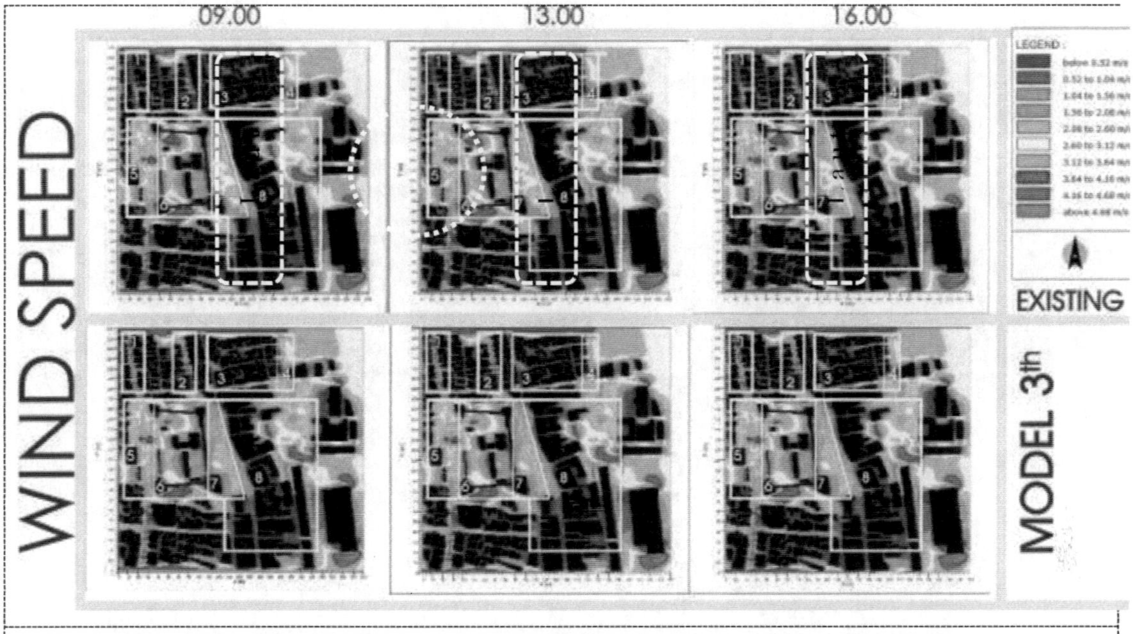

Figure 9. Result of Wind Speed Simulation Existing and Model 3th

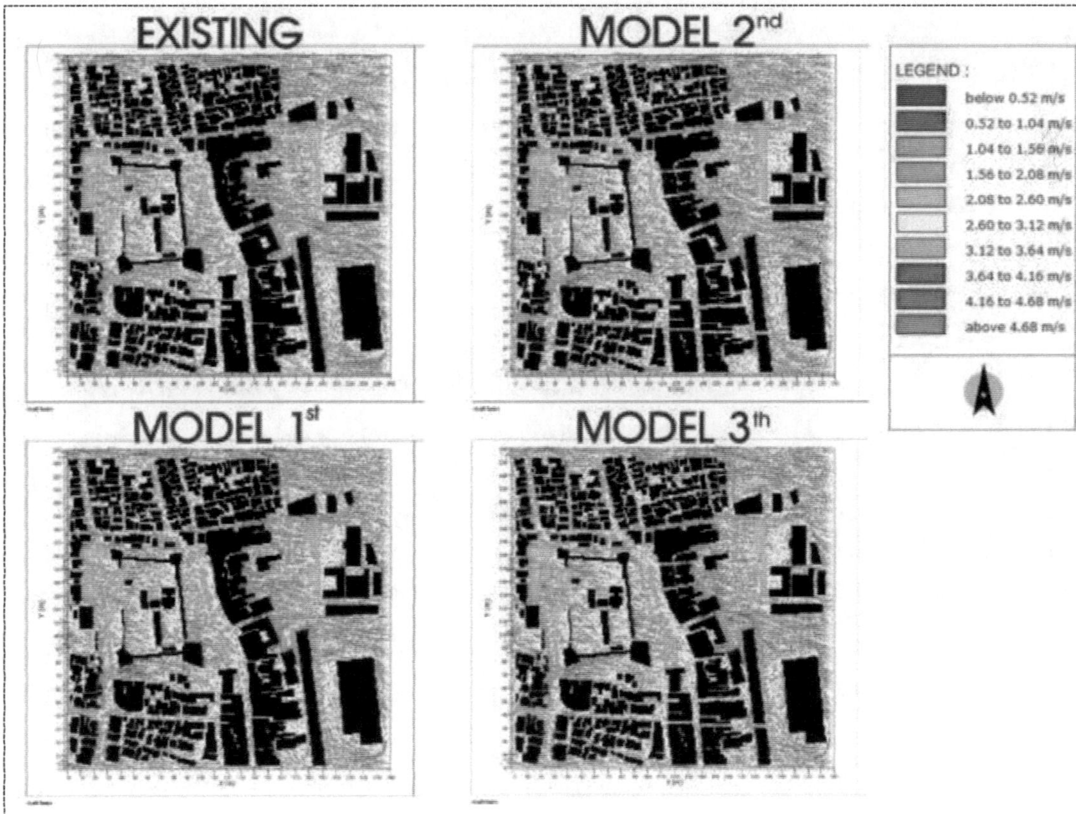

Figure 10. Wind Speed Existing, Model 1st, Model 2nd and Model 3th

IOP Conf. Series: Earth and Environmental Science **213** (2018) 012018 doi:10.1088/1755-1315/213/1/012018

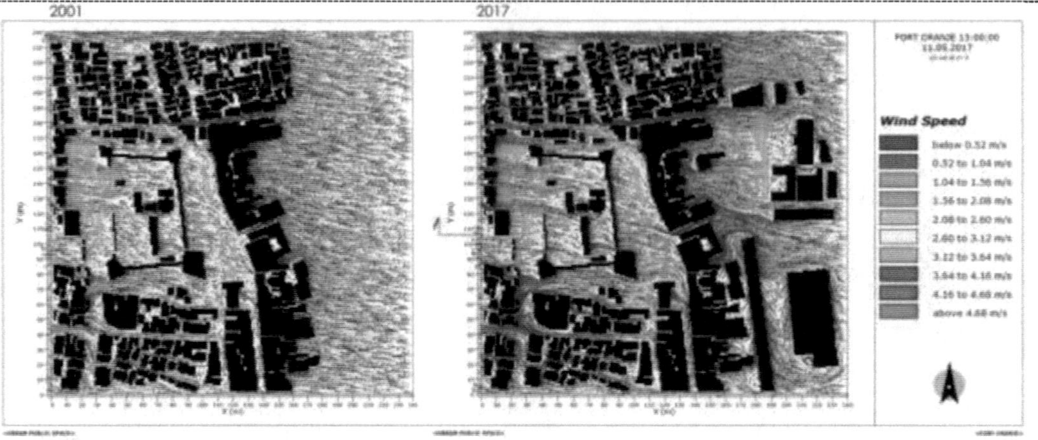

Figure 11. Air Movement of the Fort Oranje Area

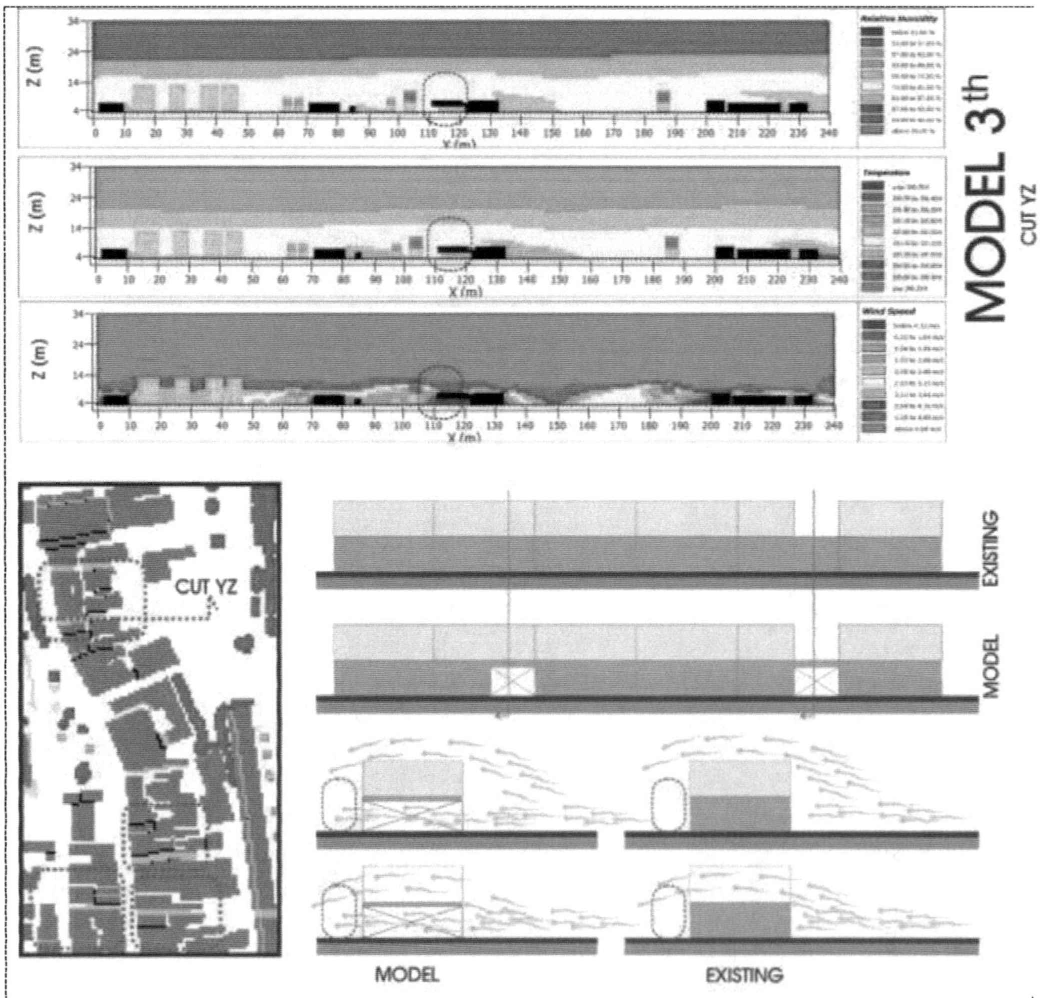

Figure 12. Wind Movement Existing, Model 1st, Model 2nd, and Model 3th

In conclusion, based on the simulation with the three models, the intervention on the building by adding the existing ventholes (see Figure 11, building intervention signs with red lines) could improve thermal area because wind movement was not hindered by the building.

4. Conclusion

The spatial configuration affected the thermal comfort of the Fort Oranje area, where the distance between buildings reduced the wind speed, increased temperature and caused low humidity. Excellent materials that increased and reflected heat from direct sunlight as well as the building height that blocked the wind movement dominated condition of the surface area. The presence of vegetation could block direct light from the sun to the surface of these materials also affect the condition. It could also reduce heat and maintain moisture in the urban spaces.

5. Acknowledgments

Authors would like to thanks to the Ministry of Research, Technology and Higher Education that funded the Inter-University Cooperation Research 2018 involving Khairun University and Gadjah Mada University. Also, authors would like to thank the survey team, especially architecture students batch 2015 and 2016 at Khairun University.

6. References

[1] Muhammad A, Marasabessy F, Kusumawanto A, and Nareswari A 2017 The effect of spatial configuration in the thermal area of *Fort Oranje* public space in ternate city *Journal of Architecture and Urbanism* **41(4)** pp 253-259.

[2] Li L, Zhou X, and Yang L 2017 The Analysis of Outdoor Thermal Comfort in Guangzhou Durring Summer, *Procedia Engineering* **205** pp 1996–2002.

[3] Mestayer P.G, and Anquetin, S 1994 Diffusion and Transport of Pollutants in Atmospheric Mesoscale Flow Fields *Atmos. Sci. Libr, Kluyer Academic Publishers.*

[4] Satwiko, P 2004. Building Physic 2 (Penerbit Andi: Yogyakarta)

[5] Younsi S, and Kharrat F 2016 Outdoor Thermal Comfort: Impact of The Geometry of an Urban Street Canyon in a Mediterranean Subtropical Climate – Case Study Tunis, Tunisia *Procedia Social and Behavioral Sciences* **216** pp 689-700.

[6] Kusumawanto A, and Astuti B 2014 Green Architecture on Cities Innovation, (Yogyakarta: UGM Press)

Reinvigorating local building envelope for translated modern building and changed environment

FX Teddy Badai Samodra[1], Irvansyah[1], Collinthia Erwindi[1]
[1]Department of Architecture, Institut Teknologi Sepuluh Nopember, Surabaya Indonesia

Corresponding e-mail: fxteddybs@arch.its.ac.id

Abstract. Different from the past, there is a current higher density of buildings resulted in less horizontal ventilation. However, a high-density environment still provides an opportunity to control solar radiation by providing shading, both at the site and on the building scale. Traditional buildings have climate as form modifier. Social and cultural aspects influence the building and as consequences the local material has also evolved to be part of the modern building in an urban area. This study reinvigorates local material in adapting to the changed environment in Surabaya as the lowland area and Malang as the highland area because of the availability of the survival building in the largest city. The field study was undertaken to identify and to evaluate the real conditions of a building envelope surface by daily measurements. An infrared thermometer was used for field measurements of the surface temperature with an additional check for the moisture content of the material with a moisture meter and solar radiation with a solar power meter for surroundings condition. Results of this study revealed that the lowland house had better material performance in controlling the climate by reducing the extreme condition than highland performance. The overhang and terrace as transitional space were able to reduce the number significantly, more than 1500 W/m^2 of daily solar radiation. In general, the problems of tropical houses were due to not only the frequency of receiving of solar radiation penetration but also the insufficient site shading to the building material.

Keywords: local material, modern building, surface material temperature, urban environment

1. Introduction

A tropical climate is characterized by relatively high temperatures and humidity. Physiological cooling is achieved from a lower wind speed to reduce heat accumulation and to increase the dehumidifying process. Cities, however, have limited space for accelerating wind because of the high density of buildings. This strategy can be harnessed to allow sustainable site planning of high-density environments, both building and human activities where solar insolation on facade simulation to assess urban typologies were very important [1-2]. The surfaces facing abundant solar intensity, East and West, have the highest insolation values [3-4]. Generally, these values increase at the upper levels of the surroundings because there are less shading from the neighboring blocks.

Traditional buildings have climate as form modifier; the building was influenced more by social and cultural aspects. On the other hand, the method for controlling the environment has been passed down through generations with improvements made from one generation to the next. In the tropical regions, the traditional building was designed in rural or open flat country terrain because of the low building density. Recently, these buildings have been constructed in various places, such as urban areas. A high-density environment controls solar radiation by providing shading even limited access for air acceleration [7]. In an urban area, because of the ecological and energy efficiency reasons, the local material from the traditional building has been adopted for the modern building in an urban area. However, the adaption of local building material to the variation of the environment and its orientation has not been found in the previous studies [1-6]. Therefore, this research reinvigorates local traditional material in adapting to the changed environment for today's building.

2. Methods

This study examined two topical urban areas, Surabaya (7.2°S, 112°E, 3 m above mean sea level) as the lowland area and Malang (7.8°S, 112°E, 575 m above mean sea level), as the highland area because of the availability of the survival local buildings in the largest city in Java Island, Indonesia. The field study was conducted to identify and to evaluate the real conditions of a building envelope surface by daily measurements on solar radiation, moisture content and their effect on surface material temperature (Table 1). The buildings were measured using the same methods: location, tools, and time. Each manual measurement was taken for approximately 3-5 minutes to have a consistent read value for every tool because the condition can change easily. The data collections had the free condition of a sensitive sensor with the height location, such as 1.5-1.7 m of height for outside/inside walls, 0 m of height for the floor and 2.5-3.5 m of height for gable (*Tutup Keyong*) and the roof parts (*Gajah and Emper*). Furthermore, Ecotect Analysis Program was used to verify and to estimate the results of field study. Figure 1 presents the profile of a building envelope in detail. The orientation rule was set with 0° as the South orientation and applied counter-clockwise. The traditional Javanese houses in both lowland, and highland regions were evaluated with the same object as thermal assessment. All materials used by a traditional building were recognized as lightweight materials.

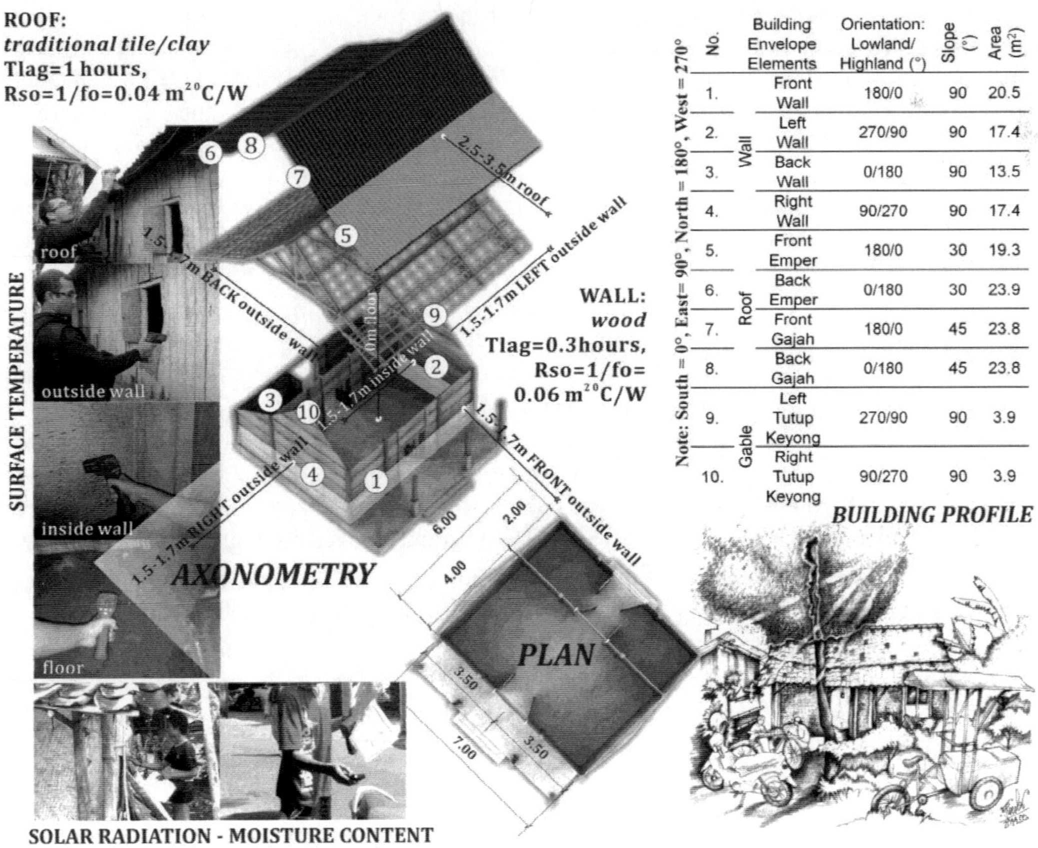

No.		Building Envelope Elements	Orientation: Lowland/ Highland (°)	Slope (°)	Area (m²)
1.	Wall	Front Wall	180/0	90	20.5
2.	Wall	Left Wall	270/90	90	17.4
3.	Wall	Back Wall	0/180	90	13.5
4.	Wall	Right Wall	90/270	90	17.4
5.	Roof	Front Emper	180/0	30	19.3
6.	Roof	Back Emper	0/180	30	23.9
7.	Roof	Front Gajah	180/0	45	23.8
8.	Roof	Back Gajah	0/180	45	23.8
9.	Gable	Left Tutup Keyong	270/90	90	3.9
10.	Gable	Right Tutup Keyong	90/270	90	3.9

Figure 1. Object measurement and profile.

Table 1. Field measurement tools.

Tools	Specification	Location	Time
Infra-red/No-touch Thermometer for Surface Temperature	Measurement range: -20°C~537°C Emissivity: 0.95 Field of view: 12:1 Resolution: 0.1°C Laser power: <1 mW	1.5-1.7 m of height for Outside/Inside Walls, 0 m of height for Floor and 2.5-3.5 m of height of Gable and Roof	7 days/ 3 times a Day: Morning (06:00-08:00), Afternoon (12:00-14:00), Evening (17:00-19:00)
Moisture Meter for Moisture Content	Range: 0% to 90%. Accuracy: ±0.1%		
Solar Power Meter for Solar Radiation	Range: 0-2000 W/m^2; Accuracy: ±10 W/m^2		

3. Discussion

3.1. Solar radiation effect on local environment and building

A building shading simulation using Ecotect Analysis was taken to identify the existing environment. Site shading gave a description of the percentage of surrounding buildings and tree shading (overshadowing) as well as the shading by their own elements (building shading). By running this program, the solar radiation distribution effects were also analyzed. A calculation of the building envelope performance, building density and its effects on the building temperature were calculated for all times of the day and followed by a surface temperature calculation, especially for un-shaded surfaces. In addition to Maestre *et al* [3], some factors are important in the solar control by shading such as different overhang and side fin typologies, latitudes of building region, orientations for shading device.

As shown in Figure 2, the percentage exposed area of the lowland site was taken in the range 06:00-18:00 and 1.50 m of height as a human standing reference. The total exposed area was indicated by the un-shaded surface of buildings and vegetation and it gave a chance for heat accumulation into the buildings. Traditional tropical house as a building object had less than 10% of the exposed area in the South, and half of the East and West. The Northside received solar radiation penetration of more than 50%, especially for the terrace of buildings because there is open space located in this area, without any protected wall. This condition was affected by the long distance of the surrounding buildings and trees as environmental shading compared to the high angle of the solar altitude and solar azimuth location in the North area.

With the same simulation time, the highland house shows slightly different performance. Different from Hii *et al* and Chow *et al* [1-2], the Western area had less than 10% exposed area, the South and North received half of them in more than 10% of the exposed area. The Eastern area had more than 50% exposed to the solar radiation because of the un-shaded environment by both the surrounding buildings and trees, even though the solar position was the same as that with the lowland house. The un-shaded or exposed areas of the neighborhoods' area had the problem of global irradiance of building envelopes. On the other hand, these open spaces still allowed wind acceleration into the buildings. In detail, Figure 3 shows the effects of environmental shading on the building envelope surface.

The roof elements of the lowland area received more than 1500 W/m^2 of solar radiation daily, whereas the building envelopes receive up to 800 W/m^2, such as the Eastern and Western Walls (Figure 3). All areas of Northern and Southern building envelopes were shaded, so the solar radiation received was less than 500 W/m^2. The 0.5 m length of the roof overhang in the East/West areas was too short to protect all parts of the wall. Therefore, only the shaded upper part (*Tutup Keyong*) received less than 500 W/m^2. The lower part of the terrace (<0.8 m), which appeared in the left of the Western Wall or the right of the Eastern Wall, received more than 100 W/m^2 because there was no wall as a protector. This condition caused a problem of heat accumulation in only sitting activities, while others activities might still have a possibility for thermal comfort. Furthermore, the terrace provided protection for openings from solar penetration.

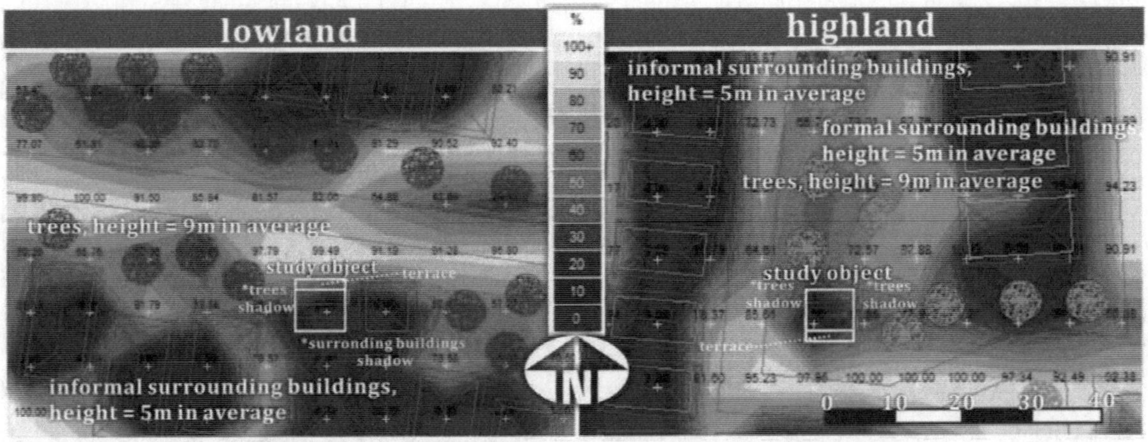

Figure 2. Site percentage exposed area in the range of 06:00-18:00.

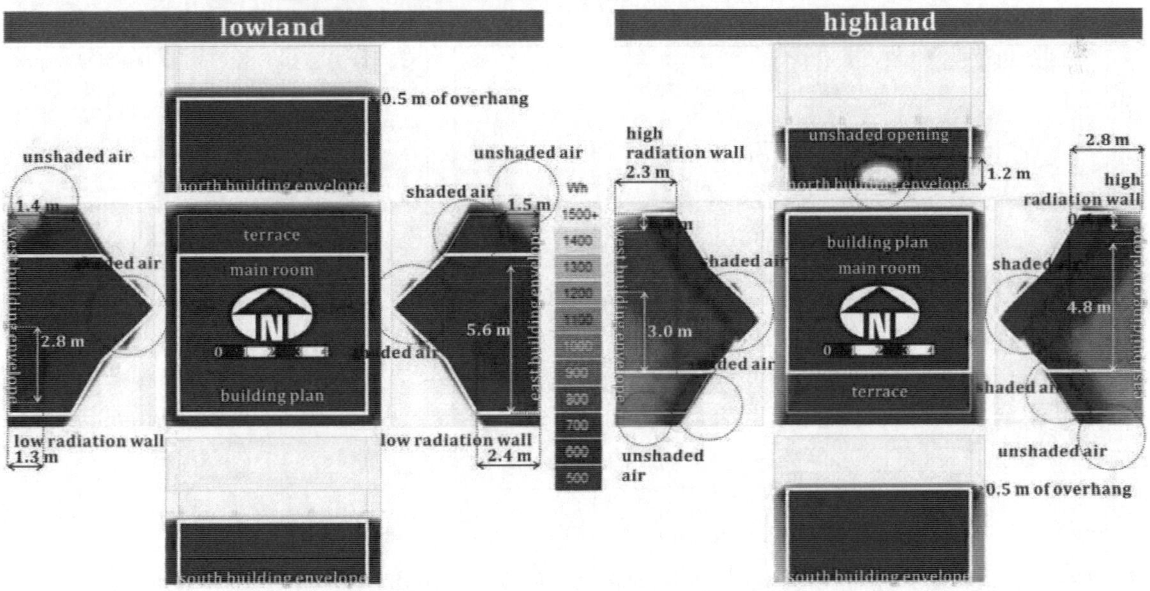

Figure 3. Average daily radiation in the range 06:00-18:00

With a higher solar radiation value, the building envelope in the highland environment had higher global irradiance. With the same condition of the roof element (>1500 W/m^2), the Northern Area received solar radiation from morning to evening. In addition, the back wall, North Wall, received high solar radiation, especially for the back opening, which had the same value as the roof, i.e. very hot. On the other hand, the wall still received less than 800 W/m^2. The front area, South, side wall and opening were protected by the terrace. The solar position in the Northern side gave an advantage. The East and West had similar characteristics to the lowland house (shaded upper part of the wall is colder/lower value). However, the value of global irradiance was higher, approximately 700-1000 W/m^2 for the un-shaded part of the wall.

3.2. Surface material temperature condition

The surface temperature measurements in lowland showed that on the first day, it was disturbed by rainy and cloudy conditions. As shown by Figure 4, the lowland lowest temperature of all building envelopes was influenced by accumulated clear conditions after rainy or cloudy periods in the first day of the measurement, which made the environment cooler before sunrise. On the other hand, the second day was the hottest day for the most of building envelopes because this day was the first clear day after the rainy condition on the first day, which allowed solar radiation to penetrate the building easily. The building roof normally had the highest surface temperature because its position is perpendicular to the sunlight. On Day-06 also, the Western outer wall had the highest surface temperature. This was influenced not only by the accumulated clear condition but also by the afternoon field measurement time (July), in which the azimuth of the solar position was 186° to 239°. Therefore, the South-Western elements, including the West wall received more solar radiation than others.

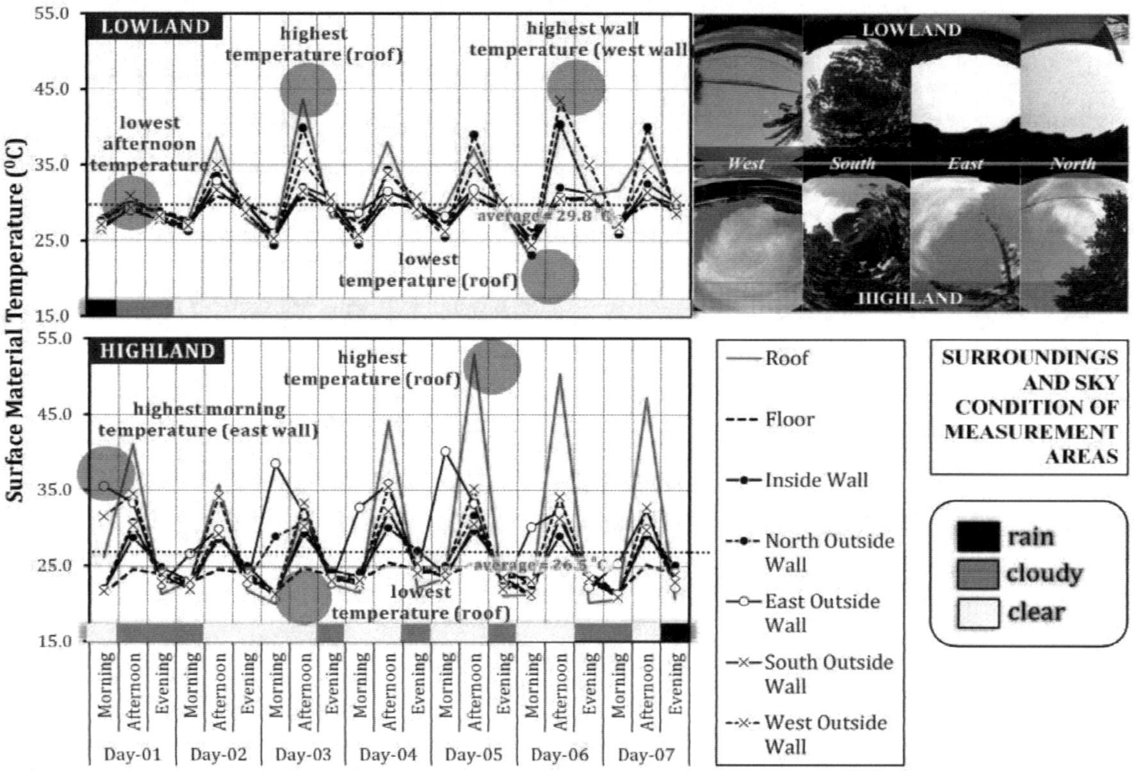

Figure 4. Measurement results of surface material temperature.

The direct solar radiation in the highland region, as indicated by the clear sky condition affected the highest temperature in the afternoon of most of the outside walls, except for the East Wall despite being under a clear sky condition (see Day-05), The highest temperature of East Wall occurred in the morning because clear sky allowed high solar radiation to penetrate the building envelope. In contrast, the cloudy sky (Day-06) affected the lowest temperature in the West Wall. For the same reason, in the morning of Day-03, the East Wall had a higher temperature than the roof at all times. In addition, the sky conditions changed from clear in the afternoon to cloudy evening. The surface temperature of the lowland and highland areas indicated that highland area had higher maximum temperature, but lower minimum temperature than the lowland area; the maximum was the roof in the afternoon of Day-05 with 53.0°C, while the minimum was the roof in the Day-03 with 19.9°C). The extreme condition of the highland region, particularly the roof element, could be affected by the higher global irradiance than the lowland region and the other elements of the building envelope. In general, the lowland building material had better material performance in controlling the climate by reducing the extreme condition than the highland has.

Based on field study above, to identify the local material, this study also results in the analysis of moisture content of critical building element, roof. Addition to Li *et al* [8], its relationships to temperature is the important factor which affect local material's thermal and moisture stress (TMS).

As presented in Figure 5, the higher surface temperature of material for all environment is affected by the lower moisture content, dealing with air temperature and relative humidity relationships. However, traditional clay material with Rso (surface resistance) is lower than its conductance (fo), it is not capable of resisting the peak high solar radiation in the daytime (see also Figure 1). As consequences, the surface material is very high when the moisture content is still in high percentage because of the high clay material porosity. Therefore, the role of wood walls for all orientation where the surface resistance is higher than the clay roof has. It is expected to have a capability of controlling the lower angle of solar radiation penetration. Potential roof material and its construction which is adjacent to accelerate the air movement has contradictory treatment in controlling solar radiation in the peak of the day [9-10].

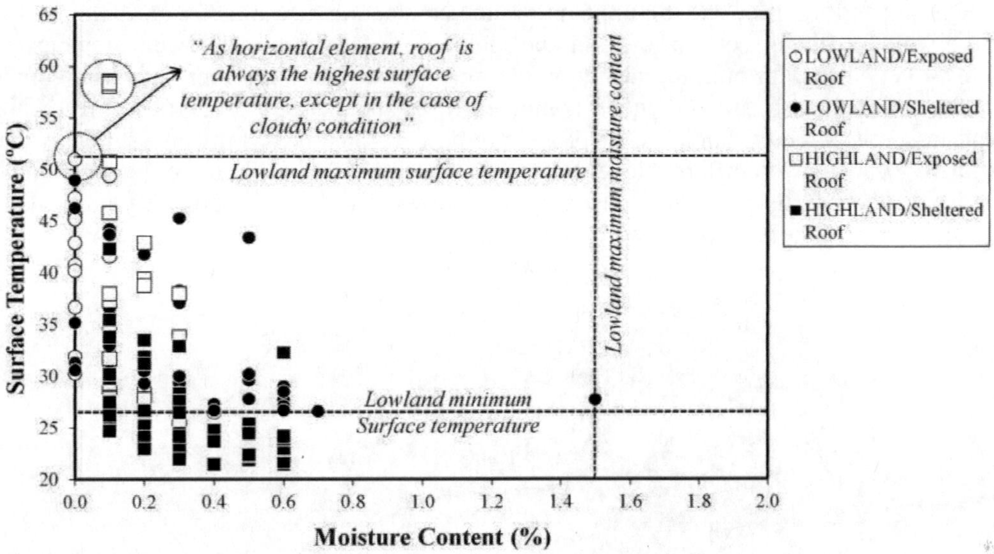

Figure 5. Moisture content and surface material temperature relationships of the roof.

3.3. Building envelope performance for translated modern building

The highland global irradiance was slightly higher than the lowland area. This was affected by the higher solar radiation, even though the time calculation was the same in the coldest month for every location. The higher altitude affects the higher value in receiving solar radiation throughout the year. An analysis of the wall and roof surface temperature (Ts) includes the outdoor temperature (To), solar radiation intensity (I), and the properties of material (α_{total}) in addition to surface conductance, as shown by Equation 1.

$$Ts = To + (I \bullet \alpha total / fo) \quad (1)$$

The high solar radiation throughout the year of tropical regions brings on the requirement for a shading design. On the other hand, some traditional buildings have been designed without a properly shaded wall. In this case, the roof is the critical element which has direct contact with solar radiation, except for providing environmental shading to the roof. The data show that the surface temperatures of a particular condition are the addition temperature of the outdoor air temperature for all building envelope elements. Therefore, the highest and lowest global irradiance also has the highest and lowest surface temperature, and if they are shaded, the surface temperature will be the same as the air temperature.

As shown in Figure 6, the lowland's lower roof slope part, front *Emper,* had a peak value of 51.6°C at 12:00, whereas the lowest element, the back wall, had a peak temperature of 30.2°C. The back wall as the South wall had surface temperatures that were the same as the air temperature all the days, because building envelopes were always shaded. Therefore, the peak temperature was not normally observed at noon but at 3pm (31°C). This also explains why the Western Wall, left wall, has a higher surface temperature than the Eastern (right wall). The Western wall had a peak temperature of 51.7°C at 4 pm. This even exceeded the front *Emper* value, which had the highest global irradiance. The Eastern wall had a peak temperature of 46.8°C at 08:00 am. In contrast to what happens with the lowland area, the highland building envelope still had back *Emper* as the highest value, even though it was only slightly different. This area had a peak temperature of 49.7°C at 1pm compared to 49.4°C at 4 pm for the Western Wall, whereas the Eastern Wall has a peak temperature of 43.1°C at 8 am.

Considering the modern building design, the detail of heat transfer is important to identify the capability of controlling changed environment from rural in the past for traditional building and today's environment for newer one. Figure 6 shows the conduction heat transfer of the highest surface temperature, wood wall, for both in lowland and highland. In addition to relevant previous findings [11-12], low-energy modern building design could be generated through local material composites. The data were taken from measurement combined with its thermal properties. Meanwhile, the calculation was conducted by Matlab-GUI programming [13]. The heat transfer resulted in the same pattern of both locations with different level of temperature in 1080 s (0.3 hours of wood's time lag). It indicates that in about 0.01 m thickness (1 cm) only, the wood wall has a capability in insulating the heat from outdoor depended on its outdoor temperature.

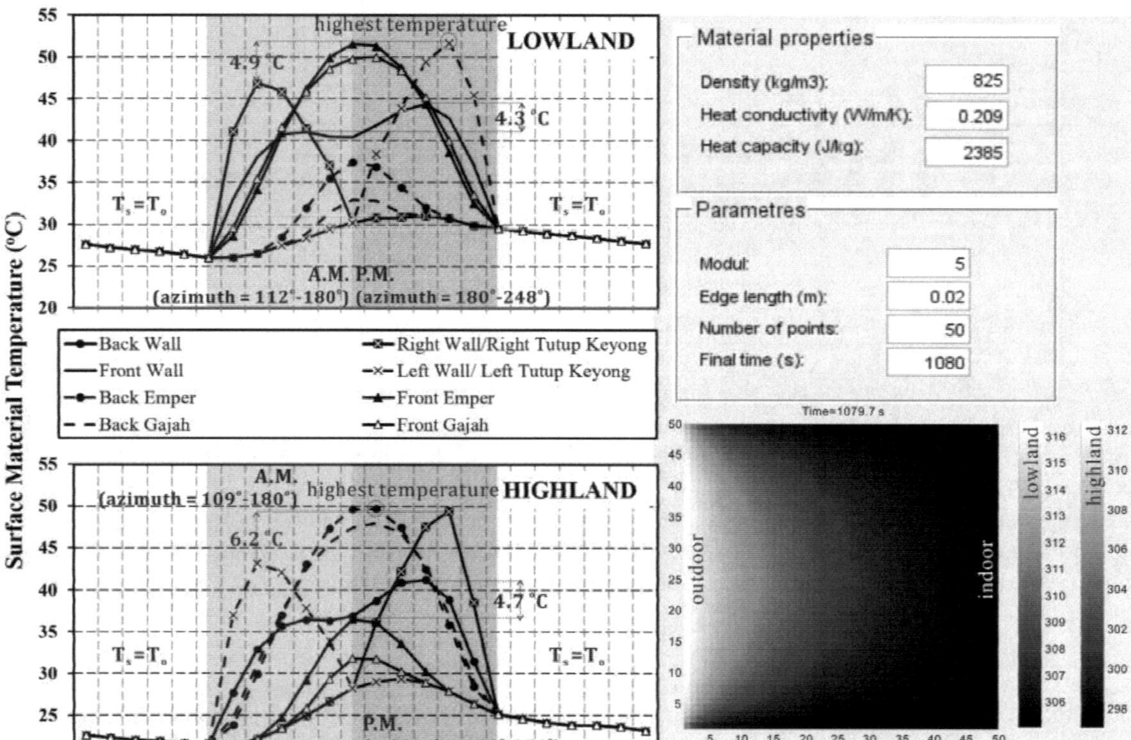

Figure 6. Potential material for future building.

4. Conclusion

This paper investigated the environmental and building to control the solar penetration into building using field measurement analysis, simulation program, and calculations for the traditional tropical building. In addition to existing findings [1,4], the surfaces facing East and West had the highest solar insolation values. However, in this research the problems were due to not only the frequency of penetration of solar radiation, but also the limited of neighborhood (site) shading, both by building and tree. The high solar radiation, 500 W/m^2, was still reached by the only short size of roof overhang (lower than 0.5 m). The terrace was one of the worthy solutions for both lowland and highland houses in reducing solar radiation into building. The orientation of the building was one of the critical factors for the walls in receiving solar radiation, daily or annually even though it was the low-rise building which had a roof as the largest area. Furthermore, the traditional house had sloped roof and separated with different orientation. Thus, the smaller angle of the roof in the Northside had a higher surface temperature, both lowland, and highland.

It recommends that the exposed aperture in highland needs protection from solar radiation with landscape, vegetation or site shading, and building shading by overhang or terrace as transitional space. This strategy will be effective to reduce of the over 1500 W/m^2 of daily solar radiation. By this evaluation, it could deliberately translate modern from traditional; from local material for global requirement have potential reinvigorated richness of culture for natural respect by the ecological and energy efficiency reasons.

In the near future, this study will extend to review the potential thermal properties integrated with the other critical urban issues, noise, and the adaptation for future building design. Furthermore, it will be integrated with controlling both noise, and thermal environment to give guideline recommendation.

5. Acknowledgments

This research is part of *Penelitian Dasar Unggulan Perguruan Tinggi* (Excellent Primary Research of Higher Education) No. 882/PKS/ITS/2018. The authors gratefully acknowledge this financial and technical support.

6. References

[1] Hii D J C, Heng C K, Malone-Lee L C, Zhang J, Ibrahim N, Huang Y C, and Janssen P 2011 *12th Conference of International Building Performance Simulation Association* 2595–2602
[2] Maestre I R, Blázquez J L F, Gallero F J G, and Cubillas, P R 2015 *Energy and Buildings* **101** 144–152.
[3] Chow T T, Chan A L S, Fong K F, and Lin Z 2005 *Applied Thermal Engineering* **25** 1908–1921
[4] Samodra F X T B 2018 *IOP Conf. Ser.: Earth Environ. Sci.* **105** 012080
[5] Vox G, Blanco I, and Schettini E 2018 *Building and Environment* **129** 154–166
[6] Alders E E 2017 *Architectural Science Review* **60** 150–166
[7] Samodra F X T B, Defiana I, and Setyawan W 2018 *IOP Conf. Ser.: Earth Environ. Sci.* **126** 012139
[8] Li N, Long J, Su L, Wang L, and Zhong S 2012 *J. Cent. South Univ.* **19** 600–608
[9] Grant E J, Black K A, and Werre S R 2017 *Architectural Science Review* **60** 137–144
[10] Wang L and Ge H 2017 *Building and Environment* **123** 363–377
[11] Shah D U, Bock M C D, Mulligan H, and Ramage M H 2016 *J Mater Sci* **51** 2991–3002
[12] Bui Q B, Grillet A C, and Tran H D 2017 *Sustainability* **9** 1444
[13] Gibala D 2012 *GUI 2D heat transfer*

Empowering traditional house for future noise control based on sound frequency propagation

FX Teddy Badai Samodra [1], Irvansyah[1], Collinthia Erwindi[1]
[1]Department of Architecture, Institut Teknologi Sepuluh Nopember, Surabaya Indonesia

Corresponding e-mail: fxteddybs@arch.its.ac.id, fxteddybs@gmail.com

Abstract. Geographical altitude represents tropical environment in Indonesia where the different wind acceleration effects the noise control. The traditional house was designed in the past when the noise was limited in the rural area, and it is changed into the noisy environment in urban space and will potentially be critical in the future. The aim of this study is to evaluate the existing traditional building noise and to upgrade its performance for empowering the future adaptation. The real-time analyzer for sound level meter was fixed in occupant reference height in order to examine the frequency bands effect. MATLAB coding is also conducted in order to evaluate the noise propagation performance. The traditional Javanese houses are the representative case of Indonesia tropical building in both lowland and highland regions. The results are highlighted that by distance and wood or bamboo wall, the sound pressure level is reduced to 20 dB and 15 dB for lowland and highland, respectively. For higher measured frequencies, more than 1 kHz, the road noise gives the major influence to the noise propagation. Meanwhile, the lowest frequency transmission, 125 Hz, has the highest consequence on the noise propagation in the road, but less in the room. It will be the consideration on empowering the local material by its coefficient selection in the future design. The traditional wall has low composite transmission loss; it can reduce insufficient rate.

Keywords: noise control, sound frequency, traditional house, urban space

1. Introduction

In urban environment, noise propagation is determined by the atmospheric absorption, diffraction and ground effect [1]. Those issues are important aside from the role of sound pressure level (SPL) on noise source and distance from noise to receiver which is still the key features. Furthermore, beside the receiver point, the terrain setting such as urban, suburban, rural and the open flat area for site barrier are the essential factors [2]. Currently, the noise environment of traditional tropical houses is affected by the growth of higher human population, building, and traffic density. Perceived noise largely disturbs the urban annoyance. Additionally, non-acoustical aspects have a significant part in the noise annoyance estimation as the response indicator [3].

A tropical environment is represented by the geographical altitude, lowland and highland. Actually, the noise propagation is not affected directly by altitude but by the air movement difference. Hence, the air flow design of the site and building may potentially affect the noise control [4-5]. The noise propagation is not affected directly by geographical setting. The site, buildings differences, and the occupant lifestyle may contribute in the noise controller. The urban thermal map indicates that the urban noise situation is also because the transportation and human movement contain the largest module of the heat source, especially in the center of the urban zones as the most crowded activities.

Compared to the similar previous studies, most of those similar findings have also analyzed traditional building performance related to the noise control. However, there were no specific results for the requirements of environment change from its natural environment to today's adaptation and potential future development [6-9]. Furthermore, most of them have also found the occupants' adaptation results in the recent environment condition [10-13]. Therefore, this study evaluates the existing traditional building noise and empowers its performance for the detailed future adaptation through potential sound frequency propagation analysis.

2. Methods

This research is located in Surabaya (District of Mulyorejo, altitude 3 m and latitude 7.2 S) as lowland representative and Malang (District of Purwantoro, altitude 475 m and latitude 7.8 S) as highland one. All locations refer to the availability of traditional houses (See Figure 1.). Recently, the both location are in an urban environment, different from the traditional houses in the past which were designed as the optimal model for a lower noise source in the rural environments. The building model for this study has been found in most Javanese houses where the Kampung type as the most used dwelling of traditional people. Actually, the building material and orientation have had many variations, but for the present time analysis, the study evaluates what is the most generally found in urban environment. The traditional Javanese houses have wood which has been used for walls (the transmission loss, TLwood =18 dB, see Figure 2).

This research conducts and sets the traditional building outdoor as noise source position and indoor as occupant receiver noise setting for on-site measurement. The traditional Javanese houses in both lowland and highland areas are 6x7 m^2 in average based on the initial field observation. The sound level meter was installed for outdoor and indoor evaluation at the equivalent period in order to see the direct result of the noise from main source to the receiver (Table 1). The sample of exposure noise was also taken at the highest traffic density in the weekday-daytime. By this detail measurement, the most critical sound frequency could be detected as a basic idea of future prediction design. Some recommendation for translated building from the traditional to the modern one will also be suggested.

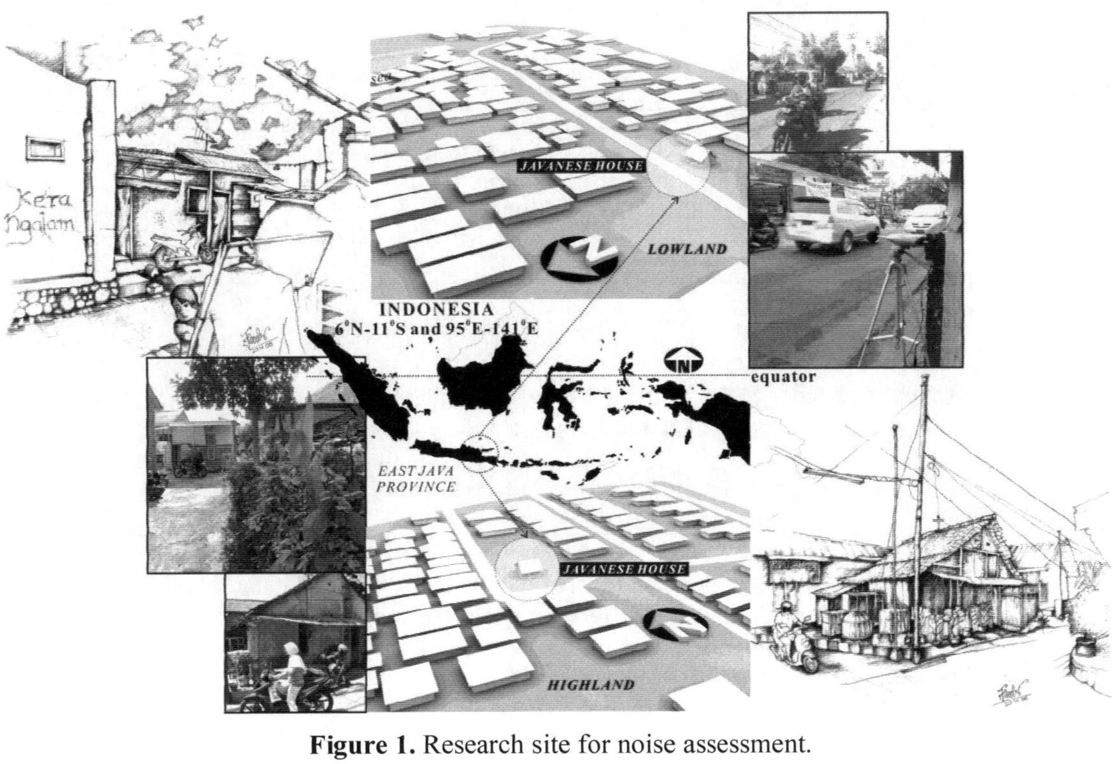

Figure 1. Research site for noise assessment.

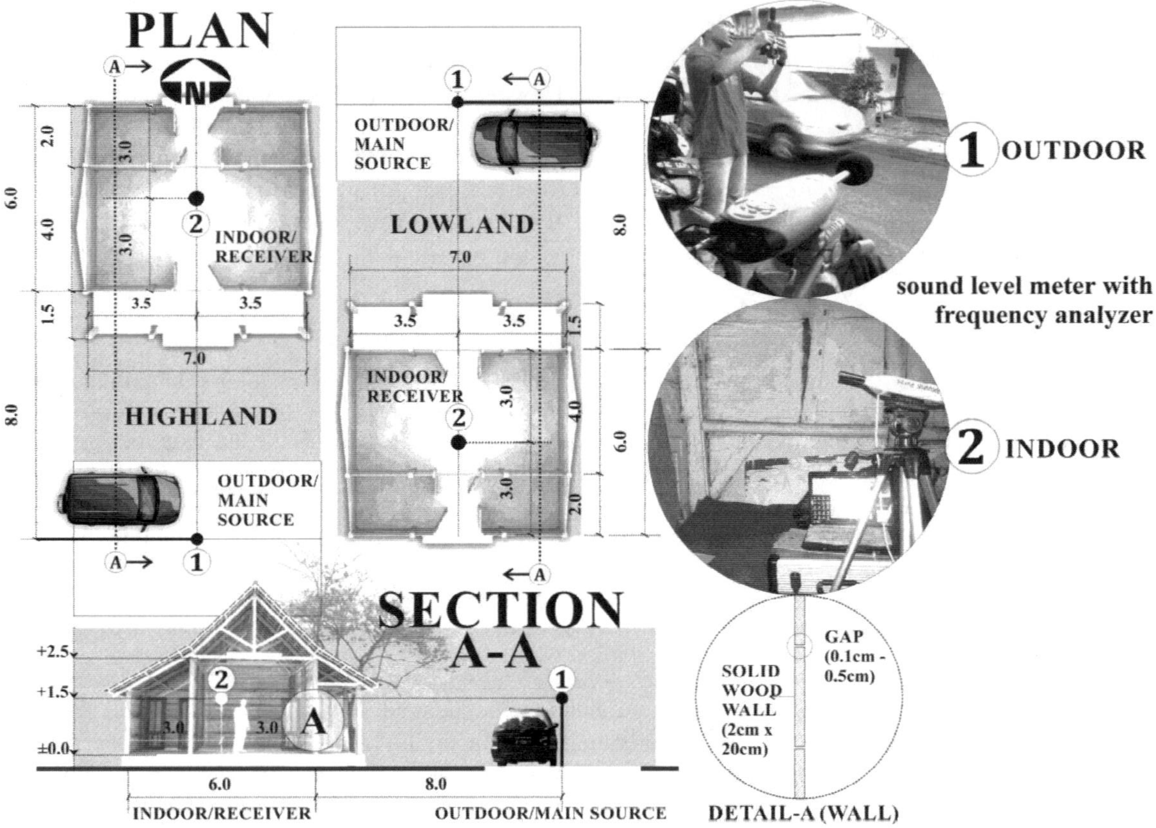

Figure 2. Traditional building profile.

Table 1. Noise measurement tool.

Tools	Range	Location	Time
Sound Level Meter for measuring Sound Pressure Level in Frequency Bands	- A Weighting - Single dynamic range: 30-137 dB(A) - Real-time frequency analysis (12.5 Hz to 20 kHz)	Main Noise Source (Road/Outdoor) and Indoor: 0.6 m to 1.5 m as reference height of occupant's sitting/standing	7 days: Morning (06:00-08:00), Afternoon (12:00-14:00), Evening (17:00-19:00), 24 hours (Weekday/Weekend), and sample for LAeq,1h in weekday-daytime

3. Discussion

3.1. Noise propagation and building as barrier

The building orientations determine the tropical house zoning in receiving road noise (Figure 3). With road as the main noise source, the distance and building wall, the sound pressure level is attenuated in 20 dB for lowland and 15 dB for highland, respectively. Though the traditional wall has low composite in transmission loss, it can significantly decrease the noise. By average distance, 12 m, and the building itself as a barrier, the noise is controlled in about 8 dBA. It may be reduced by the building barrier with closer distance from its source to the building. With the same distance, the effect of air movement for closer propagation is not significant [1]. Both sides of the building receive the same sound pressure level. Meanwhile, the difference occurs at a higher noise source at the lowland environment which has higher noise compared to the highland. In general, this phenomenon applies in all sides of traditional building. Because of blocking effect of building as a barrier, the back sides for both lowland and highland obtain the similar results.

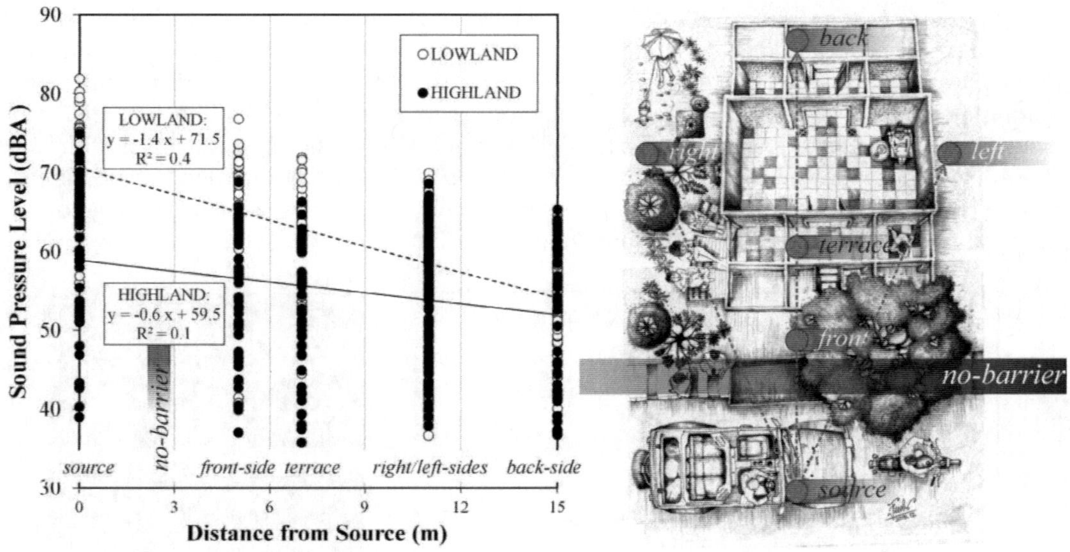

Figure 3. Distance effect on noise propagation.

The noise propagation above prevails for the most sensitive sound frequency for human, 1 kHz. This level is included mid long wave sound which effects not only on noise criteria and its comfort in acoustics environment but also the clarity of human speaking, sound intelligibility. Therefore, the field study with extracted sound frequency by sound analyzer will be discussed intensively.

3.2. Road noise effect on receiver

The high noise disturbance in the lowland site is mainly indicated by the busy traffic movement. The quite close distance to the receiver, approximately 8 m, is the critical location for survival traditional building (see illustration in Figure 4). In a week test and field measurement, the outdoor lowland reaches the peak noise in the early day (morning traffic activities time). The high SPL (around 70.0 dBA) was identified and resulted by the human actions and it is higher than in the afternoon and in the evening which is respectively 66.1 dBA and 67.7 dBA. The high opening people activities in the morning in the urban outdoor environment is the main indicator of this situation. The condition of noise source location in the building outdoor has a direct result to the receiver area where the SPL in 58.9 dBA for the indoor was also perceived in the morning and was higher than other times. The estimated noise in 12 dBA reduction from its source to the receiver is well-attenuated by its natural distance and building envelope.

This study evaluates the existence of tropical traditional building in urban environment using critical WHO recommendation as international standard (35 dB for daytime and 30 dB for night-time) compared to local, Indonesian standard (55 dB for all time) as conducted in the previous studies [5]. From Figure 5, the quadrants adapted to the referred standards can be designated as follows: The A quadrant is the most number of the results of lowland for night-time, highland's daytime and night-time and the B quadrant is the most dominant and occured in lowland daytime. Meanwhile, the C quadrant consists of the highland night-time results, but the D quadrant has no results. The leading results on A and B quadrants are the complications in accomplishment for both WHO and Indonesian standards. The dominant indication results in discomfort noise for all time both for lowland and highland buildings reached out all standards even though there are found about 50% which comply with the local standard. Actually, the noise limits by local standard in Indonesia do not provide in detail for indoor or outdoor and daytime or night-time noise condition. It sets the standard based on local allocation for the building purpose.

All conditions above have the simulation setting for the opened door (see Figure 4). Resulted by Matlab calculation, the lowland building at daytime is considered as the most critical condition even it has less strict noise criteria compared to the night-time requirement. The role of traditional materials has no significant impact to control the environmental noise because of the lightweight and typical porosity of the building material. The road noise source in highland area, particularly for the night-time, almost complies with the Indonesian standard. As the result, the building as the receiver experiences quiet situation, a potential result for WHO limits. Meanwhile, the noise level of the source and the propagation of lowland at night-time compared to highland day-time tend to be similar. In general, the typical single frequency in average as indicated by this section is the representation of the critical sound frequency deliberated in the following discussion.

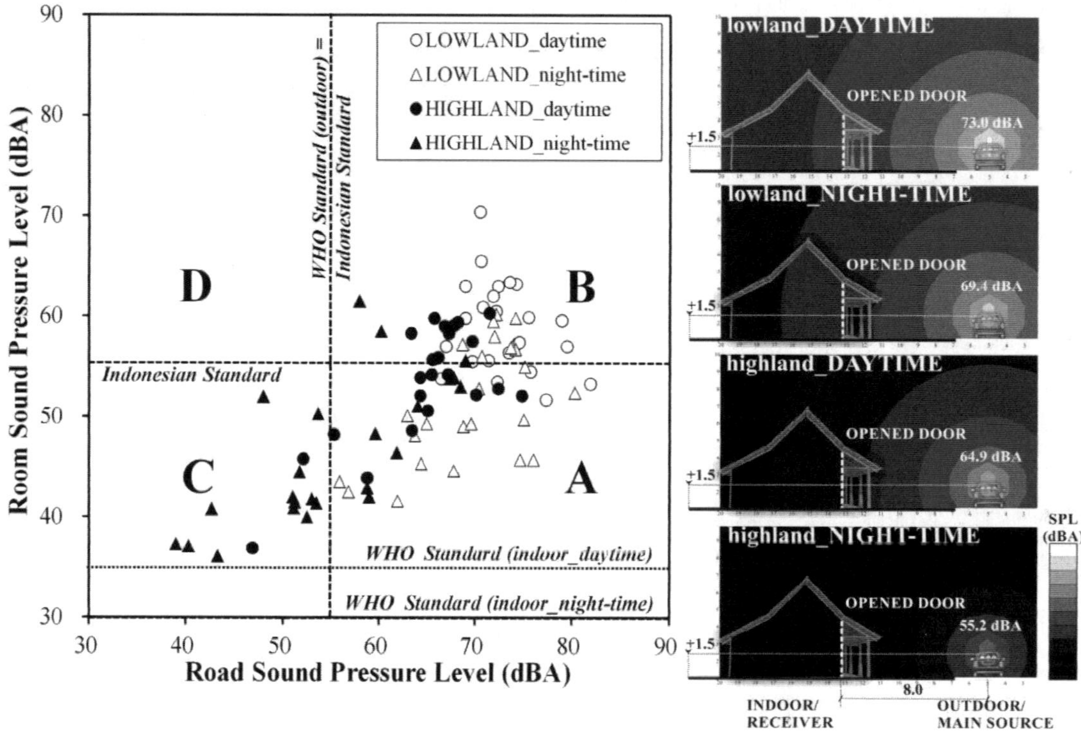

Figure 4. Effect of road noise on the receiver.

3.3. Sound frequency analysis

This section discusses the sound frequency taking on the critical and in the average method explained before. In addition to Prato *et al* [14], the investigation of low-frequency noise has become very important due to the impact sound transmission on wood-frame of traditional building. The receiver represented by the room obtains a little noise oscillation than its source after being attenuated by its distance and building wall. Resulted from the peak road noise in the daytime, the higher frequencies (more than the most sensitive band, 1000 Hz) of the noise source give the highest impact to the noise transmission received by the building (Figure 5). The 125 Hz frequency as the lowest band has less effect to the room disturbance while the highest consequence on the noise is experienced by the source area itself.

Again, analyzed through Indonesian and WHO noise standard, the most of sound frequency of the room meet to Indonesian standard while a problem is still found for the WHO has. That condition is valid for lowland while the highland building's room obtains dominantly for all the requirements. The comfort in noise adaptation of the highland building is resulted in about 25% where the higher frequencies from the noise source have a little effect perceived by the receiver. Likewise, the 4 kHz as highest frequency in this context has no significant result in both condition of the road and room noise propagation.

IOP Conf. Series: Earth and Environmental Science **213** (2018) 012020 doi:10.1088/1755-1315/213/1/012020

The analysis for peak noise sample is conducted in an hour field measurement (LAeq,1h) of lowland and highland road noise. It considers the phenomena in the peak density of the urban traffic activities at daytime (Figure 6). In general, both tropical representative locations have same noise propagation pattern in transferred sound from source to the receiver. Although the highland building receives a lower annoyance, the unpredictable grade resulted in several frequency bands affects the other disturbance because it produces fluctuated noise scale.

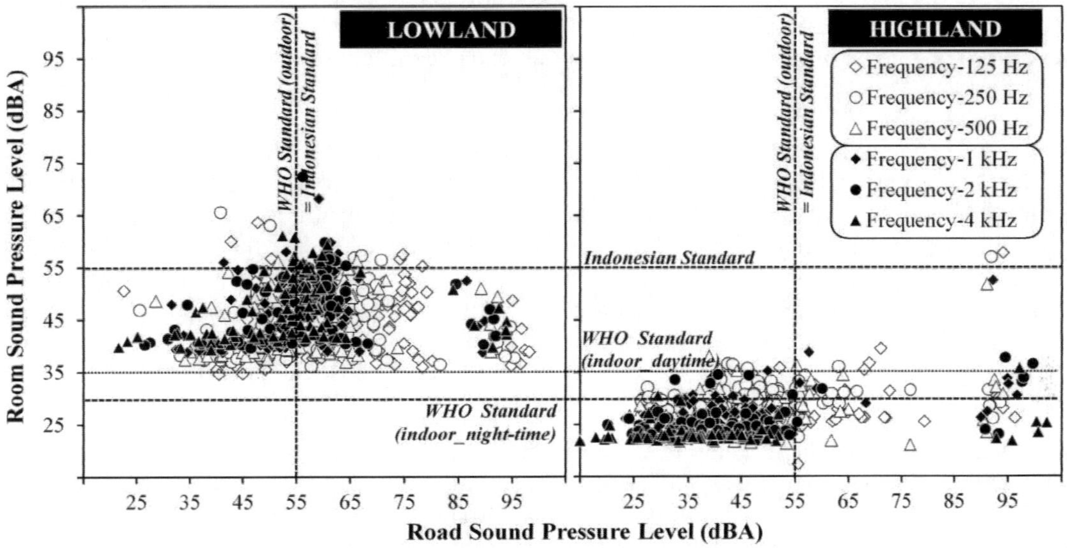

Figure 5. Effect of frequency on the receiver by source noise.

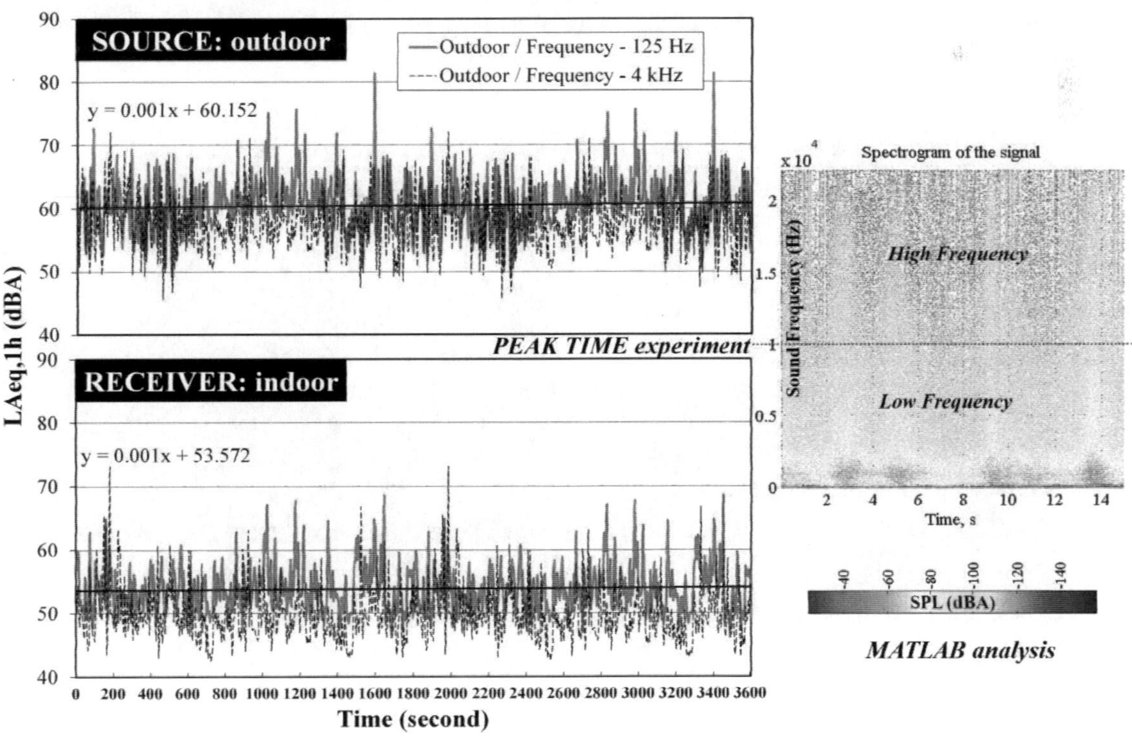

Figure 6. Sample test in noisiest time of lowland.

In general, the mid-frequency, 1 kHz, is indicated as the most penetrating frequency for human living. In this noise frequency, the lowland building obtains 10 dBA reductions similar to 14 dBA in the highland. The building aperture has opened schedule in the lowland more often than in the highland because as the strategy for occupants' adjustment, the ventilation is selected as the critical solution in hot environment. In fact, as indicated by the spectrogram of the signal produced by Matlab coding, the SPL of high frequency is higher than the lower. However, the high frequency of sound propagation is produced in shorter period and occurred in inconsistent time. This result will be valuable for predicting future noise control. With a higher wavelength of lower sound frequency, providing a larger absorber or a resonator noise control will also be very important and urgent for a denser urban environment.

3.4. Future noise control

To extend the previous research integrated with attention to energy efficiency [15], this study also intends to achieve a natural control for the future hybrid method as a compromise strategy for a better living environment. As illustrated in Figure 8, the future noise control is projected by two methods such as directing closed door and providing site barrier for opened door schedule. The empowered traditional house for future design for the most critical, lowland at daytime, in fact, is achieved by a closed-door operation. With 18 dBA transmission loss as the average coefficient of the wood wall, less than 35 dBA of receiver environment is obtained. The similar circumstance is in effect of the night-time SPL which is not more than 30 dBA. The analysis on acoustical properties is very urgent considered for the higher noise source in the future.

Based on the occupancy schedule, the possibility of the opened door for circulation activities or for providing natural ventilation of energy efficiency is reasonable. Integrated thermal and noise control is one of the cause factors of the adaptation for future urban design in the tropical area. Therefore, empowering traditional building should respect to social and economic dimension by designing site barrier. The overhang is created to accelerate air movement and to increase wind speed for physiological cooling. It is expected that the air flow for this purpose will not be interfered by airborne noise transfer, and the dealing heat with noise control is effective with this method. It is proven by the similar receiver performance in responding the noise source by opening the door as the ventilation aperture.

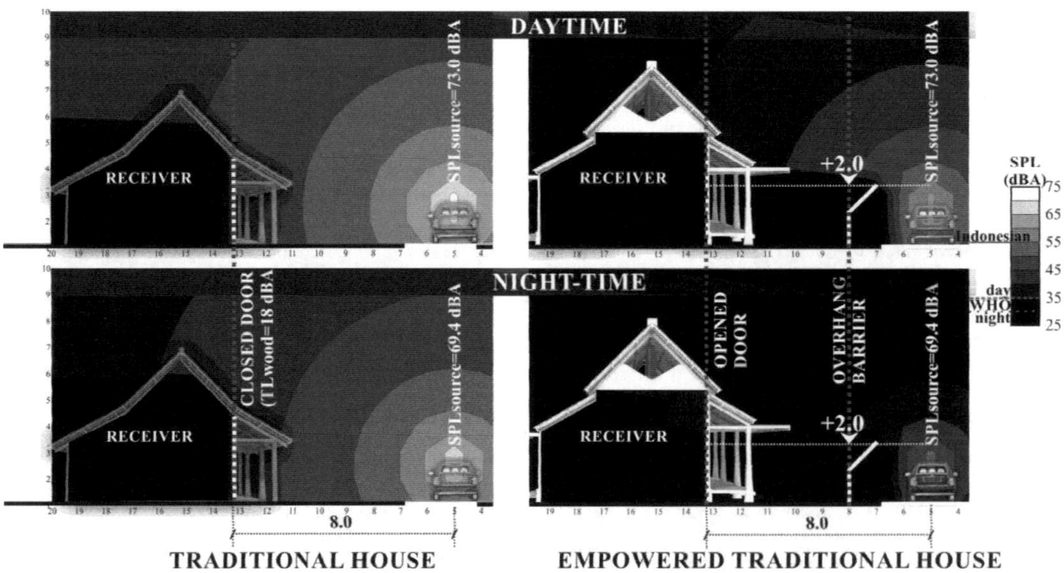

Figure 7. Empowered traditional house for future design.

In the future, the potential traditional house is also developed by accumulating the propagation distance by raising the existing floor and roof. Not only increasing wind speed for controlling tropical climate but also by the extended distance, a reduction of sound pressure level especially for the lower frequency annoyance could potential be attenuated. The noise abatement by constructing a further distance will be a little opportunity to follow the rapidity of urban growth. Therefore, the sustained idea of delivering the terrace as a transitional area could be a smart decision for the upcoming proposal of the future urban designer. The thermal environment will also be the similar challenge in urban density to the noise attenuation. The native design of traditional terrace has joined the social and physical attention absolutely (see also Figure 3).

4. Conclusion

This study finds there are about 25% of the highland tropical buildings reaching their comfort condition. Most of them area resulted by situation in shorter distance, the other indicators are the direct effect of noise fluctuation annoyance, and lack of building barrier. In low frequencies, the traffic area as main noise source has a high contribution in noise propagation perceived by the building. Because of the less wavelength and occurred in the shorter and unstable period, the highest frequency has the lowest effect on the noise propagation even having higher SPL. Furthermore, the room as receiver experiences a little fluctuations compared to its source because of the distance role and building envelope barrier, especially for the front side of the building.

The availability of transition area, terrace, could be considered as a multipurpose design idea. In an urban environment, this cultural richness provides social and physical dimension advantages. Not only for gathering activities, the terrace is also empowered as solar radiation, air temperature through wind acceleration, and outdoor glare transition. Furthermore, in this context, the building envelope as a barrier is reinforced by terrace through creating the extended distance. Designing site barrier will be the manifestation of the concept for empowering traditional design dealing with the respect to social and economic dimension. The overhang barrier integrates wind acceleration for physiological cooling through upgrading wind speed to restore thermal comfort. The efficient system as an economical intention of natural high air flow is deliberated not interfered by airborne noise transfer.

For future works, this study will cover low-frequency absorber developed from the traditional material. Prospective natural treatment of traditional house design in the future is predicted of having capacity in environmental adaptation and resilient design. By significant empowerment of frequency coefficient of material, the probable noise control from any kind and high source annoyance will be translated definitely.

5. Acknowledgments

This research is a part of Excellent Primary Research of Higher Education, *Penelitian Dasar Unggulan Perguruan Tinggi* No. 882/PKS/ITS/2018. The authors gratefully acknowledge this financial and technical support.

6. References

[1] Sakamoto S 2015 *Acoust. Sci. & Tech.* **36** 49–108
[2] Renterghem T V, Forssen J, Attenborough K, Jean P, Defrance J, Hornikx M, and Kang J 2015 *Applied Acoustics* **92** 86–101
[3] Laszlo H E, McRobie E S, Stansfeld S A, and Hansel A L 2012 *Science of the Total Environment* **435–436** 551–562
[4] Samodra F X T B 2018 *IOP Conf. Ser.: Earth Environ. Sci.* **126** 012034
[5] Samodra F X T B 2017 *Journal of Architecture and Urbanism* **41** 305–315
[6] Iannace G 2017 *Building Acoustics* **24** 101–113
[7] Tourle P 2017 *International Journal of Heritage Studies* **23** 234–247
[8] Caniato M, Bettarello F, Fausti P, Ferluga A, Marsich L, and Schmid C 2017 *Building and Environment* **120** 110–122
[9] Queirós D, Rodrigue R C, and Pereira N 2016 *Building Acoustics* **23** 181–191
[10] Oh Y K 2014 *Journal of Asian Architecture and Building Engineering* **13** 445–451
[11] Ljunggren F, Simmons C, and Hagberg K 2014 *Applied Acoustics* **85** 57–68
[12] Felcyn J, Preis A, Kokowski P, and Gałuszka M 2018 *Transportation Research Part D* **63** 72–120

[13] Koprowska K, Laszkiewicz E, Kronenberg J, and Marcińczak S 2018 *Urban Forestry & Urban Greening* **31** 93–102
[14] Prato A, Schiavi A, and Ruatta A 2016 *Building Acoustics* **23** 110-119
[15] Samodra F X T B 2018 *International Journal of Technology* **9** 246–256

The Transformation of Gegerkalong Girang Area, Bandung City: amid Educative and Religious Areas

Asep Yudi Permana[1], Indah Susanti[1], Karto Wijaya[2]
[1]Universitas Pendidikan Indonesia, Bandung, Indonesia,
[2]Universitas Kebangsaan, Bandung, Indonesia

Corresponding e-mail: yudi.permana@upi.edu

Abstract. The development of urban area is a changing process of a situational urban area into a new situational urban area in different periods of time. Those situational changes also happen in Gegerkalong Girang area, Bandung City. At the beginning of its development, this area was well known as the educational center that supported educational activities based on the activity of the oldest university and settlement in north Bandung. However, since *Pesantren* Daarut Tauhid (DT) was built and developed in 1990, this area became phenomenal. The presence of *Pesantren* DT has changed the physical formation of the environment around the area, which was essentially a place for thousands of IKIP Bandung college students to reside since the 1950s. This research also concerns on how the concepts, behaviors, and values of the area color the morphological architecture in location.

Results of the study showed that the approach of economic development and welfare for the people that was initiated by Aa Gym as the founder of *Pesantren* DT had a significant effect there. This change appears on the functions and facades of the building. The functions and the facades of the buildings, which originally had residential functions, turned into commercial functions. Thus, it affects the shapes of architecture and the environment. There are dualistic functions of the region, namely (1) educational area, and religious area; (2) college students' lodges and students' dormitories. These two functions complement each other and strengthen in increasing economic activities. activities of pesantren, trade, and education.

Keywords: religious, educative, morphological, dualistic, incremental

1. Introduction

The development of an area cannot be separated from the spatial transformation and the architectural expression that occurs in that area. The spatial transformation and the expression of architecture and environment as objects can be seen, felt, experienced, and evaluated easier than other abstract values of morality. One of the architectural expressions is depicted in the development of an area from one period to another. The development of urban area is a changing process of a situational urban area into a new situational urban area in different periods of time [1–4]. Those situational changes also happen in Gegerkalong Girang area, Bandung City. At the beginning of its development, this area was well known as the educational center which supported educational activities based on the activity of the oldest university [5,6] and settlement in north Bandung area [7,8]. However, since *Pesantren* DT [9] was built and developed in 1990, this area became phenomenal [10–12]. Less than two decades, the presence of *Pesantren* DT affects the face of Gegerkalong Girang area from a township area into a modern and popular *pesantren* area. The presence of *Pesantren* DT has changed the physical formation of the environment around the area, which was essentially a place for thousands of IKIP

IOP Conf. Series: Earth and Environmental Science **213** (2018) 012022 doi:10.1088/1755-1315/213/1/012022

Bandung (now Universitas Pendidikan Indonesia) college students to reside since the 1950s. The development of Gegerkalong Girang area today cannot be separated from *Pesantren* DT as a modern, popular, and large *pesantren* (Islamic boarding school) with billions rupiah of assets. The role of Abdullah Gymnastiar (Aa Gym), as the founder of *Pesantren* DT with his Islamic *da'wah* (Islamic proselytizing) which is packed in the concept of *qolbu* management [9–12], has become the trademark of *Pesantren* DT. His language that is very simple, easy to digest, easy to absorb, sincere, touching and soothing has become alluring and has attracted much attention from Muslims.

The interesting thing in Aa Gym's concept of *qolbu* management is that Islam is not only about the ability to explore the source of Islamic teachings but also about how the values of the teachings are practiced in everyday life [11,12]. The basic concept of *qolbu* management teaches us to synchronize the words and the deeds as well as the theory and the practice. Furthermore, the synchronization should be realized in our life.

In addition to *da'wah* as the main activities which are conducted by Aa Gym, Aa Gym also conducts economic development and welfare for the people. These activities can be seen clearly from the changes which occur in Gegerkalong Girang area around *Pesantren* DT. The functions of the area and the buildings which originally had residential functions turn into commercial functions, which obviously affect the changes in the shape of architecture and the environment.

Based on the description above, a research problem is formulated as follows: "Are the values and behaviors which are conveyed by the concept of *qolbu* management and Islamic *da'wah* reflected in the expression of the architecture and the environment?"

This research aims to analyse the morphological changes of Gegerkalong Girang area, Bandung City as the educational area and religious area, as well. This research also concerns with how the concepts, behaviours, and values of the area colour the morphological architecture in Gegerkalong Girang area, Bandung City. Morphological analysis is used to understand the developments, which occurred in Gegerkalong Girang area, Bandung City in the period of 1990 to 2018.

2. Methods

This research employed qualitative approach with descriptive and historical methods. The research site was located in Gegerkalong Girang area, Bandung City, West Java, Indonesia. Data were collected by conducting observation, interview, and documentation. The informants consisted of the occupants of Gegerkalong Girang area and the manager of *Pesantren* DT, as well as its students. Meanwhile, the objects observed were the spatial layouts of the environmental area of *Pesantren* DT.

The informants' selection was completed purposively after the observation is carried out through the grand tour. Techniques of data analysis used Spradley model [13–16]. The first stage in this model was domain analysis to obtain a general and comprehensive description of the social circumstances or research objects which are studied. The second stage was taxonomic analysis and composition to acquire the category of social situations which are studied. The last stage was themed analysis to find the linkages between categories [17,18], in order to obtain a complete picture of the object of research, so that the results become the basis for the conclusion [19].

3. Discussion

3.1. Situation and Condition of Gegerkalong Girang Area, Bandung City

The site of the study is the area around *Pesantren* DT in Gegerkalong Girang Street, SWK Bojonagara, Sarijadi District, Bandung City, West Java, Indonesia. The boundaries of the object studied are the south side of road covering Kelurahan Gegerkalong, RW 03 (RT 07, RT.06, and a part of RT.04) and north side of Kelurahan Isola Street, RW 06.

The borders of the object studied are up to Gegerkalong Tengah Street, eastward; up to Pak Gatot Raya Street, westward; up to KPAD settlement, southward. While the northward borders are UPI campus (see figure 1).

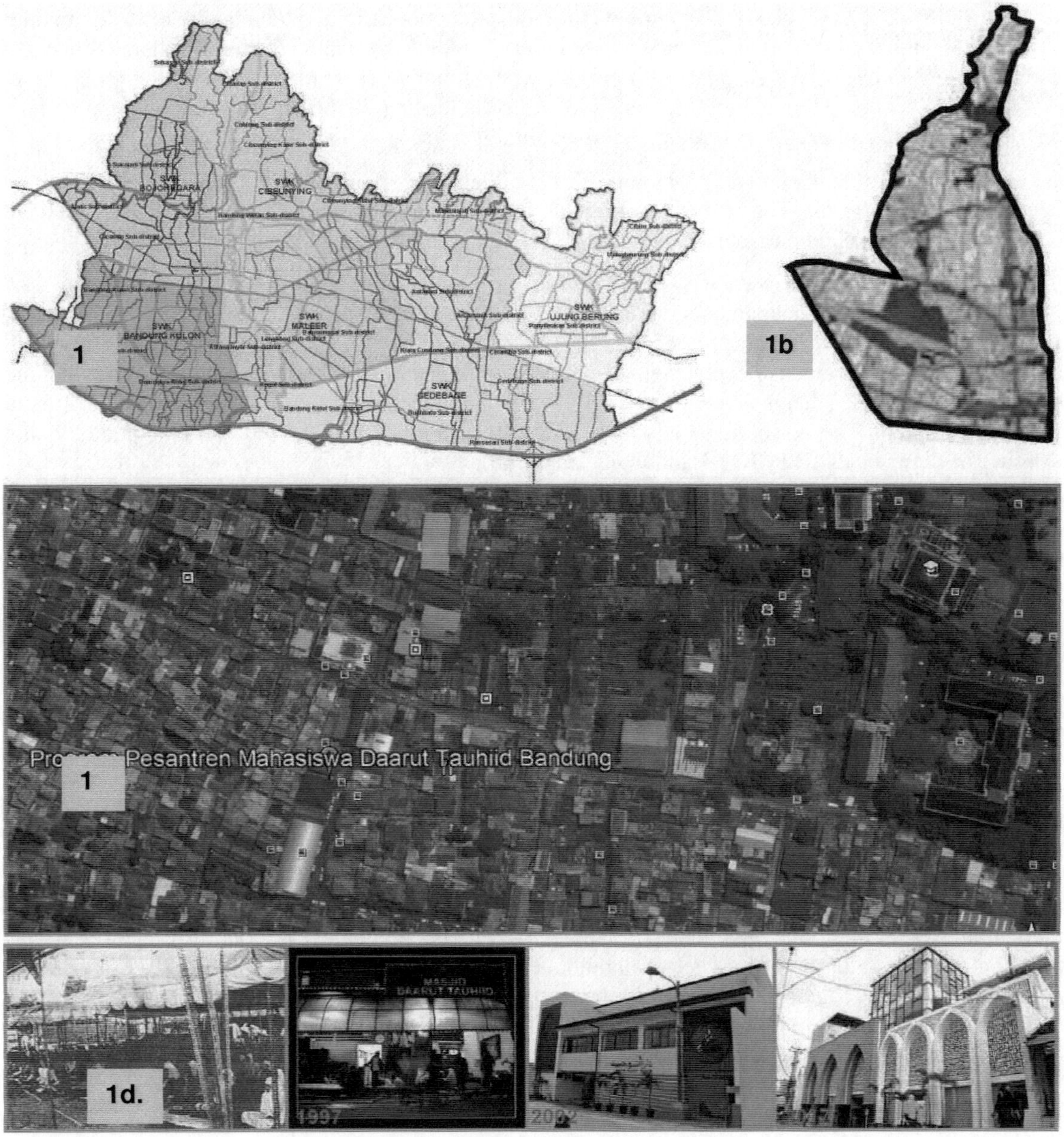

Gambar 1. Site of the Study: Daarut Tauhid Islamic Boarding School, Bandung City, West Java, Indonesia.
1a. Map of Bandung; **1b.** SWK Bojonagara; **1c.** The Area of Gegerkalong Girang Street
Source: Processed from Bandung Map, digital and Google map edition; **1d**. The development of DT mosque from its inception in 1989 to 2018

3.2. Development of Gegerkalong Girang area: pre and post-existence of Daarut Tauhid Islamic Boarding School.

The area around Gegerkalong Girang Street is a residential area. Long before *Pesantren* DT existed in 1990, this area mostly served as lodges for college students of IKIP Bandung/Universitas Pendidikan Indonesia (UPI). The existence of UPI campus provides an opportunity for local residents to rent some or all of their houses to be used as student lodges, with monthly or annual rental system. Since the presence of *Pesantren* DT, seasonal immigrants, whether they are students of UPI or students of *Pesantren* DT as well as its visitors/guests, have been increasing. The surrounding residents benefit from the overflow of migrants by renting out parts of their homes.

The existence of both *Pesantren* DT and UPI has significant impacts on the development of Gegerkalong Girang area. The texture pattern of the area and the functions of the buildings have changed, based on pre and post-existence of DT and UPI, as follows:

(1) The early period of the presence of *Pesantren* DT in 1990. Majority of the building in this area of 1990 was functioned for occupancy and partly to meet the needs of students, such as huts, food stalls, photocopy, stationery stores, and other facilities in supporting educational needs. At this period, *Pesantren* DT just occupied a rental house, which previously was a student lodge. Physically, the average building had a height of one to two floors. The texture pattern of the area was already crowded, except for the northern part, which was still an empty land in the form of paddy fields bordering the campus of UPI (IKIP Bandung at that time).

(2) The period of *Pesantren* DT development (1990-2005). The condition in Gegerkalong Girang area at this period rapidly developed, especially the physical development along Gegerkalong Girang Street. These changes included changes on the building facades, building functions, and building periods. The land around *Pesantren* DT began to be acquired to develop the Islamic boarding school (*pesantren*), which serves as not only Islamic school, but also supporting facilities, such as mosque, *Kepontren* building, residential houses for DT managers, dormitory for male students, dormitory for female students, training center building, cottages, and others. Along the street, *Pesantren* DT experienced changes in terms of its development of commercial buildings that provide the needs of the visitors of *Pesantren* DT. This condition further gave effect on changing the facades and functions of the buildings along the street that increasingly move forward to be used as a trading center.

(3) The period of *Pesantren* DT development (2006- 2018). At the beginning of this period (2006), Aa Gym followers and admirers were surprised by the news of Aa Gym's polygamy. As its consequence, the development of *Pesantren* DT was stagnated. Students who used to have regular daily or weekly visits did not have what they usually had. Moreover, the visitors who usually came on the weekday and on the weekend as the peak visits did not come to that place. This condition clearly was a significant decline in terms of its visitors. The development of the area in the last 5 (five) years has changed significantly (almost the same as the early establishment of *Pesantren* DT). The changes seem to change as the development along Gegerkalong Girang Street begins to recover. It is obvious from the buildings, which are rebuilt to support the visitors' needs.

3.3. Pesantren DT amid Planned and Unplanned Areas

The typological changes of the spatial spaces of Gegerkalong Girang result in an organic texture and spatial pattern which follow the site formation which forms a closed-linear system pattern [20]. This form is shaped from the comparison of the space and time as well as the shape of the scope. The texture pattern of the area tends to be heterogeneous and does not homogeneously follow the axial pattern along Gegerkalong Girang Street as the main axis.

The typological changes of the spatial space experienced 2 (two) forms of changing: *First*, the development of the villages around *pesantren* (the Islamic boarding school) remains incremental or patchy. It also does not form a particular block or mass unit that tends to follow the organic pattern that follows the site conditions forming unplanned and informal spaces. In this case, the relationship of buildings with others tends to be tight. This form is one of the characteristics of urban development in

IOP Conf. Series: Earth and Environmental Science 213 (2018) 012022 doi:10.1088/1755-1315/213/1/012022

the cities of the third-world countries [2,21,22]. Some of the commercial buildings along the street are new buildings replacing old buildings; while the rest are incremental or additional buildings which move forward to the curbside. In some areas, the building expands producing zero border buildings (see figure 2).

Figure 2. Above: The condition of Gegerkalong Girang Street with high traffic density in the morning and afternoon. There is a cross-circulation between pedestrians and vehicle users. **Below**: Commercial buildings that are formed from either the development of functional changes or the addition and expansion of buildings to the border of the land that results in zero border buildings.

Two, the development of *Pesantren* DT area produces a new hierarchy in which the Islamic boarding school (*pesantren)* area is split by Gegegerkalong Girang Street. It comes up with a regular, static, planned, and formal form. The pattern and mass of the buildings form certain mass units that are spread, but the identity of the mass unit of the buildings is clear. The buildings and facilities of the Islamic building school (*pesantren)* are designed in accordance with the functions that form a static space but are nevertheless linear and open. *Pesantren* DT area, which is located in the middle of the village, forms a fairly dominant block in the middle of the mass building area. The growth of *Pesantren* DT makes the mosque as a single block and as the orientation of the growth of the boarding school environment, ranging from Neighborhood Street to SMM area, Hall, Darul Jannah Cottage, female students dormitory, polyclinic, and Aa Gym's house. From this point, it can be seen that the functions associated with *Pesantren* DT become a connecting element that serves to unify and integrate the environment. (see figure 3).

Figure 3. Above: The situations of shops that are needed by students and visitors. **Below**: The situations along the housings and cottage in *Pesantren* DT that are used as vehicle parking facility.

3.4. The concept of Qolbu Management: between worldly and afterlife

The concept of *qolbu* management as a commitment between words and deeds and between theory and practice becomes an example of living the life. One of the proverbs of the *qolbu* management concept: "to change one self, change others, and change the environment" are based on three things: "Start from one self, start from a small thing [9–11]. And get started now". This proverb is not only applied as da'wah material but also practiced consistently in the educational pattern of *Pesantren* DT. This concept is reflected in the values and behavior as well as in the changes of the environment.

There are two key points in applying the concept of *qolbu* management: "get used to purifying heart" and "always have strong willing to improve oneself in any fields" [9,12]. The essence of this concept is the heart because everything must begin with self-understanding, and then, willing and be able to perform self-control after realizing who you really are. Therefore, the heart is the place to understand and control one self.

The concept of *qolbu* management is really applied in managing *Pesantren* DT, through the potential of human resources that have excellences in *dhikr*, thought, and endeavor [9–11]. Through *dhikr*, someone's mentality will be strong, full of spirit, and resilience because he/she will always remember Allah's help. Through thought, someone can be superior to others. Superior human is a thinker, and therefore he/she can use his/her mind quickly, creatively, efficiently, and effectively to recognize his/her potential and can employ it optimally. Meanwhile, through endeavor, the potential which is possessed by a human can be optimized because the potential which is physically trained through endeavor will not easily give up because everything is all left to God.

In the daily practice, all students of *Pesantren* Daarut are taught in not only reaching the afterlife (relationship with God) but also balancing the relationship with humans and the environment. All of the students applied the concept in their daily lives by keeping the environment clean. This concept is an *amaliyah* which should be practiced in daily life, so that they can obtain the balance between the worldly and the afterlife, the harmony between the spiritual and physical happiness, and the combination of modesty with modernity. The activities can be seen in the picture below. (see figure 4 and figure 5)

Figure 4a and 4b. *Akhwat* (Female students) cleaning up the environment along Gegerkalong Girang Street westward, up to Pak Gatot Raya Street;
Figure 4c and 4d. *Ihwan* (Male Students) cleaning up the environment along Gegerkalong Girang Street eastward, up to the main road (Setiabudhi Street). They swap locations every week.
Figure 5a, 5b, and 5c. The sign system is inserted in a certain place, used also to invite in good behavior

Figure 4 and figure 5 show that the concept of *qolbu* management is not only the ability to explore the source of Islamic teachings but also the ability to apply them in daily life. It indicates that there is a commitment between words and deeds as well as between theory and practices as one exemplary.

4. Conclusion

The existence of *Pesantren* Daarut Tauhid (DT) demonstrates the growth phenomenon in the setting of Gegerkalong Girang village, from a village that originally supported the fulfillment of residential area for students, to a harmonious area amid educational and religious area. However, the area does not eliminate the characteristics of a township development, which is incremental or patchy. It also does not form a particular block or mass unit that tends to follow the organic pattern following the site conditions to form unplanned and informal spaces. This change can be seen from the functions and facades of the buildings. The functions and facades of the buildings that originally had residential functions changed into commercial functions. There are dualistic functions [3] in the area that complement and strengthen each other. This area escalates economic activities and services that support the activities of Islamic boarding school, trading, and other services which support education and housing activities. The values of *qolbu* management (MQ) as Aa Gym's main *da'wah* material is a preliminary way of the heart purifying which is very practical, closed to daily life reality, and delivered in a soothing rhetoric. The basic concept of *qolbu* management is seen in figure 6

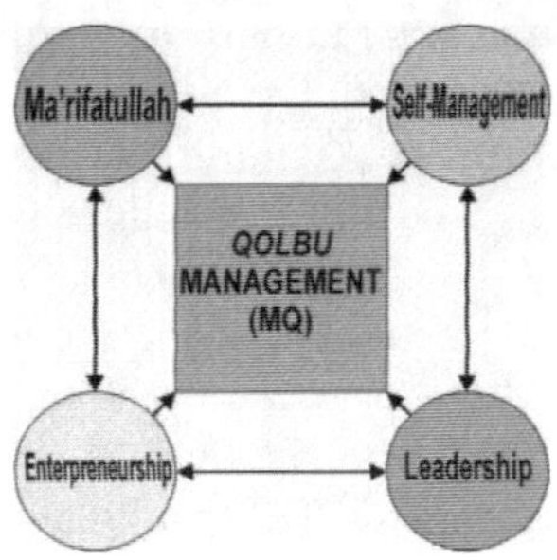

Figure 6. The basic concept of *qolbu* management

The basic concepts of *qolbu* management are divided into 4 components, namely, *Ma'rifatullah*, Self-Management, Entrepreneurship, and Leadership. These values are applied in daily life as an *amaliyah* to achieve the balance between the worldly and the afterlife, the harmony between the spiritual and physical happiness, and the combination of modesty and modernity.

5. Acknowledgment

The authors would like to thank LPPM UPI that has given the opportunity for the researcher to conduct research through its research grant funding of Research Affirmation scheme, to the leaders and the board of *Pesantren* Daarut Tauhid (DT) who have given the permission for the researcher to conduct research, and finally to all residents around *Pesantren* DT area, especially residents of RW 03 (RT 07, RT.06, and RT.04) Kelurahan Gegerkalong and residents of RW 06 of Isola Sub-district.

6. References

[1] Bintarto 1983 *Interaksi Desa-Kota dan Permasalahannya* (Jakarta: Ghalia Indonesia)
[2] McCarthy P 2003 *Urban slums reports: the case of Jakarta, Indonesia*
[3] Soetomo S 2009 *Urbanisasi dan Morfologi: Proses perkembangan peradaban dan wadah ruang fisiknya : Menuju Ruang Kehidupan yang Manusiawi* ed Graha Ilmu (Yogyakarta)
[4] Munggiarti A and Buchori I 2015 Pengaruh Keberadaan Perguruan Tinggi Terhadap Perubahan Morfologi Kawasan Sekitarnya *J. Geomatics Plan.* **2** 51–68
[5] Permana A Y 2014 *Transformasi Gubahan Ruang: Pondokan Mahasiswa di Kawasan Balubur Tamansari Kota Bandung* (Universitas Diponegoro)
[6] Permana A Y, Sumarna N and Wijaya K 2017 *Membangunan Kampung Kreatif melalui Kolaborasi Mahasiswa dengan Masyarakat. Kasus: Kawasan Balubur-Tamansari Kota Bandung* (Padang: Program Pascasarjana Universitas Andalas)
[7] Permana A Y and Wijaya K 2017 Spatial change transformation of educational areas in Bandung Spatial change transformation of educational areas in Bandung *IOP Conf. Ser. Earth Environ. Sci.* **99** 012029
[8] Permana A Y, Soetomo S, Hardiman G and Buchori I 2013 Smart Architecture as a Concept of Sustainable Development in the Improvement of the Slum Settlementarea in Bandung **2** 26–35
[9] Gymnastiar A 2002 *Meraih Bening Hati dengan Manajemen Qolbu* (Jakarta: Gema Insani Press)
[10] Hernowo and Ridwan D 2002 *Aa Gym dan Fenomena Daarut Tauhid* (Bandung: Mizan)
[11] Solahudin D 2008 *The Workshop for Morality: The Islamic Creativity of Pesantren Daarut Tauhid in Bandung, Java* vol 1 (Australia: ANU E Press)
[12] Barliana M S 2009 *Transformasi Kampung Kota: Dari Manajemen Qolbu ke Manajemen Lingkungan* (Bandung: Metatekstur)
[13] Spradley J 1980 *Participant Observation* (New York: Rinehart and Winston)
[14] Groat L and Wang D 2002 *Architectural Research Methods* (Canada: John Wiley and Sons, Inc)
[15] Moleong L 2007 *Metodologi Penelitian Kualitatif* (Bandung: Remaja Rosda Karya)
[16] Sugiyono 2016 Metode Penelitian Kuantitatif, Kualitatif dan Kombinasi (Mixed Methods) *Bandung Alf.*
[17] Strauss A and Corbin J 1997 *Grounded Theory in Practice* (USA: Sage Publication)
[18] Setioko B 2013 *Integrasi Ruang Perkotaan dalam Fenomena Pertumbuhan Kawasan Pinggiran Kota di Indonesia* (Universitas Diponegoro Semarang)
[19] Ali Ikhsan F, Setioko B and Suprapti A 2018 Ecological wisdom of Hindu-Javanese community settlement in Cetho Hamlet, Lawu Mountains, Central Java, Indonesia *IOP Conf. Ser. Earth Environ. Sci.* **99**
[20] Zahnd M 1999 *Perancangan Kota secara Terpadu* (Yogyakarta: Kanisius)
[21] Kostoff S 1991 *The City Shaped: Urban Patterns and Meanings through History* (Boston:

Little, Brown)
[22] Ardiwijaya V S, Soemardi T P, Suganda E and Temenggung Y A 2014 Bandung Urban Sprawl and Idle Land: Spatial Environmental Perspectives *APCBEE Procedia* **10** 208–13

Driving Factors of Developers in Developing Apartment in Depok

Yusup Setiadi[1], Hendrajaya Isnaeni[1]
[1]University of Indonesia Architecture Department, Depok, Indonesia

Corresponding e-mail: yusupsetiadi@gmail.com

Abstract. This research was conducted to find the factors that influenced developers in developing apartments in Depok city, especially in Margonda Street and its surroundings. Seeing the massive development of apartments, it is interesting to examine the factors that encourage developers in developing apartments in order to understand the market conditions of supply. The analytical technique used descriptive statistics, ie with one sample t-test to determine the factors that have or have no influence on developers in developing apartments in the city of Depok. In addition, it also conducted a relative rank index analysis (RRI) to determine the dominant factors on developers in developing apartments in the city of Depok. The results of the analysis show that all factors in the research variables, Population Income, Total Population, Population Growth, Total Manpower, Number of College Students, Market Price, Tax, Ease of Loan, Interest Rate, Down Payment, Land Allocation (Government Regulation), Investment and Market segmentation, were influential in developing apartments in Depok City. While RRI analysis indicated that market price and market segmentation were the most dominant factors. In addition, the number of college students, investment and land allocation were also important factors for developers in developing apartments in Depok City. The results of the study are expected to be beneficial for property developers, especially on the development of apartments.

Keywords: Development, Apartment, Supply, Demand, Factor, Developers

1. Introduction

The current apartment market is increasingly in demand for investment or housing needs[1]. Likewise that happens in the city of Depok, as a hinterland of Jakarta, the development of apartments in the city of Depok is so vibrant. Until 2017, there are 9 apartments that have been handed over and 9 apartment projects are still on marketing stage. Based on the plan, the nine projects that are being marketed will supply the number of apartments as many as 22,743 units of apartments until 2023.

The number of apartment developments is one of them driven by the presence of quite a lot of college students in the city of Depok. Increasing the number of people is unavoidable when there are educational activities in an area. This will affect the surrounding environment [2]. The large number of college students is seen by developer as a great market niche of apartment development. Because these college students will need a place to live that are close to their place of study. In addition, the proximity to Jakarta also help to stimulate the development of apartments in this city.

The proximity to Jakarta as well as the admission of new college students in each university each year, makes the city of Depok experiencing an increase in population and increased need of residence. Meanwhile, the available land is getting limited with the higher land price. The tendency that we can see is that new college students will find residential which are close to the campus and are equipped with complete facilities. Likewise with commuter workers, they will do the same because of the expensive housing in Jakarta.

Because of the increasing population, the demand for land will continue to grow, while the available land is increasingly limited. It is intensively conducted by the government and private sector to hold development especially the intensification of function in downtown, that is vertical development in this case "high rise apartment" for the concentration of residence in downtown.

As a result of the increasingly crowded conditions and the need to reside which become higher, the developers build a vertical dwelling or apartment. In the city of Depok, the construction of apartments in this city is quite massive in recent years. So, these concerns raise a question: "What factors drive developers in developing apartments in Depok City?".

Several previous researches, generally only see location as a factor affecting in developing the apartment. For instance, Krisnaputri [3] discussed the pattern of choosing the location of apartment development in Surabaya by the developer, with the aim to know the pattern of choosing the location of apartments in Surabaya based on the developer's perspective. Similarly, Nisa [4] identified the phenomenon of the development that rampantly occurred and also the factors that influence the selection of apartment construction location on the perspective of government, developers and residents. Mollah et al [5] identified the factors that affect consumers to buy an apartment rather than land. This research will try to see other demand that become the appeal of developer in developing the apartment. The results of this study is expected to be a provision of information for the people of Depok City to continue to guard the construction of existing apartments in the city of Depok. Moreover, it can be used as a guideline for local government to provide an idea of developer for preferences in choosing the location of apartment development, so that it can make the existing space beneficial to society.

In the literature there are two major factors, namely property and residential demand factors. Property is related to land used by people for various purposes: industry, commercial, residential, and special use. But the main object in this study is residential property. Residential property consists of two main categories. They are one household (single family), and multiple household (multifamily) [6].

Meanwhile, in relation to the residential demand factor, Schmitz, A., L. Brett, D [7] explains that there are at least five influencing factors: the number of labor, population or population growth, household, either traditional or students who rent a house together, revenue, and market segmentation.

Barrett and Blair [8]argue that the economic factors that directly affect the success of housing projects are demand and supply. Demand is assumed to be a function of the population, income, labor, price, tax, interest rate, down payment, and future expectations.

Furthermore, based on Putong [9], in general, the demand for occupancy is influenced by several factors, such as:

a) The wealth of society. The higher the income, the higher the demand level.

b) The rate of return on capital from owning a house. If the real rise of the value of a house as an investment increases, the demand for housing will increase

c) Acceptable income from other assets

d) Interest rate. The higher the interest rate, the larger the repayments to be paid so the lower the interest of the society in buying a house

e) Population growth and income. In the long term, population growth and income will further increase demand for housing.

Housing demand is also affected by three factors: income, mortgage rates, and taxes. As income rises, more families buy their first house or move to a larger house. People will increase their demand when they expect a higher fixed income. The mortgage interest rate will affect the demand because credit payments will be burdened with interest. In the meantime, government tax regulations do not change often, but when it happens, it will affect the demand for housing [10].

Firdaos [11] explains that the demand for residential areas is influenced by factors as follows:

a) Location

The existence of the location, whether in the center or in the edge of the city will greatly affect consumer interest in buying a home. The more strategic the location of the housing means the better and has a higher level of demand.

b) Population increase

For the reason that everyone needs a place to live as a shelter, any increase in the population both natural and non-natural (due to urbanization) will increase demand for houses.

c) Income

A person's ability to own a home is greatly influenced by the income he earns. If a person's income increases and economic conditions do not experience recession and inflation, the propensity to own a home will increase both in quality and quantity.

d) Ease of Getting Loans

In the residential property market, housing demand is also affected by government policies and financial institutions such as banking. Characteristics of the property market that requires large funds cause consumers become very dependent on the ease of funding. The ease of funding can be in the form of loan credit facility, decrease of loan interest rate, and loan repayment period.

e) Facilities and Public Facilities

Facilities here include public facilities and social facilities, including infrastructure, educational facilities, health, religion, transportation facilities, and others. The existence of such facilities builds and attracts investors who will further increase demand for houses in the region.

f) Market price

As in the case of demand and supply theory, the higher the price of goods will lead to a decrease in demand for the intended goods. If the price of a medium-sized house rises, the propensity to own a house with that price level will decrease and demand will move to the house for a lower price.

g) Constitution

Regulations on the types of land/land use rights that restrict land rights are also factors that affect consumers' demand for housing. Likewise, other regulations such as tax laws (PBB and BPHTB) also become factors that become consumer consideration in buying a house.

Eckert [12] mentioned some factors that are very important and affect the housing market demand, they are:

a) Economic factors. Changes in economic factors that affect housing demand are income, interest rate, loan policy, rental rate and house price.

b) Social factors include population density, family size, education, crime rate and age distribution.

c) Government factors involve Zoning, municipal government services and public tax rates.

d) Environmental factors. Changes in environmental factors that affect housing demand are topography, landform, soil conditions and ease of facilities such as parking, shopping centers, schools, places of worship, employment opportunities, transportation and business license services.

Based on these literature, positive factors that influence Developers in developing apartment, especially along Margonda Road, Depok are population income, total population, population growth, total manpower, number of college students, market price, tax, ease of loan, interest rate, down payment, land allocation, investment, and market segmentation.

IOP Conf. Series: Earth and Environmental Science **213** (2018) 012023 doi:10.1088/1755-1315/213/1/012023

Therefore, based on the background, problem formulation and research questions, the purpose of this study is to determine the factors that affect developers in developing apartments in the city of Depok.

2. Research Methods

2.1. Data Collection

The process of collecting data was divided into two, namely the primary data and secondary data. Primary data consisted of interview and questionnaire given to developers who are building apartment projects in the city of Depok, especially along Margonda street. In this study, there were five developers who had been successfully interviewed from a total of nine existing developers. Meanwhile, secondary data were supporting data obtained from several government agencies such as Central Bureau of Statistics and some developers related to apartment project plan.

2.2. Data Analysis

From the available data, the analysis was conducted by using the method based on the results of the review: Descriptive statistics and Inferential statistics. Descriptive statistics covers data summary in the form of tabulation data (tables). Inferential statistics involves data summary in the form of statistical estimation method and Relative Rank Index (RRI). Before performing the analysis through RRI, to determine the factors that influence developers in developing apartment, the analysis of one sample t test is operated. Then, to find out the dominant factor, RRI analysis is used. The database of this analysis is from the results of questionnaires processed through Likert-scale analysis.

The RRI formula used is (equation 1):

$$RRI = \frac{1}{nN}\left(\sum_{i=1}^{i} l_i X_i\right) \quad (1)$$

n : Likert scale highest number

N : Number of respondents

i1 : Likert scale, where i 1 scale is lowest and n 1 scale is highest.

Xi : The frequency of the selected scale scores of respondents from i = 1 to n.

3. Discussion

3.1. Factors Driving Developers in Developing Apartments in Depok City

Analysis of the factors affecting developers in developing an apartment in the city of Depok, especially around the street margonda aims to determine the factors that have influence and have no influence. The analytical technique used was likert scale analysis, with a rating scale of 1-5 (very unimportant-very important) for each category of answers on the questionnaire. The method of analysis used one sample t test.

The sample mean (t-value) set forth in this study was 3.00 on the scale of the 1-5 enters with a significance level of 5%. Then the hypothesis that came up was:

Ho: These factors have no effect on developers in developing apartments in Depok City

H1: These factors affect the developer in developing apartments in the city of Depok.

Statistical hypothesis

Ho: $\mu \leq 3.00$

H1: $\mu > 3.00$

Based on Probalitas:

If sig > 0.05, then Ho is accepted

If sig < 0.05, then Ho is rejected

Test parameters:

If t-table ≤ t-count, then Ho is accepted

If t-table > t-count, then Ho is rejected

In accordance with the litelatures described above, there were 13 factors to be analyzed: Population Income, Total Population, Population Growth, Total Manpower, Number of College Students, Market Price, Tax, Ease of Loan, Interest Rate, Down Payment, Land Allocation, Investment, and Market Segmentation. These factors then became questions on the questionnaire given to to the developers as respondents.

Table 1 Questionnaire Result with Developer

Factor	VU	U	LI	I	VI
Population Income	0%	0%	20%	40%	40%
Total Population	0%	0%	0%	80%	20%
Population Growth	0%	0%	0%	40%	60%
Total Manpower	0%	20%	20%	60%	0%
Number of College Students	0%	0%	0%	20%	80%
Market Price	0%	0%	0%	0%	100%
Tax	0%	20%	40%	20%	20%
Ease of Loans	0%	0%	0%	80%	20%
Interest Rate	0%	20%	0%	40%	40%
Down Payment	0%	20%	0%	80%	0%
Investment	0%	0%	0%	20%	80%
Market Segmentation	0%	0%	0%	0%	100%
Land Allocation (KLB)	0%	0%	0%	20%	80%

Source: author, 2018

Information : VU : Very Unimportant; U : Unimportant; LI : Less Important; I : Important; VI : Very Important

The results of the questionnaire (**Table 1**) were then processed statistically and simplified as the following summary table (**Table 2**).

Table 2 Factors affecting developers in developing apartments in Depok City

Factor	μ	p-value	Conclusion	Effect Level
Population Income	4.20	0.00 < 0.05	Reject Ho	Take effect
Total Population	4.00	0.00 < 0.05	Reject Ho	Take effect
Population Growth	4.40	0.00 < 0.05	Reject Ho	Take effect
Total Manpower	3.40	0.01 < 0.05	Reject Ho	Take effect
Number of College Students	4.80	0.00 < 0.05	Reject Ho	Take effect
Market Price	5.00	0.00 < 0.05	Reject Ho	Take effect

Tax	3.40	0.03 < 0.05	Reject Ho	Take effect
Ease of Loans	4.20	0.00 < 0.05	Reject Ho	Take effect
Interest Rate	4.00	0.02 < 0.05	Reject Ho	Take effect
Down Payment	3.60	0.01 < 0.05	Reject Ho	Take effect
Investment	4.80	0.00 < 0.05	Reject Ho	Take effect
Market Segmentation	5.00	0.00 < 0.05	Reject Ho	Take effect
Land Allocation (KLB)	4.80	0.00 < 0.05	Reject Ho	Take effect

Source: author, 2018

Based on the above table, it can be seen that all factors had an influence on developers in developing apartments in the city of Depok. The income factor of the population affected the developer because the population with high purchasing power was a big market potential. Depok demography also supported the potential income, data Central Bureau of Statistics noted that the population with the amount of income above Rp. 6 Million per month is the majority.

The other influenting factors were population and population growth. These factors were influential because the population and its growth projection greatly affected the prospects of the property market. The population in which there was potential markets should be seen as a major factor in the property business prospects, although per capita income should also be seen because not all residents were market potential.

Total Manpower was also included as factor that affected developers in developing apartments in downtown Depok. The proximity to Jakarta as well as the high price of houses in the city center encouraged developers to take this market gap.

The existence of a campus with a big name in Depok also influenced developers to develop vertical dwelling. Campuses with a considerable number of students were seen as a potential market in the development.

Furthermore, the market price was also an influential factor because this price was the main thing in property development. Competitive price was the price that could be absorbed by the market. Meanwhile, taxes, ease of lending, interest rates and down payment also affected the apartment development because with the many ease would be very helpful for the prospective buyers. Next is the investment, this factor was influential because of the potential lease of the number of students who were in downtown Depok. Market segmentation was also influential because especially in Depok, the population of the middle class was quite a lot.

The last is the factor of government regulation. In this case, the supporting location and the high coefficient of building floor (KLB) also influenced developer in developing apartment in downtown Depok.

3.2. Relative Rank Rank Index (RRI) Analysis of Factors Affecting Developers In Developing Apartments In Depok

Before performing RRI analysis, Cronbach's Alpha analysis was performed through SPSS version 21 and the data used to calculate the cronbach's alpha reability was based on the questionnaire data.

The result of Cronbach's Alpha reability test was quite satisfactory because the resulting index was 0.828. According to Sekaran [13] realibility that less than 0.60 is poor, whereas 0.70 is acceptable and above 0.80 is good. Because the index value was more than 0.80 then it can be concluded that the research instrument had been reliable or could be trusted.

Table 3 Relative Rank Index (RRI) Towards Factors Affecting Developers In Developing Apartments In Depok

Factor	Average	Rank
Market Price	1.00	1

Factor	Average	Rank
Market Segmentation	1.00	1
Number of College Students	0.96	2
Investment	0.96	2
Land Allocation	0.96	2
Population Growth	0.88	3
Population Income	0.84	4
Ease of Loans	0.84	4
Total Population	0.80	5
Interest Rate	0.80	5
Down Payment	0.72	6
Total Manpower	0.68	7
Tax	0.68	7

Source: author, 2018

Referring to RRI result (**Table 3**), Market Price and Market Segmentation were the most dominant factors. In addition, the number of college students, investment and land allocation are also important for developers in developing apartments in downtown Depok.

The dominance of price factor and segmentation were also reflected from the selling price and market segmentation which was targeted at this time, dominated by middle segment with selling price range Rp. 500 millions to Rp. 1 billion.

In addition, the number of students and investment factors were also in line with price and segmentation. In which with the target market of students and investment, then the selling price was also not too high.

4. Conclusion

Referring to the results of the analysis, results revealed that all factors, Population Income, Total Population, Population Growth, Total Manpower, Number of College Students, Market Price, Tax, Ease of Loan, Interest Rate, Down Payment, Government Regulation, Investment and market segmentation, were influential to developer in developing an apartment in Downtown Depok. However, to see the most dominant factor, based on RRI analysis results, the important factors affecting developers in developing apartments in downtown Depok were as follows:

- Market Price 1.00
- Market Segmentation 1.00
- Number of College Students 0.96
- Investment 0.96
- Government Regulation (Land Allocation) 0.96

The selected factors revealed a link between the selling price, the number of college students and the investment. These three factors indicated that the selling price of the marketable apartment was for the middle-medium segment, maximum. Students as the target market also contributed to limit the selling price which could not be too expensive, so that students as a potential target of lease could easely afford the apartment. Besides that, if the price was expensive, it would certainly compete with the choice of a landed house. This condition was seen from the selling price offered at this time that was in the range of Rp. 500 Million to Rp. 1 Billion, in which the range of selling prices was for the middle-medium segment

5. References

[1] Ardiyanto W 2017 Apartemen kini jadi hunian incaran dan investasi yang menjanjikan

[2] Allinson J 2006 Over-educated, over-exuberant and over here? The impact of students on cities *Plan. Pract. Res.* **21** 79–94

[3] Krisnaputri N A 2016 *Pola Pemilihan Lokasi Pembangunan Apartemen Di Surabaya Oleh Pengembang* (Institut Teknologi Sepuluh Nopember)

[4] Nisa I 2016 *Fenomena Pembangunan Student Vertical Apartment di Aglomerasi Perkotaan Yogyakarta dan Faktor-Faktor yang Mempengaruhi Pemilihan Lokasinya. Kasus: Apartemen Student Castle.* (Universitas Gadjah Mada)

[5] Mollah M M H, Haque I and Pasha S H A 2009 Factors influencing apartment buying decision: An analysis of company and customer *Southeast Univ. J. Bus. Stud.* **2** 109–17

[6] Kyle R C, Baird F M and Spodek M S 2000 *Property Management* (Real Estate Education Company)

[7] Brett D L and Schmitz A 2009 *Real Estate Market Analysis: Methods and Case Studies* (Washinngton: Urban Land Institute)

[8] Barrett G V and Blair J P 1988 *How to Conduct and Analyze Real Estate Market and Feasibility Studies* (New York City: Van Nostrand Reinhold)

[9] Putong I 2002 *Pengantar Ekonomi Mikro dan Makro* (Jakarta: Ghalia Indonesia)

[10] Dornbusch R, Fisher S and Startz R 2004 *Makroekonomi, Edisi Bahasa Indonesia* (Jakarta: PT. Media Global Edukasi)

[11] Firdaos A 1997 Permintaan dan Penawaran Perumahan *Valuestate* vol 007 (Jakarta)

[12] Eckert J K, Gloudemans R J, Almy R R and of Assessing Officers I A 1990 *Property appraisal and assessment administration* (Chicago: International Association of Assessing Officers)

[13] Sekaran U 2006 *Metode Penelitian untuk bisnis* (Jakarta: Salemba Empat)

Understanding *genius loci* to sustain Ume Bangka's traditional architecture based on intangible material culture

Kemas Ridwan Kurniawan[1], **Elita Nuraeny**[1]
[1]Universitas Indonesia, Depok, Indonesia

Corresponding e-mail: elitanuraeni@gmail.com

Abstract. The vernacular architecture used natural materials for its construction, such as wood, clay, stones, leaves, and animal skins. In tropical countries such as Indonesia, most vernacular architecture is made out of wood; this material is versatile, highly accessible, and affordable. Ume Bangka—a plantation house from Bangka—uses wood for its main construction. However, due to deforestation mainly by private sectors, such as major palm oil industries and private tin mining, wood becomes a scarce commodity. These activities also reduce traditional landscape areas and settlements that belongs to local people for centuries. As wood becomes unaffordable and less area to inhabit, people begin to abandon Ume Bangka as their future home and opt for modern housing type. This is very unfortunate because Ume Bangka shares a close relationship with local culture and natural landscape. This paper investigates the issue of the relationship between natural landscape and cultural values around Ume Bangka as a traditional Bangkanese dwelling. The main methodology in this discussion used literature studies on vernacular housing typology, especially in Bangka Island. Finding the *genius loci*—the spirit of a place—from the construction of Ume Bangka is the expected result of this paper that will serve as a preliminary idea in sustaining Ume Bangka's traditional house amidst globalization and proposing Ume Bangka as the cultural identity as well as the inspiration for built-environment sustainability in Bangka Island.

Keywords: Ume Bangka, vernacular architecture, material, *genius loci*

1. Introduction

Before the 19th Century, most native people lived in a dense tropical rainforest in Bangka Islands. Came as part of the Austronesians tribe called *Deutro Melayu* since 1500 BC, the ancestors of today's modern Bangka brought their lifestyle called *berume*—planting and migrating from one land to another [1-2]. Ume Bangka is their traditional stage-house made of wood and located in *dusun* (hamlet) of dense rainforest. Separated into three levels—*kolong, induk, and atap*—this modest housing unit had fulfilled the needs of people in Bangka Island. As shown in figure 1, *kolong* (base) is mainly used as storage and housing fowl or chicken coop, *induk* (main building) as their living space where daily social interaction between family members took place, and last is *atap* (roof) and space below it known as *loteng* (attic) where people kept their prized possession [3-5].

Ume Bangka has nine, twelve, or sixteen columns made of wood to support its roof and body. The uniqueness of Ume Bangka came from the way it utilized every element of trees: its trunk, bark, and even its leaves. The native people of Bangka made the tree trunk into strong posts and floor panels. The wall of Ume Bangka used the tree bark that was nailed to the wall structure. Finally, *rumbia* (or sago palm tree leaves) were clamped onto the roof structure with smaller wooden slates.

IOP Conf. Series: Earth and Environmental Science **213** (2018) 012024 doi:10.1088/1755-1315/213/1/012024

To construct the main structure, the native of Bangka Island erected columns of Ume Bangka by following the natural growth of trees: the root end against the ground, while the tip supporting the roof structure. Based on Gaudenz Domenig's book, this arrangement created strength, because "*the root end of a post is not only heavier but also somewhat thicker than the tip*" [6], thus creating a strong structure to support the entire building (Figure 1). This material knowledge was acquired by not only the indigenous people in Bangka but also various other traditional tribes across Indonesia.

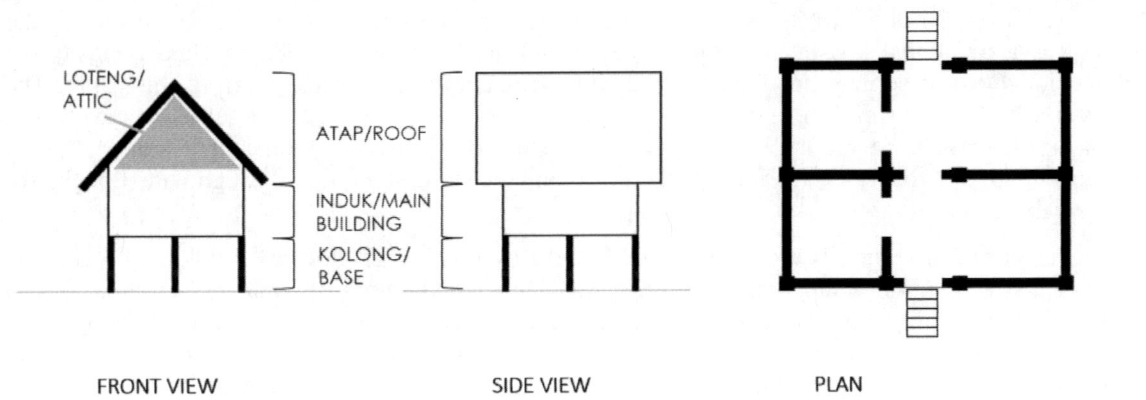

Figure 1. Typology of Ume Bangka, showing the *atap* (roof), *loteng* (attic), *induk* (main building), and *kolong* (base). The plan of Ume Bangka shows columns made of solid wood.

During the colonial and post-colonial periods, tin mining business began to rise in Bangka. Known as one of the largest tin ore producer in the world, the tin mining industry had influenced the development of politic, social, and even cultural orientation in Bangka [1-2]. The tin mining industry, then, gradually changed Bangka from an agricultural society to mining and trading community. One clear impact can be traced through the development of urban landscape and architecture in major cities in Bangka Island [2]. In the mid-19th Century, the Dutch government relocated local settlements from the forest to newer settlements alongside these new roads built to connect major mining cities in Bangka, such as Muntok and Pangkalpinang (Figure 2). The main idea behind this relocation was to win the 50-years-war against Depati Amir in the 19th century by cutting off the rebels' food source and their hidden territories [1]. Finally, in 1851, the colonial government caught and exiled Depati Amir to a south-eastern island of Indonesia known as East Nusa Tenggara, and thus successfully ending the war.

Other than ending the rebellion led by Depati Amir, resettling natives from their natural habitat was another effort made by the colonial government to gain access to tin resources. Without natives out of their woods, the Dutch colonial government was free to exploit their land to mine tin ore. Also, with more people lived closer to the city, it was easier for the Dutch government to hire more workers for their tin smelting and—most importantly—to gain control over the indigenous people [7].

The relocation of indigenous settlements from the forest to the suburban affected Ume Bangka as the local's traditional housing typology. Living far away from the forest means acquiring woods was no longer an easy task and highly expensive. Instead of woods, houses built outside the forest use recycled materials or bricks in order to save money [3,8]. Also, to preserve Bangka's water reservoir, the government limited the felling of trees; a law that forced people to switch from Ume Bangka—a traditional wooden stage house—to smaller, landed, houses [3].

However, the real menace in resettling locals to the city is that this practice made Ume Bangka—the symbol of tradition and custom in Bangka—seemed inferior to modern housing typologies. Ume Bangka is underestimated and representing 'backwardness' and 'under developed,' while the 'new' housing typology is identified as 'progressive' and 'civilized' [2]. This mindset, however, gnaws within the majority of people who are skeptical and pessimistic about the future of Bangka [9]. Lost courage to save themselves, Ume Bangka will soon perish just like other vernacular architectures that had disappeared *"to make way for new housing blocks, shopping malls, and office centers"* [10]—Ume Bangka will be gone, crumbling against time.

2. Methods

This paper investigates the relationship between the natural environment and cultural values of Ume Bangka as a traditional housing typology in Bangka Island. The importance of Ume Bangka as a cultural identity and heritage for the people in Bangka Island is the primary aim of this paper. The discussion is separated into three topics: (1) house as the *genius loci*; (2) the tension between old versus new—modern versus traditional—as a byproduct of industrialization and modernization in Bangka Island; and finally (3) the importance of finding the cultural identity through Ume Bangka for a sustainable future.

The parameter of this paper is set around Ume Bangka as a traditional housing typology, because, as Rapoport says, *"building a house is a cultural phenomenon;"* the house is *"an institution not just a structure"* that reflects as a sequence of socio-cultural factors around specific community [11]. Therefore, as the oldest housing typology exists in Bangka, Ume Bangka will be a significant parameter in studying the connection between cultural value and natural environment.

Literature research in pragmatic approach is the main methodology used in this paper. Pragmatic approach is important in this paper to allow reflective position on social and political issues following the data. Understanding values in Ume Bangka are crucial in pragmatic approach because it enables people to experience facts [12].

Discussion on Bangka—its people, natural landscape, and history—will set the overall understanding on the issue around Ume Bangka in today's modern life, emphasizing over the cultural value against social and political factors occurred in the historical timeline. Readings on the architectural history and culture in Bangka are important, as well as *Genius Loci: Towards a Phenomenology of Architecture* by Christian Norberg-Schulz (1979). Understanding *genius loci*—an ancient Roman believe that places have inner spirits that determined their essences and characters [13,14]—will help the discussion on the relationship between natural environment, built environment, and the occupants. *Genius loci* is a rhetorical device which will help people to defend places based on their values, as well as *"culturally or ecologically, against damaging forms of development"* [14]. Knowing *genius loci* will determine ways in preserving Ume Bangka for the future of sustainable heritage conservation in Bangka Island.

3. Discussion

3.1. House as our genius loci: *the relationship between our body and the natural environment*

In vernacular architecture, house means more than shelter. House is not a structure, but an institution of a complex set of purposes. Unlike the modern architecture, which placed innovation, physical appeal, and intellectual ideas as their main focus, the functional aspect of a house is far more important than its physical or concept in vernacular architecture [11,15]. House, therefore, from the vernacular architecture standpoint, *"is a human fact"* that relating *"man's way of life to the environment"* [11].

Moreover, the house is part of a larger system around its context. For example, when the forefathers lived inside Ume Bangka, living as a nomadic tribe and as gatherers, their lives were rotated around Ume Bangka. As a closed community with 10-40 houses in the woods [1,3], this indigenous people of Bangka once were capable of providing themselves without outside intervention. The indigenous people of Bangka fed their livestock under their house (*kolong*) then continued to collect edible plants in the nearby forest. When harvesting their goods, some were stored in granary while some—for the family—stored under their houses.

Socializing also took place around Ume Bangka: the *induk* part of Ume Bangka represents the idea of community gathering and social value of living *berume*. And last, the attic served as the most sacred part of Ume Bangka. The closest part to heaven, people used it to store their family heirlooms and valuables. Often used as well as a weaving station. For local people in Bangka, weaving is a tradition rich in rituals and spiritual symbols. The attic of Ume Bangka shared identity on how houses was seen as part of cosmos in vernacular architecture, where the highest part of the house is intended 'to get closer to gods' [11]. These close connection between people and nature creates Ume Bangka. Deserting Ume Bangka, therefore, similar with disconnecting from the nature that had been nurtured Bangka's ancestors for centuries old, spiritually and culturally. This *"concrete phenomena of our everyday life"* [13] is what determined Ume Bangka's character—the spirit of place: *genius loci*.

Figure 2. Ume Bangka in the 19th Century before relocation by the Dutch colonial government (left, source: KITLV) and Ume Bangka in the 21st Century after relocation (right).

To understand Ume Bangka as the *genius loci* of Bangka's local identity, we must agree that Ume Bangka should not be described as a noun. Instead, it should stand *"as a system of relations"* [13]: the relation between the physical boundaries, natural environment, and the people who lived in it. In other words, Ume Bangka shall be the complex associations between landscape, family, and community [16]; a sacred circle that linked forever, strengthening their identity. Therefore, when the Dutch colonial government removed one aspect that made Ume Bangka the living breathing cultural identity—the landscape—it disrupted the balance and alienated people in Bangka from their own identity.

"Environment affects us," says Christopher Day, *"it affects both social and personal health; body, soul, and spirit"* [17]. Day continues that different frames—or perhaps 'the relation of the built environment and the natural environment'—influence the way we think, feel, and behave. In another word, displacement will change the way we are [17]. In the case of Ume Bangka, being removed from the forest was disrupting the local custom and tradition. Once it was common for people in Bangka to work with woods, but nowadays, once abundant and commonly used as building material, becomes a burden: an expensive material only the riches can afford [3].

With less and less Ume Bangka preserved, even fewer people know how to build it. No architects nor blueprints used to construct Ume Bangka, making it difficult to recreate this particular typology in the modern day. The material knowledge, therefore, representing the character of Ume Bangka itself as mentioned by Christian Norberg-Schulz, *"character depends upon how things are made"* [13]. Thus,

Ume Bangka *"might be inherited and represent ties of kinship"* [15], further cemented the strong connection between landscape-community-family in describing the cultural identity in Bangka Island.

3.2. Modernity in Bangka: old versus new

Displacement—as discussed before—is the main issue that threatens Ume Bangka's cultural values. Even worst, the displacement disrupted the tradition and custom of the indigenous people in Bangka, slowly replacing local identity with something more Western-centric. Instead of living in harmony with nature, the connection between modern urban man and their natural environment is reduced to fragmentary relations, often ends up with excessively exploiting the soil [11,13].

Similar condition happened in Bangka Island, especially during the 19th Century as the colonial government exploited the tin resources, creating conflict between agriculture and tin mining in Bangka [3]. Illegal mining and forest cutting, as well as lack of supervision by the colonial government, had caused huge environmental damage, prompting heavy floods in traditional settlements that forced locals to move from forest to kampungs on the outskirts of the city [5].

When forced to resettle alongside the major roads in Bangka Island, the indigenous people of Bangka were exposed to the marvel of modern industrial technology. With tin mining business thrived and prospered, the urban settlement in Bangka also expanded. More roads were built, more cars were strolling on the street, and more building was connected as an object of a unified thing [18]. "*Industrialization relocated work away from homes that would, instead, turn into units of consumption, education and clean domesticity*" [16], meaning that technology had separated work, social, and home which once related to each other in the ideology of living *berume*. The issue now relies on not only being displaced from the source material but also being stripped off from their lifestyle completely.

Ignorance towards other cultures was the main problem during this relocation, especially when the idea of home between Western and native culture is completely different. In the western conception of dwelling, home is based on the notions of stability, permanency, sense of belonging, and—most importantly—being rooted in a place [19]. Unlike the indigenous people of Bangka and their Ume Bangka, dwelling means moving from one place to another [1-3]; there is no permanent location, always moving from places and continually searching for a better location to find food. Nomadic life in the industrialized society, therefore, regarded as an outdated or dysfunctional activity by Western society, and thus preventing progress and development in the future [10].

Tin mining exploitation, roads construction across Bangka, and the relocation of Bangka's native people had promoted some social changes, which then associated with rising capitalist classes—in this case the foreigners: the Dutch colonial official, Chinese and Arabs migrants who came to Bangka Island as miners or merchants, respectively [1,16]. The Dutch colonial government divided settlements in most major cities in Bangka Island based on their races and ethnicities, in the sake of gaining control over the population [1-2]. Urban segregation in Bangka Island made the locals grew jealous of others' prosperous lives and nice houses. Furthermore, Chinese in Bangka had a strong impression on the locals. While locals buried their wealth, *"the Chinese spent money on feasts"* as well as lavish weddings and funerals parties [20]. It did not help that the colonial government assumed those who lived in Ume Bangka were *"not part of the civilized society*" and therefore are not capable of handling their own laws [3]. In a sense, the local people of Bangka were ignored from the urban society.

The inferiority of the indigenous people of Bangka carried on even until post-colonial era during Sukarno's reign. As the first Indonesian president, Sukarno envisioned Indonesia as a modern country that will gain international influence. Soekarno's idea was manifested through massive infrastructures projects being constructed in Jakarta, the capital city. Bangka Island, on the other hand, was neglected. Ironically, the regional development plan for Bangka Island was crudely based on the colonial legacy during the tin mining industry. The development had been understood as a homogenous idea, instead as unique social, cultural, and historical characteristics. Just like the colonial official, traditional housing typology such as Ume Bangka is seen by Indonesian government as a symbol of 'under developed society', 'untamed' and 'old' [2,5], clearly not representing the national spirit as a modern and thriving country in the 20th century with skyscrapers that reach the sky. Woods as Ume Bangka's main material is also seen as inferior and cheap-looking material compares to stone and steel architecture which more representing the technological wonders of mankind [21].

3.3. Ume Bangka for a happier future

For centuries, Ume Bangka is the symbol that unites landscape, community, and family. As a traditional housing typology, Ume Bangka is the identity of indigenous people in Bangka Island—their culture, heritage, and tradition. Ume Bangka, therefore, is not *"a thing of the past*" [22]. However, we cannot deny that the world had changes and therefore we should not copy the past. *"It won't work*", says Christopher Day [17]. Recreating Ume Bangka as if in 1500 BC, will not work in the 21st century because modern man now seeks stability. They want to settle down and gain access to other facilities provided by their government. It is not about feeling inferior towards foreign culture, no, because culture absorbs *"'foreign' elements, made developments within the particular culture, and 'modernized' their worldviews*" [22]. If the indigenous people in Bangka Island understand their origin and their representation from Ume Bangka, adapting to modern world should not be a hard task.

Even before the Dutch came and monopolized the tin mining industry, the native people in Bangka Island had lived side-by-side with foreigners for hundreds of years. Chinese and Arabs came first as merchants and traders. Once, there were no boundaries between the locals, Chinese, and Arabs settlements; they were free to interact with each other, resulting in a rich cultural hybridity [1,3,5]. The problem in relocating the native people from their natural settlement to the city was that the colonial government did not predict the disruption they caused towards the relationship between human and their surrounding—whether towards the community or the natural environment—for the sake of having power over their subjects.

"*The identity of place is determined by location*" [13]. Some places get their identity because the location is interesting, while the man-made components might be insignificant. It could also happen the other way around: the landscape might be dull and boring but it posses a well-defined configuration [13]. In the case of Ume Bangka, its character and identity were influenced by the natural environment. When it was removed from its natural environment by the colonial government, Ume Bangka was thrown into an unfamiliar terrain of man-made construction with strict regulation. Ume Bangka was not allowed to improve and adjust; it merely just existed, torn out of its root. Without identity, all qualities are lost and thus promotes an environmental crisis which, if ignores, will implied to human crisis [13].

Continuing to preserve Ume Bangka will eventually preserve the identity and character of native people in Bangka. It does not mean that Ume Bangka must be rebuild in the forest. It is no use because people no longer live in the woods and Ume Bangka will once again be forgotten. Accepting the local heritage, tradition, and legacy will be the first step in achieving the sustainable happiness—the balance between our natural desire with the well-being of other and the natural environment—will foster sustainable behavior [23]. Sustainable behavior will lead people to understand that fixes in social conduct cannot be solve with technology, cost-benefit appraisals and taxes [17]—resolving social, economic, and political problems will take more than resettling people from the rain forest to the city, denying their tradition altogether.

4. Conclusion

For centuries, the people in Bangka Island had lived feeling inferior. They were segregated, ostracized, and alienated from their own land. First by the colonial government where the natives of Bangka Island were forcibly removed from their natural habitat. The idea of *berume* was abandoned: people were forced to change their agricultural tradition to working in tin smelting companies. Then, after independency, same ideology—new is better than old—was carried on by Indonesian government. Traditional houses such as Ume Bangka were viewed as 'underdeveloped' and 'backwardness'. In fact, vernacular architecture is more modern because it connects people, community, and landscape for a sustainable future. The interaction and connection between these three aspects make Ume Bangka suitable as a symbol of cultural identity in Bangka, also corresponding to the future of Bangka in general.

To make Ume Bangka thriving once more, we must not see Ume Bangka as a noun. Instead, we must see Ume Bangka as a living being that connects people, community, and landscape. Therefore, Ume Bangka is flexible to changes and adjustment in the future. The ideology of Ume Bangka—a nurturing system of people-community-natural environment—will be preserved, while other components might be change accordingly. First of all, Ume Bangka should adjust itself to its current location: the city. For example, revitalizing and preserving existed Ume Bangka will alert locals and government attentions regarding traditional houses. Then, reconstructing Ume Bangka are another solution to learn and pass down the traditional material knowledge and craftmanship to the younger generation in Bangka Island. In the end, people will live in Ume Bangka with pride.

This paper serves as a preliminary research and thus requires more data on how the locals live today to prove hypothesis expressed within this paper. However, current research and readings are focusing on the previous live of Bangka's ancestors and their Ume Bangka or focusing on the physical attributes of Ume Bangka. Therefore, an anthropological approach might be needed to capture the core of Ume Bangka, current habits and tradition in Bangka, and relates it to the modern world. Then, a prosperous future Ume Bangka and people of Bangka will be achieved.

5. Acknowledgments

This paper is funded by the Grant Scheme of PTUPT Universitas Indonesia in 2018.

6. References

[1] Akhmad E 2016 *Perang Bangka Tahun 1812-1851 Masehi* Pangkalpinang: Dinas Kebudayaan, Pariwisata, Pemuda dan Olahraga Kota Pangkalpinang.
[2] Kemas R K 2005 *Postcolonial History of Architecture and Urbanism: Power and Space of Indonesian Tin Mining in Bangka Islands* Saarbrücken: VDM Verlag Dr. Müller GmbH & Co.
[3] Kemas R K 2013 *The Hybrid Architecure of Colonial Tin Mining Town of Muntok* Jakarta: Penerbit Universitas Indonesia (UI-Press).
[4] Kemas R K, 2013 "Arsitektur Tradisional Rumah Melayu di Propinsi Kepulauan Bangka Belitung", *Kajian Melayu Provinsi Kepulauan Bangka Belitung nasional seminar* (unpublished).
[5] Kemas R K, Ratu A K, Evawani E, M Nanda W, and Teguh U A 2013 *Makara Seri Sos. Hum.* **17** 75.
[6] Gaudenz D. 2008 *Bijdragen tot de Taal-, Land -en Volkenkunde* **164** 450
[7] Erwiza E. 2008 *Southeast Asia: Hist. & Culture* **37** 91.
[8] Kemas R K, *et. al.* 2017 "Tipologi Arsitektur Rumah Panggung Kayu Melayu dan Tradisi Membangun sebagai Acuan bagi Kegiatan Revitalisasi Kota Pusaka di Kota Muntok", final report for Penelitian Unggulan Perguruan Tinggi (PUPT) research grant (unpublished).
[9] Kemas R K 2017 *Inter. J. Built Envi. and Scientific Re.* **1** 91.
[10] Marcel V 2006-2007 *Pers. In Vernacular Arch.* **13** 115.
[11] Amos R 1969 *House Form and Culture* New Jersey: Prentice-Hall Inc.
[12] Brian C 2009 *J. Arch. And Plan. Research* **26** 30.
[13] Christian N S 1979 *Genius Loci: Towards a Phenomenology of Architecture* New York: Rizzoli Publication.
[14] Ian T 2003 What use is the *genius loci?* Sarah M (ed.) *Constructing Place: Mind and Matter* New York: Routledge.
[15] Jenefer R 2012 *J. of Aesthetic & Art Crit.* **70** 337.
[16] Anitra N 2018 *Shared Living on a Shared Planet* Pluto Press.
[17] Christopher D 2002 *Spirit and Place* Oxford: Architectural Press.
[18] Peter T 2013 *Log* **27** 51.

[19] Didem K 2003 Migrant homes: ethnicity, identity and domestic space culture Sarah M (ed.) *Constructing Place: Mind and Matter* New York: Routledge.
[20] Mary S.H. 1995. *Archipel* **49** 25
[21] Leele V 2008 *Place and Location: Stud. in Envi. Aesthetics and Semiotics* **6** 177.
[22] Nihal P 2010 *J. Arch. Edu.* **63** 76
[23] Catherine O 2013 *Canadian J. of Edu.* **36** 228.

The spatial and settlement pattern in Mabi Hamlet, Belantih Village, Kintamani: conservation of tangible and intangible of Balinese traditional architecture in Bali Aga villages

Ni Made Yudantini
Udayana University, Bukit Jimbaran-Bali, Indonesia.

Corresponding e-mail: nmyudantini@unud.ac.id

Abstract. The Indigenous of Bali Aga villages has uniqueness in architecture, traditional norms and values. Most of Bali Aga villages are located in highland and mountainous areas. The development of economic and technology become a threat to the integrity of traditional Balinese architecture especially in Bali Aga villages. The preliminary research finding was that traditional buildings in Bali Aga villages had changed into modern style, both in spatial and settlement patterns. This happened in Mabi Hamlet, Belantih Village, Kintamani District. There is a decrease in number of traditional Bali Aga houses that threat the extinction of Balinese traditional architecture and local identity. The research aims to explore the development and changes in spatial and settlement patterns in Mabi Hamlet by focusing on Balinese traditional architecture. The importance of inventory and documentation is an effort to preserve the heritage of Balinese traditional architecture. This case study method was selected for Mabi Hamlet in Kintamani Sub-district. The interviews were conducted to village leaders, hamlet heads and traditional elders who played important roles in the implementation of customs in both tangible and intangible aspects. The literature study was conducted to find out the history of Bali Aga villages and to understand Balinese traditional architecture, its culture and customs. Studies on previous research and institutional study aim to understand the timeline, village's potency and village development. The study results recommendation for the conservation of local values, especially the legacy of traditional Balinese architecture as a local identity and a source of Nusantara identity.

Keywords: the spatial and settlement pattern, Mabi Hamlet, conservation, Bali Aga villages

1. Introduction

Bali Aga is the name of Indigenous villages in Bali that still adopt and implement their original tradition, norms and values. It is different from other villages in south part of Bali that have been influenced by the Majapahit culture while Bali Aga does not. Most of Bali Aga villages are located in the highlands of Bali Island, mountains and some are located along the Lake of Beratan in Kintamani area. Bali Aga is one of popular topics to study because its indigenous original culture and tradition including traditional governance system, architecture, and religious ceremonies implementation. They also have different dialects from general Balinese language and some village still use astronomy to decide the good day (*dewase*) for the implementation of religious ceremonies. The form of spatial and settlement pattern in the Bali Aga villages is strongly influenced by the villages topography known *hulu-teben* (upstream-downstream). It has mountain or hills/highlands as upstream (hulu), oriented for the sacred place (*kaja*) and locate village temples. There are springs for the impure zone (*kelod*) as downstream (*teben*) that lay the village cemetery and its Dalem Temple, and also public facilities, such as public bathing, parking area, community hall etc [1,2]. Budiharjo [3] distinguished three types of village's pattern that also influence the housing pattern. The first pattern is *pempatan agung* or crossroad pattern mostly found in mainland area. This pattern constructs four neighbourhoods that face to the centre of the village. The second pattern is *aling-aling* pattern (*swastika* form) that do not have" straight continuous roads" as the *pempatan agung* pattern. The third is linear pattern that is mostly found in mountains and highland areas of Bali Aga villages [3].

Many scholars who studied the Bali Aga villages, such as Covarrubias in 1974, studied the Indigenous as "original or ancient Balinese" [4], meanwhile Reuter (2002) named the Indigenous Balinese as "the Mountain Balinese" [5]. Further, Hauser-Schaublin (2004) defined that the Balinese people called Bali Aga or Bali Mula or Bali Kuna, and as an "ethnic group" [6]. Other scholar such as Muller [7] documented the Bali Aga during her fieldwork in 1980s in four regions mostly in the north, east, middle and central mountains of Bali [7]. Yudantini [8] concluded and mapped the location of 62 of the Indigenous villages spread across Bali Island from the east, middle to the west (Figure 1) [8]. One of them is Belantih Village with Mabi Hamlet as it is mentioned in Muller's fieldwork.

The rapid development and improvement of technology and communication bring impact to the communities in the Indigenous villages to have better living. There only remain five traditional houses in Mabi Hamlet that still maintain the original of architecture and norms. Other households have changed their housing pattern into modern style. This phenomenon becomes a threat to the existence of Balinese traditional architecture and the local identity. The research investigates the spatial and settlement pattern and its changes in Mabi Hamlet as a documentation concerning the Bali Aga heritage especially in the architecture field. The research sought tangible and intangible aspects related to the Bali Aga culture and tradition in Belantih Village as uniqueness of the village. The research recommended the conservation of spatial and settlement pattern of the village and served as source of basic information of the existence of Bali Aga villages in Bali.

Figure 1. Mapping of the Indigenous villages across Bali Island and Belantih Village is located near the Batur Mountain [8]

IOP Conf. Series: Earth and Environmental Science **213** (2018) 012025 doi:10.1088/1755-1315/213/1/012025

2. Methods

The research was conducted mostly a descriptive qualitative research through case study. The case study method is very useful to obtain a comprehensive study, to focus on the object under study and to gain better information regarding non-physical object and also about the behavior of the communities [9]. Direct observation on the object of case study and interviews help to obtain real data, which is used continuously in the analysis of problems related to the spatial and settlement pattern in Mabi Hamlet, Belantih Village, Kintamani Sub-district. The observation was used to investigate the spatial and settlement pattern that had been changed recently. The preliminary research found that there were five traditional houses extant in Mabi Hamlet. The interview was conducted to hamlet's leader, the householders and questions to the villagers. Most of them cannot avoid the settlement changes especially the housing pattern. The intensive literature review aims to gain understanding of Bali Aga history and its characteristic in both tangible and intangible aspects. The previous research was studied to discover the past condition of the village and to understand the socio culture life of the community. The collected data were then analyzed both through qualitative and quantitative analysis.

3. Discussion

3.1. Glance of Mabi Hamlet in Belantih Village

Belantih Village as one of Bali Aga is located in Kintamani Sub-district that has cold weather (22° C-26° C) surrounded by valley and hills. Belantih is located by coordinate between 115.2877 east longitude and -8.24246 south latitude [10]. Belantih Village is bordered to the north by Selulung Village, to the south by Belanga and Batukaang villages, to the east by Daup Village, and to the west by Catur Village (Figure 2). Belantih Village has an area of 9.06 km² (960 ha) that the land use is allocated for plantation (22.63%), dry land cultivation (62.03%), housing (2.32%), grave yard (0.11%) and other function (12.91%) [11]. Most people work in the farm especially as farmer of orange/citrus and coffee. The population of Belantih Village is approximately 2.527 people in 2017 with the density of 263 people per km² [10].

Belantih Village includes in traditional village (*desa adat* or *desa pekraman*) of *Desa Pekraman* Selulung, and administrative village of Belantih. Administratively Belantih Village has eight hamlets as follows Mabi, Sabang, Luahan, Pangkung, Kayu Padi, Tangguan, Pandan and Selulung Timur. Selulung Timur Hamlet is located in the territory of Selulung Village but administratively Selulung Timur is under Belantih Village. Likewise Sabang Kelod Hamlet is located in the territory of Belantih Village but under Selulung Village administration. This case occurs in some villages in Bali and it is called *saling seluk*. Mabi Hamlet has been observed by Carole Muller in 1980s. Mabi Hamlet has population of 95 householders with 400 people, and all embrace Hindu Religion.

3.2. Socio Culture of Mabi Hamlet, Belantih as Bali Aga village

The people embrace Hindu religion and they have 3 types of temples. They are Tri Kahyangan Tiga temples, 1 *subak* (irrigation) temple and 1 other temple. 3 priests lead these religious activities. In the culture, Belantih has art organization including 6 groups of traditional musical called *gong kebyar* and 2 types of other arts. Meanwhile the people also involve in household handicraft and only 3 household handicrafts are available in Belantih Village. The death ceremony is commemorated through *biye tanam* as a characteristic of Bali Aga tradition. *Biye tanam* means the corpse is buried for the cremation ceremony, but only the death's symbol is burned. The village also celebrates mass cremation in certain period of time that is usually 4-5 years. Belantih Village has 3 cemeteries, namely Setra Bunut, Setra Selulung and Setra Nyawah. The cemeteries are distinguished for infant's corpse in the upstream of cemetery, for noble's corpse in the northeast direction, and the rest for the corpse of common people. Most of people in Belantih village are *nyineb wangsa* (hide their caste), but only one clan is using their caste; they are Arya Kuta Waringin in Kayu Padi Hamlet.

In the governance system, Belantih Village implements a customary government system called *Apad* or *Ulu Apad*, known as an upright structure, with a total of 40 members consisting of 20 people in right side and 20 people in left side when they assemble the traditional ceremonies. Belantih Village applies the pure *ulu apad* system which means the members are selected by natural process, while the impure *ulu apad* system is bequeathed to the posterity or to the descent.

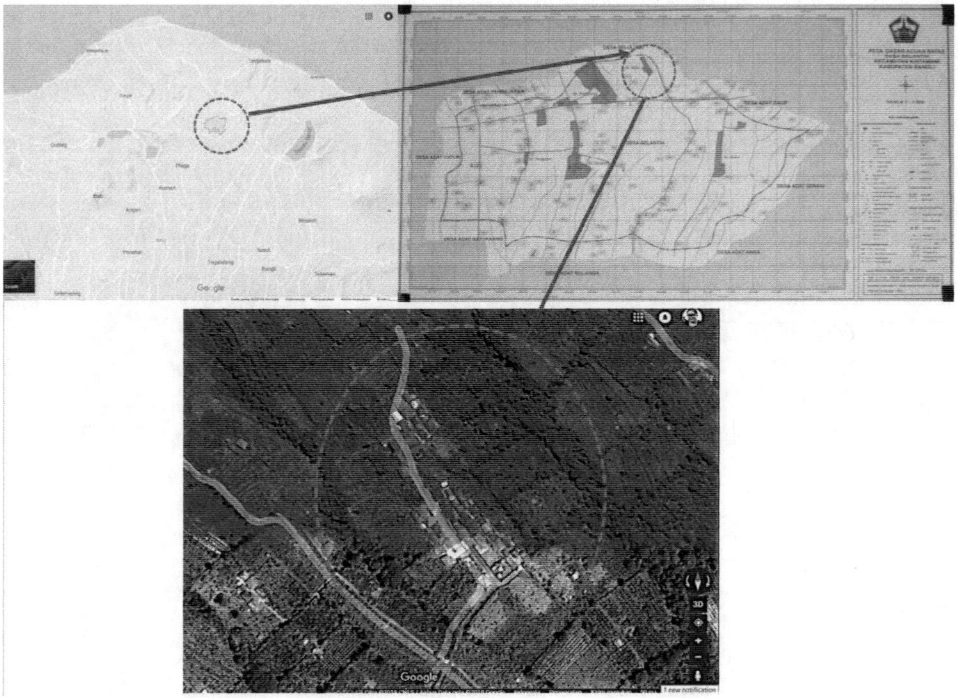

Figure 2. Mabi Hamlet in Belantih Village

3.3. The Spatial and Settlement Pattern of Mabi Hamlet

The topographic condition of Mabi Hamlet in Belantih Village influences the spatial and village patterns with having a linear pattern in the village as well as the unit of housing. Muller [7] has documented Mabi Hamlet in her book of Bali Aga Villages: Field Works in the 1980s. She presented photographs of Mabi's settlement condition with having linear pattern that followed the village's topography. The family temple was placed in the main zone as a sacred area. The settlement was having natural condition and mostly used local material such as wood for the housing structure and bamboo for the roof called *sirap* although zinc in several houses had been used (Figure 3).

Figure 3. Linear pattern of one yard of unit housing in Mabi Hamlet in the 1980s [7]
(Source: Muller, 2011).

Figure 4. The linear pattern of the settlement still extant in Mabi Hamlet, 2018

The characteristic of traditional village is divided based on the values of *Tri Mandala* concept. They are the main zone or *utama mandala* for holy place, the middle zone or *madya mandala* for settlement and public facilities, and the profane/impure zone or *nista mandala* for cemetery and services. This concept is also adopted into spatial and settlement pattern. The upstream of unit housing is placed family temple, in the middle is functioned for houses, and in the impure zone is for service function (Figure 4 and 5). One unit of the housing yard comprises of several linearly dotted houses. There are 5 to 10 of householders and the inhabitants mostly do not in one lineage. The houses are in open circumstances and no border or fencing within the neighbour. This traditional value means that the communities have good relationship to trust each other. It is also for security in the village where everybody involves in watching neighbourhood.

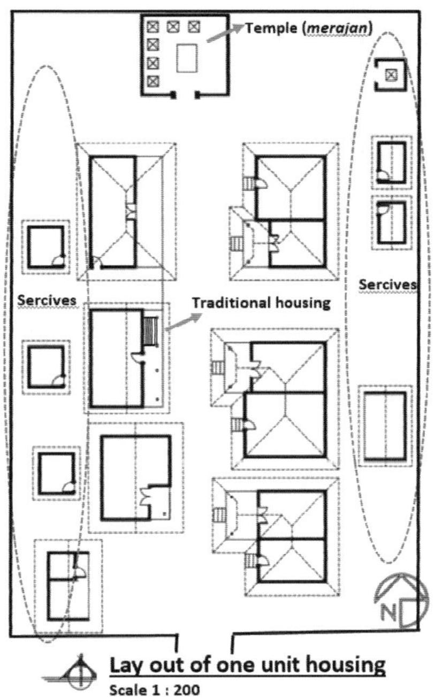

Figure 5. Linear pattern of one yard of unit housing in Mabi Hamlet.

The house pattern traditionally consists of 4 functions as follows: kitchen, bedroom, holy place and terrace or *ampik* or *terampe* (Figure 6, 7 and 8). The house has simple functions where the people do their activities in one place in their house except for livelihood. Mostly Belantih people work as orange farmers in the surrounding area.

IOP Conf. Series: Earth and Environmental Science **213** (2018) 012025 doi:10.1088/1755-1315/213/1/012025

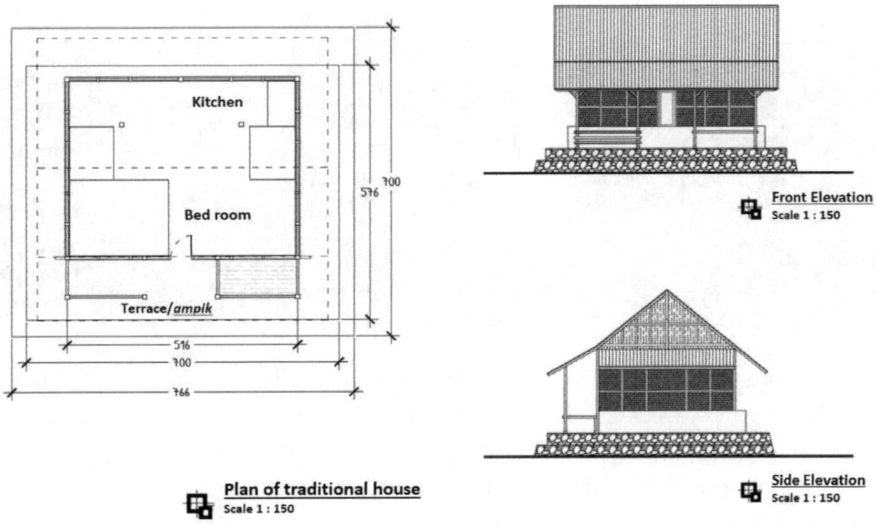

Figure 6. Sketch of floor plan and elevations of traditional housing in Mabi Hamlet, Belantih Village

Figure 7. Traditional house of Mabi Hamlet changed the roof shape and material.

Figure 8. The original traditional house in Mabi changed the roof material.

3.4. The Changes and Conservation Approaches

The changes of housing are unavoidable due to the increasing development. This also affects Mabi Hamlet and the people change their house pattern into modern style. They built new house to fulfil the demand of living to have better quality of life. The new material, such as concrete was mostly used (Figure 8). The appearance, house's shape, material, style were changed (Figure 9). Based on observation and interview, there were only 5 existing traditional houses with modification of roof shape and material. Although people changed their houses from traditional into modern style, the people still apply the settlement pattern with linear pattern. This change is also realized by Budiharjo [3] that it caused by the growing of faster tourism development. He states that Bali faces a danger. The increasing number of tourists in Bali affects the societies' economic and of course they want to change their life to modern which look simple and efficient.

Figure 9. The change of house into modern look.

Figure 10. Linear pattern is still adopted in the Mabi Hamlet.

Budiharjo [3] stated there were two points of conservation to be applied in the community to avoid more danger and huge changes of traditional values and the heritages. One approach is comprised of conserving of the ecology problems including flora, fauna, pollution and population explosion. The other approach is conservation of architecture and the built environment. Thus he emphasized that the alternate conservation of Bali's problem is by conserving the Balinese architecture including its environment.

> "The best way to avoid this danger happening on Bali is by the conservation of Balinese architectural heritages which will involve the study of the whole environment" (p. 3)

The other aspect in Mabi Hamlet, the intangible of traditional value of spatial space remain implemented in the village. Belantih people celebrate the annual festival called *ngusaba*. As it is emphasized by Reuter [12] that the people of Bali Aga in the mountains perform rituals or rituals every year to commemorate fertility for the respect of the ancestors as a mark of the turn of the season for agricultural production [12]. The people still celebrate this ritual of their tradition and believe to honour to the forefather. This ritual to honour their ancestors can be applied in their house. Although they changed the house into modern style, they still keep one holy room as a place of ancestor worship (Figure 11 and 12).

Figure 11. The place for ancestor worship

Figure 12. The offering in the front of main entrance of house, conservation traditional values

4. Conclusion

The conservation of tangible in tangible aspects in Balinese traditional architecture faces threats because of increasing development and globalisation. It cannot be denied that the people in Mabi Hamlet have changed their housing into new style and with new material. As a Bali Aga, the people still adopt original culture and tradition in their life. Comparison to other research showed that the

IOP Conf. Series: Earth and Environmental Science **213** (2018) 012025 doi:10.1088/1755-1315/213/1/012025

spatial and settlement pattern in Mabi Hamlet still implements linear pattern as it is inherited from the past. Nevertheless, the houses pattern have changed into modern form. Now Mabi Hamlet only has 5 traditional houses and that situation threatens the Balinese traditional architecture. Guidance to the people is urgently needed in order to cultivate their awareness of their traditional architecture to be conserved as it serves as reflection of Balinese identity of traditional architecture. The research recommends future research to identify, inventory and document the traditional architecture of Bali Aga for conservation.

5. Acknowledgments
I owe my gratitude and thanks to Udayana University who funded this research through DIPA PNBP Fiscal Year of 2018 with contract number 1677/UN 14.2.5.II/LT/2018. Acknowledgments are also addressed to research teams and students who have assisted the completion of this research from the beginning of the proposal to the progress report.

6. References

[1] Gelebet I N, Puja I G N A and dan Dokumentasi Kebudayaan Daerah. P I 1981 *Arsitektur tradisional daerah Bali / penyusun, I Nyoman Gelebet ... [et al.] ; penyunting, I Arinton Puja* (Departemen Pendidikan dan Kebudayaan, Proyek Inventarisasi dan Dokumentasi Kebudayaan Daerah [Denpasar])

[2] Adiputra I T, Sudaryono S, Wiyono D and Sarwadi A 2016 Konsep Hulu-Teben pada Permukiman Tradisional Bali Pegunungan/Bali Aga di Desa Adat Bayung Gede Kecamatan Kintamani Kabupaten Bangli, Bali *Forum Tek.* **37**

[3] Budihardjo E 1995 *Architectural conservation in Bali* (Yogyakarta Indonesia: Gadjah Mada University Press)

[4] Covarrubias M 1976 *Island of Bali* (Oxford University Press, Oxford)

[5] Reuter T A 2002 *Custodians of the sacred mountains : culture and society in the highlands of Bali* (University of Hawai'i Press)

[6] Hauser-Schäublin B 2004 "Bali Aga" and Islam: Ethnicity, Ritual Practice, and "Old-Balinese" as an Anthropological Construct *JSTOR* **77** 27–55

[7] Muller C 2011 *Bali aga villages : field work in the 1980s* (Walsh Bay Press)

[8] Yudantini N M 2015 *Bali Aga Cultural Landscape Challenges: Conserving the Balinese Traditional Landscape for Future Balinese Indigenous Villages (Bali Aga) and Communities* (Deakin University-Australia)

[9] STAKE R E 1978 The Case Study Method in Social Inquiry *Educ. Res.* **7** 5–8

[10] Anon DATA POKOK DESA/KELURAHAN

[11] BPS Kabupaten Bangli 2017 *Kabupaten Bangli Dalam Angka 2016* (Bali: Pemerintah Kabupaten Bangli)

[12] Reuter T A 2002 *The house of our ancestors : precedence and dualism in highland Balinese society* (KITLV Press)

Indoor thermal environment in tropical archipelago city

Mustamin Rahim[1] , Baharuddin Hamzah[2]
[1)]Associate Professor, Department of Architecture, Khairun University, Ternate-Indonesia.
[2)]Professor, Department of Architecture, Hasanuddin University, Makassar-Indonesia.

Corresponding e-mail: mustamin_rahim@yahoo.co.id

Abstract. This study aims to evaluate the indoor thermal comfort of the office building in the tropical archipelago city of Ternate, Indonesia. Survey and data collection were performed using several instruments, LSI-LASTEM Thermal Comfort Multi Logger, HOBO data logger, and HOBO data logger with the external sensor for measuring temperature, relative humidity, and wind velocity. At the time of measurement, thermal condition perceived by respondents is observed by the questionnaires of Thermal Comfort Vote (TCV) and Thermal Sensation Vote (TSV). TSV is performed by ASHRAE standard 55 and TCV from Bedford scale. The results show that indoor thermal condition is in uncomfortable level with the average temperature exceeds 29°C, RH is more than 70%, and wind velocity is lower than 0.5 m /s. More than 60% employees feel warm condition in TCV and TSV scales, but a majority of them (67%) still accept these conditions. It indicates that the employees in tropical archipelago city are quite tolerant to high temperatures. About 59% of respondents feel wind velocity is uncomfortable and 70% respondents expect to increase of wind velocity. Therefore, the new design concepts are needed to improve natural ventilation performance by combining cross ventilation with clerestory ventilation systems and planting trees around the building to reduce solar heat gain that penetrates to the building's interior.

Keywords: micro climate, thermal comfort, natural ventilation, office building

1. Introduction

Air temperatures are increasing throughout the world due to the increase of CO_2 emission in the atmosphere, especially in metropolitan cities, as the consequence of exhaust gas from industry, motor vehicle, and alteration of land surface characteristics from natural elements to artificial ones such as asphalt, concrete and steel. The increase of CO_2 in the atmosphere causes an urban heat island effect and climate change. Increasing outdoor temperatures strongly affect indoor temperatures, making it hard to achieve thermal comfort and causing the increased use of air conditioner (AC). Utilization of AC significantly affects the increase of energy consumption in the building. The building sector consumes almost one-third of energy from global energy consumption and is an equally important source of CO_2 emissions [1]. Building sector contributes up to 30% of global annual greenhouse gas emissions and consumes up to 40% of all energy, both in developed and developing countries [2].

The ISO 7730 standard defines thermal comfort as the condition of mind, which expresses satisfaction with the thermal environment [3] (ISO, 1995). Indoor thermal comfort affects the employees' work productivity. According to the Lawrence Berkeley National Laboratory, work productivity increases 7-13% in buildings with the better air quality and 15 - 50 % in buildings with the better sunlight quality [4]. Thus, improving indoor thermal comfort is very important especially by applying natural ventilation and natural lighting strategies.

Utilization of natural ventilation has been successful in many buildings in America such as school buildings, hospitals, supermarkets, offices, warehouses with the following advantages: energy efficiency, improving indoor environmental quality, lower energy equipment costs, lower maintenance costs and replacement of conditioning equipment air and electricity, enhanced thermal comfort, and integrated with natural lighting. Research results in the United States show that a naturally ventilated office building consumes less than half the energy consumed by buildings using AC [5]. Therefore, this paper will discuss indoor thermal performance on the office building in the Archipelago City of Ternate.

2. Methods

This research is focused to evaluate the indoor thermal comfort of several office buildings in Ternate City (archipelago city) by using several instruments, LSI-LASTEM Thermal Comfort Multi Logger, HOBO data logger, and HOBO data logger with external sensor for measuring temperature, relative humidity, and wind velocity. At the time of measurement, thermal condition perceived by respondents is observed by the questionnaires of Thermal Comfort Vote (TCV) and Thermal Sensation Vote (TSV). TSV is performedby ASHRAE standard 55 [6], which consist of seven scales: hot (+3), warm (+2), slightly warm (+1), neutral (0), slightly cool (-1), cool (-2), and cold (-3), and also TCV is measured by seven categories from Bedford scale [7]: much too warm (+3), too warm (+2), comfortably warm (+1), comfortable (0), comfortably cool (-1), too cool (-2), and much too cool (-3). Besides, respondents were also asked their opinions about the thermal preference and thermal acceptance whether the thermal environment was acceptable or unacceptable.

Figure 1. Five office buildings for the research objects.

3. Discussion

3.1. Climate condition of Ternate Island

Ternate is a small island in the North Maluku (Moluccas Island) and dominated by the area of volcanic mountain Gamalama with high reaches 1715 m. The population is fewer than 200,000 in the area of 111.39 km^2 [8-9]. Ternate City is an archipelago region characterized by a tropical climate with the annual average temperature is around 24-32°C, Humidity is around 77 – 83%, Wind speed is about 7 – 10 km/h, and precipitation is generally exceeded of 120 mm, except during August until October is very low of around 60-90 mm. The highest precipitation occurs on May (about 200 mm), and the lowest occurs in September (around 60 mm) (see Table 1). This condition portrays that based on annual weather averages; the temperature is not too high because rain occurs every month with different frequency or there is no summer season. However, based on the daily weather distribution especially on sunny days, the weather condition is very hot with the high temperature is up to 37°C. In this case, outdoor temperature can cause the rising of indoor temperature and application natural ventilation is very difficult to reduce temperature until comfortable level. Therefore, the new design concept is needed to improve natural ventilation performance in the building.

Table 1. Annual weather averages in Ternate during 2005–2015 [10].

	Jan	Feb	Mar	Apr	May	Jun	Jul	Aug	Sep	Oct	Nov	Dec
High Temp (°C)	30	31	31	31	31	31	31	31	31	32	31	31
Low Temp (°C)	24	24	24	24	24	24	24	24	24	24	24	24
Mean Temp (°C)	27	27	28	28	28	28	27	27	27	28	28	28
Humidity Ave (%)	83	82	81	82	82	81	78	77	78	78	82	83
Wind Ave (km/h)	10	10	10	8	7	7	8	9	8	8	7	8
Precipitation (mm)	160	122	126	147	200	176	120	92	63	69	153	142

3.2. Indoor termal performance

In Ternate City, energy resources are insufficient. It is only produces by diesel power plant, while the demand for energy keeps increasing primarily because new buildings are constructed. Based on the field observation, a cooling system in the building is generally using air conditioning. It causes the increasing of energy consumption in the building so that the supply of energy is not enough and power outages are done every day in rotation. This condition can influence the ventilation system and thermal performance in building especially in the office building.

Table 2. Indoor thermal condition [12].

Thermal Condition	Offices				
	Basarnas	Parawisata	Balaikota	Diknas	PU
Temperature (°C)	29.10	29.24	29.00	30.86	30.61
RH (%)	69.62	70.87	70.38	76.41	75.58
Wind Speed (m/s)	0.13	0.23	0.20	0.26	0.26

Table 2 portrays that the average temperature exceeds 29°C, relative humidity (RH) average is around 70 – 76% and wind speed is deficient of about 0 – 0.3 m/s. It indicates that the thermal condition in the office building is uncomfortable. The standard of thermal comfort in a hot-humid tropic area is around 24-26°C, relative humidity: 40-60%, wind velocity: 0.6 -1.5 m/s with the activity of relaxed and a thin dress [11].

Table 3 shows that thermal sensations felt by employees in office buildings during daytime of about 6% cool, 13% slightly cool, 15% neutral, 37% slightly warm, 19% warm, and 11% hot. Those conditions portray that only 15% employees felt comfortable and 85% uncomfortable: about 19% under comfortable level (-3 until -1) and 67% above comfortable level (+1 until +3). This condition can cause the increasing use of air conditioning (AC), which has an impact on the increase in energy consumption, while the energy resources are insufficient. On the other hand, it produces CO_2 emissions during its combustion process. CO_2 in large amounts is the primary cause of the global warming.

Table 3. Thermal sensation votes (TSV).

Thermal Condition	Offices				
	Basarnas	Parawisata	Balaikota	Diknas	PU
cold	0	0	0	0	0
cool	0	0	17	8	0

Slightly cool	8	0	17	17	17
Neutral	8	0	17	17	25
Slightly Warm	42	83	17	33	33
Warm	17	17	25	8	25
hot	25	0	8	17	0
Total (%)	100	100	100	100	100

Source: Analyze 2018

Table 4 shows the thermal comfort felt by employees in which more than 60% of employees feel warm condition in the Bedford scale (TCV). This condition is similar to the ASHRAE level (TSV). These state that employees are generally working under uncomfortable conditions. An office which is too warm makes its occupants feel tired and one that is too cold causes the occupants' attention drifted, making them restless and easily distracted. Even, a minor deviation from comfort affects the stress, performance and safety. Workers already under stress are less tolerant of uncomfortable conditions. Thus, indoor thermal conditions are necessary to decrease until the comfortable level in order to improve work productivity.

Table 4. Thermal comfort vote (TCV).

Thermal Condition	Offices				
	Basarnas	Parawisata	Balaikota	Diknas	PU
much too cool	0	0	0	0	0
too cool	0	0	17	0	0
comfortably cool	17	0	17	17	17
Comfortable	17	0	17	17	8
comfortably warm	33	67	25	33	50
too warm	8	33	17	17	17
much too warm	25	0	8	17	8
Total (%)	100	100	100	100	100

Source: Analyze 2018

Figure 2 indicates that the majority of respondents (69%) expect to decrease the temperature in the office building until the comfort level, and only 31% feel comfort. Although the thermal conditions are above the comfort level, a majority of them (67%) can still accept these conditions to work in the room. Only about 31% cannot accept the condition. Differences of respondent perception indicate that everyone is different in responding the thermal conditions. It may be due to habits of life in warmer conditions.

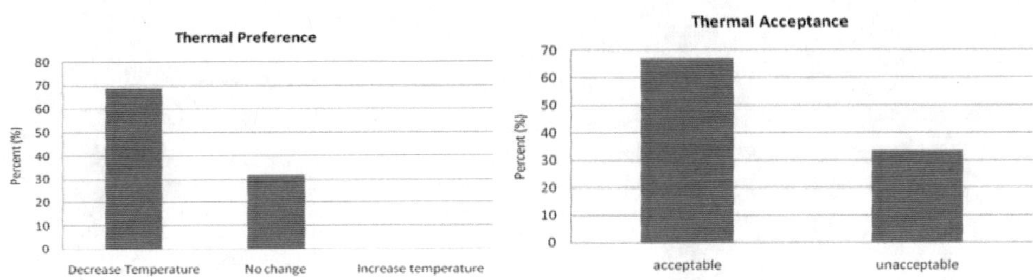

Figure 2. Thermal preference and acceptance of employees.

The average of relative humidity is generally above 70% (see Table 1). It means that the relative humidity is high and less comfortable. Relative humidity comfort is around 30-60%. Humidity vote portrays that around 60% of respondents felt dry conditions, only 32% felt comfort and 8% for slightly humid (see Figure 3). This condition can influence work productivity; high humidity can lead to a feeling of discomfort because high humidity disrupts one of the main ways our body regulates its temperature: sweat. In case the condition is too hot, the body secretes water on the skin's surface. The water cools in the breeze and lowers the temperature of the skin, but when the humidity level is close enough to the condensation point, sweating loses its cooling effect. Moreover, high humidity levels can lead to health effects.

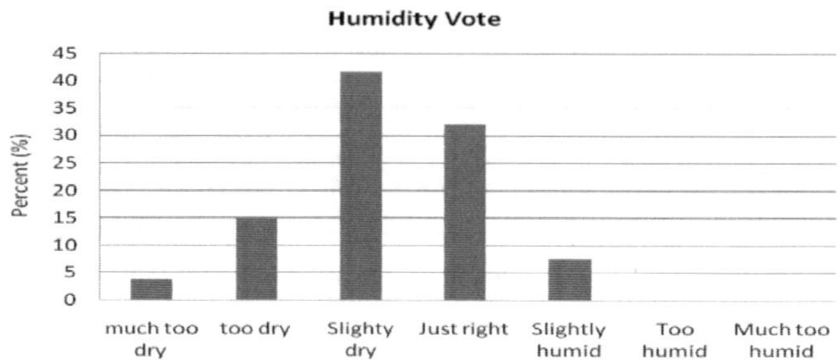

Figure 3. Employees' response to the air humidity.

3.3. Wind velocity and natural ventilation performance

Wind velocity in archipelago region such as Ternate is sufficient for the application of natural ventilation. Several previous studies suggest that natural ventilation is effective to reduce the building's temperature. For example, in Sultan Ternate Mosque, natural ventilation is used by combining clerestory ventilation and cross ventilation systems. The indoor thermal conditions can reach a comfortable level [13]. It indicates that the application of natural ventilation in the archipelago city of Ternate is effective to achieve comfort thermal level and minimize the energy consumption in offices building.

However, measurement results suggest that the wind velocity in the office building is very low. It is lower than 0.5 m/s. Based on the velocity vote, only 11% of respondents feel comfortable, 32% for slight breeze, and around 59 % feel wind velocity is very low (slightly still until much too still scale). Thus, wind velocity it is necessary to increase. About 70% of respondents expect the increase of wind velocity. Moreover, only 26% feel comfortable (see Figure 4). The distribution of air flow causes air velocity into the room not covered the entire room, so that employee sitting near the window feels comfortable. It indicates that cross ventilation system is not enough to circulate air fresh into the room due to the dimension or orientation of ventilation is not suitable to the standard. On the other hand, office buildings in Ternate City are generally designed for the usage of air conditioner, so that the natural ventilation strategy is not adopted effectively. In the application of natural ventilation, the sufficient wind speed is needed to increase air circulation in the building. Cross ventilation and stack effect can increase it.

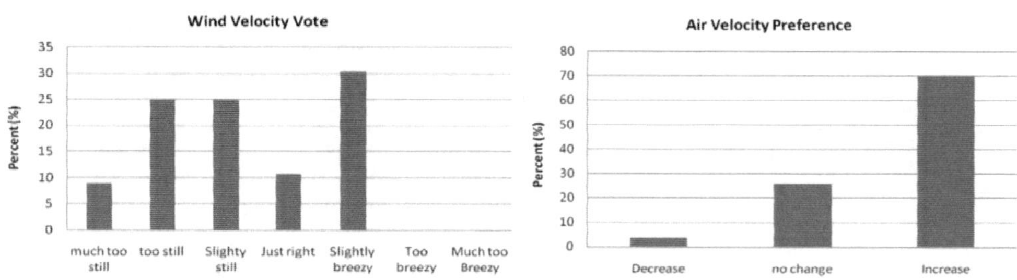

Figure 4. Wind velocity preference and acceptance of employees.

For increasing wind velocity, it is necessary to maximise the cross ventilation system, but keep in mind that Ternate Island is covered by an active volcano of Gamalama, which may emit hot dust and have a negative impact to the application of natural ventilation due to the dust pollution. Therefore, the strategy of the ventilation system in Ternate Island is should use a combination of natural ventilation and with a mechanical system. Natural ventilation is operated by the combination system between cross ventilation with open-closed ventilation systems. In normal condition, cross ventilation or open ventilation system is used, and if the eruption occurs, the natural ventilation is closed, and air conditioner can be used for preventing dust pollution into the room.

Furthermore, Outdoor temperatures are determined by land surface condition. Land surface characteristics, building materials, vegetation, water, and human activities have important effects on microclimate changes. Vegetation and water are useful means for improving microclimatic conditions [14-15]. Effect of climate change can be reduced by improving land cover, reducing anthropogenic heat, and increasing local wind circulation [16]. The proper placement of trees decreases hot temperature and then reducing air conditioning loads by 30% [17]. Vegetation is effective to decrease temperatures in the urban area so that buildings near from vegetation have lower surface temperature than those of buildings are distant from plants. Therefore, in addition to improving the natural ventilation system, mitigation of land cover or landscape arrangement around buildings is very useful to improve thermal comfort in the building.

4. Conclusion

The indoor thermal condition is uncomfortable level with the average temperature exceeds 29°C; Relative Humidity is over 70%; where more than 60% of employees feel warm condition in TCV and TSV scales. However, the majority of them (67%) still accept these conditions to work in the room. It indicates that the employees in a tropical archipelago city are quite tolerant to high temperatures due to their habits of life in warmer conditions.

Wind velocity is lower than 0.5 m/s. About 59% of respondents feel wind velocity uncomfortable and 70% respondents expect to increase of wind velocity, portrays that cross ventilation system is not enough to circulate air fresh into the room because their orientation and dimension are minimal or not suitable standard, so that natural ventilation is necessary to improve by combining cross ventilation (open-closed systems) with clerestory ventilation systems, and planting trees around the building to reduce solar heat gain that penetrates to the interior of the building.

5. Acknowledgements

We would like to express our sincere gratitude and appreciation to the Minister of Research, Technology and Higher Education, the Republic of Indonesia for providing the fund of this research.

6. References

[1] Energy-efficient buildings: heating and cooling equipment. Technology roadmap, *International Energy Agency* 2011

[2] Building and Climate Change, *United Nations Environment Programme*, 2009

[3] ISO 1995. ISO 7730 Moderate thermal environments-determination of the PMV and PPD indices and specifications for thermal comfort (2nd Ed). *International Organization for Standardization.* Geneva, Switzerland

[4] Lawrence Berkeley National Laboratory. The relation of climate and architecture. *http://rizkilesus.wordpress.com*

[5] *http://www.priceindustries.com/content/uploads/assets/literature/engineering-guides/natural-ventilation-engineering-guide.pdf*

[6] ASHRAE 2004 Thermal environmental condition for human occupancy (ASHRAE standard 55). *ASHRAE,* Atlanta US

[7] Bedford T 1936 Warmth factor in comfort at work, *report no 76. HMSO med res council*, Industry Health Res Board

[8] *https://en.wikipedia.org/wiki/Ternate*

[9] Indonesia: Administrative division (provinces, regencies and cities) - population statistics, charts and map. *www.citypopulation.de.* Retrieved 2 April 2018

[10] Annual weather averages in Ternate based on weather reports collected during 2005–2015.*https://www.timeanddate.com/weather/indonesia/ternate/climate*

[11] Satwiko P. Building physic 1. *Published: Andi,* Yogyakarta-Indonesia 2003

[12] Rahim M and Hamzah B. Evaluation of indoor thermal comfort on office building in Ternate City. *The 4th International Conference on Engineering and Technology Development (ICETD)* on October 25-28, 2017 in Universitas Bandar Lampung (UBL) Indonesia

[13] Rahim M and Marasabessy F. Evaluation of natural ventilation characteristics on the Sultanate of Ternate Mosque. *International Conference on Science, Engineering and Technology (i-SET2018)* on 2 – 3 May 2018 in University of Muhammadiyah (UNMUHA) Banda Aceh Indonesia

[14] Wong N.H. et al 2007 Environmental study of the impact of greenery in an institution campus in the tropics. *Building and Environment* **42** 2949–2970

[15] Rahim M, Yoshino J and Yasuda T. Impact of land surface changes to outdoor thermal performance and climate condition. *12th International Conference on Sustainable Environment and Architecture (SENVAR)*, Universitas Brawijaya Malang, 10-11 November 2011

[16] Moriyama. et al. The mitigation of UHI intensity by the improvement of land use plan in the urban central area, application to Osaka City. *Second International Conference on Countermeasures to Urban Heat Islands (SICCUH) 2009*, Berkeley California

[17] Ball J, Erickson D and Garbisch D.L. Urban forest. *Department of Agriculture, South Dakota 2002. www.northern.edu/notsource/HABITATS/urbanf1.htm*

Questioning Urban Symbolism in the Old City of Semarang

Yulia Nurliani Lukito, Putri Widya Pratama
Department of Architecture, Faculty of Engineering, Universitas Indonesia, Depok, Indonesia

Corresponding E-mail: yulianurliani@yahoo.com

Abstract. The aim of this paper is to analyze urban symbolism in the old city of Semarang, Central Java that is created from the physical and the non-physical elements of the city. Through historical analysis, the paper attempts to relate the physical elements that come from the structure of the city, such as buildings and transportation, with the non-physical elements that come from history and events. When Dutch colonialism came to an end, some colonial buildings in the old city of Semarang started to be neglected. However, urban symbolism created from trading, transportation and culture provides layers to the identity and culture of the city, which includes Chinese, Arab and indigenous culture. Recent effort to revitalize the city focuses primarily on physical aspects, rather than balancing both physical and non-physical aspects of the old city of Semarang. The result of this paper reveals that balancing both physical and non-physical elements of the city means taking a step further in understanding important things for the city. Moreover, understanding urban symbolism is necessary in sustaining the history and identity of the city as well as in giving direction to the development of the old city of Semarang.

Keywords: old city of Semarang, urban symbolism, colonial building, history of Semarang

1. Introduction

Kota Lama Semarang or the old city of Semarang is inseparable part of the development of Semarang city and gives important memories for the people of Semarang. Besides its physical elements, the old city of Semarang also has some historical moments and ceremonies that shape the atmosphere and memory of the city. Some of those moments were interpreted related to the physical conditions of the city such as buildings, landmarks, paths, edges, nodes, including how one building relates to other buildings.

As one of the biggest colonial city in Indonesia, Semarang has many layers of history affecting the shape of the city. Hindu-Javanese city planning principles influences the typology of the city through *alun-alun* and palace and religious buildings dominated the central square with the city center. At that time, non-native people usually live outside the royal city or the fortifications but very close to the river as the main transportation means. The oldest non-native settlement outside the wall was usually the Chinese settlement. The development of the city changed during the Dutch colonial period in the Dutch East Indies started when Vereenigde Oostindische Compagnie (VOC) occupied the Indies in

1619 and ended when the government of the Dutch East Indies surrendered to Japan in 1942 [1–3]. In the early sixteenth century, the Dutch initiated the construction of a small masonry fort near the existing city and it was located on the most strategic place near the old harbour and market located right on the river mouth. Inside the fort were some buildings for dwelling, warehouse and some military functions. In the nineteenth century, coastal colonial cities gained a new meaning as commercial centers, and the fortified cities were then no longer existed.

It meant that the old city of Semarang lost its glory and later on some marginalized people even started to live in the area since owners abandoned some of the old buildings. This situation also created the increasing crime rate and the image of the city as the center for economic and loaded with rich history was faded. In relation to its historical and architectural values, Kota Lama Semarang actually still has some economic and cultural potential. Those potential especially comes from existing colonial buildings that become a symbol of the glory of the old city. This paper aims to relate the physical aspects of the city, represented mainly by colonial buildings, and the meaning of urban symbolism in the old city of Semarang.

In order to understand the city, one should deal with both physical and non-physical conditions, including through the city's urban symbolism. Symbol is a representative of another idea or meaning and related to the signification process that includes signifier and signified [4]. Urban symbolism is related to semiotics as well as expresses through various phenomena such as the architecture, streets, and rituals in the city. The signified can be called as the domain, which dominates the dynamical changes of old city of Semarang such as during the colonial period. Those domains are the glory of trading, transport and ethnic domain reflected in the history and the existing materials that still exist in the old city of Semarang.

In the study of the cultural dimension of the city, urban symbolism concerned with the meaning of urban symbols in relation to the cultivated surroundings. There are types of symbol bearers that a city might have such as material, iconic, behavioral and discursive bearers that work as the signifiers [5]. There are some symbol bearers of the old city of Semarang but this paper discusses mostly the material bearer, which is related to the architecture of the city. Architecture is important in understanding urban symbolism and represents meanings throughout the development of the city so that people get understand both signifier and signified or connotation meanings of the city. The urban symbol in general will reflect social, cultural and historical aspects of the city and conveys a shared memory.

In this case of Semarang old city, the material bearer is a material that is potential to hold the meaning of symbol bearers or can be called as the domain. The material bearers in the old city of Semarang are related to the physical elements of its urban space such as landmarks, edges and paths as physical aspects of the city [6,7]. Architecture such as Blenduk Church can be considered as a landmark while Berok Bridge at the west edge of old city, Chinese Settlement of Pecinan and Arab settlement or Kauman as the districts and Let. Jend. Soeprapto Street can be considered as major streets or paths. Those structures are important because they create the image of the old city and urban symbolism.

Aldo Rossi in *Architecture of the City* defines architecture as forms that have continued over time to become types that constitute the history of the city or its memory, and the culture of the present [8]. Rossi criticizes the lack of understanding of the city in architectural practice and argues that a city must be valued as something constructed over time through its urban artifacts or monuments or in other words monuments give structure to the city. Functions vary over time but form remains and the meaning in the city is based on memory and monuments. Memory or urban facts withstand the passage of time and become monuments while monuments give meaning to the life of the city through memory.

A good urban space design is the design supporting the architectural value that able to contribute meaning and value to its environment [9]. By knowing the urban symbolism, hopefully it will be easier for people to appreciate and understand social, cultural and historical values of the old city. Hopefully, revealing the urban symbolism in the old city of Semarang is important because this will

IOP Conf. Series: Earth and Environmental Science **213** (2018) 012027 doi:10.1088/1755-1315/213/1/012027

help to develop the city and revitalize important meanings that come from both physical and non-physical aspects of the city.

2. Methods

The method used for this research was mostly historical and literature review to gather information of the development of the old city of Semarang. Because the old city of Semarang is a historic area, studying historical aspects of Semarang related to the development of the city was conducted through literature and archives reviews in order to understand the link between physical conditions and urban symbolism. We used qualitative method in analyzing the relation between the physical and non-physical elements of the Semarang city. First, since the research deals with urban symbolism, it is necessary to analyze the structure of the old city of Semarang through its development. Then, it is important to understand historical and cultural meanings of the city as an effort to analyze how urban symbolism is created. Lastly, information from literatures and archives, together with analysis of the old city of Semarang's map, and field research are important in analyzing and interpreting the connection between urban space and symbolism.

3. Discussion

Semarang was one of the major colonial urban centers in Java, and it gained a municipal autonomy under the Decentralization Act passed in 1903. The city was a busy port, the center of a rail networks and played an important role in connecting places of sugar industry (see figure 1). The population of Semarang between 1905 and 1920 increased from 96,000 to 158,000 and by 1930 the population reached 218,000 [2]. There was an expansion of the population because the city attracted people from the hinterland to seek job opportunities that the city provided. Semarang was home to multi-ethnic people (see figure 2) and the population in 1905 consisted of indigenous people who dominated the number of population (seventy-eight per cent), local Chinese population who dominated the commercial sector (fifteen per cent), European (five per cent) and other Asian peoples (two per cent) [2,10,11]. The Chinese built shop houses mostly near the river and those types of buildings are still existed in the Pecinan kampong, including the one that surrounded the old city of Semarang. New housing areas were developed inside and around the city to accommodate the rapid growth of urban population and some industrial buildings as well as central business districts emerged in the city centers.

Figure 1. The old city of Semarang as trading port centre. Source: http://fariable.blogspot.com/2011/01/elemen-perancangan-kota-hamid-shirvani.html

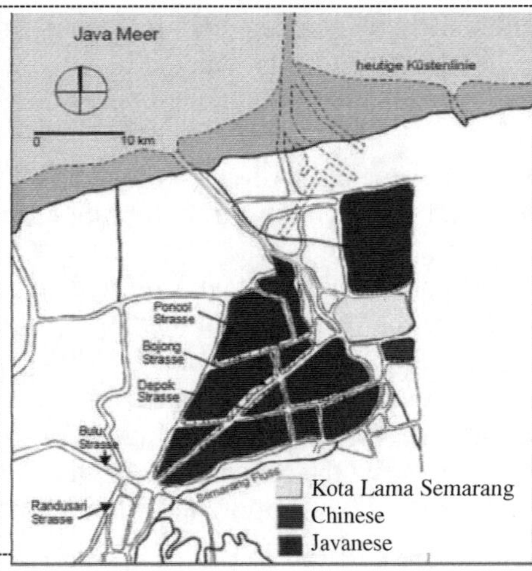

Figure 2. These two figures have been placed side-by-side to save space. Justify the caption.

During the early Dutch colonial era, the Dutch built the Vijhoek Fort to secure their business in Semarang and to provide trading facilities for VOC (see figure 3). This fort had five bastions, so that it is known as De Vijfhoek or the Five Corners bastion. The fortified wall was demolished in 1824 to make way for new commercial business area. Although the Vijhoek Fort was no longer creating a significant mark for the surroundings, however, the history of Vijhoek as an important fortress continued and gave a direction of the city development of the south of the fort. The destruction of Vijhoek Fortress was meant to accommodate the growth of the city and the further development of the area meant a consideration of surrounding areas, including Chinese, and other local settlements as part of a greater Semarang City. The invasion of Japanese Imperial army to Southeast Asia also had changed the course of urban history. After 1942, local people started to leave their buildings in the Old City of Semarang to escape from Japanese and the owners and the users left many buildings after the independence (see figure 4).

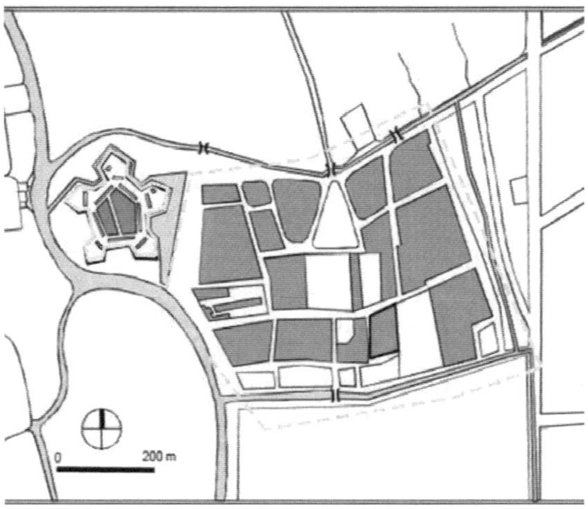

Figure 3. Vijhoek Fort in the early eighteenth century was still separated from the settlements. Source: Purwanto, 2005

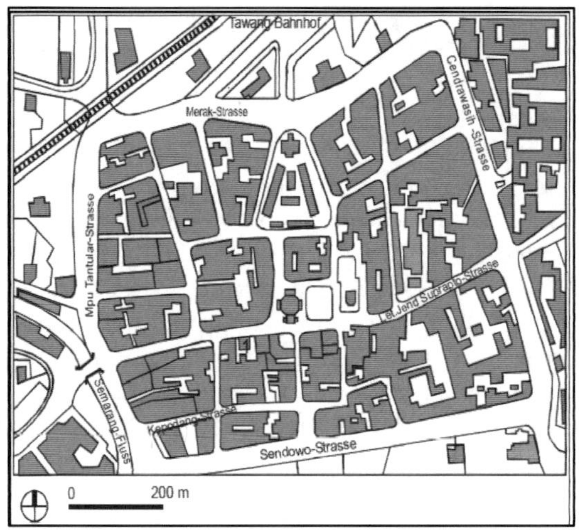

Figure 4. The old city of Semarang after the independence of Indonesia. Source: Sumaningsih in Purwanto, 2005

Below are some discussions of the old city of Semarang in relation to the development of the city and urban symbolism coming from trading, transportation, and cultural domains.

3.1. Urban Symbolism From Trading Domain

The size of the old city of Semarang is approximately thirty-one hectares more than fifty-one colonial buildings. Some of the colonial buildings are in good conditions but unfortunately a large number of those buildings are in bad conditions. The old city of Semarang had unique pattern coming from many layers of history, including during Dutch colonial period. The city was developed from a fortified city and at the end of the seventeenth century; the fortress had been torn down to accommodate the growth of the city as an important international trading port city. The city of Semarang became a multicultural and cosmopolitan city and some construction projects were done including building offices, hotels, and tramline.

By the 1700s, Javanese coastal cities, including Semarang, had central areas modeled after ideal Dutch port cities together with church, canals, and townhouses. The Dutch imposed the colonial cities' landscapes with alien cultural and physical environments [12]. However, those aesthetic features led to problems as the cities grew such as dirty canals, and heavy architecture introduced by the Dutch was proved inappropriate for local climate.

IOP Conf. Series: Earth and Environmental Science **213** (2018) 012027 doi:10.1088/1755-1315/213/1/012027

In Semarang city, the construction of Lawang Sewu, the head office of the Dutch train company, indicated how modernity had spread and the need for reliable transportation system to support for the city was increased (see figure 5). At the time, this fort area started to become a commercial center dominated by warehouses, office buildings, stores, and banks. There was also a widening of the Semarang riverbanks and the construction of a new harbour. Moreover, Lawang Sewu's unique architecture that allowed cross ventilation and the use of local roof showed adaptation with local climate through built environment. In Kota Lama Semarang, the construction of Blenduk Church marked the importance of the area and the centrality of the location. Up to now the church still becomes a landmark of the old city. The construction of European settlements and buildings was done in order to facilitate the needs of new residences and an effort to modernize and regulate the city.

Figure 5. Lawang Sewu. Source: Tropenmuseum, part of the National Museum of World Cultures.

Figure 6. Tawang Station. Source: https://kebudayaan.kemdikbud.go.id/bpcbjateng/stasiun-tawang-stasiun-yang-terindah-di-hindia-belanda/

3.2. Urban Symbolism From Transportation Domain

In 1865, the Dutch built the railway system in Semarang for 25 km and this time was also marked with the construction of the office of Nederlandsch Indische Spoorweg Mij (NIS) in Semarang (known as Lawang Sewu building). The building was designed by JF. Klinkkramer and BJ Quindag in 1902 and became one of the first modern office buildings in Semarang. In fact, Semarang city was the first city that had train and tram systems in the Dutch East Indies. Another important building is Tawang Station that created a border for the northern part of the old city, and a renowned Dutch architect Thomas Karsten built it (see figure 6). Both the construction of Lawang Sewu as the head office of the Dutch train company and Tawang Station indicated how modernity had spread and the need for reliable transportation system to support for the city was increased. Moreover, Lawang Sewu's unique architecture that allowed cross ventilation and the use of local roof showed adaptation with local climate through built environment.

The Netherlands Indies urban planning included the construction of the Great Post Road or *Jalan Raya Pos* that run across North Java during the reign of the governor general Herman Willem Daendels (1808–1811). The orientation of the old city of Semarang was also related to the Great Post Road. Previously, rivers functioned as the main transportation but with the construction of the Great Post Road and the railways changed Semarang spatial configuration from south-north orientation to east-west orientation. New settlements and commercial centers were built along the Great Post Road inhabited with multi-ethnic groups [10,13]. The Grote Postweg affected to the city's physical elements in a way the transportation became a new orientation and it created new centers. There was also the tram system run inside the old city of Semarang Between the Berok Street and the Agus Salim street. The old city of Semarang itself was designed to accommodate various kinds of transportation like bicycle, cars, horse carriage and pedestrian.

3.3. Urban Symbolism From Culture Domain

The early development of the old city of Semarang had already shown people who lived in the city came from multi-ethnic groups such as Arab and Chinese. They came to the city as merchants or live at the city periphery as farmers and came to the city to sell their goods. In addition to the trading domain that dominated activities in the old city of Semarang, cultural domain was also important in the city. The port city of Semarang was the biggest trading center during the Dutch colonial era that especially came from sugar trading. One of the main actors was the great sugar seller Oei Tiong Ham was a successful Chinese contingent entrepreneur in Semarang who built some buildings in the old city of Semarang including the one close to Blenduk Church. Part of his wealth originated from sugar industry. People of Semarang also built living quarters such as Kampung Arab and Pecinan for the Chinese. At the time the Dutch built the Vijhoek Fort, Semarang city had typology of Javanese cities with some living quarters for various ethnic groups. Some kampungs were located in the southern part of the old city of Semarang.

Since the Vijhoek fort was demolished, the Dutch housing complex was moved to the Bodjong Street and later on Thomas Karsten, a Dutch architect who practiced in the Indies, designed the new Candi housing area in the southern part of Semarang city [12,14]. Karsten's idea of creating the new housing complex was not only to provide good and healthy housing area but also promoting new social relations that was not based on race but social and economic relations. Other physical elements that became the borders of the city are Semarang River in the west, Tawang Station in the north, Cendrawasih Street in the east and Kepodang Street in the south.

The image of multi-ethnic in Semarang can be seen during Dugderan festival, an annual event to welcome the holy month of Ramadhan in Islamic calendar. People carry an imaginary animal called *warak ngendog* (see figure 7 and 8) that has the head of a dragon (Chinese), the body of a burqa (Arab influence for represent a special animal resembling the winged horse believed to take Prophet Muhammad to Sidratil Muntaha), and the tail of a goat (Javanese). The word *ngendog* itself means egg taken from the Javanese language that means the goodness that will bring peace to Semarang people who can live peacefully instead of their different ethnic backgrounds. There is even *warak ngendog* statue built on Bodjong Street, not far from the old city of Semarang. Nowadays, the government attempts to develop the area along the Semarang River to accommodate the Dugderan ceremony so that the area can be used for people to see the *warak ngendog* parade.

Figure 7. Warak Sculpture Monument in Semarang Pandanaran Park. Source: author.

Figure 8. Warak ngendog during Dugderan festival. Source: soloraya.com.

In addition to Dugderan festival, some buildings in the old city of Semarang also carry symbolical meanings for the people of Semarang. Blenduk church represents the Renaissance building style and

IOP Conf. Series: Earth and Environmental Science **213** (2018) 012027 doi:10.1088/1755-1315/213/1/012027

its location shows the centrality and the importance of the church. The church then can be assumed as the symbol of the Dutch power to shape the city and create a new central to the surroundings and do not follow local orientation. Javanese cities usually follow the orientation of the mountain and the sea, as part of the cosmology of the Javanese people. In later period, the influence of Islam in Java made religious buildings such as the city mosque had an orientation of kabah. This means that the Dutch wanted to create their own ideas of the city.

Beside the Church as the landmark of the area, there is also a symbol bearer of the old city of Semarang or the paths that are divided into major and minor paths. There is a L-Shaped building located faced north right of Blenduk Church (see Figure 9). Jiwasraya Insurance office building (see Figure 10) is another important building in the old city of Semarang. The building was designed by Thomas Karsten in 1918 with concrete-structure technology that was adapted to tropical climate through the use of corridors and a row of doors and windows to create open space, good air circulation inside the building and an area to penetrate the solar heat and to protect rainwater entering the room. The building was a symbol of modernity and adaptation to local conditions.

Figure 9. Blenduk Church. Source: Crisco 1492

Figure 10. Jiwasraya Building. Source: https://situsbudaya.id/gedung-jiwasraya-semarang/

The old city of Semarang has what Aldo Rossi mentions as forms that have continued over time to become types that constitute the history and the collective memory of the city up to the present (Rossi, 1966). Understanding of the city as something that is constructed over time through its urban artifacts or monuments is very important because those urban artifacts or monuments give structure to the city. Buildings, streets, and landmarks in the city, including the development of the old city from a port, a fortress city and colonial city withstand the passage of time and become monuments that give meaning to the life of the city.

4. Conclusion

Urban symbolism is related to the physical conditions of the city such buildings and transportation, as well as the non-physical elements such as history and events in the city. From the history of the old city of Semarang, the structure of the city relates to trading, transportation and culture domains reflected in the histories and buildings that still exist in the old city of Semarang. Those domains are also connected meanings of the city – which then becomes urban symbol.

The old city of Semarang in its trading domain means the center of economic and trades throughout the history of the city. The city also bears the idea of connecting areas in Java, trading and having significant structures related to the development of transportation. Cultural domain also gives the city its own meanings through multi ethnic culture such as reflected during the annual event of Dugderan and *warak ngendog* parade. The three domains may reflect connections between physical and non-physical aspects of the city, and how each aspect creates urban symbolism.

The result of this paper shows that balancing both physical and non-physical elements of the city means understanding things that are important for the city. Important buildings, streets and landmarks in the city are actually urban artifacts that create the structure of the city. The structure of the city shows us the signifiers or things that create meanings in the city while symbols are directly or indirectly embedded to the architecture of the city that at the end become the signified or connotation meanings of the city. In conclusion, understanding urban symbolism is essential to sustain the history and identity of the city as well as in giving direction to the development of the old city of Semarang.

5. Acknowledgments

This study was supported by research grant Hibah PITTA from the Directorate of Research and Community Engagement, Universitas Indonesia.

6. References

[1] Nas P 1993 *Urban symbolism* (E.J. Brill)

[2] Colombijn F, Coté J and Cote⁺ŕ J 2014 *Cars, conduits, and kampongs : the modernization of the Indonesian city, 1920-1960* (BRILL)

[3] Sunaryo 2005 *Morfologi Ruang Pusat Kota Jawa Periode Kolonia*

[4] Barthes R 1964 Elements of Semiology Hill and Wang

[5] Nas P J M 2011 *Cities full of symbols : a theory of urban space and culture* (Leiden University Press)

[6] Lynch K 1960 *The image of the city* (MIT Press)

[7] Rob Krier 1979 *Urban space* (Rizzoli International Publications)

[8] Rossi A, Eisenman P, Graham Foundation for Advanced Studies in the Fine Arts. and Institute for Architecture and Urban Studies. 1982 *The architecture of the city* (Cambridge Mass: MIT Press)

[9] Kasprisin R J 2011 *Urban design : the composition of complexity* (London: Routledge)

[10] Handinoto 1999 *Dimensi Teknik Sipil*

[11] Handinoto H 2009 DAENDELS DAN PERKEMBANGAN ARSITEKTUR DI HINDIA BELANDA ABAD 19 *Dimens. (Journal Archit. Built Environ.* **36** 43–53

[12] Nas P 2007 *Masa Lalu dalam Masa Kini* (Jakarta: PT Gramedia Pustaka Utama)

[13] L.M.F. Purwanto 2005 KOTA KOLONIAL LAMA SEMARANG (Tinjauan Umum Sejarah Perkembangan Arsitek) *Dimens. J. Archit. Built Environ.*

[14] Virdianti 2017 Citra Visual koridor KawasanKota Lama Semarang ditinjau dati Potensi Wisata Edukasi Arsitektur Heritage *J. Rekayasa Hijau.* **1**

Typology of Bajo Tribe Settlement in North Maluku

Mustamin Rahim, Ardi Basri, Hendra Fauzi
Lecturer of Architecture Department, Khairun University, Ternate, Indonesia.

Corresponding e-mail: mustamin_rahim@yahoo.co.id

Abstract. This study aims to identify the typology of settlement and analyses spatial pattern, infrastructure and settlement conditions by literary review and field observation. The results show that the largest of Bajo Tribe Settlement in north Maluku is located in Kayou District of South Halmahera with the 2609 population in 2016 in Laluin and 1071 in Posi-posi Villages. Typology of settlements is almost similar in both villages; during the period of 1970-1990, the settlements was deployed on shallow and potential coastal area, but in period 1990 until now, several houses is deployed on mainland area due to change of living from a fisherman to civil servant, businessmen, dealer, etc. House condition consists of three categories: traditional or temporary dwelling, semi-permanent, and permanent house. The back and below of house is used for boat mooring and storing equipment because it is easy in maintenance. A house around the main road of settlement is generally developed to permanent house with wall made of brick or zinc made of concrete and roof. However, the pattern of settlements is different; linear in Posi-posi and circular in Laluin village because it is influenced by the condition of the shallow area.

Keywords: typology, Bajo Tribe settlement, traditional fisherman.

1. Introduction

Indonesia is the largest archipelago in the world covering around 17,504 islands and 5.8 million km^2 of sea; consists of 0.3 million km2 sea territories and 2.95 million km archipelago territory [1]. Many people live in coastal areas in Indonesia, especially Bajo Tribe. They utilize marine and coastal resources as their main income. Generally they work as a boat owner, fishery labor and other marine organism farmers, fish traders, fish processors, and the equipment suppliers of fisheries production factory. Most of them live under the poverty line [2].

The number of population in Bajo Tribe is around 90,000–150,000 people [3]. They live in Riau (known as duano), North Sulawesi, Central Sulawesi, South Sulawesi, Southeast Sulawesi, East Nusa Tenggara, Maluku, and North Maluku [4]. In North Maluku, they settle in coastal areas of South Halmahera, East Halmahera, Sula and Taliabu Islands. They used to live on waters, so that; they are very difficult to be relocated in mainland area. Therefore, the settlement arrangement strategy of Bajo Tribe needs to be adjusted to their characteristics and patterns of the original settlement. According to Silas [5]; settlement development is a process of comprehensive interaction between the occupants, the location and the environment as a result of social, cultural and economic activities. The basic design of fisherman settlement should consider the house, place, building, path and drainage design which include adequate houses free of flood, sustainable waste management, easily accessible fishing boat harbors and decent infrastructure [6].

The economic conditions of Bajo Tribe are solely rely on their earning as a fisherman, so that, their income is not yet able to sustain decent life, thus supporting business is needed to increase their income. Fishing community activities must be supported by economic facilities in the form of production facilities in both housing and industrial scale [7]. The availability of economic, social and cultural facilities such as trade places and local culinary business can functionally support the coastal ecotourism-based fishermen settlement [8]. According to Holmberg [9], there are three aspects in sustainable development; (1) Sustainable economic development, supply goods from renewable natural resources, keep the balance of shopping and public debt, (2) Environmental sustainable; maintain natural resources, avoid the exploitation and utilization of natural resources un-renewable, maintain biodiversity, stability of atmosphere and ecosystem. (3) Sustainable social development: ensure social justice, provision of social and public facilities, gender equality and net political system.

2. Methods

This research is qualitative research that emphasizes on the typology identification of settlements, spatial pattern, and environmental infrastructure condition, and also analyzes the change of house characteristics and additional business. This study focuses in Kayoa district of South Halmahera covering Laluin and Posi-posi villages because the largest of settlement are located in both villages. They are the representation of Bajo Tribe settlement in North Maluku. Data collection consists of 3 steps: (1) Field observation to obtain primary data about the typology and the development of Bajo Tribe settlements. (2) Literary review: to complement secondary data about the development, economic and social conditions, and sustainable development strategy of coastal areas especially in Laluin and Posi-Posi villages. (3) Interviews: to obtain information related to primary and secondary data to support data obtained through field and literary studies. The data collected were analyzed by descriptive qualitative to describe its settlement condition and its development during last ten years.

3. Discussion

3.1. Bajo Tribes Lifestyle

The Bajo tribe is a traditional and skilled fisherman. His life entirely depends on the sea. They are often referred as sea nomads or "sea people"[10]. The term "Bajo", "Bajo Tribe", or "Bajo Man", is commonly used by eastern Indonesian to refer to these sea nomadic nomads, scattered in various regions [11].

Bajo Tribe has settled in North Maluku since 1900s. They deploy in coastal areas, especially around Sula Island, East Halmahera, West Halmahera, Taliabu, and South Halmahera. They are commonly known as marvelous sailors or wonderful fishermen. Their lifestyle is always connected to the water or sea and very difficult to be relocated in the mainland area because their livelihoods are fisherman. Therefore, they always live around of coastal areas which have potential marine resources.

Bajo tribe forefather in North Maluku is immigrant from Sulawesi. Initially, they lived in the boat *(lepa)* which is equipped with food and fishing equipment like a house on the mainland. Along with the change of mindset and the income increase, they began to build simple house as temporary dwellings by using wood materials for columns and beams and palm leaves as wall and roof. Futhermore, the temporary settlements continue to grow and develop to be permanent settlements due to the increase of Bajo Tribe population.

3.2. The Condition of Bajo Tribe Settlements

Settlement is a space or containers in which humans perform daily activity and a space can be defined as an area with a geographic limitation [12]. The limitation of a space affected not only by the surrounding physical condition but also by social and governmental system which occurred everywhere. The development of a space is required to support various human activities, to encourage benefits, environmental conservation, awareness of environmental values, and to imply the efforts of social, economic, and investment in the development of residential space environment as a container of human activity [13].

The largest Bajo Tribe settlements are located in Kayoa District, South Halmahera: Laluin and Posi-posi Villages. Bajo Tribe Settlements in Laluin village still traditionally grows and develops until now and its culture is still developed. Along with the increasing number of Bajo Tribe people, a new settlement is formed in Waidoba Island. Furthermore, the settlements development in this region continues to increase and becomes a definitive village.

Figure 1 portrays the settlements condition in Laluin and Posi-Posi Villages in 2006. The settlement condition is very simple with the roof material is generally made of palm leaves. Few of them use zinc roof. House wall uses wood and bamboo blades. The infrastructures are still limited. The footpaths are generally made of wood material and few parts of main road are made of concrete.

a. Laluin Village b. Posi-Posi Village

Figure 1. The Condition of Bajo Tribe Settlements in 2006.

Figure 2 shows the housing conditions in 2017. Few houses are still simple with wall made of wood and bamboo blades and most houses uses zinc material for roof. Houses located around main road are generally developed to the permanent house with wall made of brick or concrete and roof made of zinc. This condition indicates that their income has increased because some people already have additional livelihood, such as: fish cages, making sago cake, street vendors, fish dealer, etc.

The condition of infrastructure has increased if compared to that in 2006. Most roads use concrete material (50-60%). Pure water plumbing has been installed. As the water supply is inadequate, so, to meet the need, people use rain water or take water from other island. Electrical installation is also available in settlements area but as the power supply is inadequate, so it is only supplied for night time and also for special day such as festival day.

a. Laluin Village b. Posi-Posi Village

Figure 2. The Condition of Bajo Tribe Settlements in 2017.

3.3. The Typology of Bajo Tribe Settlements

Figure 3 portrays the typology and pattern of Bajo Tribe settlements in Laluin and Posi-posi villages, South Halmahera. It is the largest of Bajo Tribe settlements in North Maluku. Settlement of Bajo tribe in Laluin is located around shallow areas and has good fishery potential. Bajo Tribe settlements are floating settlements. They grow organically and spontaneously along the shallow areas with a linear and circular shape. This condition can be seen in Laluin and Posi-posi Villages; settlements develop without planning and develop naturally in potential coastal areas or tidal areas. They have lived in Laluin since on 1900s and 1970s in Posi-posi, and then, they continue to grow until today. Populations in 2016 were 2609 in Laluin and 1071 in Posi-posi (see table 1). They settle in coastal areas to support their activities as a fisherman and their habit of living on the sea.

The settlements typology in Laluin and Posi-Posi are similar to those in other areas. They are oriented to sea, the back or below their house are used for boat mooring and storing equipment due to its easiness in maintenance. However, pattern of settlements is different; Posi-posi is linier due to its shallow area and Laluin is circular due to its potential coastal areas. Settlement facilities in Laluin consist of roads, offices, mosques, schools, docks / harbors, markets, cemeteries, security posts, sports fields / public spaces. Posi-posi has the similar facilities to those in Laluin, but the market does not yet exist.

Initially, public facilities, using wood construction, were built on the sea areas. Now, they, such as schools, offices, and health centers, have been built on mainland area with permanent construction. Only the mosques are still built on the sea area, but they also use permanent construction or concrete material. In Posi-posi, generally public facilities are built on the sea, except for educational facilities or schools. They are built on mainland area with the permanent construction. It indicates that the facilities have been improved compared to those in 2006.

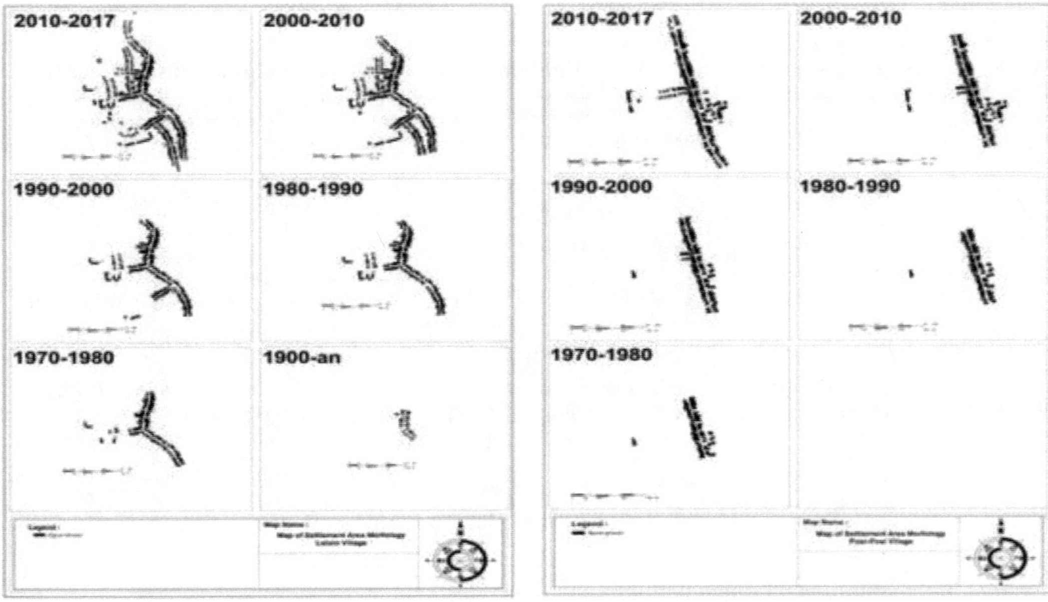

a. Laluin Village b. Posi-posi Village

Figure 3. Deployment of Bajo Tribe Settlements [14].

Table 2 shows that the house conditions consist of 3 main categories; (1) Traditional house including simple and temporary dwelling made of wood and bamboo, roof made of palm leaves. (2) Semi-permanent houses using wood and bamboo construction and roof made of palm leave or zinc. (3). Permanent houses using brick and concrete material, roof from zinc material. Generally, permanent houses are located around the main road. During the period of 1970-1990 the settlements are deployed along the shallow area and potential coastal area, but from 1990 to now, several houses are deployed on the mainland area. It is due to the change of livelihood from a fisherman to civil servant, businessman, dealer, etc.

The facilities have been improved since 2006. Pure water plumbing has been installed in Laluin Village science 2013, but the supply is still inadequate. Government (PDAM) has supplied water since 2013. But, Posi-posi gets no pure water plumbing from PDAM, so that they take pure water from another island; spring water of Tiga Lalu Island and from rain. Electrical installation is also available in Laluin and Posi-posi villages since 2015, but the supply is only for night time and for special day because the power is still limited. Therefore, it is necessary to improve environment infrastructure by sustainable development concept.

In order to improve the environment quality and the income of Bajo Tribe, the sustainable arrangement of Bajo Tribe settlement in Kayoa District consists of 3 categories: (1) Sustainable economic: development of household business; drying fish and bloater (*ikan fufu*), making crackers from fish and shrimp, fish sprouts, and cultivation of seaweed. (2) Sustainable Environment: clean system of waste disposal, sustainable fishing system: banning of fish bombs or alike, preservation and reforestation of mangrove, (3) Sustainable social development: preservation of local culture, improvement of infrastructure, utilization of rainwater and solar energy as the solution of the limitations of water and electric energy resources. On the other hand, additional business can be developed for improving their income, such as: home industry for fish canning and making sago cake, fish dealer, and street vendors. Moreover, Kayoa District is potential to develop coastal ecotourism, trade places and local culinary business.

Table 1. Description of settlement condition in Kayoa District from 1900-2017.

Year	House condition	House Function	Description of settlements
1900s	Simple	moorings boat, storage of catch & equipment	Lived in boat (lepa)
1970-1980	Temporary	dwelling, moorings boat, storage of catch & equipment	Began to settle in temporary dwelling, deploy to Posi-Posi village. Facilities: mosque and school
1980-1990	Traditional	dwelling, moorings boat, storage of catch & equipment	spread to shallow area Facilities: mosque, school, footpath from wood & stone expanse
1990-2000	Traditional and semi-permanent	dwelling, moorings boat, storage of catch & equipment, business (several)	develop to mainland area Facilities: mosque and school, footpath from wood, road from coral reef
2000-2017	Semi-permanent and permanent	dwelling, moorings boat, storage of catch & equipment, business (several)	Several houses are located in mainland Facilities: mosque, school, health, footpath from wood, road from concrete, electricity, water installation (PAM only in Laluin)

Source: analyzed 2018.

4. Conclusion

Bajo Tribe has settled in North Maluku since 1900s. They are deployed in coastal areas, especially around East Halmahera, South Halmahera, Sula and Taliabu Islands. The largest settlement is located in Kayou District, South Halmahera with the number of population in 2016 was 2609 in Laluin and 1071 in Posi-posi Villages.

The typology of settlements is almost similar in both villages; during 1970-1990 the settlements are deployed along their shallow and potential coastal area, but from 1990 until now, several houses are deployed to mainland area due to the change of their livelihood from fishermen to civil servant, businessmen, dealer, etc. their house condition consists of three categories : (1). Traditional house uses combination of bamboo and wood construction, (2) semi-permanent house uses combination of wood and concrete, (3) Permanent house uses brick and concrete materials. The back or below the house is used for boat mooring and storing equipment due to its easiness in maintenance. However, the pattern of settlements is different; Posi-posi has linear pattern while Laluin has circular pattern. It is due to their shallow area and potential of coastal areas.

In 2017, few houses are still simple with wall made of wood and bamboo blades, and most of them use zinc for roof. Houses located around main road of settlement is generally developed into permanent house with wall made of brick or concrete and roof from zinc. Most roads use concrete material (50-60%). Pure water plumbing has been installed in Laluin, but the supply is inadequate. Electrical installation is also available in Laluin and Posi-posi villages but the supply is only for night time and special day because the power is still limited.

5. Acknowledgments

We would like to express our sincere gratitude and appreciation to the Ministry of Research, Technology and High Education for providing the fund for this research.

6. References

[1] UNCLOS 1982 United Nations Convention on the Law of the Sea. Jamaica.
[2] Lasabuda, R 2013 Pembangunan Wilayah Pesisir dan Lautan dalam Perspektif Negara Kepulauan Republik Indonesia. *Jurnal Ilmiah Platax.* **1** 2 92–101

[3] Fitriana, R. and N. Stacey. The role of women in the fishery sector of Pantar Island, Indonesia. *Asian Fisheries Science Special Issue.* 2012 **25S** 159-175

[4] Nuryadin, T.L. Studi tentang kapital sosial pada masyarakat tradisional nelayan Suku Bajo. *Jakarta: Fakultas Ilmu Sosial dan Ilmu Politik Universitas Indonesia*, 2010

[5] Silas, J, 1993 Perumahan: Hunian dan Fungsi Lebihnya dari Aspek Sumberdaya dan Eksistensinya. *Fakultas Teknik Sipil dan Perencanaan Institut Teknologi Sepuluh Nopember*. Surabaya

[6] Sharvina, A.N., Faqih, M., Santosa, H.R. Sustainable Fisherman Settlement Development. *8th International Conference on Architecture Research and Design (AR+DC)* November **1-2** 2016

[7] Ekaputra, Y. Dicky 2012 Pengaruh Aktivitas Ekonomi, Sosial, Budaya Pada Sistem Permukiman Nelayan (Kajian Kawasan Nelayan Tasik Agung Kabupaten Rembang. *E-journal Unpad.* 2012 **10** 22

[8] Rusli, Santosa, H.R., Soemarno, I. Coastal Ecotourism-Based Development for Fishermen Settlement in Labuan Bajo, Donggala, Central Sulawesi. *International Journal of Development Research.* August 2015 **5** 08 5215-5221

[9] Holmberg, Johan ed (1992) Making Development Sustainable: Redefining Institutions, Policy, and Economics. *Island Press*. Washington, D.C

[10] Saman Satar. Pola Spasial Permukiman Suku Bajo di Torosiaje Laut, Provinsi Gorontalo. *Tesis UGM* 2014. Yogyakarta

[11] Anwar 2006 Kajian Pendidikan dan Kebudayaan Bajo, Tinjauan Historis dan Kontemporer, *Makalah Seminar Perumusan Naskah Sejarah.* Universitas Haluoleo, Kendari

[12] Jayadinata, J. T 1999 Tata Guna Tanah dalam Perencanaan Pedesaan, Perkotaan, dan Wilayah. *ITB Bandung*

[13] Zahnd, M 1999 Perancangan Kota Secara Terpadu: Teori Perancangan Kota dan Penerapannya. *Kanisius*, Yogyakarta

[14] Rahim, M. Basri A. and Fauzi, H. Identification of Deployment Bajo-Tribe Settlement in North Maluku. *The 4th International Conference on Engineering and Technology Development (ICETD)* on October 25-28, 2017 in Universitas Bandar Lampung (UBL), Indonesia

Hybrid Architecture and Contested Space of Colonial Interaction: Colonial Exhibitions at Pasar Gambir of Batavia and Jaarmarkt of Surabaya

Yulia Nurliani Lukito, Fildza Miranda
Department of Architecture, Faculty of Engineering, Universitas Indonesia, Depok, Indonesia

Corresponding e-mail: yulianurliani@yahoo.com

Abstract In the early of twentieth century, there was a development of technology and an expansion and cultural domination of the middle class that introduced the Dutch East Indies to modernity. Although in reality a clear dichotomy between the colonizer and the colonized was still present, there were some events that allowed local middle class to practice western culture. This phenomenon was clearly shown in the way local people mimic Dutch's lifestyles, from hybrid architectural styles and the idea of modern lifestyles spread out that basically showed a mix of both Western and local cultures. A night fair at Pasar Gambir was held by the Batavia City Government to display modern products and technology while the Surabaya government organized a similar fair of Jaarmarkt Surabaya that attracted diverse crowd ranging from the Europeans and local middle-class. Those colonial exhibitions became a rare event where both the colonizer and the colonized practiced modernity. During these period, the organizer exposed hybrid architecture adopting both European and Indies architectural elements. The night fair offered a temporary 'negotiating moment' for both the colonized and the colonizer to mingle and enjoy each other's company by creating a utopian-neutral ground and using leisure as its façade. This moment in history offered not only the way local people tried to localize the idea modernity but also a lesson of social and cultural integration. In this research, we discuss the colonial fair of Pasar Gambir in Batavia and Jaarmarkt in Surabaya that became a contested space for interaction between the Dutch and Indies people through the use of hybrid architecture.

Keywords: Pasar Gambir Batavia, Jaarmarkt Surabaya, colonial exhibition, hybrid architecture, contested space

1. Introduction

In the early of twentieth century, there was not only a fast development of technology and multi-cultural influence in the Netherland Indies that introduced colonial society to modernity but also local events that reflected how people adapted to the idea of modernity. As modern lifestyle and technology being exposed to the locals, they started adopting their colonial counterparts as a response toward modernity. Locals have been seen to model Western lifestyle due to the frequent advertisement of modern living shown in the medias, much that it became part of daily lives in the Indies [1]. Despite the contrasting difference between the colonizer and the colonized, there were some events that allowed local middle class to practice Western culture. One of the important events was at colonial exhibitions held in major city in Java, such as Pasar Gambir of Batavia, Jaarmarkt in Surabaya and Jaarbeurs in Bandung. The buildings produced exclusively for these colonial exhibitions were innovative in a sense they presented a new building style integrated both Western and Indies architecture in its design.

These temporary exhibitions attracted diverse crowd ranging from the Europeans, Chinese, and local middle-class and became a rare moment where both colonizer and colonized practiced modernity. During the flourish periods of colonial exhibitions in the Indies, roughly between 1910s-1930s, the organizer presented hybrid architecture that adopted both European and Indies architectural elements and somehow illustrated how the two contrasting cultures could live together in harmony. The colonial exhibitions offered a temporary 'negotiating moment' for both the colonized and the colonizer to accept and enjoy each other's company by creating a utopian-neutral ground and using leisure and modern products as its façade.

Before the discovery of hybrid architecture, most of the buildings styles in Netherland Indies adopted the European style. However, their European-mannered architecture style was only suitable for the climate in Europe and did not match the tropical climate in the Indies. This issue led architects of the Netherland Indies to figure out what Indies style architecture might be. The discussion was prominent among the Netherland Indies architects, such as Thomas Karsten, Wolff Schoemaker, and Henri Maclaine Pont who were some Dutch architects and planners who involved in the search of suitable architecture for the Indies at the time [2]. Some analysis conducted upon the Javanese architecture had led them to discover how local culture became the element that embodied the building and how it became the main characteristic of Indies architecture. It was not long after the realization of richness of Indies architecture that drove Dutch architects and planners to create modern innovations that combined both European and Indies styles. The mixture between both contrasting cultures had presented as a new style that expressed cultural hybridization that they believed to be the birth of the new and appropriate Indies style. The idea of hybrid architecture, as a combination of both local and Western architectural principles was spread throughout the colony that considered suitable for the tropical climate and adoption of modern architectural style. It was in the temporary colonial exhibitions buildings, such as in Pasar Gambir of Batavia and Jaarmarkt of Surabaya that offered illustration how architecture played a great role in mediating modernity to local people and even helped to create a contested space for colonial interaction.

This paper examines primarily architecture during colonial Indonesia or the Netherlands Indies, because the practice of architecture during this time also showed an adaptation to local condition. It is in the effort to understand that the hybrid architecture present during this period was loaded with cultural, historical, and local context that this paper will be settled. This paper proposes to contribute to the variety of approaches of sustainability, especially as part of architectural education and practice through history. As the case studies were two colonial exhibitions held during the Dutch colonial era in the Netherlands Indies because hybrid architectures in those two temporal exhibitions presented some innovations and relations to tropical architecture.

This paper attempts to analyze how hybrid architecture mediated changes in the relationship between the colonizer and the colonized and the colonial exhibitions created a contested space for unusual colonial interaction. The duality in architecture – the strict segregation between the colonizer and the colonized culture – was challenged by ephemeral exhibitions during the Dutch colonial period when the idea of modern lifestyle started to develop in the colony.

2. Methods

The research conducted for this research was based on qualitative methods. Primary data were obtained from collection of colonial studies and literature reviews. Qualitative methods, such as interviews, document, archive reviews and survey of colonial exhibitions and architecture were used in order to uncover motivations and development of hybrid architecture during the Dutch colonial period in the Netherlands Indies. First, literature from books, journals and local archives for the colonial exhibitions were collected to gain information for the purpose of holding colonial exhibitions and the strategy of using local architecture. The focus of the analyses was the Pasar Gambir colonial exhibition in Batavia and Jaarmarkt in Surabaya. These chains of events then led us to the findings of colonial exhibition as a contested space with the organizer of the exhibition made use of hybrid architecture. The hybridity that we found in colonial exhibition acted as proof on how the Dutch Indies struggled to find ways to design a building that was suitable for tropical climate but also attracted the local people. Then, analyses of colonial exhibitions were carried to determine the significance of the use of hybrid architecture and how the exhibition became a contested space that used local images as a way of spreading the idea of modernity.

IOP Conf. Series: Earth and Environmental Science **213** (2018) 012029 doi:10.1088/1755-1315/213/1/012029

3. Discussion

3.1 Colonial Exhibition of Pasar Gambir in Batavia

During the 1920s-1930s the annual Pasar Gambir fair was held in Koningsplein Batavia or nowadays Lapangan Merdeka, Jakarta. The Chairman of the Pasar Gambir Committee Meyroos proudly mentioned what the popularity of Pasar Gambir meant for the Dutch colony and how the colonial exhibition corresponded with two elements of modernity, light and leisure. Pavilions at Pasar Gambir were ephemeral in which each year came with new architectural forms yet often used Indies vernacular architecture as references. J.H. Antonisse, the architect of Pasar Gambir, brought together local and Western architectural forms that in the end invented hybrid architecture that had never presented before. Pasar Gambir was not only successfully delivered the tradition of the nineteenth century world fairs in the West but also gave way to hybrid architecture that dealt with locality to a greater extent.

Figure 1. The 1927 Pasar Gambir

The 1927 Pasar Gambir adopted Joglo-house-style for the roofs. The design seemed simple and the pyramid shaped roof was accentuated at the end of every side with greater pitch roof. The architectural composition followed three divisions of traditional houses and was purposely made symmetrical. The lower part of the pavilions consisted of square columns that extended to the body part. Due to the elimination of solid base and pavilions, Pasar Gambir had spacious and open areas and the roofs dominated the architectural compositions. The area within one building seemed broader than the vertical oriented pavilion and the connector between two pavilions looked more horizontal compared to last years' layout. The roof was made from palm leaves.

Other type of pavilions was set lower yet looked more elaborated. The roofs of this second type had three levels with moderate and taper angles and ended with a long antenna. The body was solid but left with some openings. The walls made good ventilations as well as fantastic decorations.

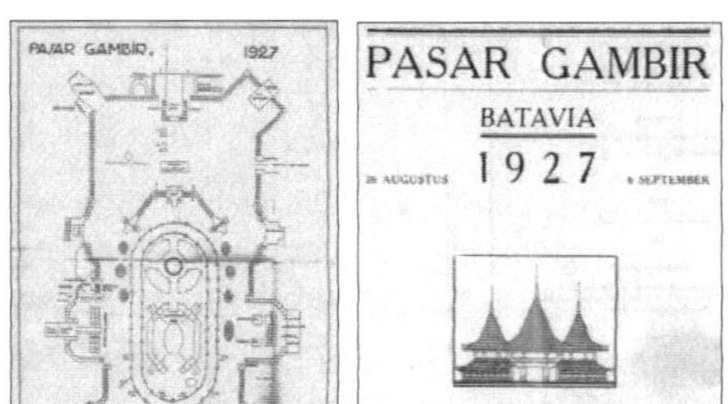

Figure 2. Program and Plan of the 1927 Pasar Gambir. Programma Van Den Pasar Gambir 1926.

The image from the 1927 Pasar Gambir's program showed the symmetrical arrangement of the pavilions. The main entrance was the tallest building, stood aside twin buildings. The middle lower part was open that show the main gate. The First Aid Post was located at the west side of the fair close to the street and a bicycle park was located on the southeast area. *Droog Aquarium* Pavilion was on the north area showing J.H. Soumokil collection from his underwater findings in Molucca. The agriculture and livestock department looked very open to the visitors, showing the agriculture and the horticulture products and some demonstration. The native industry pavilion was bigger than the previous year and a small exhibition hall was added to the pavilion.

Figure 3. The bird eye view of the 1927 Pasar Gambir.

Slightly different from previous years, the 1927 Pasar Gambir had more varieties. Besides the plain rectangular plan, some buildings had rotated rectangular orientation and cross plan. The exit area was located at the center of north side, besides the theater hall. The main restaurant and the water fountain were at the central axis not far from the dance floor and the main restaurant. It became a bit different compare to the last year where more linear structures made the dance floor more open to public.

3.1.1 Colonial Exhibition of Jaarmarkt in Surabaya

The term Jaarmarkt was derived from the Netherland's word *Jaar* from year and *Markt* from market. It was an annual fair conducted by the Netherland Indies to showcase local and imported goods as well as using entertainment as leisure for marketing purpose. J.E Jasper held the first fair in 21 May in 1905 with the financial help from the government [3]. A Dutch architect Cosman Citroen was assigned to design the buildings of *Jaarmarkt.* Citroen was involved with various committees and councils [4] including a member of the board of the *Soerabajasche Jaarmarktvereeniging* (Surabaya Annual Fair Association) from 1923 onwards and he was also asked to design a building complex for the ninth Annual Fair in 1923 and to continue the new Surabaya Town Hall in Ketabang. He held the position of president of the Vereeniging "*het Oudheidkundig Museum*" (Archaeological Museum Association) in Surabaya for some years [5]. Citroen's projects and experiences showed that he was involved in various projects and aware of both modern and local architectures. In addition, Citroen was one of Dutch architects practicing in the Indies who got honorary mention at the Paris International Exhibition with the theme of International Exhibition of Modern Decorative and Industrial Arts held between 28 April and October 1925. Citroen's works were showcased for the first time for modern decorative arts together with J.H. Antonisse and B.J.K. Cramer, two Dutch architects involved in the design of Pasar Gambir as well as Thomas Karsten and PAJ Moojen who supported modern architecture and local adaptation in buildings.

Citroen was known as an architect who considered tropical climate of the Indies in his design. His design for Lawang Sewu in Semarang for instance used stacking roof or double gable system that allowed fresh air to come in and supported cross ventilation and natural light in the building.

In 1925, Citroen produced a general layout for Jaarmarkt where buildings were arranged almost symmetrically with the main entrance was located alongside the main artery in order to capture the crowd's attention. The side entrances were placed on the axis of northern and southern part of the front elevations, which were used for service access.

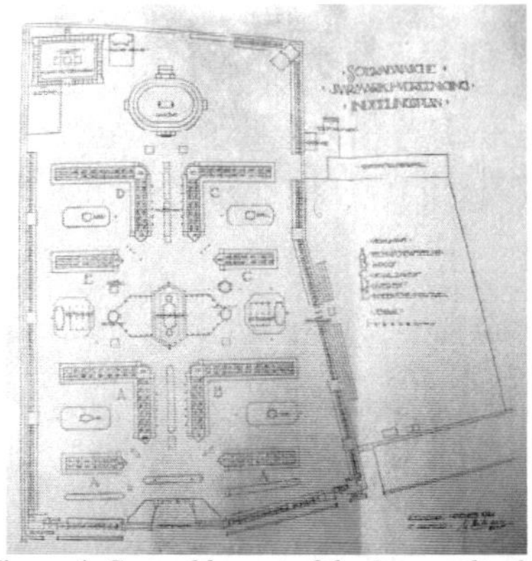

Figure 4. *General lay-out of the Jaarmarkt of 1925*

In line with the temporality of the event of Jaarmarkt, Citroen made his design to be non-permanent in order to decrease the budget. At the time Surabaya was known as *Kota Pelabuhan* or main trading harbor in Netherland Indies, and Citroen imbued the idea of trading harbor into his 1930 Jaarmarkt design. The building appearances were composed with pylons that similarly resembled the prow of a ship and built using wood as its main material. Citroen used teak wood as the main material for composing the main and side entrance, the music tents, and the building on the corner of the site were an expression of Citroen's talent in wood construction [6].

Figure 5. *1930 Jaarmarkt design*

Despite Citroen's effort to create a design that resembled the deck of a ship, the building also showed a similar feature as that of a fort depicted through its symmetrical rectangle form. Since the event was only eligible for those who could afford the entrance fee, Jaarmarkt designed to be surrounded by these wood structures, which created some sort of an enclave that distinguished the fair from the rest of the city. The method of enveloping a space was similar to what we can find in a fort without the trench that surrounded it. However, the design was able to maintain its hybridity in a sense that the European styled *fort* was balanced by the Indies styled slanted roof for the lower level in which it provided an access for the rain to flow through.

Figure 6. Jaarmarkt layout (1925) in bird-eye view in comparison with a VOC Fort in North Batavia

Jaarmarkt attraction was related to the events that were being offered to the visitors. The fair exhibited many varieties of entertainments ranging from local entertainment such as *Wayang, Ludruk, Reog, Gandroeng, Barong, Keroncong, Paseran* to European entertainments such as fireworks, dance halls, and musics [3]. However, in Jaarmarkt there was also an exhibition called *Kampung Pertukangan,* which were built to showcase the craftsmanship of Javanese and Madura people as well as important products, such as beautiful patterned cloth Batik. The design of Kampung Pertukangan adopted the Minangkabau style of Rumah Gadang for the central part with the roof pointed upward and did not take Javanese architecture as its main reference.

Figure 7. *Comparison between Kampung Petukangan exhibition building and Minangkabau style Rumah Gadang*

3.2 Contested Space

These colonial exhibitions provided a *stage* for their visitors to practice modernity but also in a sense of celebrating the colonial achievements on regulating the land and illustrated the future for the colonised natives [7]. As modernity emerged along with the exposure toward Western lifestyle, local people had shown to be more interested in lifestyle (compared to achieving their own independence). The sense of wanting to be a part of the modernity had influenced local people to believe that their identity was related to their culture and they were aware of modern culture. The term *cultural citizenship* is the idea of how the position of ethnic minorities identified to achieve liberation [8]. The occurrence of *cultural citizenship* occurred when the colonized started embracing the modern culture of the colonizer that they saw in some colonial events like colonial exhibitions. The idea of being a part of modern culture influenced local people to modify their lifestyle that was previously related to local traditions to be adapted to modern conditions. Local people started embracing modern lifestyle and accepted the idea of hybrid architecture or the appearance of *dandies* that acted as fashion ambassador whom they modeled after their colonial counterparts. They tended to seek refuge among the crowd but at the same time became a part of the crowd in the event like Pasar Gambir and Jaarmarkt that celebrated modernity. The sense of duality that hybrid architecture or the dandies had – between local and modern culture – could only be practiced in colonial exhibitions as a modern but temporary event. Similar phenomenon also occurred, for instance, in the nineteenth century Paris where there was the presences of *flaneur* who drifts amongst the crowd yet also alienated themselves because of their individuality [9].

The appearance of the 'modernized local fair' at Surabaya's Jaarmarkt indicated the importance of both local fairs as places for interaction between the Dutch and middle-class Indies in the Dutch late-colonial era. As public spaces, colonial fairs became a stage for the negotiation of economic, social and cultural relations that could not be directly spoken or easily created because of the strict segregation in colonial lives. The hybrid character of the architecture appeared, although in different degree.

4. Conclusion

Hybrid architecture that was used for colonial exhibitions showed innovations in architecture that combined both Western and Indies architectural styles. The use of hybrid architecture can be a lesson in using local consideration in creating sustainable response toward tropical climate. Colonial exhibitions also acted as a contested place where the idea of the hybridity also happened in interactions among visitors and became the identity of the space despite the contrasting difference of colonial lives. From this knowledge of the past, diversity and temporality can be a point of architectural adaptation to changes in society. Although in reality a clear dichotomy between the colonizer and the colonized was still present, there were some events that allowed local middle class to practice western culture. This phenomenon was clearly shown in the way local people mimic Dutch's lifestyles, from hybrid architectural styles and modern lifestyles that basically showed a mix of both Western and local cultures. Pasar Gambir and Jaarmarkt Surabaya not only displayed modern products and technology but also attracted diverse crowd and created a rare event where both the colonizer and the colonized practiced modernity. The night fair offered a temporary 'negotiating moment' for both the colonized and the colonizer and became a contested space for interaction between the Dutch and Indies people through the use of hybrid architecture.

5. Acknowledgments
Authors thank PITTA grant from DRPM Universitas Indonesia that supported the research conducted for this paper.

6. References

[1] Mrázek R 2002 *Engineers of happy land : technology and nationalism in a colony* (Princeton ;;Oxford: Princeton University Press)

[2] Veneendal M 2015 *Building Modernity.* (Utrecht University)

[3] Dyokta N A and Nasution 2013 Jaarmarkt di Surabaya Masa Pemerintahan Hinda-Belanda. **1**

[4] Zeeuw W 2001 *Cosman Citroen 1881-1935 (transcript)* (Rotterdam NAI)

[5] Lemei W 1935 *Indisch Bouwkundig Tijdschrift Locale Techniek*

[6] Santoswana 2010 *Analisis perjanjian kerjasama operasi sarana perkereta apian di wilayah Jabotabek* (Universitas Indonesia)

[7] University of Sydney. Department of Indonesian and Malayan Studies. J and Koninklijk Instituut voor Taal- L V (Netherlands). A D M I 2006 *Review of Indonesian and Malaysian affairs.* (Dept. of Indonesian and Malayan Studies, University of Sydney)

[8] Nordholt H S 2011 Modernity and cultural citizenship in the Netherlands Indies: An illustrated hypothesis *J. Southeast Asian Stud.* **42** 435–57

[9] Walter B 2002 *Arcade Project* (Belknapp Press)

Improving the Quality of Urban Lives through Public Art in Taman Suropati, Jakarta

Yulia Nurliani Lukito, Rima Aisha Zahra
Department of Architecture, Faculty of Engineering, Universitas Indonesia, UI Campus, Depok, Indonesia

Corresponding e-mail: yulianurliani@yahoo.com

Abstract. A successful public space requires good infrastructures and atmosphere so that it can successfully invite and engage its users in various activities. The placement of public art in a city park may stimulate and invite active contacts rather than just passive observation, thereby promoting social interaction. The purpose of this study is to discuss the use public art as well as its role in improving social interaction in public space. The aesthetic quality in a work of art can transmit impulses resulting particular suggestions in human's imaginative life and even attract activities that in turn create a great public space that sustains stronger communities. As a case study is public art in Taman Suropati, Menteng, Jakarta, which is known as the first garden city in Indonesia. In 1984, six statues from six pioneer countries of ASEAN were placed in this garden. This garden functions not only as a green open space and a recreational area for nearby residents but also as a place for some art communities to meet and hold their activities. The phenomena of a park that enhances community's activities in Taman Suropati becomes the focus for the study especially how the idea of public art is initiated, sustained, and affected urban lives. This study also suggests some possibilities of improving urban life's quality by using public art located at public spaces as the medium.

Keywords: Taman Suropati, public art, garden city, public space, urban life

1. Introduction

Social interaction is one of the most fundamental human needs, and living in a healthy society means the need of communication and interaction are fulfilled. A similar realization is included in Sustainable Cities and Communities, in the United Nation's 2030 Sustainable Development Goals (SDG) established in 2015 [1]. A city can be seen as a potential hub for preserving social and economic growth which is driven by the subsistence of ideas, culture, science, and supported by productivity and social development of the citizens. Thus, a good living environment must be an environment that is adequate and supportive to generate social interaction. A city's potentiality in supporting social interaction is shown through the public spaces.

Public art is an artistic expression of artist and community that can be found in various forms, materials and processes, both installed permanently or temporarily in a public space [2]. The practice of public art includes a collaboration of stakeholders, including the authority, artists or designers, and community members in order to promote a sense of belonging. Public art may contribute to the enrichment of public realm, the success of place making and the development of cultural and social capital [3]. Public art usually relates to local culture and establishes history, memory, the sense of notable event, as well as reflecting the belief system and at the same time, maintaining the city's identity.

IOP Conf. Series: Earth and Environmental Science **213** (2018) 012030 doi:10.1088/1755-1315/213/1/012030

Every installation of public art has its rationale towards improving the quality of the living environment and the urban community. Public art plays an important role for the improving the quality of urban life in the following manner [4]:

- Reflecting and expressing city's identity
- Celebrating significant people and places
- Telling the stories and histories of a place
- Contributing to a sense of pride and belonging to the community
- Improving the attractiveness of an area.

Arguably, in developing countries like Indonesia, there is a relatively less awareness of the role of public art in improving the quality of urban life compared to Western countries. People still question the connection of public art to urban landscape and to extend of public art contribution to the improvement of the quality of urban life. This lack of awareness leads to some debates of the public art's meaning, placement and appreciation. Moreover, the practice of public art should respond to a particular setting of a public place, such as its use and significance history [5].

The aim of this paper is to analyze public arts located in a city park in order to investigate the aesthetic and functional values of public art and its relation to the particular setting of the public space. This paper takes Taman Suropati as the case study, a garden city and a public space widely known as a relic of Jakarta's development in the early twentieth century. Located in Menteng, the central part of Jakarta, Taman Suropati is also a home for art communities. There are some statues from Association of South East Asia Nations or ASEAN countries that makes this city park has its own political statement added to the long history of the development of Jakarta city starting from the Dutch colonial period. It is this writing's quest to discover the association of historical background, spatial qualities, and the narrative in Taman Suropati as a public space with the presence of public arts and community activities that in the end improving the quality of urban life through social interaction.

2. Methods

The research used mainly literature research and on site observation. Qualitative methods such as interviews, document and archive reviews, and field survey of public art were the main methods used for this research in order to uncover motivations and expectations that drive the development of public art and public space projects. On site observation that included interviews and visual documentations become the primary way of getting firsthand information for the analysis. As a case study is Taman Suropati in Jakarta that has expressive artworks and emphasizes the importance of locality as criteria for successful public art implementation [2]. The criteria included highly visible spaces and free accessed by the public. The review of the public art projects and public space were carried out firstly, an extensive review of the literature from books, journals and local archives for the art projects was conducted to gain information for the purpose of enhancing the relationship among ASEAN members. Then, information from the literatures was combined with the results from on-site observation in order to analyze two important aspects of the public art: aesthetic and functional. Lastly, analysis of the public arts and urban space was carried out to determine the significance of public art in improving quality of life also including the consideration of the local contexts of the public space.

3. Discussion

To be able to analyze Taman Suropati and its function as a public space, we first have to acknowledge its historical background and the most apparent narrative of the place, including its public art. The narrative of the place should be combined with an analysis of physical elements in the park [5]. These two points are considered substantial in Taman Suropati's sustainability in allocating its site for public interaction and activities.

3.1 History of Taman Suropati

Taman Suropati is located in Menteng, a site which was a part of the first garden city project built during the Dutch colonial era and an expansion of the city of Batavia or nowadays Jakarta. The area of Menteng was originally *particuliere landerijen* or private land – a kind of property that was commonly

IOP Conf. Series: Earth and Environmental Science **213** (2018) 012030 doi:10.1088/1755-1315/213/1/012030

owned by noble society since the seventeenth century in the Netherlands Indies [6]. The Dutch architect Pieter Adrian Jacobus Moojen was assigned to design a satellite town for Batavia in 1910. Allegedly, influenced by the growing ideology of a garden city utopia in Europe emerged and popularized by Ebenezer Howard, Moojen designed Menteng in a form of a garden city, making it the first ever prototype in the Netherlands Indies. His design was implemented into a radial constellation that covered residential blocks, gardens and boulevards, including connections with other residential areas. The broad Nassau Boulevard (Imam Bonjol Street) that bisected the Menteng residential area next to the central square (Taman Suropati) was intended to connect Tanah Abang to the west and Meester Cornelis to the east. The Java Street (HOS Cokroaminoto Street) was a north-south connector from Kebon Sirih to the southern fringe of Batavia.

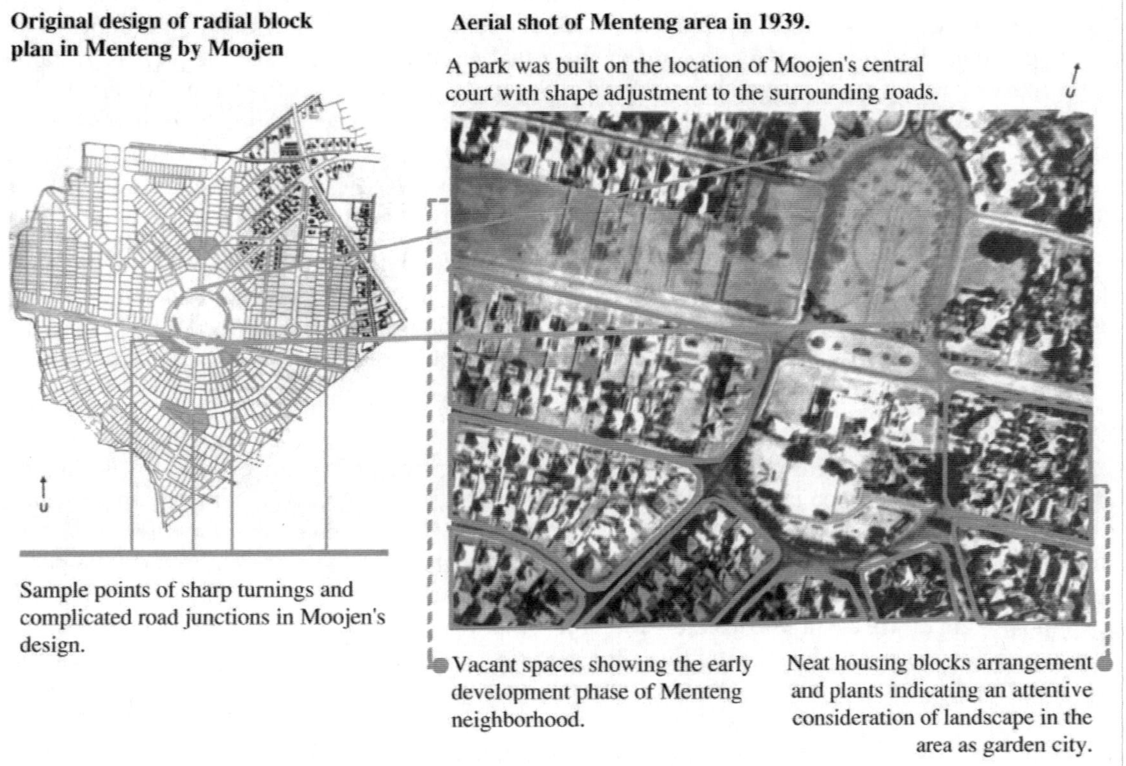

Figure 1. Comparison of Menteng neighborhood from its original design and early development.

Moojen's design shows a number of angles at the end of the streets that were considered too sharp for car manoeuvres and confusing. Moojen's plan was later modified in 1918 by F.J. Kubatz who changed the street pattern to become more realistic for vehicles to make turns and added a small pond (Situ Lembang) to the east of the central park area. Throughout the development of the area, the big park that was existed in Moojen's design was never really built. But on the approximate location laid a smaller park positioned in the middle of Menteng. This park was originally called Burgemeester Bisschopplein, named after the first *burgemeester* or the city mayor of Batavia, G.J. Bishop (1916-1920) [7].

In 1920, an article on Taman Suropati was written and it illustrated how the park was cultivated with trees and flowers [8]. Figure 1 shows an aerial picture of Menteng taken in 1939 that showed a landscape of a newly grown park surrounded with blocks starting to be filled with houses. There are two other parks in Menteng beside Taman Suropati namely Taman Situ Lembang and Taman Menteng. Taman Situ Lembang was actually a water storage reservoir for the neighbourhood while Taman Menteng was originally a sports stadium for local football club Persija. Thus, Taman Suropati is the only park that is originally intended to function as a public park.

From its origin, the plan of Menteng garden city was intended for the elites. The site fit the purpose because the closeness of Menteng to the center of capital and government, and the idea of utopia were a great appeal to its potential buyers. Up to the present, Menteng area is still known as an elite area and the prestige of Menteng is transferred to the urban elite of Jakarta. With the prestige of the area, and the utopian ideal of a garden city, the existence of a public park is necessary to implement the function and image of the city [6].

During the 1990s, the image of Taman Suropati was similar to other city parks in Jakarta for being rather neglected and sometimes associated with negative image of the city. The city parks also had to compete with modern recreational space such as mall. In 1997, a senior violinist Agustinus Hesti Dwiharso who had a vision to preserve folk music from different regions in Indonesia started to use the park for his musical activities. Moreover, Agustinus believes that city parks are important public spaces and people should make a good use of the park that motivate him to make the park as his home base for musical activities. Then, he initiated Taman Suropati Chamber in 2007 that consisted of street musicians and artists and used the park as a meeting place. Instructors for violin, cello, percussion or guitar were since then available on weekends and yoga community came during weekend. Later on, Taman Suropati Art Community was created that involved divisions of music, crafts, sketch, photography, and theatre, still with active collaboration with street artists.

The history of Taman Suropati tells the importance of Menteng and the park in the development of Jakarta. The park bears the idea of public space with symbolic arts and cultural activities that the park offers. The rich meaning of Taman Suropati is also related to physical elements inside the park that in turn influences the image of the place and users' behavior. Public arts at the park add artistic elements and consecutively affect activities in the park.

3.2 Public art and activities at Taman Suropati

In 1984, the local government decided to use the park for more prominent events. Six sculptures designed by various artists coming from the six pioneer countries of ASEAN were placed in Taman Suropati and substituted previous animal figures at the park. The sculptures are known as ASEAN's Monument of Peace and bear the idea of support and participation for the regional league. The public art project in Taman Suropati also means that the government recognizes the park as an important space and is worth to deliver political statement of peace and friendship among ASEAN members.

When one walks the path from the south to the north of the park allows a view of six prominent artworks embodying the symbolic friendship and diplomatic ties between Indonesia and ASEAN countries. These six sculptures include (see figure 2): "Rebirth" by Luis E. Yee Jr. from Philippines (A), "Harmony" by Awang Hj Latif from Brunei Darussalam (B), "Fraternity" by Nonthivathn Chandhanaphalin from Thailand (C), the "Peace-Harmony and One" by Lee Kian Seng from Malaysia (D), "the Spirit of ASEAN" by Wee Beng Chong from Singapore (E), and "Peace" by Sunaryo from Indonesia (F).

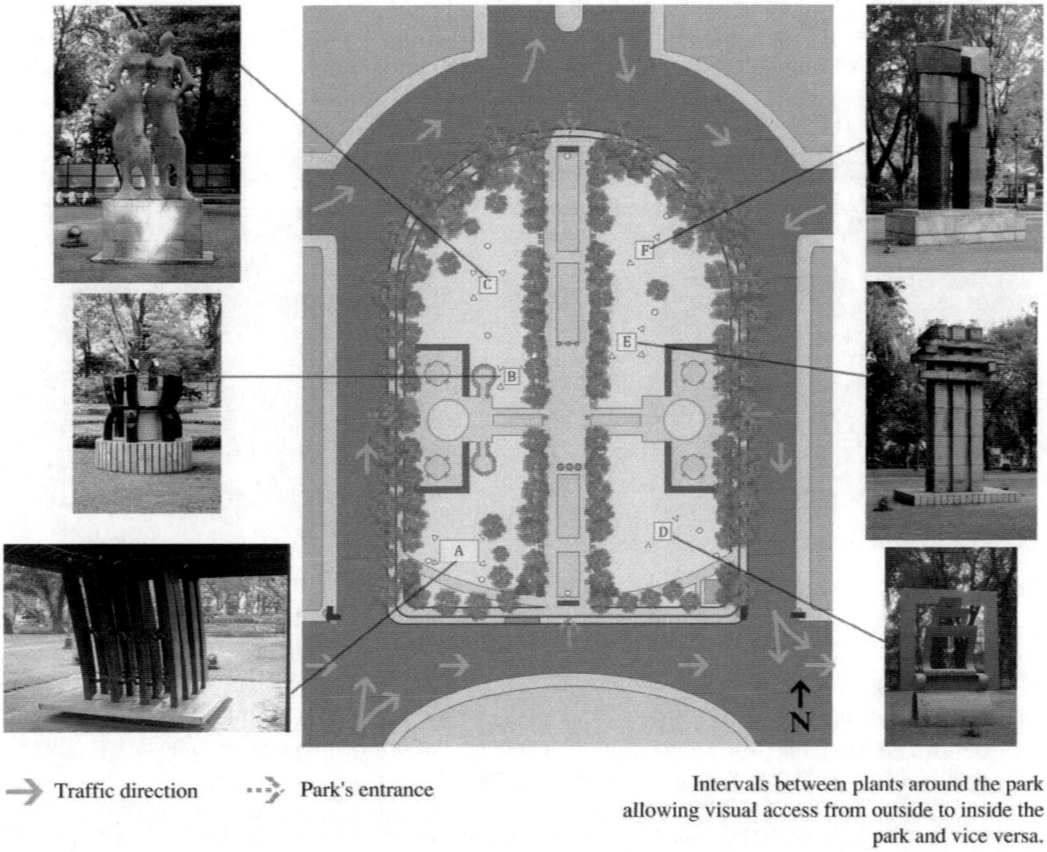

Figure 2. Artworks at Taman Suropati.

An axis divided the park into two symmetrical areas, leaving a clear influence of European garden. There are three plazas at the park – west, central, and east plazas – with four accesses to the park (south, west, north, and east). The west and east plazas have two identical water fountains while in the middle stands the central plaza as the biggest plaza. People usually use the central plaza for communal activities, such as tai chi and yoga exercises as well as community gatherings. Trees and plants at the park are arranged with intervals, giving visual access from outside to inside the park and vice versa. With this kind of arrangements, it is possible to gain visual access to the artworks from almost every direction. Although not symmetrical, the sculptures were placed in balance – three on the left and three on the right – and in accordance with the park's access and axis. Those artworks are put on the grass with low platforms.

These sculptures become a set of static artistic element of the park. The strong sense of artistic value in the park does not only come from these sculpture, but also from musical activities. The sounds from violin and guitar instruments, for instance, create unique atmosphere and artistic identity of the park [8]. Moreover, the activities of Komunitas Kota Seni Taman Suropati also emphasize the function of the park to support community activities related to arts.

Komunitas Kota Seni Taman Suropati holds a violin course every Sunday in Taman Suropati. The class is divided into several times, starting from 10.00 AM to 17.00 PM. Usually, the violin class involves some teachers who tutor several groups, marking each 'classroom' with plastic seats around the plaza located on the west side of the park. This group of musicians becomes an informal institution and the class is open for all ages. It is also common to see elementary school children sit next to middle age men because they take the same class. There are also other art groups such as sketch, theatre, crafts, and other visual artists in the park. They usually hang out or have an informal meeting at the park as the musicians play their instruments. Beside regular activities, some art communities

IOP Conf. Series: Earth and Environmental Science **213** (2018) 012030 doi:10.1088/1755-1315/213/1/012030

also organize special events at the park. The special events that include discussions, performances, and exhibitions. Those events are usually open for public and those events attracts people who live far from the park.

Figure 3. Private violin lesson

Figure 4. Discussion in an event, "Nyandang Sandang" on 5 May 2018

Public arts at Taman Suropati can be understood as two different kind of artworks based on motives of their movements. First, public arts at Taman Suropati are a set of display and inanimate objects. Second, public arts at Taman Suropati are animate objects and become influential elements for public activities [8]. In inanimate objects, the movement becomes a consequence of external physical forces while in animate objects the movement is the result of intention and desire. These two distinctive practices of public arts give different impacts to the park's users.

The sculptures, which have fixed places at the park may be treated as art displays. There is visual reception from people in the park when they see the public arts. As the result of passive interaction between the public arts and the park's users, users perceive the public arts as a part of the idea of the park. In this way, the existence of the public arts are relative to their own position in the park, other physical elements in the park, and the image of the park. Since the position of those public arts are not central but spread in the park, no single public art looks dominant. Since there is no particular access or path to reach the artworks, the artworks seem to be understood only through visual connection.

Community's activities are influenced both directly or indirectly by the presence of the public art. Public arts indirectly give influence to the users, especially art community, for the presence of those public arts give significance artistic value to the park. Public arts, together with the plazas and other objects at the park, also mark the plaza and influence how people perceive their places. For instance, people at the central plaza will most likely see the artworks from Brunei Darussalam the tall artworks from Singapore so that when they do their activities those artworks become part of the nearest surroundings. In addition to this, not only physical objects located at the park but also the sounds of music give influence to the atmosphere of the park. The dynamic sounds of music successively add channels of experience, enabling users to directly react at its presence. By these two dissimilar degrees of influence, we can assume that community's activity is closely related to the presence of public arts that contributes to the creation of unique qualities of the park and in turn affects how the users perceive and use the space.

As the violin course takes place in the west plaza of the park, the sound can be heard from all directions. The unique violin sounds create a background sound for other activities, together with natural sounds of water from the fountain, birds chirping, and people chattering; all those sounds subdue the sound of the passing vehicles around the park. The presence of the specific sounds of music has affected the atmosphere of the entire place. The sound of nature combined with the unique sound of the instrument result in a synergy of space and event that produce a certain kind of artistic quality of the park.

Figure 5. Section drawing and situation of west, central, and east plaza

Public art has some historical, aesthetic and functional values [9]. Taman Suropati is loaded with the historic meaning of the development of Menteng area and ASEAN public arts encourage the connection between the societies with their history and create a collective memory. Public art conveys the basic impression of art that is to beautify spaces and improve the visual quality. Public arts at Taman Suropati do not perform as functional objects such as street furniture. However, the existence of public arts in Taman Suropati give additional functional value for the park as a landmark and borders of activities. The success of public space in Taman Suropati arises from collaboration with the users of the place who articulate what they value about it and public artworks stimulates active interaction rather than just passive observation from the users [10]. In short, public art has helped to improve the quality space in the urban environment through fostering social interaction.

The appropriation of the park by the local government through public arts plays a significant role in creating a public perception of the park [11]. The rich historical context of the park overlaps with the narrative of the place as meaningful public space that influence how people response to the idea of Taman Suropati as a public space. The appropriation of the park as the place for art communities' activities is related to the accessibility of the park for public. Taman Suropati has become a medium for people to encounter urban space while the implementation of public art has encouraged the participation of public in public space. Up until today, the community is still dominated by middle to lower class society who actually live outside the luxurious residential of Menteng. In contrast to big houses in Menteng, people who use the park are actively use the park for as a regular venue of their routine activities and allow various background of society to join. In Taman Suropati's case, the rich historical background of the place is added with the utilization of space by street artists. As a result, the communities who occupy the park and have variety in their backgrounds, contribute in branding the landmark as an art place. The artworks placed in the park are actually not too influential in affecting the park's users. Nevertheless, the artworks help to strengthen the whole narrative of the park as a set of art displays. Besides improving the attractiveness of the area, public art also reflects the city's identity, and celebrate the history and the uniqueness of local activities. People gather and interact in Taman Suropati with art as the medium of interaction. Public activities, discussions and collaborations held at the park reflect a lively social environment.

In a wider understanding, public art in Taman Suropati has its purpose in the urban community and helps to improve the quality of urban life. Following Hamilton City Council [4], the park reflects and expresses city's identity as both loaded the idea of garden city in the park's history and telling stories of its own as the place for art community. Taman Suropati also celebrates significant people and places that are related to art and contributes in giving a sense of belonging to the community. The identity of the park as important place in the history of Jakarta and as a place for art communities to do their activities improves the attractiveness of the area and the quality of urban life.

4. Conclusion

The analysis of public art and activities in Taman Suropati has shown an apparent case of an active public space that is, directly or indirectly, powered by the influence of public art. In Taman Suropati, public art is a promising agent to revive and sustain the function of public space as the place for community collaboration.

In Taman Suropati, there is quite a successful implementation of public art that generates artistic activities in the city. One of the reasons is because various stakeholders have a significant collaboration as well as the effect of physical elements of the park that create good access, visual connection and unique atmosphere. The role of public art in benefits not only the city's development but also its community's activities. Thus, further efforts to ensure collaboration of stakeholders, accessibility of public space and sustainability of public art and artistic activities at the park should be maintained. In order to sustain the uniqueness of the park, respecting local history and promoting collaboration of communities through the arts should be continued. In the future, it will be helpful for city planning to include potential treatments and utilization of public art in the grand design to support the growth of its community. The awareness of public art as a significant agent that can influence the life in the city should be related to the understanding of the history of the city and the enhancement of physical elements that surround the public art.

5. Acknowledgements

The research of this paper was conducted with the support of PITTA grant from DPRM Universitas Indonesia.

6. References

[1] UNDP 2016 Sustainable Development Goals

[2] Chang T C 2008 Art and Soul: Powerful and Powerless Art in Singapore *Environ. Plan. A* **40** 1921–43

[3] Anon National Policy Statements

[4] Anon Public Art - Hamilton City Council

[5] Irish Museum of Modern Art 2008 What is public art? (Irish Museum of Modern Art)

[6] Moore C W, Mitchell W J (William J and Turnbull W 1988 *The poetics of gardens* (MIT Press)

[7] A H 2001 Kota Taman Pertama di Indonesi (Yayasan Cipta Loka Carak) pp 9–51

[8] N A 2005 Asal Usul Nama Taman Suropati

[9] Ozsoy A and Bayram B 2007 The role of public art for improving the quality of public places in the residential environment *Sustainable Urban Areas* (Rotterdam)

[10] Harrison C and Wood P 2003 *Art in theory, 1900-2000 : an anthology of changing ideas* (Blackwell Pub)

[11] A M 1962 *The perception of causality*

The Comparison of Structure Construction "*Atap Empyak*" with Roof Construction of "*Atap Rumah Bolon*" Jangga Dolok

Ch. Koesmartadi, Yulita Titik S., Gregorius Adiprakoso
Soegijapranata Catholic University, Semarang, Indonesia.

Corresponding e-mail: ch.koesmartadi@unika.ac.id

Abstract. It is amazing that Indonesia has hundreds of traditional houses that can be categorized as a wealth of archipelago architecture. One of which is the architecture of "*Atap Empyak*" house in Central Java. The construction of the roofing house of *Empyak* is a unique construction where the whole field of the entire roof serves as a structure, so that the roof construction is enough to rely on the strength of field and pedestal beam as the landing and pedestal of the two areas of the roof of *empyak*. Architectural formation system allegedly has resemblance to the roof structure system of traditional houses in the archipelago of Nusantara. There are allegations of some similarities of the roof principles of traditional houses in Indonesia referring to the towering high roof architecture. One of the allegations is the architecture of *Ruma Bolon* Jangga Dolok North Sumatra which has the same principles of structure and construction with the roof of the house *empyak*. The *Ruma Bolon*'s roof allegedly has similarity in terms of the principle of field structure as the roof structure. The similarity of the construction lies in the *usuk/urur* construction as the connection of the two joists, the arrangement of intestinal/urinal concrete as the forming of structural strength so that it does not require horses buffer roof.

Keywords: without *kuda-kuda*, *Nusantara* roof, roofing house, *usuk, urur*

1. Introduction
As mentioned before, the House of Roof *Empyak* has a uniqueness that is the roof of the field structure standing by mutual lay down each other's roof. The uniqueness of the structure and construction of the roof of the *empyak* can be a source of inspiration in searching and pacing steps against the structure and construction of other roofs scattered in the Indonesian archipelago. One of which is the structure and construction of the roof of *Bolon Batak Toba Sumut*. After the comparison, similar as well as distinguishing principles can be the reference material on the tracking system of custom house roof form in the Indonesian archipelago.

2. Methods
The method to prove of these allegations was done by comparing techniques by putting forward all the similarities and differentiation between the two, so that the results of the match obtained some principles adding to the wealth of archipelago architecture.

3. Discussion

3.1. Empyak in Java Architectural Space
The character of a humid tropical country of Indonesia has the nature of scorching heat and high intencity rain that cause covering and shading to become important [1] Covering or shading is the act as putting

yourself under the shade. This action is not to hide or to escape, but the act of not being directly exposed to the rain or the heat of the sun. The term of this roof construction is 'shelter' as Paul Oliver stated in his book 'Shelter, Sign & Symbol'.

3.1.1. Empyak as Shelter

This shelter is possibly made by the presence of *empyak*. In determining the amount of *empyak* the field and the designation of the building were carried out. This construction suggested that *empyak* became a shade that produced a volume of space that could be utilized by people to run activities. Now it appears that a linkage has been expressed by *empyak* that is linking for use with shade. *Empyak* refers to shelter because of the function of *empyak* as a shade.

a) The position of *empyak* in a certain level facing of the soil produced the volume of space for the implementation of a human activities. Here the floor did not make up the volume of the room space, but the roof that did it.
b) *Empyak* served as shade - shelter, here human activities could be held with comfort. The distance between roof and the soil became the volume of the room formed. Based on the research conducted by the author, there were other studies conducted by A. Rubbi Magzaya from the Postgraduate Program of Architecture Engineering UGM in Journal of Architecture Tesa in 2012 with the title "Empower construction technology (emphasis on structural stability)". Magzaya states that: "*Empyak* is a dialogue between man and his environment.
c) Factors that influence *empyak* construction:
    1) The roofing house of *empyak* is not developed anymore.
    2) Roof geometry and elements of *usuk* are very dominant.
    3) The change of society leaves the nature of mutual cooperation.
    4) Nowadays, it is easier to build a house using new materials and technology.
    5) The purpose is efficiency of time and materials.
    6) It needs wider place to build *empyak* construction.
    7) The theory of *empyak* construction is not developed.

The architecture of the roof houses *empyak* in general is still in the house type of *limasan*. However, based on the construction system, the roof can still be categorized as the architecture of the house of earthquake-resistant *limasan*. Its ability to muffle comes from the structure of the frame system where the logs are arranged by applying the form of a cube limas.

3.1.2. Construction of Empyak Roofing House

As construction and structures of skeletal structures, the structural strains of the Javanese *griya* (house) get their own nickname [1]. The framework of the *griya* is called *balungan* (skeleton), while the roof is called *empyak*. This means basically a Java *griya* is formed from two main structural components of the building that is the component *balungan* and the components of the *empyak*. The separation of the Javanese structural compartment is determined by the action of preparing the rumble or substance of the component. For the *balungan* component as the building framework, the measurement is done by applying *petangan* (prediction of a good day), such as the measurement guidelines and the division of building component.

To quote Frick's view, *empyak* is the traditional Javanese roof cover plate [2]. The main ingredient of the roof of *empyak* is bamboo construction. The size of each *empyak* is the roof area of the house. Assembing *empyak* begins with making *jidaran empyak*. *Jidaran empyak* is a series of *gapets* that have been shaped *empyak* but still empty (not filled with *usuk*). Used charging (called "*ngragum*" = stringing) on the *jidaran* can be handed over to others in mutual assistance. *Usuk* is tightly held carrying with a distance of 75 cm. Finally battens can be installed with the distance as needed by layering the roof (*welitan* or *genting*).

To lift up *empyak* at the time of building, a house needs many people in mutual assistance. The empire roof construction explains that the hanging roof construction is more ancient than the intercropping that cannot be assembled with the oil. The roof of intercropping usually uses roof construction consisting of the *usuk* and battens with *welitan* layer of shingles or tiles. Empires are done indirectly on the premises, but assembled first by *blandong* (carpenters) in accordance with the procedures that have been submitted, newly installed on the house in mutual assistance. According to Frick [2], *empyak* as a roof assembly with various functional tasks is also a ritual with the boundary between the roof (as the god of goddess or ancestral sacred place) as cover of house (micro-cosmos) and sky (universe or macro-cosmos).

Empires as roof assemblies have a variety of functions / tasks [2]. The *empyak* serves as a house ceiling and the oil also has the function of an additional layer to prevent the leakage (base sheeting). In addition to the bonds, strong empirical assemblies can also serve as construction stabilizers, such as batten and *usuk*.

3.1.3. The roof of the house empyak

The principle of the roof of the empire is to unite the two fields of the empire and the two cuddles mutually lie against each other forming the *limasan* roof. Empire field is a form of field structure that mutually strengthens themselves and receives loads on the load. Thus, other elements, such as *wuwungan* and *ander dudur,* function as binders four layers of *empyak*. Thus, the roof of the house *empyak* does not use the triangle horses wood construction as a buffer roof load (Figure 1)

Figure 1. The proces forming *empyak* roof.	
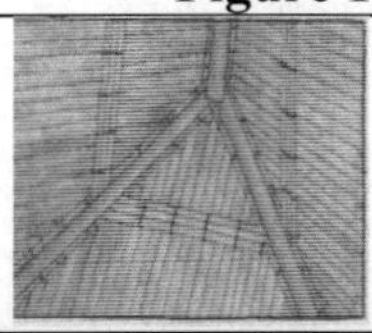	**Figure 1.a**: there is upper horizontal wood *suwunan* (between 2 *empyak*) and *dudur* (between *kejen* and *empyak)*
	Figure 1.b: Vertical wood (*ander*) together with horizontal wood (*gonjonya)* to form *empyak* construction work entirely

3.2. Principles of Structure and Construction of Roof Ruma Bolon Jangga Dolok

In a building construction its performance can be seen when viewed intact in a single structure of strong rigid and beautiful structure. The shape of roof loomed on both sides of the roof, as well as wooden construction buildings to arrange construction as a whole. The construction of the arrangement is done by arranging and connecting with each other. In order to achieve this, it is necessary to connect a construction by knowing construction relationships, such as *gapit*, *pen* and hole construction or by combining with the addition of other elements that strengthen with iron plate. In the *gapit* system relationships there is no weakening, as both rods remain intact if the construction relationship with perforation on a plane or construction rod weakens the strength of the construction.

The purpose of the building structure is to make sure the building will stand under load and pressure on it: weight, wind pressure, style due to climate change, and possible earthquake shocks. The builders want to make sure the buildings will not collapse, and they also hope the building is not damaged, because it can kill people and be costly to fix it. They want to make sure the building will not move. If a house slips from the hillside or if the skyscraper collapses due to the wind, the building has failed from its destination, even if its structure is not damaged [3]. The structure of the building and the concept of architecture are two sides of the coin that cannot be separated. The structure of the building cannot be ideal without paying attention to form, function, building system, material and so on. Conversely, the concept of architecture cannot be done without the selection of the right structure system for a building. Regarding earthquakes, building structures are not the only determinant of the safety of building users. But structural systems can also form and be shaped by architectural concepts. Thus, the linkage of building performance under the earthquake is necessary [4].

In ancient times most forms of *autotoknos* structures were very limited according to the experience and techniques of carpentry as well as by metaphysical factors (custom, primbon, religious challenge etc.), based on the form, width of the span, as well as building materials used traditionally, such as the Batak home structure Toba [5].

3.3. Traditional Houses Rumah Gorga and Batara Siang House.

In Batak Toba there are two types of traditional houses namely Rumah Gorga and Batara Siang House. Gorga's house is decorated with carvings all over the house, while Batara Siang has more limited decoration. Both Batak houses are very distinctive in their roof structure with the tumpu on both inclined plane at the edge of each plane and connected with the pull rod (hill/*nok*) without the triangle horses construction and reinforced with cross wood on both slopes.

At the Gorga Jangga Dolok house the role of the wood board acts as a connecting beam, inter-pole fastener, stability to earthquake wobble. The entire construction is made of special wood at least 50 years old and before cutting it is required to plant first as a replacement tree. In this house the position of the racang board serves as the stability of the building against the lateral force on the column pole, so the hole on the columns of each column are leased with the intention of facilitating the installation, inserting the wood boards into several cumns, in case of difficulty mounting the stone can be raised [6].

The construction of traditional horses in Indonesia is very rich associated with different architectural and cultural diversity. But in principle the construction of horses can be classified on two principles: stage roof construction and horse construction. An example of stage roof construction and construction of horses is construction of roofs of Batak Karo (West Sumatra).

The construction of traditional horses in Indonesia is very rich in relation to the richness of different architectures and cultures. But in principle the construction of the horses can be classified into two principles: the construction of the roof of the stage and the construction of the horses, such as the roof stage construction of Karo Batak roof house [7] (Figure 2).

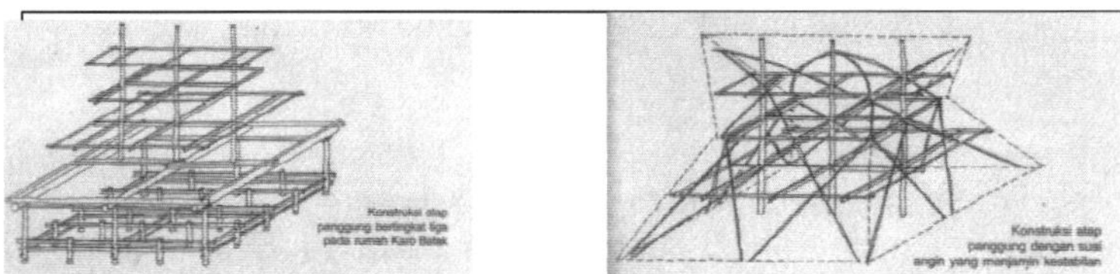

Figure 2, a, b: Three-story concrete roof construction on Karo Batak *ruma* [7]

3.3.1. Previous Existing Houses

Figure 3, a, b, c: Toba Batak House previous existing houses

Early Batak house first had a lot of population in the past, along with the development of the era, the population of the Batak house decreased (Figure 3). So this study is expected to reveal its potential. This paper discussed a series of process of establishing customhouse until the bias data obtained as complete as possible.

3.3.2. Somba and Si Bong Bong Ari

Somba is the base of the construction that will determine the location of Si Bong Bong ari (triangle barrier *pegulan*) (Figure 4).

Figure 4, a, b: *Begu-begu* that is mounted on Sibong-Bong Ari front (front part of triangle horse construction)

After Si Bong Bong Ari stood then adjusted the position for the placement of the month (*nok*) (Figure 5).

Figure 5, a: Measurement with rope to determine the length of the triangle wood

After the book rests on both Si Bong Bon Ari then do the binding and then continued the next job.

3.3.3. Urur (usuk)

Figure. 6 a, b: The *usuk* is fitted with a alternating hose based on the lower makeon and on the hill at the top. The top left 15cm from the book and the rest of the pedestal to the makeon in the puncture to the already bunched and punched.

IOP Conf. Series: Earth and Environmental Science **213** (2018) 012031 doi:10.1088/1755-1315/213/1/012031

The dirt is mounted on the top (*nok*) at the top and bottom on the Tombonan Urang lisplang board as a *tumpian urur*) which is then bonded with a rattan rope tied to the Createon (round beam as rattan handle) (Figure 6 & 7).

Figure: 7.a, b: Installation of *Usuk (Urur)* and *sandals* Slippers, strap as a buffer and the worker's footrest to bind	

3.3.4. Installation of Ijuk Fibers

Bamboo ladder to help install the fibers from the outside	Tomboman, as pedestal thatuk tied on the makeon	Installation of *ijuk* fibers

Figure. 8, a, b, c: principles and order of installation of lais-lais and fibers

3.4. The *Comparison of Structure and construction Empyak Roof House and Ruma Bolon*
This pairing is in search of the similarity of construction of Batak Toba house with the roof of the house. In the discussion of the roof, many found alleged suspicions that the condition and shape of the roof of a house Batak Toba Batak resemble the roof of the house *empyak.* In order to obtain the results of these allegations then this study conducted evaluation on the construction of both roofs (Figure 9-20).

	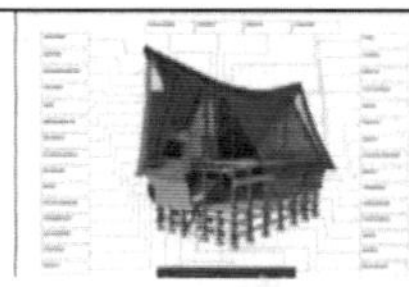 
Existing house data roof of *empyak*	Existing data *Ruma Bolon* Batak Toba

Figure 9.a, b: The existing data of the roof house of the tuna fund of the Bolon tap. Each image shows the shape of the roof that becomes the object of research. Both types of roof forms become the object of research, so it is expected that the matching process will obtain optimal results.

The shape of the roof *empyak* consisting of two layers and two *kajen* closes tightly and forms a pyramid roof	Both areas of the roof leaning against the horisontal beam (*nok*) at the top and the bottom pedestal

Figure 10, a, b: If the *empyak* is a two-sided plane and two *kajen* only as pyramid formers, it is not the case with the ruma roof. The second bolon of roof ruma bolon rests on a Tombonan board tied by a rope rope with a round beam (makeon). A field with a steep angle the load will be at the bottom.

The lower pivot area is on the beam	The base of Bolon house is on a *Ropu* (rattan rope) strap that holds the load at an angle above 60 degrees that holds evenly the *tombonan* board

Figure. 11. a, b: both the same support images work on the roof but with different shapes.

The shape of the roof space *empyak*	Space in a Bolon house

Figure. 12. a, b: The rooftop bottom space, seen from below shows that each of *usuk* (*urur*) is constructed directly without a buffer supported on both ends of the field.

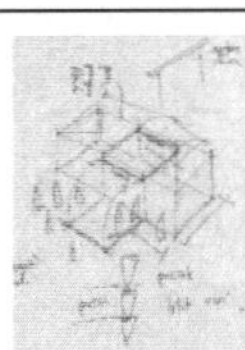 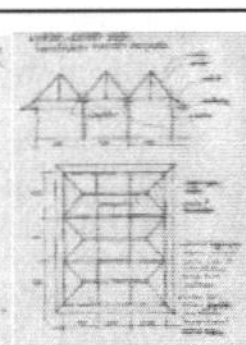	 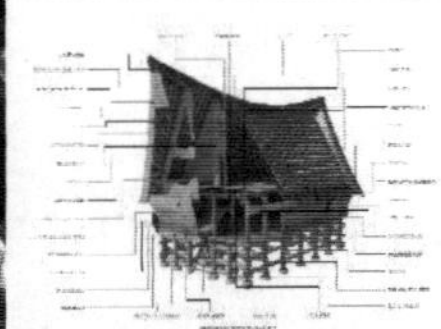
The system of roof construction *empyak* done with raft system outside the location so that to stay upward it is arranged into a *limasan* roof.	Roof Bolon house is done with one by one stem where in the place of the house was built.

Figure 13.a, b, c: actually with the advancement of the roof kompas *ruma bolon* bias in the program into the system roof of the *empyak*.

Empty roofing console system built by the console itself so that the bias covers the wall	Another case with the bolon house roof. *Tumpua* there is pulled by strands of rattan rope, so as well as a console.

Figure. 14, a, b: the element of shade bias through the bolon house design without altering the alleged functions of the empire in the house.

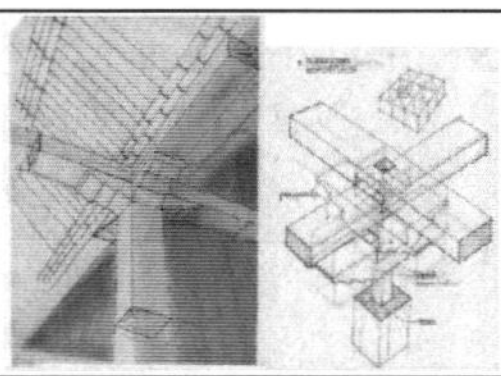	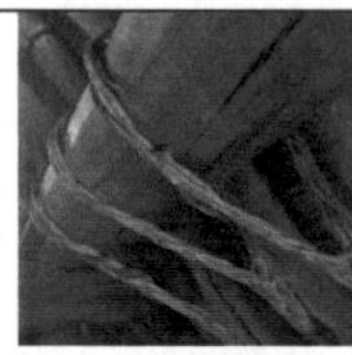

Figure 15. a, b: the two constructs mentioned above are unique and having no equivalent to any other construction, but both constructs are of great importance.

	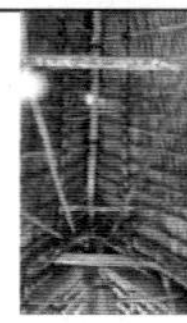
Space underneath *empyak* roof	Space underneath *Bolon* roof

Figure 16, a, b: the lower chamber of the two cases shows a similar function.

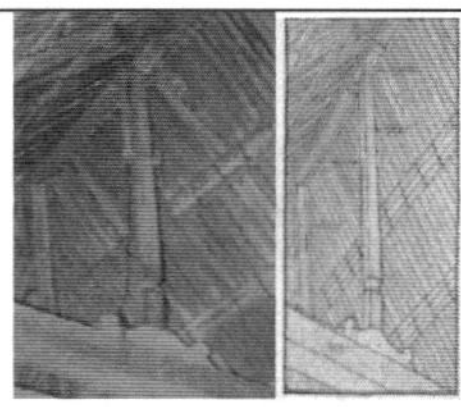	
Vertical wooden column *suwungan*	Vertical wooden column *bukulan*

Figure 17, a, b, c,: buffer roofs and house bolon, both of them are not triangle horses wood construction.

Empyak before constructing	Construction roof *bolon* house

Figure 18.a, b: the intact condition of the roof and the roof of the Bolon House indicates that the roof area intactly works on the system of drafting in such a way as to form the construction field.	

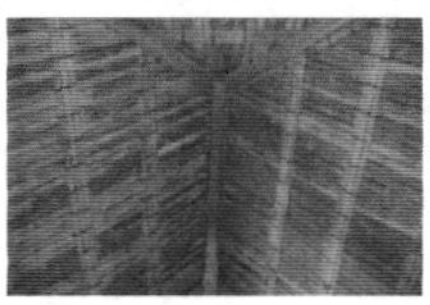	
Suwunan (*nok*) in *empyak* construction act as a base upper part of *empyak*	The shape of *bukulan* (nok – horizontal beam) act as base of roof layer

Figure 19.a, b: place of construction of upper field of suwunan (*empyak*) and *bukulan* (bolon house), both rods serve as pedestal of construction field of both types.

Article I. 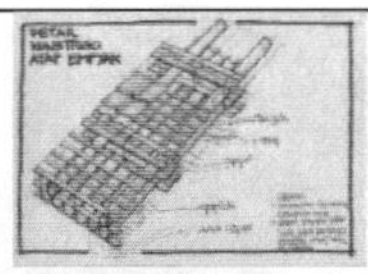 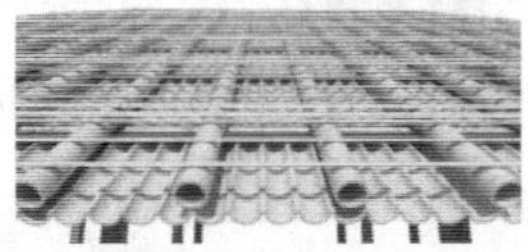	
Empyak consists of a semicircular meeting compact, *gapet, gaplok, gapet*, the new uppermost *gapet* serves as a *reng*	The arrangement of *usuk* consists of a whole wooden trunk arranged like a bucket and tied to one another with horizontal rods

Figure 20: a, b, c: The arrangement of the ions is the bond of some bamboo stems and the gaplok causes the oil to not curve, otherwise the bolon house is only reinforced by the long wooden timber and is combined with a curved saddle-like bookmark.

4. Conclusion
Based on the discussion of the two rooftops of the archipelago it can be concluded as follows:

1. The result of the discussion of the pairing of both types of roof found that construction system relied on the stem as the framework of the construction field.
2. The roof of each pedestal was equally pedestal, but the house bolon had naturally curved bent with wooden arch then the curved roof field in accordance with the arch of the book.
3. The roof *empyak* was arranged in a layered good half bamboo and *gaplok* was bamboo intact so that the stiffness occurred.
4. The roof of the house Bolon consisted of a layer of wooden rods that lined uric side form the side of the roof following the arch of the book. *Bukulan* was made of pokki logs that were more than 50 years old placed under the tree in front, so that the arch followed the wooden stem that curved a lot in the back.
5. Both roofs had a principle as a field construction system with a reinforcement on its own stem rod.
6. It seems that the principle of *empyak* at the bolon house becomes the answer to the match.

5. References
[1] Prijotomo J 2006 *(Re-)konstruksi arsitektur Jawa : griya Jawa dalam tradisi tanpatulisan* (Surabaya: Wastu Lanas Grafika)
[2] Frick H 1997 *Seri Strategi Arsitektur 1. Pola Struktural dan Teknik Bangunan di Indonesia* (Yogyakarta ;Semarang: Penerbit Kanisius ;;Soegijapranata University Press)
[3] Salvadori M 2008 *Seni Konstruksi* (Pakar Raya)
[4] Anon 2014 *Prinsip-prinsip desain arsitektur tahan gempa: lengkap dengan bahasan gempa bumi dan metoda praktis evaluasi kerentanan terhadap gempa Noor Cholis Idham* (Yogyakarta Andi)
[5] Frick H and Purwanto L 2007 *Sistem Bentuk Struktur Bangunan* (Kanisius & Soegijapranata University Press.)
[6] Schefold R, Nas P and Domenig G 2003 *Indonesian houses* (KITLV Press)
[7] Frick H and Setiawan P L 2001 *Seri Konstruksi Arsitektur 4. Ilmu Konstruksi Struktur Bangunan, Cara Membangun Kerangka Gedung. Ilmu Konstruksi Bangunan 1* (Yogyakarta; Semarang: Kanisius; Soegijapranata University Press.)

Architect's vision, creative process and architecture for empowerment: learning from creative process of Indonesian architects

Tulus Widiarso[1,2], Himasari Hanan[3]
[1]Lecturer, Graduate Program, Department of Architecture, Trisakti University, Jakarta, Indonesia,
[2]Student, Architecture Doctoral Program, Bandung Institute of Technology, Bandung, Indonesia
[3]Lecturer, Architecture Doctoral Program, Bandung Institute of Technology, Bandung, Indonesia

Corresponding email: twidiarso@gmail.com

Abstract. In the course of architect's career, a vision of work is developed. The firm's orientation where the architect works becomes one determining factor of the vision. The vision may change over time based on their professional experiences. At some point, some architects decide to focus only on a particular project, such as architecture for empowerment. This research was conducted by analysing theoretical and empirical studies of architect's creative process. The study was done through a qualitative approach. The participating architects were chosen purposively. Exploration was conducted through deep retrospective interviews. This study aims at: (1) describing the relevance between the architect's vision and his creative process; (2) describing the proportion of combination between problem solving process and the process of self-expression in the creative process in the architecture domain; (3) describing the proportion of combination between problem solving process and self-expression process in architectural creative process for empowerment. The main findings are: (1) every architect develops a personalized creative vision. The architect's creative vision will vary the creative process, although it may not always be the dominating factor of the creative process; (2) in the creative work process, the architect may not always determine the problem solving domination or the self-expression domination as his identity; (3) in the case of architectural design for empowerment, the architect's creative process is dominated by problem solving, even it is possible for the loss of self-expression.

Keywords: architect's vision, creative process, architecture for empowerment.

1. Introduction

Creativity is a complex cultural process involving the interrelation between domains (culture or a particular science), a field (control domain system) and a creative person which produces valuable and novelty creative work. [1]. Architect is a creative person in the field of architecture. The creative process undertaken by the creative person is usually a mix of problem-solving-oriented process [1] [2] [3] [4] [5] [6] [7] and a self-expression process [3] [5]. The attempts to uncover the joint proportions between the two processes in a person's creative process on a particular domain are significant supports for the creative science. However, this has not been most researchers' interests. In the context of stakeholders' (clients, users and others) empowerment, it is essential to understand whether or not more problem-oriented processes are required.

In the context of today's business institutions, especially in the field of creative industry, the term vision and creativity are often used interchangeably. In the context of the architectural firm, the creative vision of the firm is identical with the creative vision of the architect's principle [8].

In his research on some architectural firms in the United States, Coxe [9] categorized those firms based on their services' characteristics. The architectural firm's characteristics of services are very influential to the shape of the architects' visions. However, Coxe [9] has not yet revealed the relevance between the architect's vision and the blended proportion of problem-solving orientation and self-expression orientation.

This paper was analysed based on an empirical exploration study on architect practices, aiming at: (1) describing the relevance between the architect's vision and his creative process; (2) describing the proportion of combination between problem solving process and the process of self-expression in the creative process in the architecture domain; (3) describing the proportion of combination between problem solving process and self-expression process in architectural creative process for empowerment.

2. Methods

The study presented in this paper was analysed by using a qualitative approach. The participants were 10 architects who have received both national and international awards and hold essential role in Indonesian architecture. A retrospective interview regarding the creative vision and creative process of work were asked to the participants. The transcript of the interview was analysed by using the procedure as shown in Figure-1

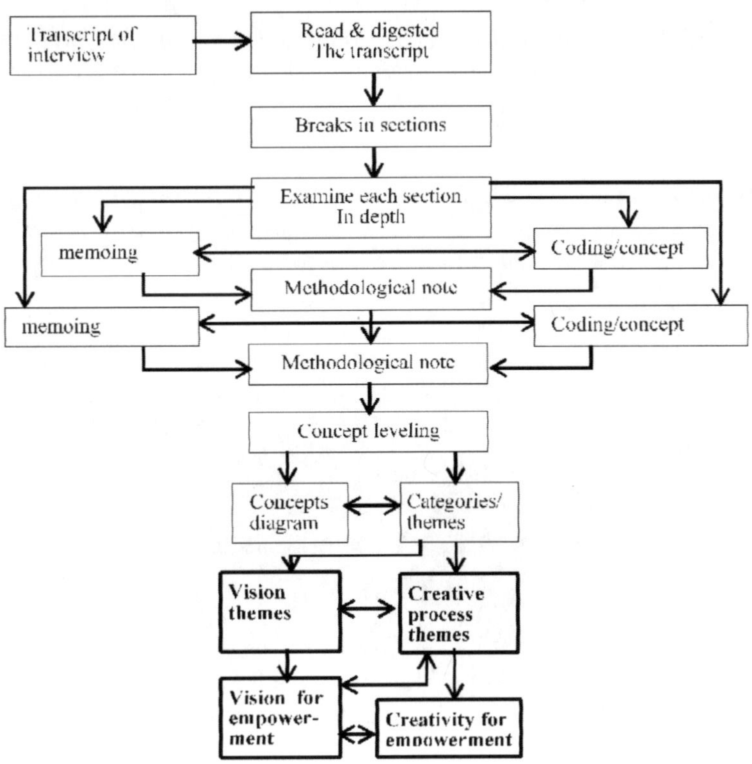

Figure 1. The process of data management and analysis of interview results constructed from Corbin & Strauss [10]

The themes that emerged from the interview transcript result were analysed and interpreted based on relevant creative science theories.

3. Result, Interpretation and Discussion

The interview results of the participating architects show that all architects describe the firm where they work as a "strong idea firm". In accordance with the Coxe [9] categorization, the "strong idea firm" is a service-oriented bureau, focusing on providing specific expertise or innovation and uniqueness. The fundamental value of architectural service-oriented bureau is being specific, unique, and innovative. In this study, the specific services in the bureau are services on the architecture and interior design. Other design aspects such as structure, mechanical-electrical, landscaping and others are distributed to third-party consultants. Based on the volume of workload, the bureaus are considered to be in a small and medium category. It is acknowledged by the fact that there is no firm whose participants are among the "top ten BCI-Asia". This is in line with the Coxe's finding [9] that the "strong idea firm" generally grows as a small and medium-sized bureau. Nine out of ten participating architects, who are the Principal Architect of their firms, provide reasons that: by focusing solely on architectural and interior design, a unique and innovative works are more possibly created. One participating architect who is not the Principal Bureau suggests that he feels comfortable creating such a unique and innovative results.

3.1. Architect's Vision, Creative Vision and Firm Orientation

Vision is very important strategic horizon to build short and long-term organizational goals and objectives [8]. Creativity in organizational context is simply defined as using imagination in designing and actualizing the organization's production. Institutionalizing creativity involves the modification of a future-oriented organizational culture as well as the organization's vision. It should be realized that a creative vision requires the articulation of vision and mission and its implementation through appropriate management strategies.

Coxe [9] distinguished the architect's firm orientation into: (1) strong idea firm – the orientation is to give specific expertise or innovation and uniqueness; (2) strong service firm – the orientation is to provide a reliable service on complex projects; (3) strong delivery firm – the orientation is to provide efficient services on regular projects for product-oriented clients. The orientation of a "strong idea firm" leads to the creation of the architect's vision for a unique and innovative work. All participating architects indicate that they possess a clear vision of work. A number of 6 participants expresses their visions of the work clearly (explicitly), while the others do not express it directly. Nevertheless, it is elaborated during the interview. Despite of all the visions to achieve innovation and uniqueness, each architect expresses the value of his vision differently (Figure-2).

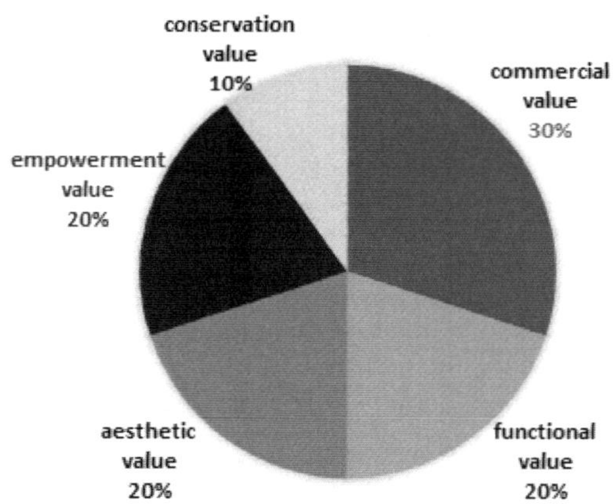

Figure 2. The Vision-Values of the Participating Architects

3.2. *Creative Process and Architect's Vision*

Creative process is done by a creative person. He/she combines problem solving process oriented and self-expression process oriented [1] [2] [3] [4] [5] [6] [7]. Vision, which is the expression of architect's value, is essential for every architect. It is found in the empirical practice of the participating architects. In this study, three participating architects shared the same visions, commerciality. However, it is essentially different from each other. First, the strength of commercial vision is built from the business domains; second, the power of commercial vision is built based on the effectiveness and efficiency of the design; and third, the power of a commercial vision is built based on their distinctive work. The implementation of the creative process vision as a form of self-expression is agreed by three participating architects. However, regarding to the creative process, they are very concerned about how the problem of architectural design should be well answered.

The main component of creative process is the process towards problem-solving and self-expression. Several researches have explored the complexity of the creative process. It can be described as [7] [11] [12] [13] [14]: both conscious and subconscious heuristic processes towards problem solving and self-expression process supported by strong internal and relevant external conditions.

None of the participating architects guarantee that creative process is always dominated by problem solving process or vice versa. Each project case has its own specific creative process. External conditions cannot be controlled by the architect's internal strength. In many cases, external conditions precisely dominate the architect's internal strength.

3.3. *Creative Process and Architecture for Empowerment*

The proportion between problem solving and self-expression cannot be considered as the characteristics of the architect's creative performance. This depends on the external conditions of the project unless they choose to deal only with controllable external conditions projects. Or else, the architects deal only with projects with external conditions that are beyond their control capabilities. In architectural projects for empowerment, it is necessary to trace whether external factors are beyond the control of the architects or vice versa. Here is a theoretical insight to note.

The architecture of empowerment is built to respond to the needs of low-class society, respect the value of humanity, and take responsible act for their lives [15]. The society's participation is emphasised for not only creating average achievement but also providing valuable contributions in social or political values. Additionally, this can open bigger chances in architecture [16]. The architecture for empowerment practice is architect's platform of creativity and innovation elaboration. The agents of initiator for the society of empowerment are: government, private, non-governmental organizations (NGOs). Architect who takes the role in community empowerment is known as architect community [17]. The architect community serves as a supporter providing important information on architectural techniques and facilitating dialogue for various parties.

In actualizing the benefits of community participation, especially in the context of empowering underprivileged communities, many architects encounter various obstacles while trying to engage their clients during the design process. In regards to the community participation in today's architectural design, there has been an essential implication for the role of architects and professional design teams. In the context of participatory design for empowerment, development does not limit only to build buildings. Further, the development process may inspire the society to be skilled and knowledgeable people. This may occur through various stages of participation in design and construction. Broadly speaking, the participatory design process for empowerment is shown on Figure 3 below [18].

The model chart on the left side is the participation of the community, the ways (means) of contributions in development, as well as the responsibility upon the completion of the program (end). The right side are architects as designers who are working on the process and product design aspects. In the production aspect, architects as design product providers need to provide designs that guarantee the efficient and effective development. In process aspect, design and design process should encourage the implementation of development that enhance community capacity (building capacity) and community empowerment. The synergy between the community and the architects and other stakeholders (stakeholders) embody three empowerment components: efficient and effective development, development processes that provide community learning, and development processes and outcomes to empower the society.

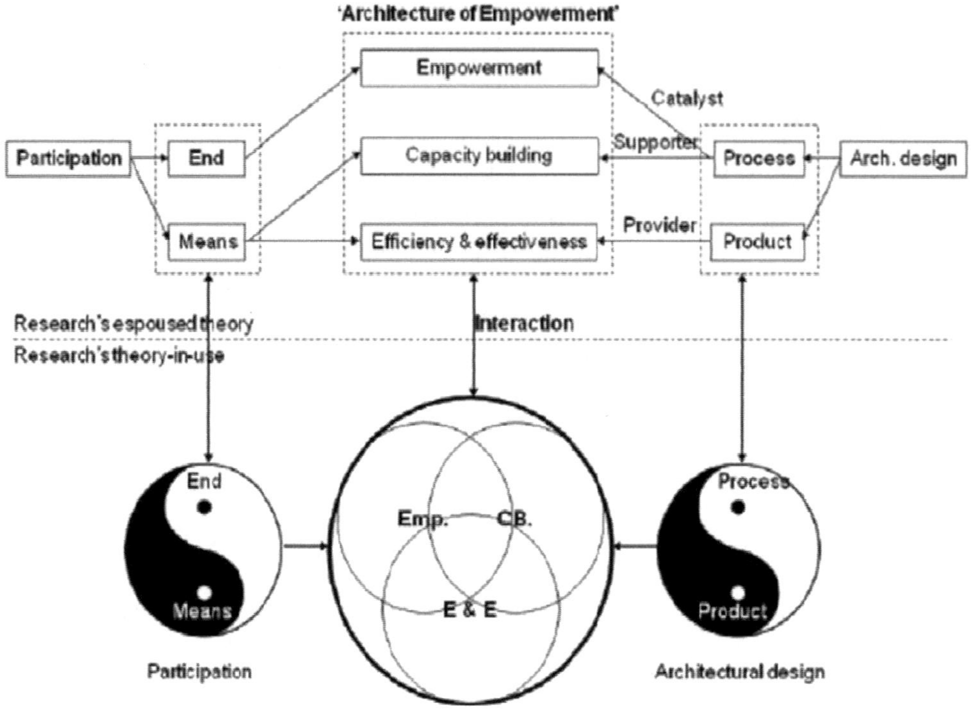

Figure 3. Model of Participatory Design Process for Community Empowerment [17]

Based on the Tovivich's model, the main components of the development process of empowerment are design and participation. To create the development for empowerment, the architectural design has to align both the process and the result. On the other hand, empowerment participation has to be aligned between its means and its end.

Figure 3 shows clearly that the core of the architecture for empowerment is the interrelation between community participation and the architects' contribution. Those are: (1) empowerment; (2) building capacity; and (3) efficiency & effectiveness. Architects need to answer all the three issues addressed by the process and the design result. Theoretically, in architectural design for empowerment, problem solving process dominates the creative process.

The two participating architects in this study involve themselves mostly in architectural design for empowerment. Among their works which receive award-winning appreciation are the works for empowerment. The first participating architect focused on providing problem solving facilities for the community in the creative design process for empowerment. The second one described his work in the design process for empowerment, fused himself in an empowered community, and totally performed problem solving oriented creative process.

4. Conclusion

Generally, the empirical exploration studies of the creative process of architects, especially in the context of community empowerment result in the following conclusions:

- Every architect develops a personalized creative vision. The architect's creative vision will vary the creative process, although it may not always be the dominating factor of their creative process;
- In their creative process, architect may not always determine their problems solving orientation domination or self-expression orientation domination as their characteristics. Not all of the external conditions can be controlled by the architects; in fact, many project external conditions are predominantly control the architects;
- In the case of architectural design for empowerment, the architect's creative process is dominated by problem solving even it is possible for the loss of self-expression.

5. Acknowledgments

The authors would like to thank to the participating architects, Bandung Institute of Technology as the institution where the authors carry out the research, as well as Trisakti University as the research funder.

6. References

[1] Csikszentmihalyi, Mihaly, (2013). *Creativity: The Psychology of Discovery and Invention*, Harper Perennial Modern Classic Edition Published, New York, 278-323.
[2] Gardner, H. (1999). *Intelligence reframed: Multiple intelligences for the 21st* ***century.*** New York: Basic Books.
[3] Mumford, M. D. (2003). Where have we been, where are we going? Taking Stock in Creativity Research, *Creativity Research Journal*, **15(2-3)**, 107-120.
[4] Newell, A., dan Simon, H. A. (1972). *Human Problem Solving*, Englewood Cliffs, Prentice-Hall, New York, 41-76.
[5] Runco, M. A., & Pritzker, S. R. (Eds.). (2011). Encyclopedia of creativity, two-volume set, second edition (2 ed.). Boston: Academic
[6] Torrance, E. P. (1974). Torrance tests of creative thinking. Scholastic Testing Service, Inc
[7] Wallach, M. A., & Kogan, N. (1965). Modes of thinking in young children: a study of the creativity-intelligence distinction. New York: Holt, Rinehart & Winston.
[8] Karaman A, Kok S B, Hasiloglu S B, Rivera M 2008. Vision, creativity, strategic innovation, and transformational leadership *Problems and Perspectives in Management* 6-2 pp 104-109
[9] Coxe, W. (1987). *Success strategies for design professionals: super positioning for architecture & engineering firms*. New York: McGraw-Hill.
[10] Corbin J and Strauss A, (2008). *Basic of Qualitative Research, Techniques and Procedures for Developing Grounded Theory*, Sage Publication Inc, California.
[11] Dewey, John, (1934). Art as Experience, New York: the Berkley Publishing Group,
[12] Guilford, J. P. (I 956). The structure of intellect. *Psychological Bulletin, 53,* 267-293.
[13] Kolodner, J.L. and Wills, L.M. (1996). Powers of observation in creative design, *Design Studies*, **17**, 385- 416.
[14] Amabile, T.M., Conti, R., Coon, H., Lazenby, J. and Herron, M. (1996). Assessing the Work Environment for Creativity, *The Academy of Management Journal*, **39**, 1154-1184.
[15] Serageldin, I. (1997). The Architecture of empowerment: people, shelter and liveable cities, London ; Lanham, Md, Academy Editions : Distributed to the trade in the United States of America by National Book Network Inc
[16] Till, J. (2005). The negotiation of hope. in Blundell J, P., Petrescu, D. and Till, J. (Eds.) Architecture and participation. London: Spon Press.
[17] Valladares, A. (2012). The Community Architect Program: Implementing participation-in-design to improve housing conditions in Cuba, *Elsevier Ltd.*
[18] Tovivich, S. (2010). *Architecture for The Urban Poor, The 'New Professionalism' of 'Community Architects' and the Implications for Architectural Education: Reflections on Practice from Thailand*, PhD Thesis, the Faculty of Architecture, Silpakorn University, United Kingdom, 43-88.

The adaptation strategy of dwelling in the riverside settlement of the Arut River in Pangkalan Bun City, West Kotawaringin Regency, Central Kalimantan

Edi Purwanto, Edy Darmawan
Department of Architecture Diponegoro University, Semarang, Indonesia,

Corresponding e-mail: edi.purwanto@ft.undip.ac.id

Abstract. This research entitled the adaptation strategy of dwelling in the riverside settlement of the Arut River in Pangkalan Bun City was conducted with the consideration that the settlement has a similar typology to settlements in the Kalimantan which are generally next to/on the rivers. This settlement, located both on the river and in the transition points between lands and rivers, is familiar with damp environments. The people dwelling in thisplace, dealing with damp environments and striving to continue their life, must adapt to such enivironment. Therefore, this research tries to discover what kind of strategy done by these people to live in this kind of settlement.
The research method used is mixed method. Mixed method is a combination of quantitative methods based on interview data and questionnaire, and qualitative exploration method that is based on field observation in the exploration of adaptation strategies settled in Arut River settlement.
The research finally discovered the adaptation strategy of dwelling in this riverside settlement particularly the adaptation strategy to deal with tides and climate problems related to wind movement, precipitation, temperature, humidity, and other factors related to socio-economic condition of the people.

Key words: adaptation strategy, riverside settlement, continue their life.

1. Introduction
In general, developed traditional settlements in Indonesia have their own uniqueness when it is related to the geographical location. There are settlements located on the hillsides, over the river, are on the banks of the river. Initially, the existing settlement was linked to the livelihood of the community, for example, the settlements on the river of Kalimantan. Many settlements is located on the river/ near the river because of the most people livelihood, firstly, is fishermen. In the past, this riverside was the grown settlements centre with village background. Aside from being the source of life for the people, the river is also used as inter-island transportation on Kalimantan. From the past until now, the rivers in Kalimantan has an important function in the economic development and has become a vital part related to the activities of the community as a whole, where various activities of the community are undergone, both settlement activities and other social activities. So the existence of settlements along the banks of the river is not something new. The geographical condition of the country in Indonesia which has many rivers as the life orientation makes the river as a place to live and seeking livelihood, even its existence has become a tourist spot now [1],[2],[3].

In the subsequent development, the settlements existence on the riverbank experienced dynamics mainly related to environmental conditions. Some rivers have been considered disturbing the community life as it causes flooding, especially during the rainy season. In addition, flooding disturbance, climate factors also contribute to environmental disturbance for the settlements [4].

A study, entitled settling adaptation strategy in the urban waterfront settlement of Pangkalan Bun was deliberately conducted as the settlements had similar typology to the settlements on the banks of Kalimantan. The settlement experiences the phenomenon of natural disturbance affected by the flood from the Arut River, especially during the rainy season and climate disturbance. With the phenomenon of natural disturbance, to maintain the survival of life, inevitably, the residents have to adapt to their environment. What we want to explore through this research is the process and form of adaptation.

Adaptation and change are inseparable for living. Adaptation is applied in every living things ti survive in a constantly changing environment. Adaptation is a human responsive behavior towards the environmental changes. Such responsive behavior allows them to organize certain systems for their actions or behavior, in order to adapt to the situation and conditions. Those behaviors are related to the necessities of life, having previously passed certain circumstances and then developing a strategy and a certain decision to deal with the next situation. Thus, adaptation is a strategy used by humans in their lifetime to anticipate changes in both physical and social environment [5]. Adaptation strategies can be defined as the ability of a person to apply a set of ways to deal with the many of issues which is surrounding his or her life. This problem-handling strategy is basically the ability of the community to manage all its assets [6].

2. Literature review

2.1. Natural phenomena and physical environmental changes as the causes of adaptation

Physical environmental changes can be considered as a pressure for families or communities to make adjustments so that the residents' needs can be adequately fulfilled. Briefly, these adjustments are expressed in terms of on-site adjustments and adjustments by residence movement. If a settlement area is disturbed by a natural phenomenon, then certain the residents will make adjustments to nature. Sea level rise directly or indirectly impacts the lives and livelihoods of the residents, will be answered in accordance with unique natural adjustment processes. Various adjustments to stress are revealed in accordance with the cultural and habitual form that the residents possesses. The natural phenomenon and physical changes in the environment in question are floods and climatic conditions [7].

2.1.1. Flood

Flood is one of the most frequent disasters in the last decade. The impacts of this flood include damage to residential areas and infrastructure damage. Indonesia is one of the countries considered to be vulnerable to climate risks, which is affecting the occurrence of floods [8].

Flood is a frequent phenomenon and directly could impact community activities. Areas that are often affected by floods are no exception dense residential areas in the river border due to overflow of the river. Floods occur due to two events namely floods and puddles that occur in areas that are often flooded and the second is the flood that hit an area due to river water runoff [9].

The factors causing floods on Kalimantan are very diverse. Some of them are forest destruction caused by forest exploitation, illegal logging, land fires, and so on. Then, there is the mistake of the area designation, with the real evidence that the number of water catchment area is now clearing, and it leads many expansion of open land. In addition, many large rivers on Kalimantan contributed to the cause of the flood. The character of the river on Kalimantan is generally a tidal river and influenced by the rainy season and the conditions of the upstream and downstream.

2.1.2. Climatic conditions

The tropical climate into several classifications based on average daily temperature and the difference between day and night temperatures. In this grouping, only cities or regions with average daily air temperatures of 28° C or more are included in tropical climatic categories. A prominent feature of the tropical climate is the high average daily temperature compared to other climates. The problem posed by this climate is the warming caused by solar radiation. The implications of solar radiation to the surface of the earth depend on the falling angle, depending on the conditions of the cloud that can block the radiation emission and depending on the difference in surface character so as to have a difference in terms of reflection and absorption of the radiation [10],[7], [11].

In many ways, architectural design is less attention aspect of tropical climate in solving problems in various settlement areas, both in urban and rural areas. Problems caused by humid hot tropical climate such as high rainfall, temperatures generally above the comfort tolerance, stinging solar radiation especially with global warming issues, high humidity and relatively slow airflow for gaining the thermal comfort, is not anticipated by a residential area planner or designer. Though, it can disrupt the comfort and stability of residents in a residential area. For the enivironment adaptive architectural design is expected to reduce the microclimate, thereby generating thermal comfort in occupancy in the tropical city. Convenience can be obtained from aspects of the building itself, and from the environment outside the building [12].

There is a balance relationship between microcosm in the architectural building where humans live with the macrocosm that is the universe where the architecture building is related to nature and tropical climate. Architectural buildings in Indonesia are mostly influenced by nature and tropical climate, namely the influence of sunlight and the orientation of buildings, wind and air space, temperature and protection against heat, precipitation and heat humidity [13].

The architectural design, especially architecture of houses must adapt to the tropical climate in Indonesia [14]. The embodiment of adaptation is manifested in the design of the building which has the following characteristics (figure 1):

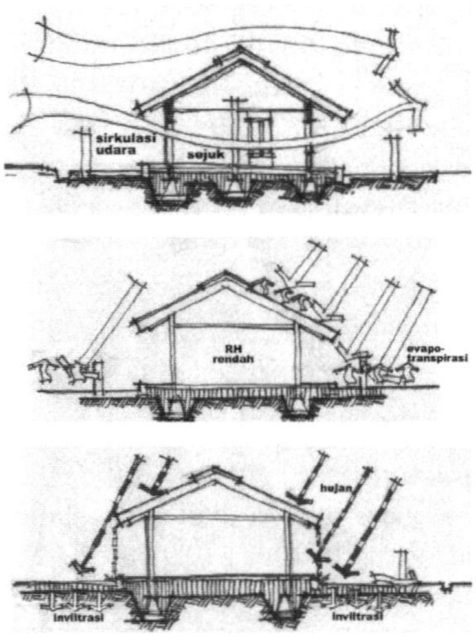

Maintain the air movement within the house where living, thus it is able to move heat and humidity out the building by convection, so that the air in the house is always relatively cool and comfortable

Open space provides an opportunity for evapotranspiration, to suppress excessively high humidity

In the rainy season, open spaces are useful in maintaining hydrological cycles by suppressing run-off and keeping infiltration on green land. On a broad scale, it could prevent floods and maintain a balance of freshwater and saltwater pressures to prevent sea water intrusion.

Figure 1. Architectural design adapting to tropical climate

Local wisdom manifests in the form of Indonesia architecture that is adaptive-responsive to the tropical-humid environment of the tropical archipelago. The original architectural form of Indonesia presents a number of appropriate characteristics to the tropical-humid nature situation. This characteristic is an adaptation and response to the local environment (Wikantari, 2008). One of the characteristics of tropical-humid in traditional houses is the use of stage structures near riverbank (figure 2) [15],[4],[16].

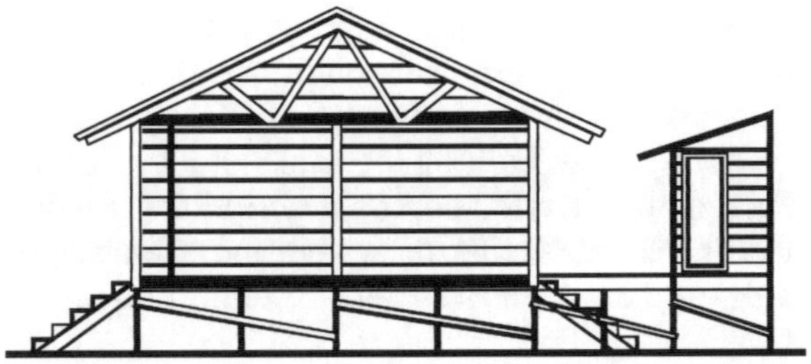

Figure 2. Sample of house on stilts house model

2.2. Characteristics of settlements on the banks

River is the lifeblood of the people who have inherited in developing Kalimantan, so that cities in Kalimantan basically grow and develop from the embryo of the riverside settlement. The cities in Kalimantan are now developing very rapidly, but pay less attention to the potential of riverside settlements and pay more attention to the growth of land settlements [17].

Settlements on the banks of the river consist of many variations of houses and housing. Such variation will be more pronounced in a heterogeneous society. For example in Kalimantan society there are various forms of houses, although it can be said that almost everything is a house on stilts. In indigenous communities Kalimantan, Dayak tribe, main form of the main is the elongated *rumah panggung* (stage house) extends to the left and right inhabited by a number of families. Although architecturally, the house of stilts has almost the same shape but the similarity in some places is somewhat different. For example, there is a call by the name of "lamin" there is also a call as "betang" house. Among the Malays who mostly live in the lowlands or beaches, most of the houses are also houses on stilts.

The main materials used to create a house on stilts are wood that can be easily obtained from the surrounding environment. It is just in the present, because of the technological developments of house on stilts, the wood has been using non-timber materials, such as floor mounted wooden pegs, it is affecting the existence of wood as the main materials of house building. The settlements form for the riverside area has the shape of a house on stilts. This is due to the influence of natural components in the form of tidal surface of the river (figure 3). As for the settlements away from the river, there are some which the shape is house on stilts, yet because the existing land is a swamp sediment, then the existing resident houses form the house on stilts pattern [4].

Figure 3. Rows of Housing on the banks of the Arut River

Community activities on the banks of the river make the river as the center of everyday life, such as livelihood and as a water taxi. The river is made by the surrounding community as a means of connecting with the surrounding village/town. In general, the construction of residential buildings on the banks of the river has a form of house on stilts. This stilt house is suitable to be built in swamp area, because this area is risky to flood which will rise every time and also the influence of tidal affected by the river. In accordance with geographical conditions, the houses are built on poles by the river, or over the river. The residents' houses are constructed of forest wood which is widely found in Central Kalimantan. Originally, houses were built on the banks of the river, facing the river so that the river into the front yard [4]. According to Kertodipoero's description [18] about rivers and settlements on Kalimantan, houses stand on poles, all facing the river, and each house has wooden stalks. He mentions villages along the banks of the river as a "station", connecting one village to another village, and anyone passing through it can stop.

2.3. Adaptation strategy

Hardesty, in[19] put forward the notion of adaptation: "Adaptation is the process through which beneficial relationships are established and maintained between an organism and its environment". Meanwhile, cultural ecologists define that adaptation is an adjustment strategy that humans use during their lives to respond to environmental and social changes [19].

When humans are confronted with a solid situation, which can be perceived as situations that threaten their existence, humans will adapt. This means that there is an interactionist relationship between the environment and humans. Environment can affect humans, humans can also affect the environment [20]. Because of their mutual influence, there is an adaptation process of the individual in responding to the pressures coming from the environment. To some extent, humans have the flexibility that allows individuals to adjust to their environment. This adaptability has a value for their sustainable life.

The occurrence of adaptation can not be separated from perceptual understanding [21]. According to Bell, perceptions may vary due to physiological processes. In terms of human interaction with the environment, humans will always try to obtain harmony with the environment. It is possible by the present of cognitive ability to make certain reactions to the environment that contain certain things that interest them in fulfilling their needs. How is the process of relationship with the environment that occurs since humans interact through sensing up to the occurrence of reactions, depicted in the scheme of perception as follows:

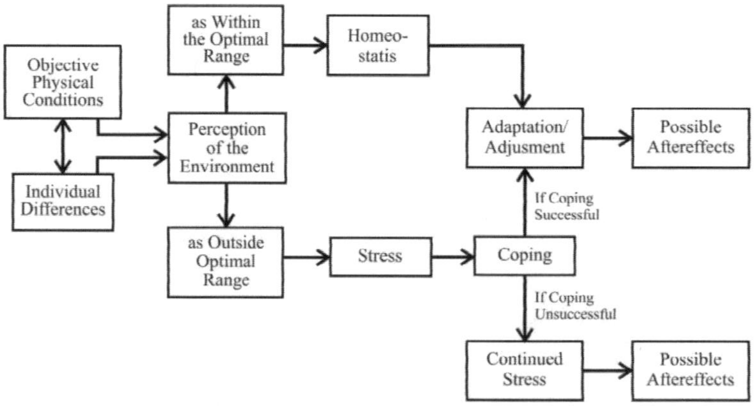

Figure 4. Perception of the environment scheme

Figure 4 explains the results of human interaction with the objects resulting in human perception of the object. If the perception is within the optimal limit, then the human being is said to be in a static homeo state, that is a well-balanced state and usually always wanted to be defended by every human being because it creates a pleasant feeling. Conversely, if the object is perceived as beyond the optimal limit, then people will experience stress, there is an increase in energy, so it must be done coping to adjust the environment on their condition. Human adjustment to the environment is called adaptation, while the adaptation of the environment to humans is called adjusment. In terms of human interaction with the environment, humans will always try to obtain harmony with the environment. [22] distinguishes between adaptation behavior and adaptation strategies. Adaptation behavior is, behavior aimed at overcoming problems encountered or to obtain something desired and this is different from adaptation strategy. Adaptation strategies are defined as the patterns of various efforts planned by humans to meet the minimum requirements they need and to solve the problems they face. The patterns in question here are patterns of behavior or action.

3. Methods

The research method used is mixed method [23]. Mixed method is a combination of quantitative methods based on interview data and questionnaire and qualitative exploration method that is based on field data (field observation) in the exploration of adaptation strategies settled in Arut River settlement.

The boundaries of the study area are settlements along the Arut River that belong to the administrative area of Mendawai and Raja villages. The reason for determining the area of this research is because this area became the starting point of the growth of Arut river settlement and became the forerunner of Pangkalan Bun city (figure 5). The number of home population in the identified research area is 2,580 units with relatively similar homogeneous level, thus the number of samples set as many as ± 10% or as many as 260 houses and its occupants as respondents.

Figure 5. Boundaries and research location

Components of the object of observation are: : [i] dimentional of buildings, [ii] building structure, [iii] roof cover used, [iv] wall material used, [v] flooring material used, and [vi] distance between buildings. Interview data is needed to know the social, cultural, economic and perception backgrounds of respondents as residents. The observations and interviews were then analyzed to produce findings in accordance with the objectives of this study: to reveal adaptation strategies in settlement on the banks of the Arutriver, especially to tidal discharges and climate disturbance related to wind, rain, and temperature, humidity and other factors based on socio-economic conditions of the community.

4. Discussion

4.1. Geographical condition, administration, and population of Pangkalan Bun city

Pangkalan Bun City is a part of South Arut Subdistrict of West Kotawaringin Regency which is geographically located between 02'-44'-03"South Latitude and 111'-42'-04" East Longitude. The administrative boundaries of Pangkalan Bun City are as follows:

North : Medang Sari Village
South : Kumpai Batu Atas Village
West : South Arut River
East : Kumai Sub-district

The total area of Pangkalan Bun city recorded 821 km^2 consists of 9 villages (Mendawai, Mendawai Seberang, Raja, Sidorejo, Madurejo, Baru, Raja Seberang, Natai Baru, Pasir Panjang), which includes 39 RW and 149 RT.

The population in the city of Pangkalan Bun at the end of the year 2016 is 98,691 people. The composition of population by sex is known that the population of male sex is 51,024 people and the population of female is 47.667 people. Based on the data, the largest population is in Baru village which is 24,324 people and at least in Natai Baru village is 2,132 people (Kotawaringin Barat Regency in Figures, 2017).

The pattern of settlements in downtown Pangkalan Bun is divided into two parts, namely North Arut River and the South side (figure 6). North of the Arut River, the settlement is relatively low density, that is a linear pattern along the riverside. The area includes Village Mendawai Seberang and Raja Seberang. While in the South River Arut is the heart of the city that became the center of the activity so the settlement is high density.

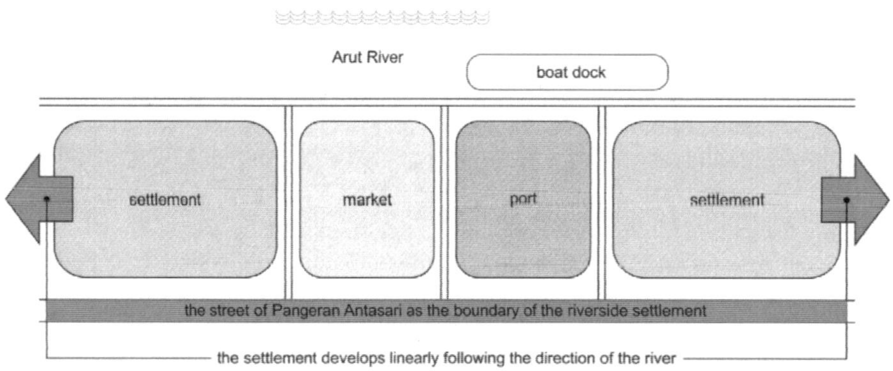

Figure 6. Early area structure of Arut river settlement pattern development

4.2 Research findings

4.2.1. Adaptation strategy based on dimensional of buildings

In the study area there were 57.30% small type houses with an area of <36 and 36-45 m^2, the remaining 45-54 m^2, 54-60 m^2, and > 60 m^2. Based on this type of field observation,it shows that more on the edge and facing the river. Not much information is obtained from the citizens, but the contention appear that small types are more existing on the riverbank because of the flexibility factor. This means that small type has a lighter weight and certainly has a small impact if the condition of the soil that supports the building is not hard ground. In addition, small type of house has a smaller wall area so that if there is wind blowing from the direction of the river does not have a bad impact when compared with large type of house with a large wall area [16]. Thus citizens have anticipated such conditions, and this is one form of sttle adaptation strategy by the river (figure 7, table 1).

Table 1. Dimensional of buildings

NO.	DIMENSIONAL OF BUILDINGS	TOTAL	%
1	< 36 m^2	81	31,15
2	36 – 45 m^2	68	26,15
3	45 – 54 m^2	43	16,54
4	54 – 60 m^2	15	5,77
5	> 60 m^2	28	20,39
	Total	**260**	100,00

Figure 7. Small type building example

Residents adaptation to residential design with relatively small dimensions has succeeded in building the physiological comfort of the inhabitants achieved through an understanding of the climate and the human responsive system [24].

On the other hand, according to the residents' explanation, the figure of the building on the banks of the river with a small dimension in addition to saving development costs, is also efficient in maintenance considering the economic condition of the residents who earn an average of 1-2 million rupiah / month.

4.2.2. Adaptation strategy based on building structure

There are 76.92% of buildings using a stage structure with wooden building materials, and the rest using concrete. It shows that the use of stage structures for residential buildings above the river has a specific purpose (table 2).

Table 2. Based on the structure used

NO.	BUILDING STRUCTURE	TOTAL	%
1	Stagestructure with wood material	200	76,92
2	Stagestructure with concrete material	51	19,62
3	Non Stage	9	3,46
4	Floating	0	0,00
	Total	**260**	100,00

Figure 8. The use of a stage structure for residential buildings on the riverside

According [4],[16], the *arsitektur panggung* (stage architecture) of riverside dwellings also tried to adapt to water level elevation conditions. With the hous on stilts shape, the rise or fall of the river water surface will not interfere with the activities in the house in daily activities (figure 8).

Residential architecture of house on stilts shape, is an eco-responsive wisdom to the local environment, or appropriate and responsive to the topography of tropical areas that are hot and high humidity [25].

The structure of the stage is related to the tectonics of engineering and the form of arrangement, connection, the circuit consists of various elements and materials that form a construction and building structure [26]. In residential architecture in the area of research, the structure and construction in the form of column frames and beams with dynamic jointed connection arrangement.

Architectural building of residential house in research area is building with construction system connection (peg-hole and pen-hole), it is dominated by wooden/bamboo building material. Tectonics deals with the technique and the arrangement form, connection, the circuit consists of various elements and materials that form a construction and building structure. The wisdom of the geo-adaptive geo-responsive tectonic system, which is appropriate and responsive to the Indonesian archipelago, which tends to sedation and earthquake vibration forces, as well as the force of wind blowing [27]. The determinants of the design and construction of inhabited comfortable buildings are how to relate to the conditions of climate [28].

4.2.3. Adaptation strategy based on the use of roof covering

Based on the data obtained from the respondents, the roof cover of most buildings used is zinc material (86.92%), the rest using tile, asbestos, wood. The results of the interviews show that the reason for the use of roofing material from zinc because it is considered cheaper, lighter and more durable though has the disadvantage of having a catalyst to heat (figure 9, table 3).

Based on the analysis, in addition to cheap, lightweight and more durable, the use of zinc makes the building lighter considering the ground conditions at the bottom / riverside is not hard ground. The use of a saddle roof shape with the terry model is possible because of its correlation with the tropical climate. The shade of a saddle-shaped gaze is a steep angled saddle and a wide-ranging diaphy reveals the role of shade. The roof shape thus reflects eco-adaptive eco-responsive, which is appropriate and responsive to the vertical burden of tropical rain and lateral force of wind blowing [16].

Architecture is a shadow. That is the understanding of architecture in Indonesia. For these needs, namely the presence of shade or shade, then the presence of a parade of shade or shelter that the answer. Diversity of roof formation is related to the important and honorable role of the roof. Through the review of the tropics containing the dry season and the rainy season, of course become more stable and real that the first and main elements of Indonesian architecture is the roof. With this main role and position also the structure of the building make the pillars supporting of the roof as the main pillars [16].

IOP Conf. Series: Earth and Environmental Science **213** (2018) 012033 doi:10.1088/1755-1315/213/1/012033

Table 3. Based on roof used

NO.	Roof Covering	TOTAL	%
1	Tile	4	1,54
2	Asbestos	5	1,92
3	Zinc	226	86,92
4	Fibers/Leaves	2	0,77
5	Shingle	23	8,85
	Total	**260**	100,00

Figure 9. The use of zinc as a roof cover (left) wooden board as a ceiling to reduce heat (right)

4.2.4. Adaptive strategy based on used wall

Based on data from the respondents, as many as 81.54% of houses using wooden boards as wall building materials, the rest using brick and bamboo walls. With such large amounts indicating that the use of wooden boards for residential walls has a special purpose. According to residents interviewed, initially the use of wooden boards for the walls was done because at that time the supply of wood is very abundant and at a cheap price. The current condition of residents still maintains the existence of wooden board as a wall building material because it still survive its durability and easy in its maintenance (table 4).

The use of wooden wall building materials for walls presents a light-porous membrane. Thus raises eco-adaptive eco-responsive in the formation of the inner baffle and the outer casing of the field to a tropical-humid environment. Wall insulation is made using wooden board arrangement by leaving the gap between them [16].

Table 4. Based on used wall

NO.	WALL	TOTAL	%
1	Bricks	48	18,46
2	Wood/Board	212	81,54
3	Bamboo	0	0,00
4	Zinc	0	0,00
	Total	**260**	100,00

Figure 10. Coloring on wooden wall

The purpose of the use of porous inner and outer sealing walls is to preserve the air circulation system in the dwelling house, to induce heat and humidity out of the building by convection, thereby keeping the air temperature in the house cool and comfortable [14].

Outdoor space with indoor space limited by dinsing should be of particular concern, thus the wall must be capable of providing cross-ventilation media [29].

Some homes provide a combination of wooden window board openings in the form of windows with the aim that the air circulation more smoothly in the room. Even some houses paint a wooden wall with some kind of color to beautify the look of the house. The above explanation provides assertion that from the beginning the community already has adaptation strategy to humid tropical climate especially in the selection of environmentally friendly wooden wall building materials to facilitate indoor air circulation to be more comfortable. In addition, wooden building materials are lighter when compared to bricks (figure 10).

4.2.5. Adaptation strategy based on used floor

Similar to the use of wooden boards for house walls, the use of wooden boards for home floors is also included in large numbers (80%), the rest using stucco and ceramics. Based on interviews with respondents, the use of wooden boards for the floor has a purpose other than lightweight, wooden boards are reducing heat so as to help lower the temperature in the room. This is supported by the structure of the stage, where the underside of the wooden board floor has air circulation so that the accumulated heat can be removed. According to some respondents, some houses coated wooden floorboards with ceramic/plaster to facilitate cleaning. But there is also a wood floor plaster with ceramic floor patterned plastic sheet with the same purpose of ease of cleaning but still aestethic (figure 11, table 5).

Table 5. Based on used floor

NO.	FLOOR	TOTAL	%
1	Soil	1	0,38
2	Plester	16	6,15
3	Ceramic	35	12,46
4	Wooden Board	208	80,00
	Total	**260**	100,00

Figure 11. Wood floor coated with plastic

The explanation above confirms that the use of wooden planks for the floor has been thought by the community from the beginning, they have tried to develop adaptation strategies for their homes, especially those related to the environmental conditions in which they live in tropical climates.

The purpose of the use of apertured wood floors and porous is to keep the movement of air in the house where living, so that it can move heat and humidity outside the residential building by convection, so that the air in the house is always relatively cool and comfortable [14].

4.2.6. Adaptation strategy based on distance between buildings

Based on the opinion of the respondents to the distance between the residential buildings, 24.62% said that there is no distance between buildings / mutually adjacent and as much as 65.00% said the distance between buildings ranging from 1-2 meters. This phenomenon illustrates that the condition of settlements in the study area has a fairly high building density (figure 12, table 6).

Based on the respondents' explanation, the distance between adjacent buildings has the purpose of social relations among citizens, namely the creation of a more familiar communication with neighbors. In addition, promenade deck that formed between two buildings functioned as a means of transportation as well as an open space for the interaction of residents.

Table 6. Distance between buildings

NO.	BUILDING DISTANCE	TOTAL	%
1	No distance	64	24,62
2	Has distance 1 – 2 m	169	65,00
3	Has distance> 2 m	27	10,38
	Total	**260**	100,00

Source, Field survey, 2018

Figure 12. Distance between buildings is near
Source: Field survey, 2018.

At the front of the buildings row, there is a promenade deck, other than as a means to walk while enjoying the river scenery is also used by the community as a place to interact. According to Lenzholzer and Brown [30] and Khatibi[31], humans who are part of nature take the most benefit from natural facilities in the environment so as to create harmonization between the buildings they occupy with the environment (figure 13).

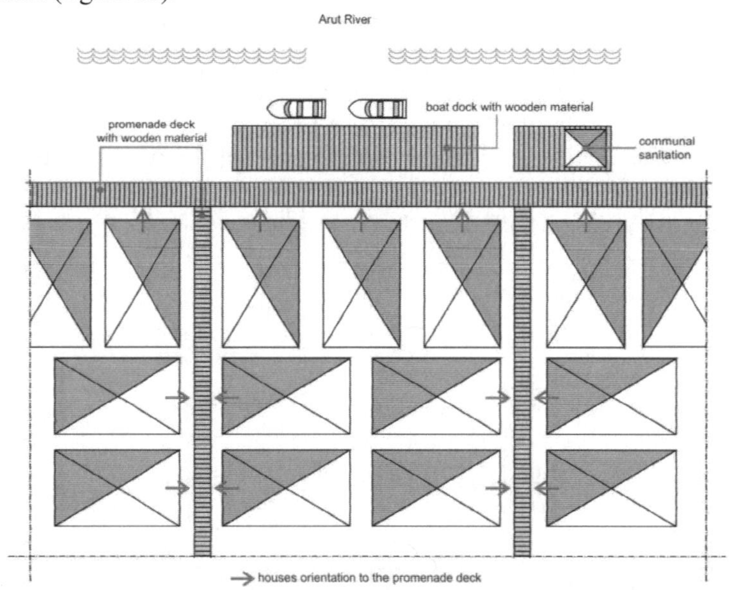

Figure 13. Houses orientation to the promenade deck

5. Conclusion

This research succeeded to reveal the adaptation strategy in settlement caused by natural factors such as Arut River flood disturbance especially during the rainy season and climate disturbance related to wind movement, rain, and temperature, humidity, and other factors related to socio-economic condition of the people. These natural factors affect the physical shape of residential buildings and the environment they create.

Adaptation strategies are referred to as patterns of various efforts planned by humans to be able to meet the minimum requirements they need and to solve the problems they face. The patterns in question here are patterns of behavior or action. Such patterns or actions are human responses to the physical of residential buildings and their environments are based on: [i] dimentional of buildings, [ii] building structure, [iii] roof cover used, [iv] wall material used, [v] flooring material used, and [vi] distance between buildings.

6. Acknowledgments
The authors would like to express much gratitude to the Faculty of Engineering Diponegoro University for the encouragement through the fund granted for this research in 2018 through strategic research programme.

7. References
[1] Rahmadi, D.K 2009 Permukiman Bantaran Sungai :Pendekatan Penataan Kawasan Tepi Air (Riverside Settlements: Upgrading of Waterfront Areas Approach). *Buletin Tata Ruang* (Spatial Bulletin) September – October 2009 Edition
[2] Rahmawaty S. and Khadiyanto P 2014 Kualitas Lingkungan Permukiman di Tepi Sungai Kelurahan Pelita, Kecamatan Samarinda Ilir (Quality of Environmental Settlements on the Riverside Pelita Village, SamarindaIlir District). *J. Teknik PWK* **34**
[3] Hamidah, N. ,Rijanta, R. , Setiawan, B. , Marfai, M.A 2016 AnalisisPermukimanTepian Sungai Yang BerkelanjutanStudiKasus: PermukimanTepian Sungai Kahayan Kota Palangkaraya (Sustainable River Side Resettlement Analysis, Case study: The Kahayan River Settlement Palangkaraya City). *J. Inersia*, **121**
[4] Purwanto, E. Gultom, J 2013 Typo-Morphological Study of Settlement at Riverside, Case: Settlement at Arut Riverside in Pangkalan Bun at West Kotawaringin region, Central Kalimantan Province. *J. Tesa*, **112** 77-91
[5] Helmi A. and Satria A 2012 StrategiAdaptasiNelayanTerhadapPerubahanEkologis (Fisherman Adaptation Strategy to Ecological Change). *J. Makara, SosialHumaniora.* **161** 68-78
[6] Suharto, E 2002 *Coping Strategies danKeberfungsianSosial: MengembangkanPendekatanPekerjaanSosialdalamMengkajidanMenanganiKemiskinan*(Coping Strategies and Social Function: Developing a Social Work Approach to Review and Handle Poverty). Seminar Paper, IPB Bogor 17 December 2002
[7] Purwanto E 2010 The Re-Designing of Responsive Coastal Settlement Toward Tropical Climate (A Case Study: Semarang Coastal Settlement). *J. Pembangunan Wilayah & Kota.* **63**.47-54
[8] Utomo A.P., Suprayogo D., Sudarto 2014 Estimasi Sebaran Daerah Rawan Banjir Bandang Sub Daerah Aliran Sungai Sumber Brantas Kota Batu: Aplikasi Model Genriver & Sistem Informasi Geografi (Estimated Area Distribution of Banjir Bandang Sub Area Brantas River Basin Source Brantas Batu City: Application of Genriver Model & Geographic Information System). *J. Tanah dan Sumberdaya Lahan.* **12** 01-14
[9] Sugiyanto and Kodoatie R 2002 *Beberapa Penyebab Banjir dan Metode Pengendaliannya* (Some Causes of Flood and Control Methods). Yogyakarta: PustakaPelajar
[10] De Wall, H.B 1993 Recomendation for Building in Tropical Climate, *J. Building and Environment.* UK. **28** 271 – 285
[11] Purwanto E 2012 Redesign City: Adaptation Steps In Creating Sustainable Cities. *J.Komposisi* **101** 1-10
[12] Naing N 2008 Peran Arsitektur Rumah Mengapung di Danau Tempe Dalam Mewujudkan Kota Tropis (The Role Architecture of House Floats on Lake Tempe In Creating Tropical Towns). *Proceedings of the National Seminar: The Role of Urban Architecture in Realizing Tropical Cities*. University of Diponegoro. 6 August 2008
[13] Frick H. and Suskiyanto B 1998.*Dasar-Dasar Eko Arsitektur*(Eko Architecture Basics), Kanisius Press
[14] Indarto E 2008 Arsitektur dan Perubahan Iklim Dalam Tantangan Proses Desain Arsitektur (Architecture and Climate Change in Challenges of Architectural Design Process). *Proceedings of the National Seminar: The Role of Urban Architecture in Realizing Tropical Cities*. University of Diponegoro. 6 August 2008
[15] Purwanto, E. 2009. “Rumah Panggung” for The Settlement With Sea Level Rise Problem in The Fishermen Settlement of TambakLorok Semarang. *Journal Of Coastal Development* **132** 67-80

[16] Purwanto E., Apriyanto H., Yarjani 2016 *Elements of Local Wisdom Traditional House as the embodiment of Archipelago Architecture in Sukamara Regency, Central Kalimantan Province*. UNDIP Press
[17] Muchamad, BN, 2010, *The Concept of Floating Construction as a Model for Sustainable Development in Wetland Area*, Proceeding of International Seminar SENVAR **11**, ITS, Surabaya
[18] Rochgiyanti 2011 Fungsi Sungai Bagi Masyarakat di Tepian Sungai Kuin Kota Banjarmasin (River Function For People on the Riverside of Kuin Banjarmasin), *J. Komunitas*. **5** 2
[19] Rahmasari L 2011 Strategi Adaptasi Perubahan Iklim Bagi Masyarakat Pesisir (Climate Change Adaptation Strategy for Coastal Communities). *J. Sainsdan Teknologi Maritim*. **101** 1-11
[20] Kumalasari N.R. and Yuliastuti N 2013 Bentuk Adaptasi Terhadap Banjir di Kampung Purwodinatan dan Jurnatan (The Form of Adaptation to Flood in Purwodinatan and Jurnatan Village). *J. Teknik PWK*. **23** 457-467
[21] Bell, P. A., Greene, T., Fisher, J.D., and Baum. A 2001 Environmental Psychology (5th edition). Belmont, CA: Harcourt College Publisher
[22] Bennet J.W 1976 *The Ecological Transition, Cultural Anthropology and Human Adptation*. Oxford: Pergamon Press
[23] Grout, L. And Wang, D 2013 *Architectural Research Methods*. 2nd. Wiley.
[24] Olotuah, A.O 2015 Climate-Responsive Architecture and Sustainable Housing in Nigeria. *Global Journal of Research and Review*. **24** 094-099
[25] Wikantari R 2008 Citra Kearifan Arsitektur Lokal dalam Pewujudan Kota Tropis (The Image of Local Wisdom Architecture in Tropical Town Realization). *Proceedings of the National Seminar: The Role of Urban Architecture in Realizing Tropical Cities*. University of Diponegoro. 6 August 2008
[26] Frampton, Kenneth. Cava, John, Editor 1995 *Studies in Tectonic Culture: the Poetics of Construction in Nineteenth and Twentieth Century Architecture.*Cambrirge, Mass.: MIT Press.
[27] Bakhtiar, Waani J.O., Rengkung W 2014 TipeTeoriPadaArsitektur Nusantara Menurut Josef Prijotomo (Type of Theory in Archipelago Architecture According to Josef Prijotomo). *J. Media Matrasain*. **112** 32-47
[28] Zareaian, S. And Zadeh, K.A 2013 The Role of Climate Factors on Designing and Constructing Buildings (From Urbanization Architecture Approach). *Bulletin of Environment, Pharmacology and Life Sciences*. **31** 197-200
[29] Karol, E.and Lai, V.V.C 2014 Climatic Design and Changing Social Needs in the Tropics: A Case Study in Kuching, Sarawak. *Sustainability*. **6** 6278-6292
[30] Lenzholzer, S. And Brown, R.D 2013 Climate-responsive landscape architecture design education. *Journal of Cleaner Production*. **30** 1-11
[31] Khatibi, S.2012. Vernacular Architecture Values (Humid and Moderate Climate in Iran). *International Journal of Architecture and Urban Development*. .**24** 39-42

Building Task Concepts of the Vernacular Settlement in Tamkesi Village

Reginaldo Christophori Lake, Benediktus Boli, Ubaldus Djonda, Yohanes Nurak Siwa
Universitas Katolik Widya Mandira, Kupang, Indonesia,

Corresponding e-mail: egilake@yahoo.com; egilake@unwira.ac.id

Abstract. Tamkesi village in Timor Island is a vernacular village designed and constructed by the local people based on local wisdom and knowledge. The Tamkesi settlement was developed on the basis of culture of Southern Biboki ancestral traditions, either in terms of the order of space outside the building or the order of space inside the building. The mass building of Tamkesi people contains 6 masses, either occupancy masses or non-occupancy masses. Tamkesi vernacular settlement is very significant to be analyzed because of the phenomenon of its local wisdom concept. This research, therefore, was focused on the process and concept of building task architecture. Such concept is believed to contribute to today's architectural designs. This study used literature review and field notes as its methodology. The employed reading tool in this study is based on the relationships which exist in function-form-meaning of vernacular architecture. The building task analysis was used as an identification tool of a concept which influences function-form-meaning of a case study. The result indicates that the building task of the study object is a phenomenological concept of the existence of a vernacular settlement which has been survived until this time based on the local natural condition and social symbol of cultural settlement. The contribution of the process to discover the concept of building task of Tamkesi vernacular architecture is the understanding about the value of culture and the symbols of tradition as the concept of local identity of today's architecture.

Keywords: building task, traditional vernacular architecture, Tamkesi village

1. Introduction

Tamkesi village is a vernacular village [1]. A vernacular village is designed by the first generation of the villagers and further developed by their descendants [2]. The development of village's space as a whole refers to the physical order established by the founding generation through customary rituals, in which the ancestors were invited to be present and give the directional signs to the villagers [2]. The understanding of such village spatial phenomenology is evident in Tamkesi village, which can be described in the socio-symbolic spatial concept of the settlement inner space, that is, brotherhood relationship of the ethnic/tribe, spiritual relationship, relationship of the obedience to the tradition, and the special relationship of humans and nature [3].

The establishment of retidential space of Tamkesi village is very unique, which is composed of grave part (*naukel*), traditional farm (*lele*), customary forest (*nasi*), and a new village (*kuan feu*) [1].

In the settlement tradition, Tamkesi people talk about the expressions which indicate the category of "in-village" and "outside-village". The socio-culture of Tamkesi reveals that Usboko is "an inner tribe/the guardian of in-village", because Tamkesi people acknowledge Usboko tribe as the revelation of *Uis neon* (as the descendant tribe of Tamkesi *sonaf* Kaiser). Some people claim that *Fai San Nionu* tribe which is located in the east of Tamkesi village and Belsinkone tribe which guards the west part as "outside-tribe/ the guardian of outside-village". Based on the local phenomenological concept, this reality indicates the existing of spatial awareness among the villagers in terms of the relationships of settlement spatial orders through actual experiences on such reality [1].

This spatial perception is based on the tasks and functions of each tribe in Tamkesi village. Tamkesi people say that the West part and East part are outer village because those areas are known as the territory used to collect tributes by *Fai San Nionu* tribe and *Bel Sinkone* tribe to be offered to Kaiser (king) which resides in 'in-village' (*Usboko* territory). The living experience on such spatial concept was finally crystallized and giving birth to typical expressions of Tamkesi village [1].

The settlement phenomenon in Tamkesi village is interesting to be observed, especially the meaning of "*building task*" concepts of Tamkesi people settlement based on their ancestral cultural customs.

This paper aims to explore the phenomenon of concept and function of elements which establish spatial form of Tamkesi village as an internal inspiration for Tamkesi village and the motivations for other villages to maintain their typical characteristics and identity through the preservation architectural settlement elements. The research problem was formulated as, "What is the meaning of spatial physical form based on the theory of *building task* function of the vernacular settlement of Tamkesi village?"

2. Methods

The methodoly used is literary review, that is, the elaboration between *building task* theory [4] and the *ordering principles* of function-form-meaning (Pursal, 2010), supported by the field notes (*logbook*) produced from the results of research process based on on the *physical trace*, and equipped with ethnographic verbal data [2]. The field notes used as the data of this paper were taken in 2010 and 2014, used in the thesis entitled, "Konsep Ruang Dalam dan Ruang Luar Arsitektur Permukiman Tradisional Suku Atoni di Kampung Adat Tamkesi di Pulau Timor" (The Concept of Indoor Space and Outer Space Architecture of Atoni Traditional Settlements in Tamkesi Customary Village in Timor Island) and the book with the same title, completed in 2014. The data on the concept of settlement phenomena in the thesis field notes [5] that was examined through the *building task* concept has never been used as the materials for research paper due to its variation of contents. Thus, the data is deemed relevant to be used in a research paper with the specific topic on the phenomenon of *building task* concept in vernacular settlemen in Tamkesi village.

2.1. Case Study

Tamkesi traditional village is located in the hamlet III of Southern Biboki Usboko, North Central Timor District (TTU). Geographically, TTU district is located between 9^0 02' 48"LS-9^0 37' 36" LS and between 124^0 04' 02"BT-124^0 46'00"BT [6]. The borders of Tamkesi traditional village: North side with Tautpah and Sapaen villages, East with Oekopa and Oriabesi, South with T'eba and West with Tokbesi [1].

Specifically, Tamkesi village which consists of 3 hamlets is located in RT. 2 Usboko. Tamkesi village is a home to 18 families [1].

2.2. The Analysis Procedures

Literature review was done on substantial literature and the literature on methodology. The substantial literature contains the information about Tamkesi villange, among others, a descriptive book [1], and the research report [3]. In literature review, the data was read again with the focus on the spatial aspects of Tamkesi vernacular settlement, either on the scope element or its function and meaning. The reading of Tamkesi vernacular spatial case was purposively based on the the principles of searching *building task* concept [4] and *ordering principles* of function-form-meaning [7].

IOP Conf. Series: Earth and Environmental Science **213** (2018) 012034 doi:10.1088/1755-1315/213/1/012034

The overall picture of this study can be illustrated as follow: (see figure 1 below).

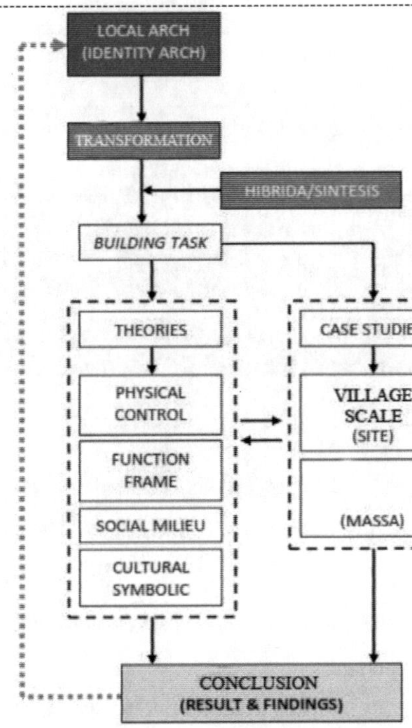

Figure 1. Reseacrh design.

The analysis consists of two main parts, the first, physical analysis of the study object, and the second, theoretical analysis. In the first analysis, the object was redrawn in three dimensions, and then parsed based on the scale (scope) of the *site* and the *mass*. This analysis was done in order to know in details the *building task* concept of all elements forming Tamkesi settlement architecture. In the second analysis, the relationship of function-form-meaning aspects according to the ordering principles was operationalized. Although the three aspects cannot be separated, the scope of this study focuses on functional aspect which is related to form and meaning. This analysis is based on the theoretical analysis.

3. Discussion

3.1. First, the Building Task Analysis of the Village Scale (Site)

The elements which form the scale of the village (*site*) consist of open space and mass.

3.1.1. Open Space

The open space in this context is the outer space, which is located around the building, and the opposite of interior space. In the compound of Tamkesi vernacular settlement (Sonaf Biboki), there are some open spaces, namely: settlement courtyard (open courtyard), graves and circulation [8].

a. Settlement courtyard (open courtyard)

The pattern of Tamkesi vernacular settlements is formed in cluster *(tanean)* with the arrangement adjusted to hill contour conditions in the form of a linear pattern, so that the houses (*ume, lopo and sonaf*) grow in a circle on the hillsides. The arrangement or hierarchy of settlements in Tamkesi is also influenced by the customary provisions, characterized by the special or sacred zones (places) that need to be taken into considerations [3].

Sonaf is a name for the king's residential house (palace) which is regarded as a sacred house or a house with some restrictions. In the daily life, the open courtyard (settlement courtyard) is usually used for social acivities, a place to dry up agricultural products, even as children's playground, see Figure 2) [9]. However, in a special open courtyard, there is *haumonef* (a pole

IOP Conf. Series: Earth and Environmental Science **213** (2018) 012034 doi:10.1088/1755-1315/213/1/012034

with three branches having a flat stone as its altar to place offerings), used for spiritual activities (customary rites). It means an offering place to show dependence, respect and gratitude to the Almighty (*Uis neon*) and the ancestors (*Usif*) [1].

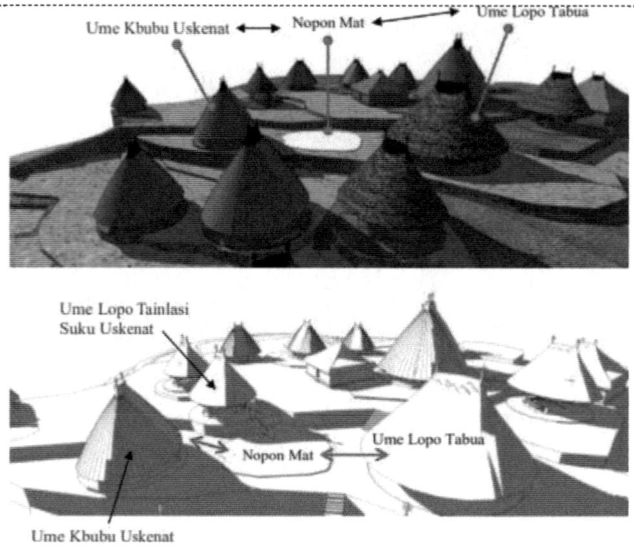

Figure 2. Open courtyard characterized by the existence of nopon mat and haumonef.

b. Grave (naukel)

The stone grave (*naukel*) which is circularly arranged (*nopon ni nopon*) is one of the megalithic traditions in the Tamkesi customary settlements (see Figure 3). Most graves are special tombs for *Usif* (king), and *amaf* (tribal chief) buried in the vicinity of the Tamkesi customary settlement.

The tomb is made of circular cobble stones. The tradition to put tombs around Tamkesi vernacular settlement is an expression of respect of Tamkesi people to their ancestors and a belief in the spirit which guards Tamkesi customary settlement [10].

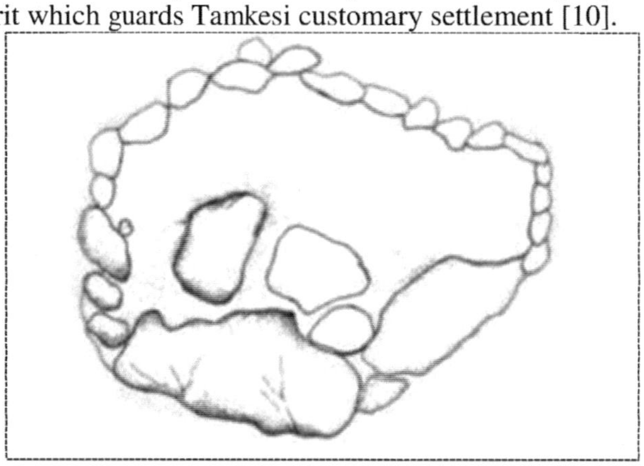

Figure 3. Nopon ni nopon as a binder/datum on megalithic culture in Tamkesi settlement.

c. Circulation

Tamkesi vernacular settlement is a close settlement, on the basis of regulations of customary traditions. It is a fortress palace, with only one main gate as the entrance and exit access to the settlement. Having only one access (circulation) is believed for security and denfense purpose. The circulation pattern symbolizes that the Tamkesi vernacular settlement is central to the activities and life of the Tamkesi-Biboki community, since the beginning (birth) to the end (death) [1].

The Tamkesi settlement pattern is generally oriented towards the North-South, with the south direction as the the entry toward the circulation, the northern direction as the peak of the ritual, marked by the presence of *sonaf mnasi* (the residence of King/Kaiser) . This pattern also indicates that the position (orientation) of the settlement is similar to the position of a sleeping man, that is, the head rests on the mountain (north direction) and the feet are on the river (south direction).

The concept of cultural settlement in Tamkesi has 4 circulation zones in the settlement pattern, namely: (i) The *eno naikah* zone (first door/ front zone) is the entrance zone (entrance gate), occupied/guarded by *taslulu* tribe and *tkafun* tribe whose title is *meosopan* (warlord/fenching); (ii) The *eno tnana* zone (second door/middle zone) is a zone to bring tributes to *Uis neon* and Kaiser, occupied/guarded by *Uskono* tribe, *Uskent* tribe, and *Bel Sinkone* tribe whose title is *aat* (Kaiser spokesman/ press secretary of *Usif/* king and *amaf-amaf/* tribal chiefs); (iii) The *natna* zone is called the seventh ladder, as the center of Kaiser's residence. This zone is intended for the descendants of the king (*Usboko* tribe); (iv) The *eno kotin* zone (back zone), consisting of rocks, hill cliffs and Biboki customary forest, is a defense area at the back circulation.

The consequence of the existence of these four types of zones/circulation is the sustainable efforts by the tribes on customary rules and settlement traditions as well as the responsibility to own and maintain the territory of living space in Tamkesi.

In addition, the hierarchy of zone designation in the settlement culture of Tamkesi custom is based on the concept of sacred area (private/*fafon*) to the profane (public/*munini*). The settlement culture in Tamkesi customary village believes that the higest area or Biboki hilltop as private space (sacred/*afon*) while the hillside section as public space (profane/ *munin*) (see Figure 4 & 5).

IOP Conf. Series: Earth and Environmental Science **213** (2018) 012034 doi:10.1088/1755-1315/213/1/012034

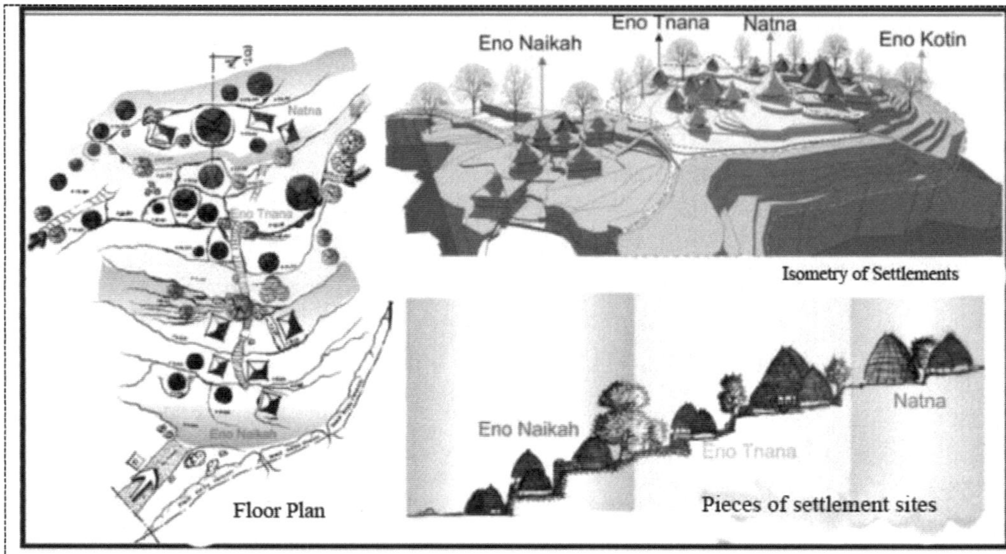

Figure 4. Circulation zones of Tamkesi vernacular settlement.

Figure 5. The upper hierarcy is a hill/mountain. The bottom hierarcy is the river. This is the concept of top-down orientation of Tamkesi vernacular settlement.

3.1.2. Mass.

The total number of mass buildings in Tamkesi vernacular settlement compound is 6 units. In general, the mass bulding can be categorized into two based on its functions, namely occupancy mass and non-occupancy mass. Generally, the occupancy mass in Tamkesi consists of *Sonaf Mnasi*, *Ume Mbubu, Ume Lopo* and *Ume Kbat*. *Sonaf Mnasi* is Kaiser palace (the residence of Usboko tribe), *Ume Mbubu* (the house for mother/ women tribe), *Ume Lopo* (the house for men tribe, which functions as a place to welcome guests, and barn), and *Ume Kbat* (the house to gather all members of the tribe as one family, either women tribe or men tribe, see Figure 6). However, the non-occupancy mass consists of *sonaf nenobiboki,* and *hauteas/haumonef*. It is the mass which is considered spiritual (sacred) [1].

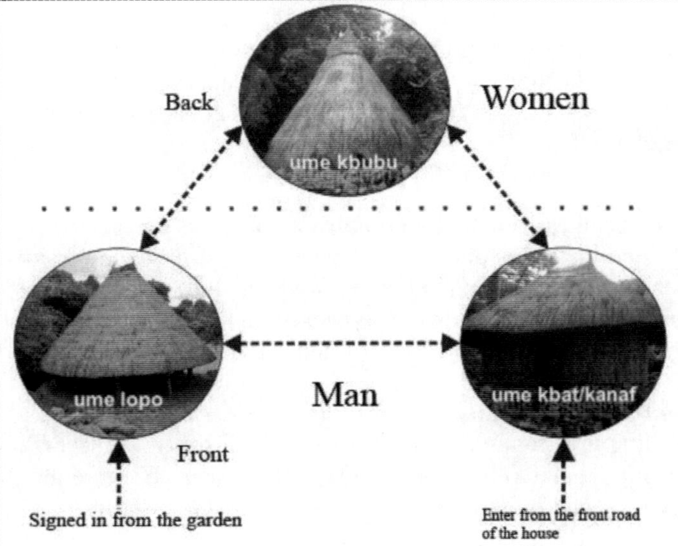

Figure 6. Spatial formation of the residential houses in Tamkesi.

From the above description, Tamkesi vernacular settlement complex as an environmental model accommodates various functions. The following is the identifiable functions of the vernacular complex of Tamkesi that is classified based on the theory of "task-building" functions by Christian Norberg Schulz:

a. Pragmatic Function

The spatial physical form of Tamkesi settlement shows pragmatic functions namely functional frame and physical control.

- Physical control

(1) The pattern of complex arrangement of Tamkesi settlemen in cluster form is an adaption effort to the relatively flat topological condition. The hilly condition in the forms of footprints, however, stimulates the growth of mass organization. Surface land was made terraces in order to create suitable soil conditions for activities and the placement of building; (2) The circulation with stone pavement was made in ladder form in order to facilitate the access to the areas with different height. This is an adaptation the typology; (3) *Sonaf* and *Ume Mbubu* (masses) are houses with limited open space. This is an adaptation to the South Biboke nature which is cold throughout the day.

- Functional frame

(1) The open space in the form of the settlement courtyard has daily functions as a children's playground, to drying up the harvest products, and the entrance and exit circulations of Tamkesi settlement; (2) The courtyard around *naukel* and *haumonef* is used to perform customary rituals asking blessings and protection from the Lord and the ancestors; (3) In everyday life, *sonaf mnasi* (Kaiser/king palace), *ume mbubu* and *ume lopo* function as the place to sleep, welcome the guests, cooking, while non-occupancy masses (*sonaf nenobiboki* and *haumonef/ hauteas*) are the masses which are considered sacred (spiritual place and the place to keep heirloom items).

b. Symbolic Function

The spatial physical form of Tamkesi settlement also shows symbolic functions which cover social milieu and cultural symbolic.

- Social milieu

(1) Tombs (graves) were made with a unique shape and its existence indicates a hierarcy, which means that the tombs around Tamkesi settlement are the tombs of the ancestors and the tribal chief (*amaf*) while ordinary people were buried outside the Tamkesi settlement; (2) As a sacred site, *sonaf nenobiboki* can only be entered by Kaiser's wife.

- Cultural symbolic

(1) The cluster pattern of Tamkesi vernacular complex symbolizes unity, leading to a single point, that is, *Uis neno* (the Lord) and the ancestors; (2) *Sonaf mnasi* as a palace, *ume mbubu* as the symbol of women, while *lopo* as the symbol of men; (3) *Ume mbubu* symbolizes a woman (mother). The condition of a short dimension of the door symbolizes the respect of visitors or residents when entering the house in which they should bow down as an expression of honor to the figure of a woman (mother).

3.2. Second, the Building Task Analysis of the Building Scale (Mass)

In architecture, the building function is closely related to the form and space. If the discussion is about function, it is then certainly related to form and space. This happens because the function is a picture of an activity that requires form and space.

The building mass of *sonaf mnasi* (Kaiser's palace) was used for the building task analysis of building scale because it is more representative for in depth analysis. Beside its function as Kaiser's residence, *sonaf mnasi* also functions as a religious and sacred place in South Biboki area. In terms of form, *sonaf mnasi* is almost the same with other *sonaf*. The floor plan is quite oval and elliptical; the patterns of structures, construction, walls, roof, and other materials used are generally the same (Figure 7).

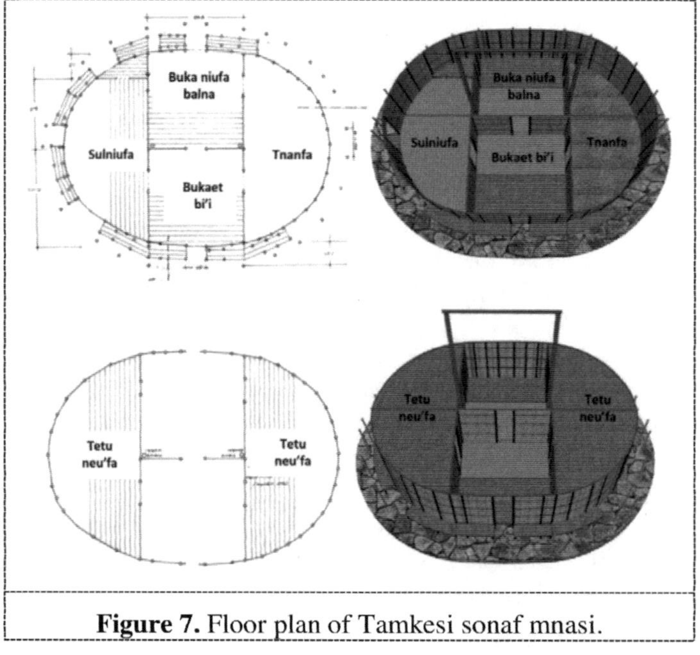

Figure 7. Floor plan of Tamkesi sonaf mnasi.

IOP Conf. Series: Earth and Environmental Science **213** (2018) 012034 doi:10.1088/1755-1315/213/1/012034

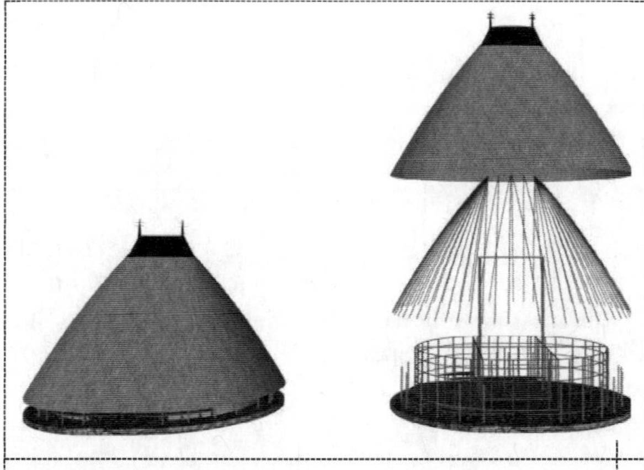

Figure 8. The isometry of Tamkesi sonaf mnasi.

Figure 9. Inner space of Tamkesi sonaf mnasi.

With in terms of the structure, *sonaf mnasi* is a bulding with the frame structure which relies on the pedestals of main poles (two main poles: *liurai* pole and *sonbai* pole) and supporting poles connected to each other by the main bars and supporting bars.

From the visualization analysis of Tamkesi *sonaf mnasi* (Kaiser's palace), the function of that building can be identified based on the theory of building task function of Christian Norberg Schulz:

a. Pragmatic Function

The physical form of Tamkesi *sonaf mnasi* (Kaiser's palace) shows pragmatic function including functional frame and physical control.

- Physical control

(1) The roof of *sonaf mnasi* which almost touches the ground gives a wide space to the roof in order to avoid the heat of the sun and the rain. In addition, it decreases the circulation of cold air from outside to inside the building, so that the inner space becomes convenient for activities; (2) The elevation of wooden flooring (the concept of stage house) is a control effort to avoid flooding, rain, and soil surface humidity. It also facilitates the cleaning of the house and allows the air circulation through the slit of the floor; (3) The physical condition of *sonaf mnasi* which is completely closed and less open is used as a barrier to protect the inhabitants from cold weather and providing privacy for the residents. Although closed, the cracks of the plank walls and bamboo blades facilitate the air circulation; (4) The roof shape with a slope of more than 60° expidates the flow of rain water. The use of *alang-alang* materials provides a good protection against the local climate, creating thermal comfort and good insulation against noise. In addition, the relatively light *alang-alang* materials provide a relatively light construction loads which consequently reduce the risk of collapse during the earthquake (Figure 8 & 9).

- Functional frame

(1) The terrace is made in the form of a veranda that can be used as a seat. It is located on the edge of *sonaf*, precisely in front of the side door and the front door to accommodate the reception of common guests (ordinary people); (2) The function of the living room (*buka nifa balna*) is used as a central sacred space, characterized by the presence of a pillar of *liurai* and *sonbai*. It is usually used just for customary rituals; (3) *Buka niufa/sulniufa* (the room for Kaiser/king's seat) is the room for Kaiser and his family. It is a very sacred and private place. The room is also often used as a meeting place for the tribal chiefs (*amaf*); (4) In daily life, *sonaf mnasi* functions as kaiser's palace, a place for living together with his immediate family, to rest and keep the valuable heirlooms of the king.

b. Symbolic function

The physical form of Tamkesi *sonaf mnasi* (the palace of Kaiser/king) reveals symbolic functions that include social milieu and cultural symbolic.

- Social milieu

(1) *Sonaf mnasi* is the residential palace of Kaiser/king Usboko (tribal king). On a cultural scale, it shows a hierarchical difference from ordinary houses. In terms of dimension, *sonaf mnasi* has a larger size than other houses, due to various activities. *Sonaf mnasi* is meant for the family activities (women and men tribes of the king's descendants), and used also as the place to keep heirlooms, and a meeting place for the main tribal chiefs of Tamkesi. Thus, it requires a large circulation (space); (2) The spatial pattern of *sonaf mnasi* has strata (position) division. The outside terrace of *sonaf* is intended for ordinary people, *buka niufa balna* as a central sacred space for the Kaiser/king, *buka niufa bahana/ sulniufa* as an assembly place for tribal chiefs and *tetu* (ceiling) as a place to keep sacred food tributes; (3) The door was constructed in a low level, so the visitors (guests) must bow down as a sign of respect to the kaiser/king; (4) The social value of *gotong royong* (working together) and *musyawarah* (dialogue) is expressed through the construction process of *sonaf mnasi*. For *sonaf mnasi* is a palace that is well respected by all people of South Biboki.

c. Cultural symbolic

(1) There are two sacred poles in *sonaf mnasi* namely *liurai* pole and *sonbai* pole as symbols for men and women (sun and moon). This symbol is an expression of the ancestor and the Creator; (2) The materials for the construction of *sonaf mnasi* should be taken from a customary forest. The gathering process was preceded by a traditional ritual in order to ask permission from the spirit who guards that place. It is a symbol to keep harmony between human beings and the nature; (3) the carvings in the poles and entrance door of *sonaf mnasi* in the form of nature motif become the identity of Tamkesi people.

4. Conclusion

Based on the analysis of building task concepts of Tamkesi vernacular settlement, some points can be taken as a conclusion: (1) The building task concept of Tamkesi settlement was formed based on the open space (open courtyard, circulation and grave), mass occupancy (*sonaf mnasi, ume mbubu, lopo* and *ume kbat*) and non-mass occupancy (*sonaf nenobiboki* and *huateas/haumonef*). In the occupancy scale, the building task was materialized through the environment of Tamkesi vernacular settlement. Both pragmatic function and symbolic function are interrelated and influence physical form of Tamkesi vernacular settlement. Physical control of Tamkesi vernacular settlement was influence by weather, tribal organization and topography. On the other hand, the functional frame is reflected on the existence of various masses with different functions accompanied by all supporting space equipments. Symbolic function is social milieu which is influenced by the viewpoint about social strata concept, while cultural symbolic is related to the belief concept (religion). (2) The bulding task concept of mass scale is materialized in the spatial physical form of *sonaf mnasi*. Both pragmatic and symbolic functions influence spatial physical form of the mass building. Physical control appears at the surface, either in the roof, wall, and floor in order to create a comfortable room to be occupied. Likewise, the functional frame is revealed at the division of rooms inside *sonaf mnasi*. Symbolic function is expressed in the form of social milieu, the division of room functions in *sonaf mnasi* based on different social strata, and the tasks of customary stakeholders. The dimension of the door was constructed in a small size so that the visitors need to bow down during the entry as a sign of respect to the hosts (residents). Likewise, the values of working together and dialogue during the construction process of *sonaf mnasi* are meant for the common ownership of *sonaf mnasi* by the people of South Biboki. The cultural symbol is closely related to the belief in the ancestors and the Creator.

5. Findings

The shape, space and pattern of Tamkesi vernacular settlement is influenced by physical and non-physical factors. The physical factors include climate, weather, and topography. The non-physical factors are social system, tribal organization, belief (spiritual factor) and daily activities.

6. References

[1] Lake R C 2015 Gramatika arsitektur vernakular suku Atoni di kampung adat Tamkesi di pulau Timor

[2] Purbadi Y D 2017 " CONTINUITY AND CHANGE " DALAM ARSITEKTUR VERNAKULAR KAJIAN FENOMENA LOPO DI DESA KAENBAUN (Kupang: Program Studi Teknik Arsitektur Fakultas Teknik Universitas Katolik Widya Mandira) pp 69–82

[3] Lake R C 2016 TATA SPASIAL ARSITEKTUR TRADISIONAL SUKU ATONI DI KAMPUNG TAMKESI PULAU TIMOR *ATRIUM-Jurnal Arsit.* **2**

[4] Norberg-Schulz C and Schulz C N 1966 *Intentions in architecture* (MIT press)

[5] Lake R C 2014 Konsep ruang dalam dan ruang luar arsitektur tradisional Suku Atoni di kampung Tamkesi di Pulau Timor *E-Journal Grad. Unpar Part D-Architecture* **1** 61–74

[6] TTU B 2017 *TTU Dalam Angka 2017*

[7] Pursal. 2010 *Arsitektur yang membodohkan* ([S.l.]: CSS Publishing)

[8] Salura P and Lake R C 2014 The Architectural Language of Inner and Outer Space as Observed among the Atoni Tribe in the Tamkesi Kampong on Timor Island *Int. J. Acad. Res.* **6**

[9] Tallo A J 2013 STRUKTUR RUANG PERMUKIMAN SUKU ATONI BERBASI BUDAYA, STUDI KASUS: KAMPUNG ADAT TAMKESI, KABUPATEN TIMOR TENGAH UTARA *Tesa Arsitektur, J. Archit. Discourses* **11** 16

[10] Amsikan Y G 2006 MANFAAT KEARIFAN EKOLOGI TERHADAP PELESTARIAN LINGKUNGAN HIDUP Suatu Studi Etnoekologi di Kalangan Orang Biboki *AKADEMIKA* **4** 1–14

Does ICT make city compactness higher? Evidences from compact city attributes in Yogyakarta City's districts.

M. Sani Roychansyah[1], Sushardjanti Felasari[2]
[1]Universitas Gadjah Mada, Yogyakarta, Indonesia
[2]Universitas Atma Jaya Yogyakarta, Yogyakarta, Indonesia

Corresponding e-mail: saniroy@ugm.ac.id

Abstract. From discourse to implementation compact urban development has grown rapidly in nearly two decades. In the era of information communication technology use and popularity of smart city concept, inevitably compact city has entered a new phase: Compact City Version 2.0. In Indonesia, the smart city concept is also flourishing in some cities, even though they are still in early stages of exploring some appropriate strategies. This study aims to look at the effect of the use of ICT in the Yogyakarta City's districts, proven in previous research to be categorized as districts with high compactness, especially on the attributes of density of population, concentration of activities and social welfare. This research continuously examines how far the influence of ICT is able to give effects on making urban compactness higher. Three districts in the city of Yogyakarta generally with high to very high-density score of compactness were taken as objects of this research. Furthermore, the compact city attributes and its indicators were explored in more detail associated with the use of additional criteria, the ICT usage. The result informs that ICT was able to improve performance for a better compactness. Some of the attributes that were previously weak, such as local public transport services and range of public services were capable of being upgraded by the use of ICT. Meanwhile, attributes that had good performance previously also became stronger. This suggests that ICT with various applications that are appropriate for the life of the urban residents significantly help the performance of the city to be more compact. Welcome to the Compact City Version 2.0!

Keywords: compact city attributes, urban compactness, high-density district, information and communication technology, Yogyakarta City

1. Introduction

Although it has been initiated more than three decades ago, sustainability still faces major challenges, especially its realization in the urban level. Cities in developing countries, for instance, still face various fundamental problems. This issue is quite difficult in terms of where the initiation of sustainable cities supposed to start. Nonetheless, discussions related to compact city as one of opportunities toward sustainable cities are responded positively by the cities in developing countries. They are optimistic that some of key attributes either academically or practically in the compact city have already existed naturally in their cities, including, those with high population density or various activities concentrated in their area.

In the few existing studies, population density and concentrated activities indeed have been one of advantages of cities in developing countries to accelerate the realization of a compact city. There are also indications that support the use of technology, especially information and communication technologies that may also push the speed of realization in a city to be more compact. This indication recently tends to be more intensified with the increasing proliferation of concept and implementation strategies related to smart city, where ICT becomes one of the backbones of its manifestations.

In this case, ICT has provided a new picture that compactness has been determined by not only the physical performance, but also the city's performance in using resource (technology) in their city management. This research aims to provide a preliminary picture related to this new paradigm. Index of compactness today should not depend only on the input of physical city, but the most important thing is outputs or achievements. One of them may be through ICT usage.

1.1. Compact City as Sustainable City: Smart Requirement

Sustainable city is possible not through a formal or normative process, but on capacity and uniqueness of each city. Using various strategies to achieve sustainable performance of the city, any city is able to adjust this purpose based on its own potential and capacity.

Figure 1. Smartness through ICT is one of attempts towards achievement of compact sustainable city

The question is whether a sustainable city must have a compact urban performance, or automatically compact. In this case the degree of compactness is also believed as a representation to illustrate an achievement of sustainability. From other perspective, compactness is a requirement prior to becoming a sustainable city. Scenarios built by Jenks et al. [1] illustrated that a sustainable city is a city that has a certain level of degree of compactness, because it concerns with some essential consideration of urban space, like balance, efficiency, and effectiveness.

Thus, as shown in Figure 1, the compact sustainable city can be achieved through a variety of strategies, such as smartness, health, resiliency, green, creativity, inclusivity, and so forth. This paper is set to concentrate on achieving a compact city by smartness strategy, through the use of ICT.

1.2. Transformation of Compact City

As revealed by Dantzig and Saaty [2], first compact city is solely determined by form or morphology of the city. For the next development, compact city is more associated with set of indicators directly related to the urban form, physical performance of the city. At this stage, these indicators have been linked to the assurance of sustainability achievement. While at recent time, in the third stage, many indicators have undergone more various and complex processes. They are not only associated with sustainability itself, but also unique or special aspects owned or being developed in the city.

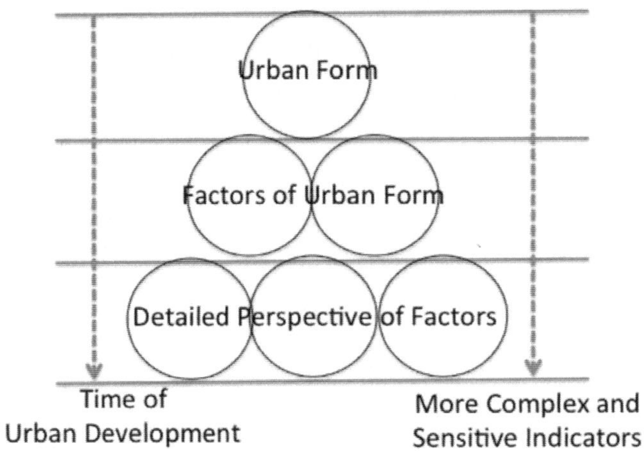

Figure 2. Trend of compactness indicators

As shown in an illustration in Figure 2, the development of compact city concept should not be separated from the history of urban development and the influence of existing industries, ranging from use of machinery to electronics. Today the world has entered the final stage of the third stage of industries, namely use of computer technology and cyber physical era. Thus, the compact city also needs to be approached through the indicators, which are more sensitive in the realization of compact city in the era of ICT.

2. Methods

2.1. Urban Compactness Index in Yogyakarta City (2015)

This study used urban compactness index (UCI) measurement in 14 districts of Yogyakarta City in 2015. Yogyakarta (see Figure 3) is an important medium city in Indonesia, located in the middle of Java Island with about 480 thousands inhabitant. The variables consist of 5 variables: densification of the population, concentration of activity, intensification of public transport, size and access to the city, and socio-economic welfare respectively [3]. These variables were developed through several related indicators as shown in Table. Generally, the UCI of Yogyakarta City and other cities in Indonesia have strong results in densification of the population, concentration of activities, and socio-economic welfare.

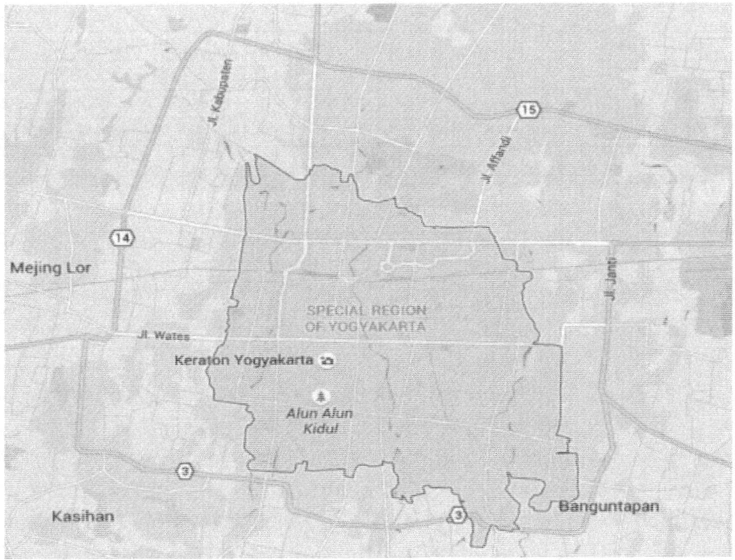

Figure 3. Map of Yogyakarta City

Based on the analysis as shown in Table 1, Gondomanan has the highest index of urban compactness (79.63) and Kotagede has the lowest (54.04). This showed that population density, mobility, and activities in each district have become key achievements of UCI in Yogyakarta City. Furthermore, urban compactness index was grouped into three ranks of UCI (high, medium and low). From each rank, a district took as representative of the rank, and these three districts were used as a case study. They were Gondomanan District as representation of high rank, Gondokusuman District as representation of medium rank, and Kotagede District as representation of low rank of UCI.

Table 1. Urban Compactness Index (UCI) of Yogyakarta City
(*Source: Roychansyah, 2015*)

No	Districts	Popdens	Actons	Pubtrans	Citysize	Sosecwel	UCI	Class of
1	Kotagede	81,97	43,91	48,91	14,81	80,62	54.04	
2	Ngampilan	88,33	29,85	54,17	27,78	79,39	55.9	

3	Mantrijeron	84,28	28,47	54,17	43,33	74,04	56.86	Low
4	Umbulharjo	87,94	45,11	54,17	43,33	78,41	61.79	
5	Tegalrejo	83,97	54,15	54,17	41,67	75,25	61.84	Medium
6	Kraton	91,73	39,43	54,17	64,17	74,42	64.78	
7	Wirobrajan	80,19	50,82	54,17	58,33	80,59	64.82	
8	Gedongtengen	91,01	44,23	54,17	68,33	74,75	66.50	
9	Mergangsan	92	45,13	54,17	58,33	86,08	67.14	
10	Gondokusuman	85,63	82,75	54,17	50	79,15	70.34	
11	Pakualaman	94,82	77,64	54,17	54,17	77,34	71.63	High
12	Jetis	81,16	85,8	59,55	55,83	79,53	72.37	
13	Danurejan	92,66	83,88	54,17	60	85,85	75.31	
14	Gondomanan	90,59	96,18	54,17	79,17	78,05	79.63	

2.2. Development of ICT Related Indicators in Compact City

Principally, the new set of indicators was added as set indicators of smart city considered to be suitable to strengthen the condition of compactness, from the view point of ICT usage. Each attribute was elaborated with several ICT related indicators as presented in Table 2. Some of indicators were developed from smart city studies pioneered by Giffinger, et al. [4] when he measured rank of European medium city size for their smart city performance.

Table 2. Development of ICT Related Indicators

No.	Attributes of Compactness	ICT Related Indicators
1	Population densification	Rate of HP user in the area/total population Rate of mobile internet user/total population
2	Activity concentration	Rate of creative industry or IT related/total industry Rate of online shop located in the
3	Public transport intensification	Rate of resident using internet based transport service/total population Rate of resident using GPS or geo-location
4	City size and access consideration	Number of telecommunication antenna in the area Number of hotspot or wi-fi
5	Social economic welfare target	Rate of elderly who use gadget/total population Rate of facilities where serve online or phone/total population Rate of communication budget in

3. Result and Discussion

The calculations depicted the three districts as representation of various class of UCI that were measured, namely Gondomanan, Gondokusuman, and Kotagede. All showed improved performance of UCI when indicators related to the use of ICT were added. Even in some weak attributes, such as public transportation intensification and size and access consideration, the indicators related to the ICT use were very helpful in improving the performance of district compactness. More results of these calculations are shown in Figure 4.

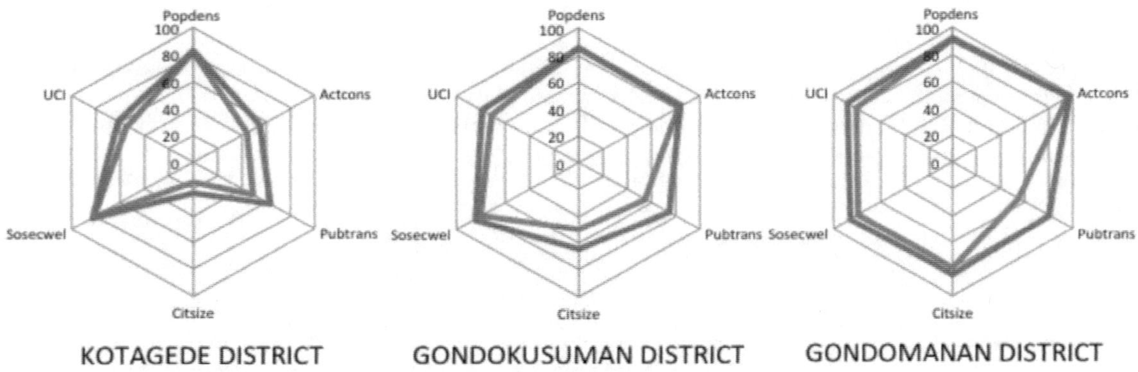

Figure 4. Result of measured districts (Kotagede (low), Gondokusuman (medium), Gondomanan (high))

This result also showed that ICT as well as new media strategy in triggering and accelerating development of the regions seem to be very promising. For example, Kotagede District located on suburban of Yogyakarta City looks was helped by the use of ICT in improving non-personal transport services. Innovations in car taxi and motorcycle taxi service that use user-friendly applications to support ICT sector attract the public.

Participations of public or community and private sectors in increasing provision of hotspots or access points in urban areas also became key factor to reach higher UCI. This phenomenon can be seen from the size and access in the urban area, from low to high level when ICT related indicators were taken into account. Currently, the availability of access points has become requirement from the markets when investors attempt to open new business and if they want to smoothly continue to gain public favour, especially from the younger generation.

4. Conclusion

This study confirmed compact city position and achievement of sustainable development in the city. This study emphasized that the compact city was a representation of sustainable cities that can be approached from the unique aspects of cities or the themes that had become trends in the worldwide today. A smart city strategy is an example that may be one way to achieve a better compact city in the future, through the intensification of existing use of ICT (information and communication technology).

This study also indicated the needs for new approach in viewing the compact city. Complex resources of city formed the compact city as a product and its process required a more comprehensive approach, including the application of ICT in the city. This study also proved that the elaboration of ICT indicators in the new set of indicators in measuring the urban compactness index had produced better results. This has obviously been due to the current position of ICT as a key role that receives a lot of attention as a priority in the urban development. This study, however, still requires a lot of development, especially to see the important findings of negative impacts regarding the use of ICT, its media and data, as well as its target for urban efficiency and effectiveness.

5. References

[1] Burton E, Jenks M and Williams K 2003 *The compact city: a sustainable urban form?*

(Routledge)
[2] Dantzig G B and Saaty T L 1973 *Compact city: a plan for a liveable urban environment* (WH Freeman San Francisco, CA)
[3] Roychansyah M S, Ishizaka K and Omi T 2005 Considerations of Regional Characteristics for Delivering City Compactness *J. Asian Archit. Build. Eng.* **4** 339–46
[4] Giffinger R, Fertner C, Kramar H, Meijers E and others 2007 City-ranking of European medium-sized cities *Cent. Reg. Sci. Vienna UT* 1–12

Historical Buildings Study in Purwodinatan Semarang as Urban Conservation Guidelines

Rina Kurniati, Nurini Nurini, Diah Intan Kusumo Dewi, Retno Susanti
Department of Urban and Regional Planning, Engineering Faculty
Diponegoro University, Semarang, Indonesia

Corresponding e-mail: adalerina_66@yahoo.com

Abstract. Conservation of historical buildings in an area becomes an attraction to see the past and present sustainability for a region. Purwodinatan urban village as one of the oldest kampong in Semarang city has important historical value for the development of settlement in Semarang city, therefore need to be preserved. This area has an important role with the economic activities that take place in it. The purpose of this study so that historical buildings in Purwodinatan Urban Semarang can be treated, protected and developed as an object of conservation and has value for the local community. The research method used is quantitative research. Analytical techniques using evaluative analysis include weighing and scoring techniques including aesthetics, plurality, scarcity, privileges, historical roles, strengthening of the surrounding area as well as the characteristics of the local community. The results of this study provide academic benefits for the conservation of buildings and the environment of history for a city. And the practical benefit is it can be used as guidance to determine the direction of conservation policy of building and historical environment in Semarang city.

Keywords: historical buildings, historical areas, urban conservation.

1. Introduction

Conservation is a way to keep, hold, manage, and utilize an object whether it is a place or thing to become a flashback of the present and the past [1]. Urban buildings in a region are an attraction for that region. Preserved urban areas which contains historical building have the harmony between the present and the past for that region [2].

Purwodinatan as one of the oldest kampong on Semarang City have an important historical value towards the development of Semarang City. Purwodinatan initially developed from the existence of Chinese immigrants since 1416, they stop by to trade, then eventually settle and establish a settlement [3]. Following the trading activity, the place keep developing then European and Arabic immigrants started to establish a settlement on the region also. This settlement becomes the initial seed for the settlement development on Semarang City. Because of that historical role, then this region's conservation should be kept.

Architectural characteristic of buildings on this region is mostly Middle-east and Chinese style, but most of the buildings have the function changed and no longer maintained. Moreover it also has an important historical role on local and regional scope. Therefore by this study it's expected to bring direction and recommendation towards buildings that should be preserved and protected as an appreciation towards the regions' past history.

Purwodinatan region have many buildings with both tangible and intangible values that should be preserved. Moreover, the pluralism of the people is seen from the buildings that resemble the diversity of the people living on that region. Based on those things that have been fit with the set requirement from UU No. 11 year 2010 about cultural heritage, hence Purwodinatan region is selected as the research region.

The research area is a part of Purwodinatan areas which consist of Kampong Jurnatan, Purwodinatan, Petolongan, Pekojan, Bustaman, Bustaman Gedong, Gedong Mulyo, Kertobangsan, and Pesantren of 33 Ha wide with physical border of road and river. The majority of people on this region work as merchants with shopping complex located along the Jl. H. Agus Salim, Jl. MT. Haryono, and Jl. Pekojan. Here are the borders of the research area:

- North : Jl. H. AgusSalim
- East : Jl. MT Haryono
- South : Jl. Petudungan and Semarang River
- West : Semarang River

Figure 1. Delineated of Study Area

2. Methods

Building and historical environment conservation should be based on motivation and the goal of doing conservation action. Physical conservation methods consist of widely known conservation technique such as initiative awareness, legal basis, concept planning, funding realization and organization. Based on the aspects, conservation can be divided to physical and non-physical conservation. The explanation of each conservation methods can be elaborated as follows:

2.1. Areal Physical Elements

There are many form of direction on physical conservation action for a historical region. Usually the direction forms will be customized with the existing problems and expected to reduce the problems.

2.1.1. Land Use

Historical area can be developed by activate new activities that are still compatible and able to trigger economical growth [4]. The effort to bring new function for old building on historical area can be done as long as not ruining the typical characteristics of that area itself [5]. The direction of land usage should be a usage with functions that are not ruining the historical characteristics of the area.

The forms of that usage are in form of suitable dynamic activities such as development of tourism service sector and museum.

2.1.2. Building

Some activities such as renovation, adaptation, beneficiation, and protection of building face. Conservation with building situation that undergone small to medium change consist of activities such as restoration, renovation, adaptation, beneficiation and protection of building face. Rehabilitation or Demolition with medium level change consists of activities such as restoration, renovation, adaptation, beneficiation, reconstruction, replication, demolition and rebuilding.

2.2. Areal Non-physical Elements

Conservation effort of urban building and environment non-physically needs a set of suitable criteria. Gerds set five criteria derived from the experience of European countries on doing the conservation activity. Those criteria are as follows:

Table 1. Building and Environment Conservation Planning Criteria

No	Conservation Criteria	Implementation
1	Initiative and awareness	Government, private and society motivation
2	Legal base	The existence of policies that regulate conservation actions
3	Planning concept	Organization between involved departments, studies, specialized education to provide experts on the conservation sector, design steps and person in charge
4	Realization and organization	Executor organization, approach system and execution support.
5	Funding	Government subsidy and special programs

Source: Gerd, 1975

2.3. Conservation Properness Assessment

On conservation properness assessment, then the building elements based on high, medium and low potential classification. Every element on a particular potential level then directed to further conservation actions. Direction forms focused on physical actions. Direction of physical actions on building classified into four classes, which are preservation, conservation, rehabilitation and reconstruction.

Table 2. Potential Building Element Classification Table

Potential Building Element Classification	Physical Conservation Direction	Allowed Physical Change Level
High Potential	Conservation	Very Small
	Restoration	Small
Medium Potential	Conservation	Small
Low Potential	Rehabilitation	Big

Source:Gerd, 1975

Table 3. Building Assessment Table

No.	Assessment	Note
1.	Score< 10	Low Potential
2.	Score 11-15	Medium Potential
3.	Score>16	High Potential

Source:Gerd, 1975

High Potential preservative or restorative treatment to bring back the original form of building or building elements that have high value and have changed a lot. The return of those elements advised to usethe same or similar material to produce building value as the original situation.

Medium Potential conservation effort done to preserve the building situation and direct its development course. Conservation effort also connected with building conservation effort that can be done by routine or periodic maintenance.

Low Potential rehabilitation effort can be done by replacement of broken parts to be able to perform again. On rehabilitation effort the low-potential building element are possible to be added with new elements customized with the building function.

3. Discussion

3.1. Urban Conservation

Conservation expert it has a series of definition with various implications. The term conservation refers to the charter of International Council of Monuments and Site (ICOMOS) on 1981 also known as the Burra Charter. Conservation is a place or object management process concept to preserve the cultural meaning within it. Conservation activity including the whole maintenance activity that fits with local situation and development effort for further utilization. If associated with region then sub urban or region conservation includes a preventive action for social change activity or inappropriate utilization and not just physical efforts.

The existence of historical places gave the feeling connection about the past and share life experience to the people also gave identity for that region. Hence, places with historical value need to get conserved. Conservation on urban scope not only refers to buildings and physical aspects but also on people's activity which is socio cultural activities. Moreover, through conservation it's also expected to bring another benefit such as economical increase for the residents and society. Thus, conservation can't be separated with efforts to bring back the face of the preserved object.

3.2. City Conservation Form

Dynamic conservation concept not only keeping the originality and maintenance of an object, but also able to add economical value or another benefit for the owner or society. This activity requires cross-sectoral, multi-dimensional and discipline and also sustainable effort. Conservation as a way to add future cultural heritage to be remembered by the next generation. The forms of conservation, is: (a). Restoration. (b). Preservation. (c). Conservation. (d). Reconstruction. (e). Consolidation. (f). Revitalization. (g). Reparation.

3.3. Conservation Steps

There are some steps on a conservation activity, those steps can be seen on Burra Charter that explain the sequences of doing conservation as a conservation plan which consist of: a. Step 1: Starting Cultural Significance, is an effort to understand and assess the cultural meaning of the building and the place value with certain assessment criteria i.e. attractiveness, historical and scientific value, or demonstrative, associational relationship, formal quality and aesthetical value; and b. Step 2: Conservation Policy, is the search for the best methods to keep the values on the building usage and future development (Kerr, 1982).

On Indonesian Constitution No. 5 year 1992 about Cultural Heritage Object, which is a man-made object, static or non-static, as a unity or group, or the parts or the rest of it, aged minimum 50 (fifty) years, or represent the typical era style and represent the era style minimum 50 (fifty) years, also considered as having an important value for history, science, and culture. Moreover, assessing a historical region or building can be done by research based on Catanese [6] building criteria.

Table 4. Cultural Heritage Region/Building Assessment [5,6,7,8]

No	Criteria	Definition	Benchmark
1.	Aesthetics	Associated with the building's aesthetical and architectural change (building style, building facade/covers, ornament/element, and material)	The change of building style, roof, facade/cover, ornament/element, and also building material and structure
2.	Exceptionality	Have a characteristic that can be represented by age, size, building shape and other factors	The existence role is able to increase the quality and also building's character and image
3.	Historical Role	Associated with history of either the region or the building itself	Associated with historical event as symbolical ties of past and present event
4.	Scarcity	Building shape, style, and elements and the use of different ornament that not exist on other building	Is a rare building and not exist on another place
5.	Building Characteristics	Have important role on the creation of building's character	Have typical characteristics such as building age, size/wide, etc
6.	Strengthen the region image	Have important role on the regional character creation	The existence role can fit with regional function and increase the quality and character of the region

3.4. History of Pekojan to Chinatown Region

The more Tionghoa people settle on Pekojan to Chinatown region makes them build temples. After a temple was build on 1771 they planned on making another temple on the riverside near the chili farm. To build that temple a significant amount of money has been accumulated, the workers are being brought from China. After more than a year, finally the temple have been built, then named Tay Kak Sie, or widely known as big temple. [9]

Figure 2. Tay Kak Sie Temple, 1910 [10]

On 1770s, Semarang City becomes wider, the native kampong that now known as Tudungan become more crowded. This is because Tudungan area becomes the passing way for people from east Semarang. The habit of the people at the time, if natives comes out of their house most of them bring Keris and using tudung (hood). Even for the natives and Tionghoa these things are considered as luxurious items, so the place named **Tudungan** or **Petudunga**n.

Figure 3. Petudungan Area [11]

Figure 4. Pekojan Area [11]

The prosperity of Tudungan area brings impact to the kampong next to pesantren kampong (located besides Tudungan) resided by Moor people or at the time called as Kojan people. Hence, the kampong is named Pekojan.

On 1811, news spread about the arrival of Britain on Semarang and predicted there will be a big war broke on Semarang. Hence, some of Tionghoa people at the time escaped to another area. After the situation have become quite safe, each of them go back home. On 1816, Semarang region once more ruled by Netherland. But on 1821 the Netherland government make a pass policy or passenger letter called passenstelsel that requires someone who wants to go to another place must bring the passenger letter. This makes the mobility of Tionghoa people become limited, while their activity is travel for trade.

Not long after the passenstelsel, released the wijkstelsel constitution which set that Tionghoa people should resides at their environment. Before the stelsel, the living place of Tionghoa people are scattered but then they live in a place set by the Netherland government which is Chinatown that still be known until now. But on 1906 and 1915, passenstelsel and wijkstelsel consecutively revoked by Netherland Government.

On 1835, Javanese-Tionghoa trade increasingly develops until the arrival of Tionghoa merchant. Tionghoa people already existed on Semarang created a congress with the office stood next to Tay Kak Sie Temple, which at the time still a chili farm. Therefore, that area is named Gang Lombok (Chili Alley) and the Tay Kak Sie Temple also called Klenteng Gang Lombok (Chili Alley Temple). Gradually, the congress function that at first only as the direction place for Tionghoa merchants that just arrived, developed into a place to arrange light weight case for Tionghoa people, such as birth, death, marriage, business, organization and special tax registration for Tionghoa people.

On 1876, Tudungan and Koja area that heading to Lloji (Netherlanders place at the time) have a cemetery for Tionghoa people that believed to be the oldest cemetery on Semarang. At the time, Netherland government planned to build a railway that passed through the area. Therefore, Netherland government agreed with Tionghoa people to relocate the cemetery, so the railway can be built.

3.5. Activity Analysis

Daily activities done by the people of conservation region mostly are trading and praying. This is because there are place of worships in the form of Tay Kak Sie Temple and Pekojan Mosque, also shopping complex corridor such as printing house and household material trade. To be more clearly, will be discussed into 2 parts that is non-conservative and conservative activities:

3.5.1. Non-Conservative Activities

Non-conservative activities are activities that commonly happened on various regions, and don't need to be preserved because it commonly happened everywhere. On Purwadinatan conservation region, there are 2 major non-conservative activities, namely praying and trading. The praying activity on the conservation region focused on Tay Kak Sie Temple, with Tri Dharma religion, and also focused on Pekojan Mosque with Islam religion.

Figure 5. Praying Activities on Tay Kak Sie Temple [13]

Figure 6. Praying Activities on Pekojan Mosque [14]

Trading activities done along the Jl. K.H. AgusSalim and Jl. Pekojan. The traded goods are various, including foods to building materials, household, stationery, printing, etc.

Figure 7. Non-conservative Activities on Purwodinatan Region

3.5.2. Conservative Activities

Conservative activities are activities worth to be conserved. Because of its typical and scarcity element. Conservative activities at this Purwodinatan conservation region are in form of culture/important commemoration that usually done on this region such as:

a. Medicine God Commemoration Parade

This commemoration also known as *Jut-Bio Kongco Poo Seng Tay Tee,* held annually every 1st day of imlek month. Medicine God's dooly paraded from the temple to Kranggan, followed by liong and barongsai performance heading to Marina Beach Semarang [14]. This commemoration believed to remove disease and illness. This commemoration need to be preserved that includes the protection of physical building of Tay Kak Sie Temple, and non-physical protection for this *Jut-Bio* commemoration activity. This conservation that involves physical and non-physical activities is called **dynamic conservation**.

Figure 8. Medicine God Parade on TayKakSie Temple [14]

Figure 9. Commemoration of Admiral Cheng Ho's Arrival [15]

b. ***Kongco Sam Poo Tay Djien* Arrival Commemoration**
This *Kongco Sam Poo Tay Djien* arrival commemoration event held to commemorate the arrival of Admiral Cheng Ho to Semarang as one of the figure on the history of Semarang City. This commemoration shows barongsai and liong show, and Gods' worshipping ritual on the altar of Tay Kak Sie Temple.[15]. This belong to dynamic conservation, with the protection on the Temple and the commemorations within it.

c. **Trading Activities near Pekojan Mosque**
Pekojan Mosque used to be inhabited by Indian and Pakistani people that then called as Kojan people to shift and spread Islamic religion. The arrival of these people shaped the Pekojan kampong and doing trading activities that still exist until now, trading prayer beads, gemstones, and typical Kojan cloth as the hereditary activities since their arrival [16]. These activities can be conserved because existed near the Pekojan Mosque as the cultural heritage building, hence belong to dynamic conservation.

d. **The Culture of Distributing Indian Porridge near Breakfast.**
This unique culture introduced on PekojanJami Mosque at Ramadan, that heredited by Kojan People since long [16]. This unique culture also belong to dynamic conservation.

Figure 10. Distribution of Indian porridge at Pekojan mosque [18]

Table 5. Evaluative Analysis of Buildings in Chinatown

Building Name	Photo	Building Age	Building History	Evaluative Analysis (Quality)					
				Aesthetics	Plainness	Scarcity	Speciality	Historical Role	Strengthen Nearby Area
Tay Kak Sie Temple		245 years	The building used as worship place since it's built until now	The architecture on this building have typical Chinese ornaments	The building don't have a common ornament, hence the plainness on this building don't exist.	the building architectural differ from other buildings nearby	The building not dominating on nearby environment and the building architectural differ from other buildings nearby	Since built the temple already used as worship place, up until now still used as worship place and religious tourism	Elementsonbuilding as a whole quite creating Chinatown feel by Chinese ornament architectural
Score				1 2 **3**	**1** 2 3	1 2 **3**	1 2 **3**	1 2 **3**	1 2 **3**
Score Total				16					
Masjid Jami' Pekojan		167 tahun	This building used as a place of worship from the beginning built up until now, but originally a surau	The Architecture in this building have ornaments typical of India and Java	The building don't have a common ornament, hence the plainness on this building don't exist.	the building architectural differ from other buildings nearby	The building not dominating on nearby environment and the building architectural differ from other buildings nearby	Since built the temple already used as worship place, up until now still used as worship place and religious tourism	Elements on the building as a whole create enough atmosphere Pecian with architectural ornaments India
Score				1 **2** 3	**1** 2 3	1 2 **3**	1 2 **3**	1 2 **3**	1 2 **3**
Score Total				16					

IOP Conf. Series: Earth and Environmental Science **213** (2018) 012036 doi:10.1088/1755-1315/213/1/012036

....Continued Table 5

Building Name	Photo	Building Age	Building History	Evaluative Analysis (Quality)					
				Aesthetics	Plainness	Scarcity	Speciality	Historical Role	Strengthen Nearby Area
Building 1			The building is used as a house from the beginning built to date	The architecture on this building have typical Colonial style	The building don't have a common ornament, because of this the plainness on this building don't exist.	the building architectural differ from other buildings nearby	The building not dominating on nearby environment and the building architectural differ from other buildings nearby	Since the beginning of the building has been used as a residence until now still used as a residence	Elements on the building as a whole less create a pecian atmosphere with architectural colonial
Score				1 2	3 1 2	3 1 2	3 1 2 3	1 2 3	1 2 3
Score Total				10					
Building 2			The building was originally used for shelter but now it is just an empty building	The architectur e on this building have typical Chinese ornaments	The building don't have a common ornament, because of this the plainness on this building don't exist.	the building architectural differ from other buildings nearby	The building not dominating on nearby environment and the building architectural differ from other buildings nearby	Since the beginning of the building is used as a place of worship, but at this time only a vacant building	Elements on the building as a whole create enough atmosphere pecian with chinese ornamental architecture
Score				1 2	3 1 2	3 1 2	3 1 2 3	1 2 3	1 2 3
Score Total				12					
Building 3			The building was originally used for shelter but now it is just an empty building	The architecture on this building have typical Chinese ornaments	The building don't have a common ornament, because of this the plainness on this building don't exist.	the building architectural differ from other buildings nearby	The building not dominating on nearby environment and the building architectural differ from other buildings nearby	Since the beginning of the building is used as a place of worship, but at this time only a vacant building	Elements on the building as a whole create enough atmosphere pecian with chinese ornamental architecture

3.5.3. Urban Conservation Direction

SWOT analysis can be implemented by analyzing and sorting out many things that affect the four factors, then implement it on SWOT matrix, where the application is how the strength can take advantage from existing opportunities, the next is how the strengths can deal with existing threats, and the last is how to overcome the weaknesses that are able to realize the threats or creating new ones. The goal of SWOT analysis it as an instrument to determine the potentials and problems of Chinatown area and knowing which the area to protect or change. (Table. 4)

Table 6. SWOT Analysis

	Strength	**Weakness**
	1. Cross-ethnicity relationship harmonization formed 2. Have cultural heritage worship place in form of Tay Kak Sie temple 1771 and 250 years old Pekojan Mosque 3. Have strong history about Chinatown and Pekojan area 4. Have historical role for Semarang City	1. Don't have the supportive infrastructure as tourism area 2. The unclear cultural heritage status establishment, creating the frequent situation of conservation region threatened to be damaged or destroyed by human actions
Opportunity 1. Located on Semarang city that have the image of city with high attraction rate 2. Located near the old city region that are more developed and be the best practice/reference on tourism area development	Using power to utilize the opportunity and otherwise utilizing opportunity and make it as strength.	Opportunity used to suppress various existing weaknesses or in other words eliminate weaknesses with opportunity
Threats 1. Conflict about utilization and maintenance of historical buildings.	Using power to avoid threats.	The threat should be able to cover the existing weaknesses on itself with strength and opportunity

Source: Analysis Result 2017

Buildings on Purwodinatan areas have the medium potential to be conserved related with gradual or frequent maintenance effort.

Table 7. Potential Buildings to be Conserved

No	Photo	Physical Conservation Direction
1.	 Tay Kak Sie Temple	Conservation efforts that can be done on *Tay Kak Sie* Temple is small-scale conservation as building maintenance and slow down the weathering process since the building have old age. By this conservation effort the the building can last long because *Tay Kak Sie* building have history with urban kampong and Semarang city and as the strengthener for regional image.

2.	 Pekojan Jami Mosque	Conservation efforts that can be done on Pekojan Jami Mosque is small-scale conservation by building maintenance to slow down the damage for the building. Pekojan Jami Mosque have the building style that influenced by the existence of Kojan people and this building contains history thus need conservation.

4. Conclusion

The conclusion of Purwodinatan building conservation study is as follows:

a. Purwodinatan is a kampong that have characteristics that is cultural diversity by Tay Kak Sie Temple and Pekojan Jami Mosque.
b. Settlement density on Chinatown and Pekojan region also the habit of drying clothes in front of the house reduce the aesthetical value of the region.
c. There are some buildings that can be conserved with medium and high potential. Furthermore there are cultures that need to be conserved such as God's commemorations and distributing Indian porridge.
d. Purwodinatan is worthy and able to be a conservation region in terms of the role of Purwodinatan area towards Semarang Regency and City, ad religious bond associated with Tionghoa people.

4.1. Physical Urban Conservation Guidelines

Physical urban conservation guidelines offered for Chinatown and Pekojan region conservation are:

a. Doing conservation studies towards Chinatown and Pekojan region Semarang that can be used as a guidance to set the conservation actions that will be done.
b. Socialization or counseling to the people about the importance of cultural heritage building and the regulation and cultural heritage protection policy to make the people care about the existence of the region as cultural heritage.
c. The government are expected to provide the budget as compensation for historical/ancient building owners in form of incentives and disincentives.
d. Forming a data collection and ancient/historical building construction institution to fulfill the conservation principles, to be able to do renovation supervision done by owners so it won't cause the change in building shape that can eliminate the originality of the building

4.2. Non-physical Urban Conservation Guidelines

As for the non-physical urban conservation guidelines offered for Chinatown and Pekojan region conservation are:

a. Government Role
Creating conservation policies and execute the cultural heritage regulations on Chinatown and Pekojan region Semarang.
b. Academics Role
Give input to the related stakeholders based on the study done on cultural heritage region conservation.

5. Acknowledgments

The authors strongly acknowledge the historical buildings study in Purwodinatan Semarang especially members of the urban conservation team 2017 who have helped provide data and assessment in this study. The authors are also grateful to Urban and Regional Planning Department of Diponegoro University and all parties who have assisted the completion of this research.

6. References

[1] Tanselle, G. Thomas. 1998. Literature and Artifacts, Charlottesville, VA: Bibliographical Society of the University of Virginia, *ISBN 1-883631-06-8, OCLC 39223648*

[2] Larkham, Peter. 1996. Conservation and the City. Routledge. New York.

[3] Joe, Liem Thian, 1933, Riwayat Semarang dari Djamannya Sam Poo sampai Terhapoesnya Kongkoan, boekhandel Ho Kim Yoe, Semarang.

[4] Shirvani, Hamid. 1985. The Urban Design Process. Van Nostrand Reinhold: New York]

[5] Budihardjo, E. 1985. Arsitektur dan Pembangunan Kota di Indonesia. Bandung: Alumni.

[6] Catanese, Anthony J. Snyder, James C., 1989. Second Edition. Urlan Planning. Erlangga. Jakarta.

[7] Nurmala. 2003. Panduan Pelestarian Bangunan Bersejarah di Kawasan Pecinan-Pasar Baru Bandung. *Tesis.* Tidak dipublikasikan. Bandung: ITB

[8] Hastijanti, R. 2008. Analisis Penilaian Bangunan Cagar Budaya http://saujana17.wordpress.com/2008/analisis-penilaian-bangunan-cagar-budaya .html. Accessed on November 5, 2017

[9] Widodo, J. 1988. Chinese Settlement Changing City an Achitectural Study Of Urban Chinese Settlement in Semarang. Lauven University, Belgium.

[10] Semarang.nl

[11] Google street view

[12]Surya, Johan. 2015. Event Tahunan–Memperingati Kedatangan Sam Po Tay Djien. www.anakkopi.id. Accessed on Sunday, November 5, 2017

[13] Adikurnia, Muhammad. 2016. Menengok Sisa-Sisa Peninggalan Salah Satu Masjid Tertua di Semarang. www.travel.kompas.com. Accessed on Sunday, November 5, 2017.

[14] Radar Semarang. 2016. Jut Bio Dipercaya Hilangkan Wabah. www.radarsemarang.com. Accessed on Sunday, November 5, 2017.

[15] Loo, Petrus. 2017. Meriahnya Peringatan 612 Tahun Kedatangan Laksamana Cheng Ho (Sam Po Tay Djien) di Semarang. www.klayapan.com. Accessed on Sunday, November 5, 2017.

[16] Viva. 2016. Masjid Pekojan dan Kisah Kampung India di Semarang. www.viva.co.id. Accessed on Sunday, November 5, 2017.

[17] Solo Pos. 2015. Bubur India Tersedia di Masjid Pekojan. Dalam www.solopos.com. Accessed on Sunday, November 5, 2017.

Vertical Greenery System as the Passive Design Strategy for Mitigating Urban Heat Island in Tropical Area: A Comparative Field Measurement Between Green Facade and Green Wall

Ratih Widiastuti[1], Chely N. Bramiana[1], Bangun I.R.H[2], Bintang N. Prabowo[1] Mirza Ramandhika[1].
[1]Architectural Study Program, Vocational School, Diponegoro University, Semarang, Indonesia
[2]Department of Architecture, Faculty of Engineering, Diponegoro University, Semarang, Indonesia

Corresponding e-mail: ratihw@arsitektur.undip.ac.id

Abstract. During the last decade, greenery aspect in the building envelope starts to increase their presence in the building design. Vertical greenery system became the one of alternatives that potential to energy-saving in the building. However, in tropical area, researches related to vertical greenery system are enigmatic since reducing heat in tropical area are crutial for maintaining space in comfort condition. In this study, a comparative thermal analysis between green facade and green wall will present. A model called house miniature set up with climbing plans as model of green facade. While for green wall, data measurement was conducted in the office building with green wall as building envelope. The results showed a high potential for energy savings during peak time in the afternoon for green wall (1.5%) and green facade (1.3%) in comparison to the bare wall. While for interior facade are 2.6% for green wall and 2.2% for green facade. Effect of vertical greenery also found in the outdoor air temperature around greenery system. Results show at the hottest time of the day, the green wall and green facade reduced outdoor temperatures by 3.0°C and 1.2°C, respectively. Based on the analysis, compared to green facade, the cooling effect produced by green wall is faster. Possibility, the substrate properties in the green wall also help to decrease temperature. The reduction can be due to the foliage density as the leaves of green wall is denser than leaves in green facade.

Keywords : building envelope, green facade, green wall, vertical greenery system

1. Introduction

Over the past decade, due to population growth, energy consumption has dramatically increased. More than half of human life is spent in building and cause the increasing demand for building functions and indoor environmental quality. This increases the energy consumption and becomes the factor of global warming.

Based on data from European Commission, buildings are responsible for approximately 40% of energy consumption and 36% of CO^2 emissions in the EU [1]. While according to Energy Information Administration (EIA) [2], the global totals of primary energy consumption and CO^2 emissions have grown by 85% and 75% from 1980 to 2012, with average annual increases of 2% and 1.7%. In this case, heating and cooling loads become the largest energy use. On the other hand, significant energy saving can be achieved in building if they are properly designed, constructed and operated. This condition then encourages building planners to create sustainable design in the building.

Among other innovative technologies to improve energy saving in the building, vertical greenery systems are considered as design trends in the field of architecture and construction that increase building thermal performance. Many building planners used vertical greenery sytems to enhance the quality of urban landscapes around buildings. Within this concept can provides multiple eco-system services in the building and urban scale and in the end resulting in mitigating urban heat island.

Research conducted by Perini et al., [3] devided vertical greenery system in the two terms :

- First, green façade, when the systems use climber plan as greenery element. The plants root directly in the wall or using wire or terallis as supporting element.
- Second, green wall, when the systems use small schrubs such as mosses and grass as greenery element. For applying in the building envelope, this system uses planter element and also needs special treatment for watering. Building planners can make special pattern by arranging the vegetations based on the colour.

Basically, vertical greenery systems have four key factors that influence the system as passive cooling design for energy saving : the shadow produced by vegetation; the insulation provided by vegetations and substrats; the evaporative cooling through evapotranspiration and barrier effect to wind [4]. From the previous studies, shadow effect produced by vertical greenery system has the highest impact in the reduction of building wall temperature and resulting in saving energy consumption [5]. Especially when the vegetation has higher percentage of leaves densities [5][6][7].

To arrange and summarize all the key factors, **Table 1** and **Table 2** showed the previous studies related to vertical greenery system in tropical area. There are nine studies were found, six related to green facade and three for green wall.

This research relates specifically to the comparison between thermal performance of green facade and green wall. In this regard, it must be taken into account that the each of study objects clearly has different design and different vegetation. Results from this study are expected can help architect and building planners to make more appropriate decicions in the bilding design.

IOP Conf. Series: Earth and Environmental Science **213** (2018) 012037 doi:10.1088/1755-1315/213/1/012037

Table 1. Most significant previous studies on the use of green facade as passive design cooling in building in tropical countries.

Authors	Publication year	Location	Method	Period of observation	Plants species	Building storey and material	Control treatment	Surface temperature reduction (°C)	
								Interior	Exterior
Laopanitchakul et al. [6]	2008	Thailand	Experimental	Summer	*Mann bali* (local name)	One storey concrete	Bare wall	Depending on the systems	Depending on the systems
Wong et al. [7]	2010	Singapore	Experimental	February; April; June	Climber plants	One storey concrete	Bare wall	-	4.36
Rashid et al. [8]	2010	Bangladesh	Observation	-	-	Six storey	-	-	-
Sunakorn and Yimprayoon [9]	2011	Thailand	Experimental	Summer	*Thunbergia grandiflora*	One storey concrete	Bare wall	-	-
Rahman et al. [10]	2011	Malaysia	Experimental	Summer	*Psophocarpus tetrogonobulus*	Multiple storey brick.	Bare wall	0.38-1.28	3.97-10.7
Safikhani et al. [11]	2014	Malaysia	Experimental	April; Mei; June	*Thunbergia grandiflora*	One storey plywood	Bare wall	4.5	-

Table 2. Most significant previous studies on the use of green wall as passive design cooling in building in tropical countries.

Authors	Publication year	Location	Method	Period of observation	Plants species	Building storey and material	Control treatment	Surface temperature reduction	
								Interior	Exterior
Wong et al. [7]	2010	Singapore	Experimental	February; April; June	M3: *Hemigraphis repanda*; M6: *Phyllanthusmyrtifolius*; *Tradescantia spathacea*; M7: Mosses	One storey concrete	Bare wall	-	Depending on the systems
Widiastuti et al. [12]	2016	Indonesia	Observation	October	*Phalaenopsis sp*., *Dracaena warneckii* and local climber plants	Three storey brick	Bare wall	2.1	-
Safikhani et al. [11]	2014	Malaysia	Experimental	April; Mei; June	*Thunbergia grandiflora*	One storey plywood	Bare wall	6.5	-

Note : M Model

IOP Conf. Series: Earth and Environmental Science **213** (2018) 012037 doi:10.1088/1755-1315/213/1/012037

2. Methodology and Materials

2.1. Climatic condition

The location of experimental site is in Semarang City, Indonesia, can be seen in **Figure 1.**, at south latitude 6°58'0.0012" and east longitude 110°24'59.9904" under tropical monsoon climatic condition defined according to Koppen-Geiger climate classification [13] as the climate that has a driest month which nearly always occurs at or soon after the "winter" solstice for that side of the equator with rainfall less than 60 mm, but more than 1/25 the total annual precipitation.

2.2. Experimental set-up

In order to get green facade data, a building miniature called house miniature has been built in Architecture Department of Universitas Diponegoro to carry out the experiment, can be seen in **Figure 1**. The dimension is 1m x 1m x 1m, and can be considered as one real scale building, illustrated one storey building. The comparison scale between model and real building is 1:4. The fondation is made from rainforced concrete slabe and in the base of building is equipped with rail from steel that the model can be spin 360°. The roof constuction is gable roof with asbestos as roof covering material. The length of overhang is 50 cm, possibility can create shadow in the building walls. While for the walls are constructed from brick and have inlet oulet with pososity 30%.

a b

Figure 1. Study object. a). House miniature; b). Detail of house miniature

A green facade is installed on the east facade. Climber plant attached directly in the facade used mesh trallis, while the plant roots was planted in the planter boxes, can be seen in **Figure. 2**. The facade covered by green facade orientated to face the east according to sunlight direction. *Passiflora flavicarva* and *Pseudocalymma alliaceum* are selected as greenery element because it is ease to climb and presents well adaptation to micro climatic condition, can be seen in **Figure. 3**. As data comparison for green facade is the exsisting model without green facade.

Figure 2. Model of green facade

a b

Figure 3. Climbing plants for green facade application. a). *Pseudocalymma alliaceum*; b). *Passiflora flavicarva*

On the other hand, an *in situ* measurement was done in the office building of PT. Pertamina to get thermal data of green wall. This green wall previously was used as study object in the previous research [12]. The green wall was using geotextile system that consists of an aluminium structure, a PVC panel and felt layers. The plants are growing in the plant pockets with automatic watering system from irrigation pipes in the top and side of green wall.

The green wall is installed on the external non insulated wall. While the wall construction is brick and the thickness is 15 cm including internal and external plasters. The green wall is using various types of plants as greenery element such as *Phalaenopsis sp.*, *Dracaena warneckii* and some of local climber plants, can be seen in **Figure 4**. Bare wall for data comparison of green wall is a wall above the green wall.

Figure 4. Model of green wall

2.3. Data measurement

Both of green facade and green wall are using bare wall as control treatment. Each of data measurement from vertical greenery system will compare with data measurement from bare wall and analyze the difference of thermal profile.

The experiment consisted of three data measurements :

- Measurement of surface temperature on the interior and exterior facade.
- Measurement of outdoor air temperature near vertical greenery system.
- Measurement of outdoor air humidity near vertical greenery system.

Data measurement was done in October from 05.00 AM to 06.00 PM. The interval of data collecting are every 1 hour. For data measurement of surfaces temperature, the surface of measured wall devided into 9 point. They are three in the mindle, three in the bottom and three in top. The analysis was based on the average of hourly data measurement from each measured points.

3. Discussion

Because there are limited of test room with bare wall, present of data measurement between green facade and green wall are separately. Both vertical greenery systems has different models size, with the result that, it is invalid to compare humidity and air temperature inside test room. The analysis only restricted in comparison of surface temperature and outdoor humidity and air temperature near the vertical greenery systems.

3.1. Surface temperature

3.1.1. Green wall

Accoding to data measurement, surface temperature of green wall is lower than surface temperature bare wall. Respectively, the difference between the average of surface temperature are 3.17°C for exterior facade and 5.38°C for interior facade. In fact, exterior surface temperature of green wall is lower than exterior and interior of bare wall.

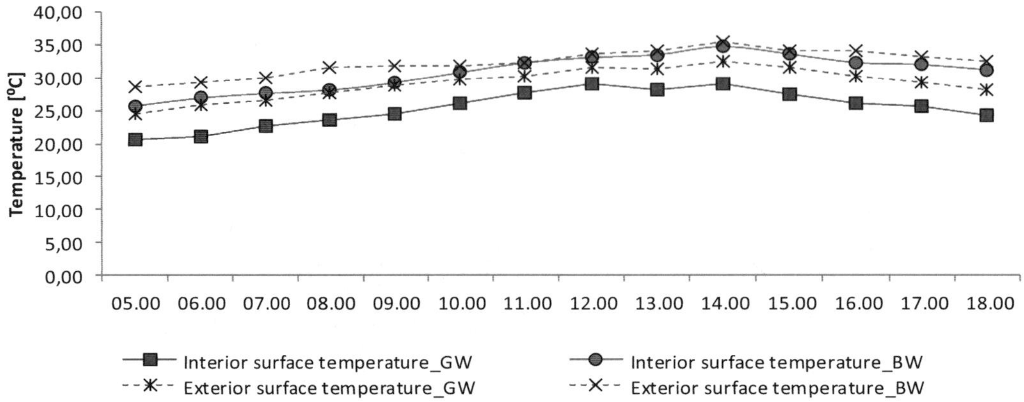

Figure 5. Surface temperature profile of green wall

3.1.2. Green facade

It is noticeble that surface temperature in the bare wall model is considerably higher than surface temperature in green facade. In general, the average of surface temperature difference between bare wall and green facade are 4.4°C for exterior facade and 2.9°C for interior facade. The lowest difference between interior surface temperatures and exterior surface temperatures, respectively are 1.3°C at 1:00 PM and 1.5°C at 10:00 AM.

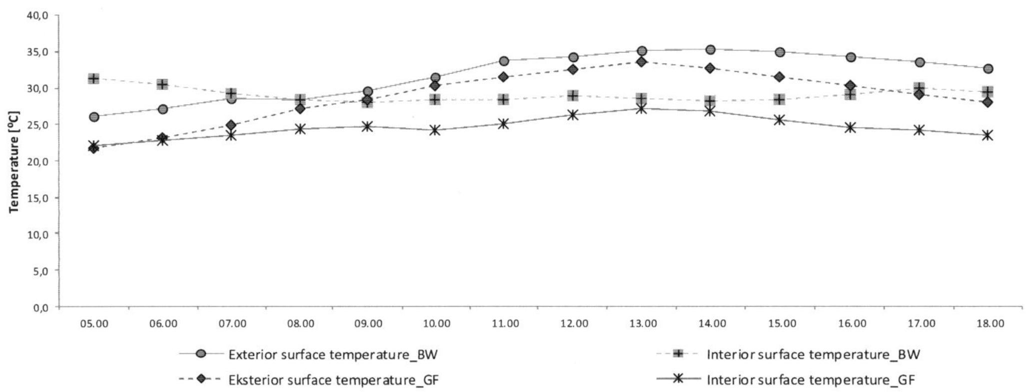

Figure 6. Surface temperature profile of green facade

3.2. Analysis surface temperature

In the analysis the difference on the thermal performance between VGS and bare wall were not analysis.

The results showed that both the green facade and green wall reduced the temperatures especially in the afternoon. During the morning, direct solar radiation fell on the east part of the facade. Plants reduced temperatures using natural evapotranspiration and cooling effects.

In the afternoon when outside temperatures are at their highest, vertical greenery systems have a significant effect on temperature reduction. During the afternoon when direct solar radiation hits the plants on the west side, cooling effect and the shade provided by the vertical greenery systems effectively reduce temperature.

The results showed a high potential for energy savings during peak time in the afternoon are 1.5% for green wall and 1.3% for green facade in comparison exterior facade to the bare wall. While for interior facade are 2.6% for green wall and 2.2% for green facade.

At that time, green wall can reduce temperature better than the green facade,especially in the afternoon until evening. The reduction can be due to the foliage density as the leaves of green wall is denser than leaves in green facade. In addition, the substrate properties also help to decrease temperature. In green facade, the cooling effect is directly due to the shading effect and evapotranspiration from the leaves of the climbing plants. There is no substrate to contribute to the cooling effect due to evaporation of moisture from the substrate as in the green wall.

3.3. Air temperature

From **Figure 7.** of data measurement on the trends of the ambient temperature, the aim are to observe the potential of vertical greenery system in producing a cooling effect on the immediate external environment.

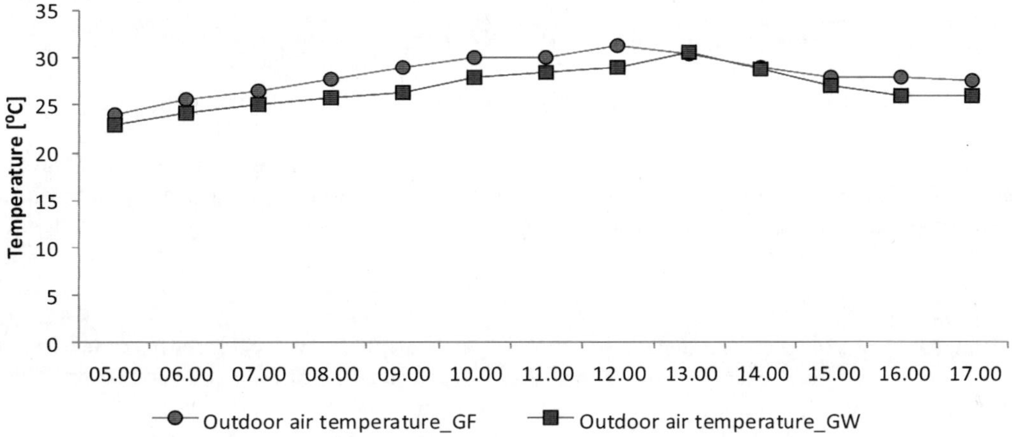

Figure 7. Profile of outdoor air temperatures near green facade and green wall

At the hottest time of the day, the green wall and green facade reduced outdoor temperatures by 3.0°C and 1.2°C, respectively. Possibility, the ambient temperature maybe affected by air circulation, yet the factor not always occured during data measurement. Though green facade and green wall are covered by well-distributed greenery, the green wall has thicker greenery near that may block the air circulation and trap heat. The substrate surface of the green wall also had a significant effect on temperature reduction around greenery application.

3.4. Air humidity

Although installing vertical greenery systems have a positive effect on indoor temperatures in hot and humid climates, they have a negative effect on humidity control.

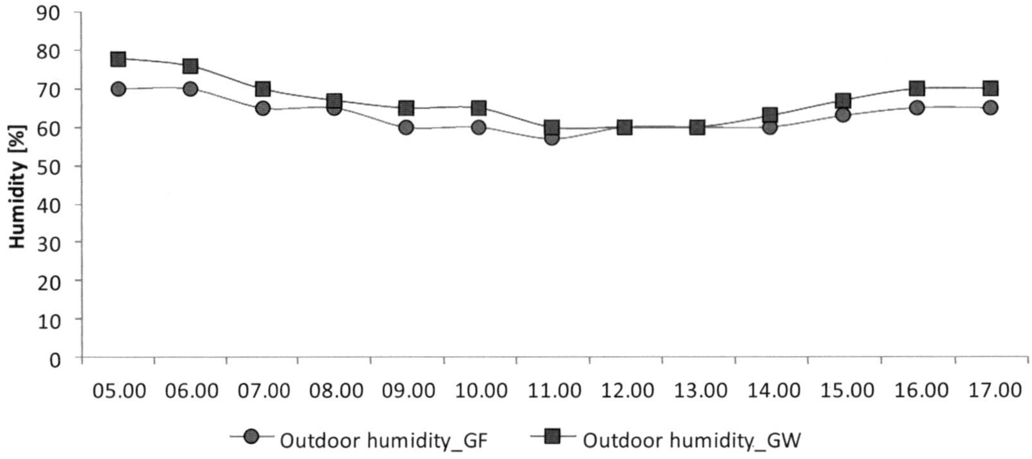

Figure 8. Profile of outdoor humidities near green facade and green wall

The green wall increased humidity around facade more than the green facade because the leaves densities and substrat in the green wall interfere with evaporation resulting the increase of humidity. Calculation on the average of humidity around the greenery systems are 67.2% for green wall and 63.% for green facade.

4. Conclusion

As continuation of a long term research in order to study the potential of vertical greenery systems as a passive cooling design for energy saving in the buildings in tropical countries, two in situ observation have been down to collect the data of vertical greenery system.

The effects of vertical greenery system on the building thermal profiles are depend on the specific parameters of vertical greenery system. From the results obtained in the data measurement, it is found that green wall can produce cooling effect better than green facade. The reduction can be due to the foliage density as the leaves of green wall is denser than leaves in green facade. Substrate in the green wall also contribute in the cooling effect. In the green facade, the cooling effect is directly due to the shading effect and evapotranspiration from the leaves of the climbing plants.

The temperature reduction is most prominent around noon when it is the hottest weather. This is a significant reduction in surface and outdoor air temperature that will lead to a corresponding reduction in the energy cooling load. A cooler ambient temperature means that the air intakes of air-conditioning are at a lower temperature, translating into saving in energy cooling load.

5. Acknowledgement

We gratefully thank to PT. Pertamina Indonesia and Architecture Department, Faculty of Engineering, Universitas Diponegoro for giving the licence to conduct this research using vertical garden application.

6. References

[1] European Commission. Energy efficiency building.
[2] U.S.E.I. Administration, International Energy Outlook 2013, in, 2013
[3] Perini, K., Ottele, M., Haas, E.M., Rossana, R. R 2012 *Vertical gardens, a process tree for green facades and living walls*. Urban Ecosyst

[4] G. Perez, L. Rincon, A. Vila, J.M. Gonzalez, L.F. Cabeza 2011 *Green vertical systems for buildings as passive systems for energy savings*. Applied Energy, 2011. **88** 4854-4859
[5] G. Perez, J. Coma, I. Martorell, L.F. Cabeza. 2014. *Vertical Greenery Systems (VGS) for energy saving in buildings: a review*. Renew. Sustain. Energy Rev., 2014. **39** 139-165
[6] Laopanitchakul, V., Sunarkorn, P., Srisutapat, A 2008 *Climbing-Plant on Solid Wall for Reducing Energy in Tropical Climate*. Sustainable Building Conference 08, Soul, Korea
[7] Wong, N.H., Tan, A.Y.K., Chen, Y., Sekar, K., Tan, P.Y., Chan, D., Chiang, K., Wong, N.C 2010 *Thermal evaluation of vertical greenery systems for building walls*. Building and Environment, 2010 **45** 663-672
[8] Rashid, R., Hamdan, M. bin Ahmed, Sayem Khan, Md 2010 *Natural Green Application technology on Building in Dense Dhaka City is provide a sustainable, energy saving, comfortable and healthy environment*. Annual Asian Business Research Conference, BIAM Dhaka Bangladesh
[9] Sunakorn, P., Yimprayoon, C 2011 *Thermal Performance of Bio Facade with Natural Ventilation in the Tropical Climate*. Procedia Engineering **21** 34–41
[10] Rahman, A.M.A., Yeok, F.S., Amir, A.F 2011 *The Building Thermal Performance and Carbon Sequestration Evaluation for Psophocarpus tetrogonobulus on Biofaçade Wall in the Tropical Environment*. World Academy of Science, Engineering and Technology **52**
[11] Safikhani, T., Abdullah, A.M., Ossen, D.R., Baharvand, M 2014 *Thermal Impacts of Vertical Greenery Systems*. Environmental and Climate Technologies
[12] Widiastuti, R., Prianto, E., Budi, W.S 2016 *Evaluation the interior façades performance of vertical garden*. International Journal of Architecture, Engineering and Construction. **51** 13-20
[13] Kottek, M., Grieser, J., Beck, C., Rudolf, B., Rubel, F 2006 *World map of the Koppen-Geiger climate classification updated, Meteorol. Z.* **15** 259-263

Spatial pattern of Islamic settlements of kampung Kauman Malang in the perspective of typical Mzab

Nunik Junara[1,2], Sugeng Triyadi S[2]
[1]UIN Maulana Malik Ibrahim, Malang, Indonesia,
[2]ITB, Bandung, Indonesia

Corresponding e-mail: nunikjunara@arch.uin-malang.ac.id

Abstract. Kampung Kauman is a representation of the Muslim settlement in Indonesia, especially in Java. Its location behind Masjid Jami' in the city center makes it one of the destinations to stay. The existence of the Kampung Kauman has been synonymous with the religious activities of its people. One of Kauman that still exists is in Malang, East Java. Kauman was formed along with the establishment of Masjid Jami'. In the beginning, people who lived in Kauman were Dutch until the colonial government established territories based on the race. Many moved to occupy the area around the Ijen Street, and Kauman was reserved for the Muslim population.
The present kampung Kauman is not only inhabited by Muslims, but there are also non-Muslims. There have been social and physical changes in the environment. Settlements are increasingly crowded, but religious life continues to influence the community activities. Mostly, non-Muslims live outside the kampung with their trading activities.
This study is field research, in which the phenomenon is tried to be analyzed based on facts in the field. This study aims to determine the pattern of settlements in kampung Kauman Malang. The result in this study; the spatial pattern is similar to the Islamic spatial planning principle which is the basis for the Muslim settlement according to Buchair namely Typical Mzab

Keywords: pattern of settlement, the principle of Islamic city, Kampung Kauman Malang

1. Introduction

Kampung Kauman Malang is located in Kelurahan Kauman, Kecamatan Klojen, Malang (Figure 1). Precisely in RW 3 Kampung Kauman, it consists of 6 RT. The number of residents based on survey results in October 2014 was about 1399 with 279 families. Almost 98% of the population is Muslim.

The development history of Malang cannot be separated from the function of the city as the center of trading and the spread of Islam. Babad Malang mentions the arrival of Islamic scholars on the orders of Sultan Sadad of Gedah State to spread Islam while trading. This incident occurred in 1293 Saka, or 1371 M. Babad Malang also mentions that these preachers failed to Islamize the king of Majapahit, but obtained permission to spread Islam [1]. Based on the Dutch colonial Law (Wijkenstelsel) in 1914, which governs the pattern of ethnic-based settlements, the Kampung Arab was located behind the mosque of Alun-Alun adjacent to the Kauman area (Figure 2).

IOP Conf. Series: Earth and Environmental Science **213** (2018) 012038 doi:10.1088/1755-1315/213/1/012038

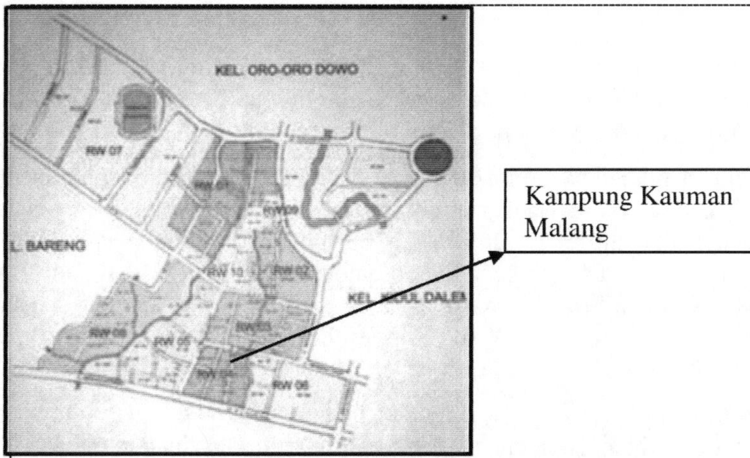

Figure 1. The Location of Kampung Kauman Malang

The history of the entry of Islam according to historians from the State University of Malang, Cahyono [2], there are several stages related to the entry of Islam in Malang Raya. It is based on relics that can still be traced by date.

Stage 1. The existence of Muslim settlements in Gribig Malang Timur was around the year 1500 as a part of Syech Manganti, the uncle of Sunan Giri, da'wah. Gibig area (Gribig) is a stepping stone for the kingdom of Demak to conquer the Sengguruh kingdom which is a Hindu kingdom.

Stage 2. The historical moment of the struggle against the colonial

Sub 1. The Trunojoyo Era, around the early 1600s, fugitives of many Trunojoyo followers who settled in Selokurung Ngantang formed Muslim settlements, including Kraeng Galesung from Makassar was buried there.

Sub 2. Untung Suropati Era, in the late of 1600 took advantage of Malang as a stepping stone (after losing to the VOC) and successfully conquered Pasuruan to become Pasuruan regent. In 1767 Pasuruan conquered VOC, and many Untung Suropati's grandchildren ran away to Malang Tengah (Singosari, Bungkuk, and surrounding areas).

Sub 3. Diponegoro Era, Prince Diponegoro's followers ran away to Gunung Kawi, West Malang. Jugo's grandfather was buried in the Blitar Kesamben, and Raden Imam Sujono established a hut for the spread of Islam in Kawi mountain.

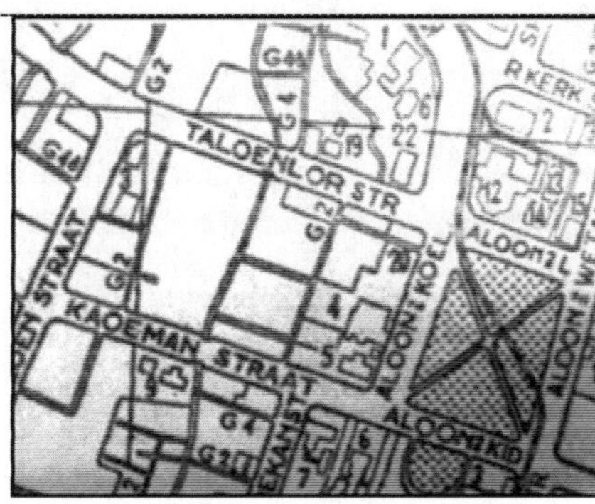

Figure 2. The map of Kauman Malang 1938 [3]

IOP Conf. Series: Earth and Environmental Science **213** (2018) 012038 doi:10.1088/1755-1315/213/1/012038

Kauman region itself cannot be separated from the existence of Talun region that has mentioned in Sengguruh inscription, as a new opening area. In the colonial period after the construction of the Jami' mosque and Alun-Alun, Kauman was a part form of Dukuh Talun.

Another historical factor related to the formation of Malang is because of the concept of the Javanese Islam formed by the Dutch colonial government around the 1800s. In the beginning, Kauman behind the Jami mosque is a settlement for Dutch citizens in Malang. As time goes by, and the efforts of the separation of indigenous and non-indigenous settlements, Ijen region was established for the Dutch settlement. Gradually Kauman was abandoned by Dutch citizens and began to be inhabited by Muslim immigrants from Java, Madura, and Arab. Similarly, several other areas around the Alun-Alun Malang such as Chinatown and Embong Arab is part of the region separation.

Kampung Kauman Malang is a Muslim settlement formed since the construction of urban centers in the Dutch colonial era, which is intended for Muslim religious leaders to be closer to the Jami 'mosque. To date, Kauman Malang is not only inhabited by Muslim but also non-Muslim, especially Chinese descendants who occupy large roadside areas. Culturally, the Kauman community is a merchant society; it could be seen from the composition of the livelihood that most of the traders with various independent efforts is undertaken by the community in Kampung Kauman.

Based on survey results (2014), there are two mosques, three mushola and one church in Kampung Kauman. The mosque is Masjid Jami Malang and masjid LDII. While mushola among others, mushola Bin Yasin, mushola al Ikhlas and mushola Muslimat. Religious activities were not only done in the mosque and mushola but sometimes also done in people's homes in turns. Based on the results of interviews with residents, the composition of Islamic adherents is the majority of Nahdatul Ulama, followed by Salafi identified with Arab descendants, Muhammadiyah and LDII. Although there are masjid LDII, the congregation is mostly from outside the kampung.

2. Principles of Spatial Management of Islamic City

Islamic city not only can be seen from the physical characteristics of the building but also can be seen from the social and cultural expression of the people. Some things that become a typical marker of Islamic cities according to Saoud [4] include:

a) Citadel, the fort is used as guard and defense of the city. The fortress is usually regarded as a barrier and is a natural defensive device.

b) The City Center (Royal City / Quarter City) is the place where the imperial palace was located. The city center serves as a landmark that is both contras and aligned with the city's identity.

c) Central urban Complex is the center of Islamic city civilization in the form of mosques and schools, also trade centers/markets and certain places provided to traders and artists.

d) Core City, at least have two main characters namely a combination of local, ethnic and religious, and serves as a district barrier.

e) Suburbs and suburban residences, where non-permanent settlers live while working in one place, such as a "caravan" spreading along the highway.

Based on the points above can be suspected that in the center of Islamic cities there are mosques as a central.

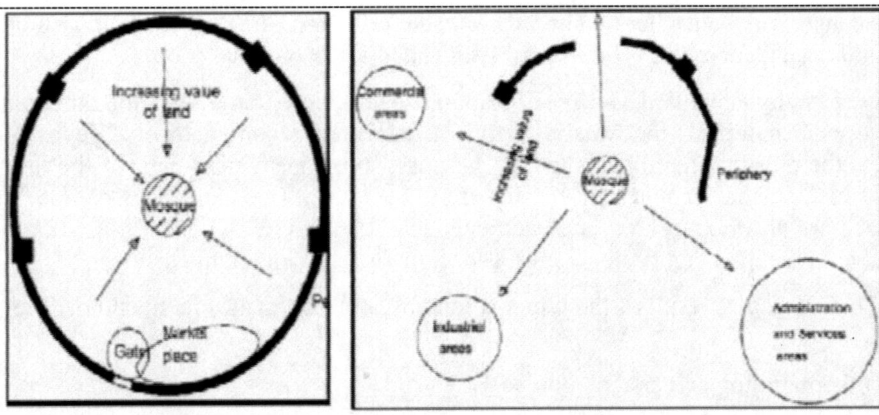

Figure 3. The concept of centrality in Islamic cities [4]

In the picture above (Figure 3), it can be seen that the mosque as the center and parts of the structure of the city followed. This consideration is made to show the existence of the mosque as a center of civilization, not only about worship, but also with education, socio-culture, and all the society's order.

In other information presented by Bouchair in Saoud [4] it is explained that Islamic city could be described with this following form (Figure 4).

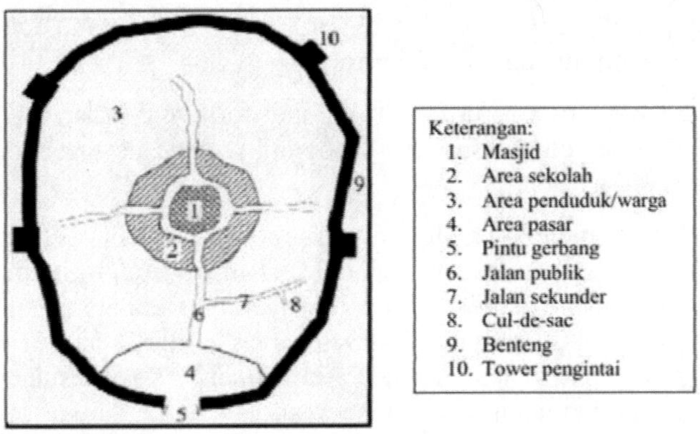

Figure 4. Schematic Diagram of Typical Mzab Settlements [4]

Related to the settlements' expansion, the market area was placed on the side of the gate to facilitate access for incoming traders goods. In addition, in the Islamic city, dimension of the road within the city is also arranged. The main street of the city (thoroughfare / thariq almuslimeen) is used for all, opened for public, whether Muslim or not. The main road has a minimum width of 3.23 - 3.5 meters (7 cubits) as determined by Rasulullah SAW. The Messenger of Allah does not specifically regulate the height of the road, but the Fuqaha agree to use the base of the man who drives the camel without obstruction, so the size is seven cubits as well. While the cul-de-sac is set high and the width 1.84 - 2.00 meters (4 cubits) based on Ibn Ar-Rami explaining that the width is adequate for the load / passing through it, while three cubits are too narrow for it (Ar-Rami, 1982).

Earlier Hakim [5] has written about the principles of urban planning in Islam which mentions:

a. Harm: everyone is encouraged to exercise full individual rights in all actions, but no harm or harm has been made, included the matter concerning the placement of locations within the city that cause smoke, dust, smells, etc.

b. Interdependence, between the inhabitants of the city and the structure in which they live, considering interdependence; this is what was then called as "ecological sense."

c. Privacy, every family is entitled to privacy from noise and views, as well as the other annoyances. It is seen by the main nature of the Muslim family that protects women from all eyes of foreigners. Privacy affects the physical buildings such as the position of the window above the human eye, the height of the road, so that people can not see into the house. The placement of door and window between houses are not face-to-face but interdependent, especially along the lane (alley). Similarly, avoid being able to see the neighboring pages, especially where women move.

d. The higher building, as a result of the window landing, not diminishes its neighbors' rights to get air and sunlight.

e. Respect to other building owners.

f. Selling someone's ownership to neighbors, rejection advice can be from neighbors themselves, previous owners, or family.

In the discussion of the road, Hakim [6] explains that one important element in Islamic city planning is the presence of two types of roads. The first is the "trough" (the continuous path) and the "culdesac" (the culminated path). The continuous road is as serving activities in the public area while the cul de sac road is for more private activities.

Saoud [4] in research on Muslim cities in the Middle East states that some factors influence the formation of Muslim cities. In addition to the influence of the topography and morphology of the previous city, Muslim cities reflect the general socio-cultural, political, and economic structure of the newly formed society. In general, the relevance follows:

a. Natural law, this first principle of adaptation of building form and urban planning to the climate is expressed to respond to weather conditions and topography, such as the concept of the courtyard, terrace, narrow and closed road garden.

b. Trust in religion and culture, beliefs in religion and practice form the center of the cultural life of the people so that the mosque is the center of spatial and institutional hierarchy. The belief in this culture separates public and private life through separate spatial rules between usage and territory. The city is designed with narrow streets and cul de sac to separate public and private domains and also considering the use of land to separate areas of men and women. As the result, activity in public areas and major roads is more economic than in residential areas.

c. The design principles derived from sharia law, the Muslim city, also reflect the laws of sharia in regulating public and private area relationships, and between neighbors and community groups.

d. Social principles, social groups within a city is usually based on the perspectives of inherited similarity, ethnicity and culture. It takes a social venue to bring them together regarding solidarity, social order, defense and religious celebrations.

3. Discussion: The Spatial Pattern of Kampung Kauman Malang

Based on the survey results in Kauman Malang, there are some elements of space commonly used by the Kauman community. The spatial elements found in the Kauman settlements will be reviewed according to the concept of the Islamic Town Arrangement from Buchair [4].

3.1 Masjid/mushola

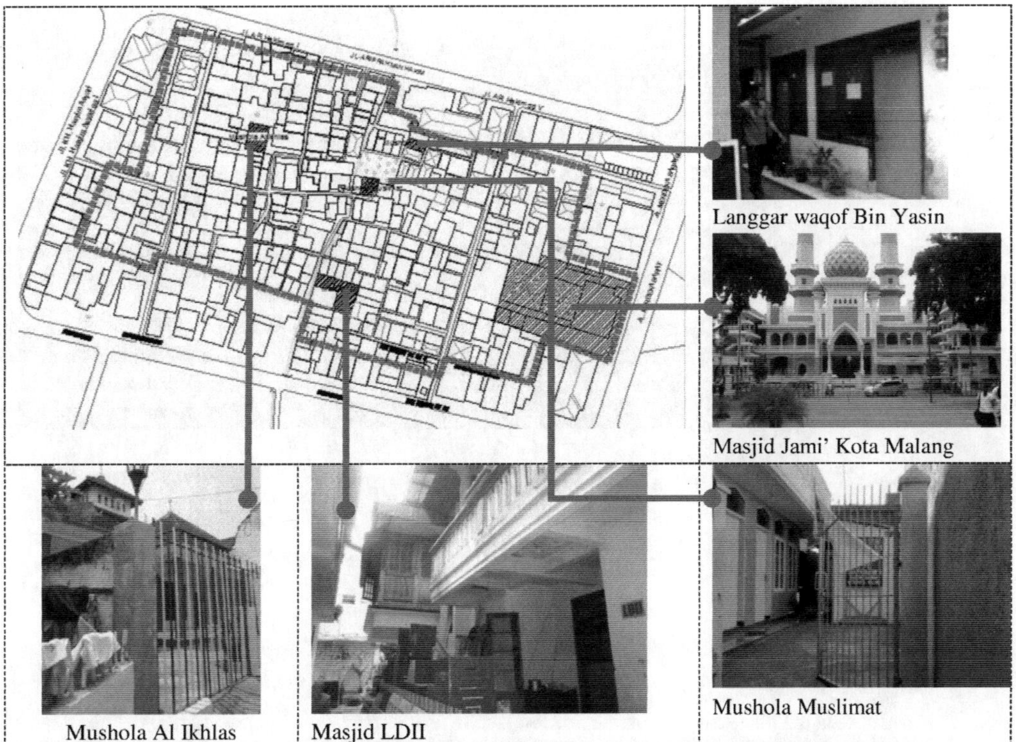

Figure 5. The position of the mosque / musholla as the center of religious activity

As the center of religious and social activities of the community in Kauman, besides as a place of worship, mosque and praying room is a means of interaction between individual communities through recitation activities, commemoration of religious holidays, RT/RW meetings and deliberations among communities (Figure 5).

IOP Conf. Series: Earth and Environmental Science **213** (2018) 012038 doi:10.1088/1755-1315/213/1/012038

3.2. Formal and Non-Formal Education Areas

Educational facilities in Kauman consist of Formal and Non Formal education. The new formal education is Kindergarten and Primary School (Figure 6).

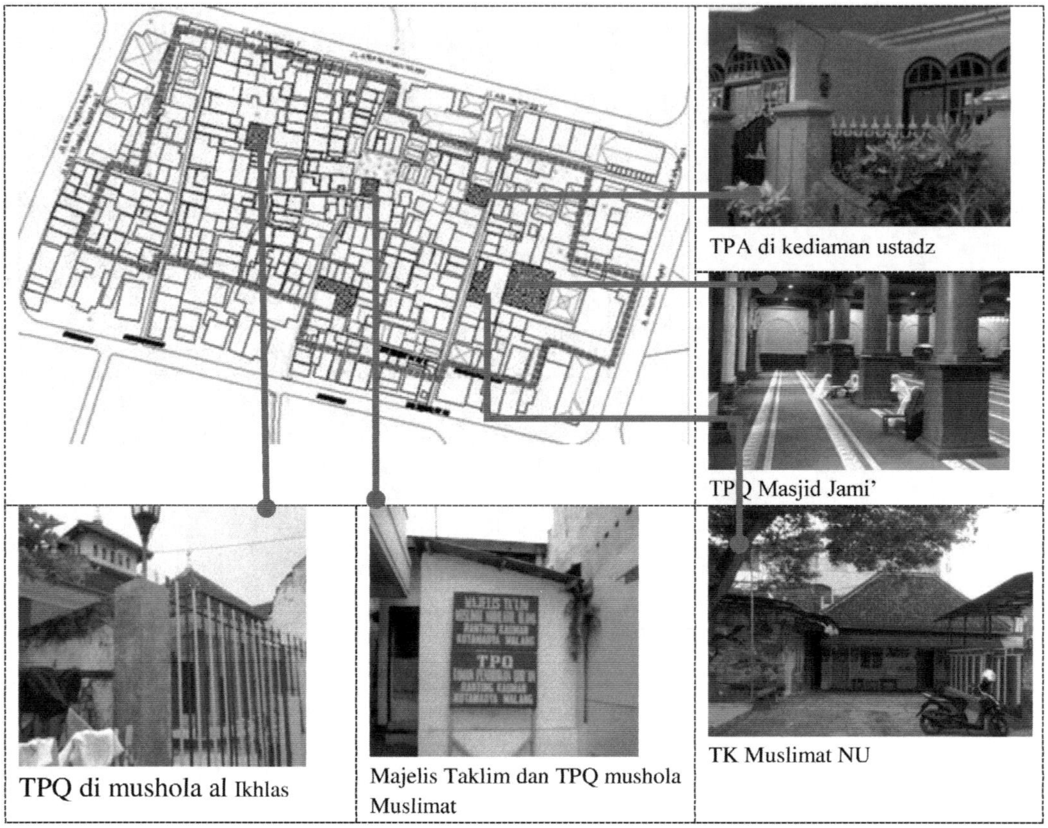

Figure 6. The position of Educational Facilities

IOP Conf. Series: Earth and Environmental Science **213** (2018) 012038 doi:10.1088/1755-1315/213/1/012038

3.3 The Settlement Area

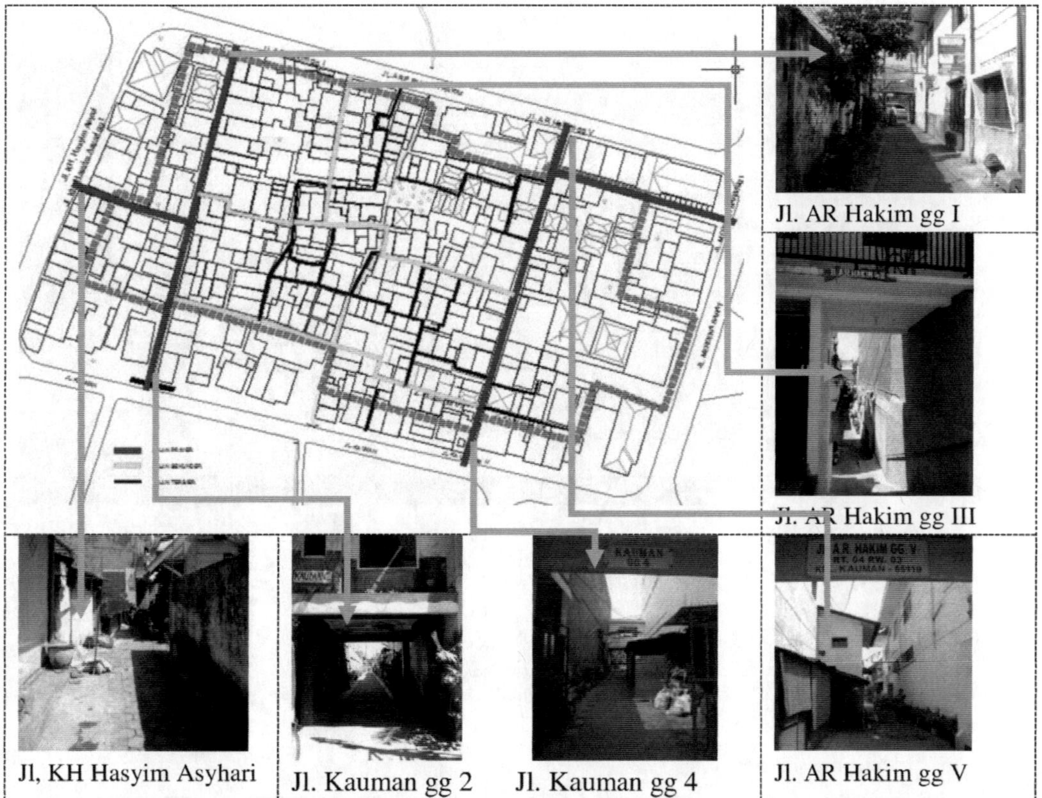

Figure 7. The position of the entrance to the kampung

The settlement area is located within a neighborhood surrounded by shops along the road that surrounds the Kampung Kauman. The entrance is accessible via eight driveways. All entrances are used as guard post facilities, by building a guard post on it (Figure 7). There is a gateway at most entrances, especially from the direction of Kauman road, Jl Merdeka Barat, and Jl.AR Hakim. Though there is no gate from Jl. KH Hasyim Asyari, but inside, there is a portal which two-wheeled vehicles and four wheels cannot pass it.

IOP Conf. Series: Earth and Environmental Science **213** (2018) 012038 doi:10.1088/1755-1315/213/1/012038

3.4 Market Area

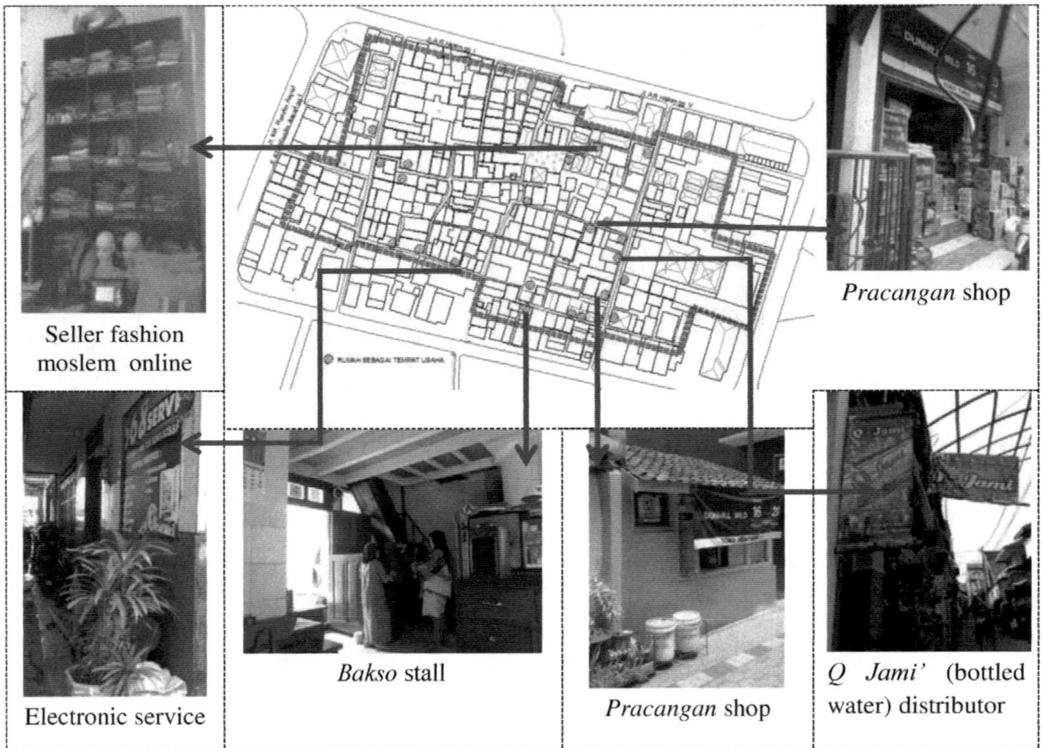

Figure 8. some home-based businesses conducted by Kauman residents

The market is identical with trading center where the seller and the buyer meets to trade all sorts of goods. In Kampung Kauman, the location of the market (Figure 8), Pasar Besar Malang, as an economic center, is about 1 km. But not too far from kampung Kauman, there are several modern shopping centers such as Sarinah Plaza, Ramayana, Mitra II, Malang Plaza and Gajah Mada Plaza. The area outside Kampung Kauman settlement is shops dominated by Chinese, especially along Kauman street and AR Hakim street.

Inside the kampung itself, there are several home businesses which is occupied by Kauman residents as a part of their household economic improvement. Various businesses is undertaken, among others, opened a shop *pracangan*, room rent/*kos-kosan*, food stalls, salon, selling *kripik*, batik, food entrepreneurs and other trading business.

Residents also use the road as a means of trading at certain hours.

3.5 The Gate

The gate here, in the form of double left and right doors that can be opened in and out, is one of the residents' security tools. This position is due to the high privacy and protection of citizens against the security. In addition to the gate, there is also a security post on it, such as a tower that can be used to look around (Figure 9).

The gate of jl.Merdeka Barat

The gate of jl. AR Hakim gg V

Security with the portal

Guard post that is above the gate jl. Kauman gg 2

Guard post located above the gate jl. AR Hakim gg III

The gate of jl. Kauman gg 4

Visible gate post which located enters from jl. AR Hakim gg V.

Guard post jl AR Hakim gg V saw from the inside.

The guard post located from jl. Kauman gg 4.

Figure 9. Gate Condition in kampung Kauman Malang

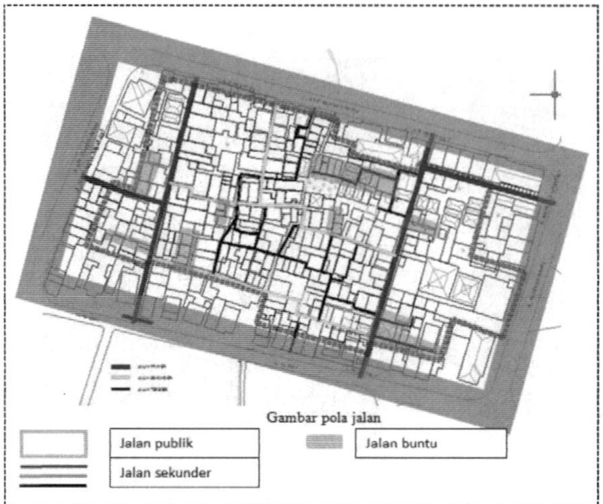

Figure 10. Public road

The public road is a passable road by anyone, opened for public and all walks of life, either on foot (Figure 10). The width of the public road is about 6-8 m and is located around kampung Kauman. Along public roads lies the means of trade, offices, and other public facilities included churches and schools, which can only be accessed from outside. At a glance, the placement is like a castle that limits the settlement of people from outside.

Secondary road is a residential street located within the village area. It consists of 2 types differentiated by width. The first is 3-4 meters width, and the second is 1-2 m width. In this residential area, a 3-4 m wide street is located at the entrance to the residential area. It can only be passed by people walking, bicycle, or motorcycle. This road is more opened to be traversed by citizens and outsiders. To some extent, citizens may pass through the road with motorcycle. Then, to go inside, it is recommended to turn off the engine and push it to the destination.

Secondary roads with 1-2 m width seem to be more intimate and private (Figure 11). This road connects to a deeper residential area and serves as a road to the opposite of the kampung. When outsiders pass through these streets, they feel like foreigners.

Figure 11. Secondary Road image with a width between 1- 2 meters

Cul de sac or a dead-end, with 1-2 m width, is a private road that is only passed by residents, neighbors and relatives who live there, included guests who visit (Figure 12).

Figure 12. Some Condition of Street in Kampung Kauman

From the results of the above discussion, the application of the Islamic spatial planning concept from Buchair is applied in the Kampung Kauman. The dense settlements and the relatively small radius of settlement is adjusted to the structure of the settlement space. If the concept is depicted on a city scale, then this settlement is minimized in the RW and RT spatial clusters. The interaction among residents occurs in the streets of the kampung as part of the public space. They also build a playground for children, places to sell, drying furniture, and various other activities.

There is a significant difference associated with the position of the mosque as the center of people activity. On a settlement scale, the Jami mosque is located on the edge of one side of Kauman, not in the middle of the settlement. The hierarchy of community activities tends to be more crowded on around the Jami 'mosque. Similarly, shops that surround Kampung Kauman serve as a fortress that limits the settlement area from external influences. It is due to the absence of direct access that connects shopping to residential areas. The massive wall boundary behind the shops does not allow interaction with the community behind it. Placing the gates in every access to the settlement area, most of them are also equipped with two guards, such as towers that keep security around the settlement (Figure 13).

Some of the factors that affect the layout of settlements include:

1) Activity factors, which form the pattern of settlements and positions of places of worship and the flow of the citizens circulation
2) The safety and comfort factor, which arranges public facilities located in areas of community-accessible settlements
3) Religious and cultural factors, which lie behind the rules applied in activities in the environment
4) The privacy and security factors, leading to a hierarchy of roads and elements of environmental safeguards

Figure 13. Scheme of application of Islamic spatial planning principles in Kampung Kauman

4. Conclusion

Based on the analysis that has done, there are similarities and differences from the application of Islamic spatial planning principles in settlements in Kampung Kauman. It has been seen in the position that includes:

1) Masjid/mosque
2) Schools and non-formal education
3) Settlement area

4) Shopping area
5) Access to the street, such as public street, secondary roads/settlements, and cul-de-sac roads
6) The existence of environmental protection devices (gates, portals, guard posts)
7) Some influencing factors include:
    a. Activity factors,
    b. Security and comfort factors,
    c. Religious and cultural factors,
    d. Privacy and security factors.

5. References

[1] Ardi 2010 Malang Tempo Dulu 2003 *Malang Tempo Dulu*

[2] Cahyono D 2013 *Wan Wacarita* (Malang)

[3] Handinoto & Soehargo P H 1996 Perkembangan kota dan arsitektur kolonial Belanda di Malang *Surabaya Lemb. Penelit. dan Pengabdi. Kpd. Masyarakat, Univ. Kristen PETRA*

[4] Saoud, Muazaz Y. D 2010 Aspects of Decline in Contemporary Islamic Cities *Eur. J. Sci. Res.*

[5] Hakim B S 1982 *Designing in Islamic Cultures 2: The Aga Khan Program for Islamic Architecture. Arabic-Islamic Cities: Building, Planning, and Principles* (Massachusetts Institute of Technology)

[6] Hakim B S 2007 Generative processes for revitalizing historic towns or heritage districts *Urban Des. Int.* **12** 87–99

Contribution of balcony of *Rusunawa* Surabaya on the use of space

Rika Kisnarini, Johanes Krisdianto, Iwan Adi Indrawan
Dept.of Architecture, Institut Teknologi Sepuluh Nopember, Surabaya, Indonesia.

Corresponding e-mail: r.kisnarini@gmail.com

Abstract. Balcony is actually not included as a basic functional space in the dwelling. However, for walk-up flats like *rusunawa* in Surabaya in which most masses are in double loaded, balcony can be the only space directly related to outer space where fresh air can be obtained. It must be noted that some of them are not real balconies, because they are covered with walls and glasses. Previous research has proven that balcony of *rusunawa* Surabaya is very effective in improving the thermal comfort of internal space. *Rusunawa* Surabaya, although it should be occupied temporarily for only 5-10 years, in reality many families still inhabit for longer period. Even many of them made changes on their space because their needs also increase spatially. Balconies often play an important role in terms of space adaptation. The purpose of this study is to find out how important the role of this balcony according to the residents of *rusunawa* in Surabaya. This research was descriptive explorative research. Data collection technique applied was field survey by spreading questioners on 110 respondents who were determined by stratified random sampling. The data collected was about whether the balcony was really desired by the occupants; what activities were carried out on the balcony; and the effectiveness of balcony in spatial adaptation. The result that is expected to assist provider of *rusunawa* ascertains that balcony is definitely required by the occupants.

Keywords: contribution, balcony, use of space, *rusunawa*, Surabaya.

1. Introduction

Outdoor spaces have been essential parts of a house because much of free time of household members were spent in the garden or terrace. Most families give high priority to enjoying outdoors as a way to improve health [1], on which health is the primary requirement for human being [2]. Easily accessible, high quality green open areas facilitate people to spend more time outdoors that has a positive effect on life expectancy, health, happiness and general wellbeing [3]. Outdoor space is indeed important and required by houses to supply fresh air to refresh the air of internal space [4][5]. This reality is enforced by [6] which said that the availability of natural ventilation and outdoor private living spaces play important roles in resident perceptions of liveability. For vertical dwelling, the chance of connection to the outdoor space is more available by the existence of balcony. According to [7] a balcony is even found to have also a positive effect on the value of a property irrespective of the quality of the view. Moreover, according to [8] balcony also plays a decisive role in the energy behaviour of the buildings since they may intervene and influence almost all of the mechanisms of how buildings interact with its environment. Even to be more creative, the balcony can also have additional function as potted desert garden [9] that can turn a simple balcony into a tropical paradise.

Outdoor space of skyscraper apartments of large cities in developed countries are created by providing roof garden completed with full greeneries, cafes, pools, gyms, barbecues etc. Such facilities are impossible to be created for *rusunawa*, a walk-up flat for low-income households, which even each unit space area is very limited. However, in a tropical city like Surabaya, the easiest way to get thermal comfort is through cooled outside air [10] and balcony is the exact intermediary. This is real, as only a few of residents capable of providing air conditioner. So it can be imagined how important the procurement of balconies on *rusunawa*, an apartment unit that addressed for the poor families. This demand is rarely done though in fact it is highly required [11]. Balcony is a platform on the outside of a building, enclosed by walls or balustrades, supported by columns or console brackets. The platform is projecting from the wall of a building, usually above the ground floor. Balconies are typically small and are not used as social spaces or for entertainment purposes [12].

Figure 1. Simple balcony (source: Homedit Interior Design & Architecture)

Although the space area is small, the balcony of *rusunawa* in Surabaya is not only used for enjoying outside air and outdoor view as intended, but also for conducting household activities such as drying, cooking, even sleeping and doing business activities [13]. These activities cannot be done inside the unit because the unit space is too small. Normally, balcony is designed to be located in front and it can greatly contribute to the look of an apartment. [14]. For *rusunawa*, especially that building mass is of double loaded, the balcony is designed to be backward of the unit as the entrance into the unit is at the corridor side. This condition often results in problems especially to the performance of the building, as the occupants use the balcony for storing and drying laundry pieces. Therefore, this research is meant to investigate whether the occupants desire the balcony.

2. Methods

This research is actually a whole study conducted in 2 years. In the first year (2017) the research focused on contribution of balcony on thermal comfort of the internal space with simulation method using Ecotect 2011. In the second year (2018), the focus of contribution is on the use of space of the balcony according to the residents of *rusunawa*, by field survey method using questioners. Result of research in first year indicated that contribution of balcony on thermal comfort of internal space was positive, meaning that the existence of balcony can reduce the temperature of internal space. By applying sun devices the reduction of temperature was more significant. This was in accordance with [15]. In this second year, the research aims to strengthen the importance of balcony existence based on the users' opinion, so that in the future design the inclusion of balcony for *rusunawa* is no longer doubted.

The step taken in second year research was initiated by observing the balcony types that exist in 11 locations of *rusunawa* spreading out in Surabaya. This observation was done in order to reconstruct the balcony types. Next action was preparing questionnaires in which the questions focused on among others: whether balcony existence is desired, which type of balcony is desired; what activities are carried out by the residents on their balcony; how effective the role of balcony in solving the space shortage, etc. Before data collection in field research it is important to prepare the surveyors. Surveyors must master exactly what is meant in each question. Thereafter is data compilation and processing. The last step is analysis and synthesis that finalised by conclusion. In more detail, steps of research can be seen below.

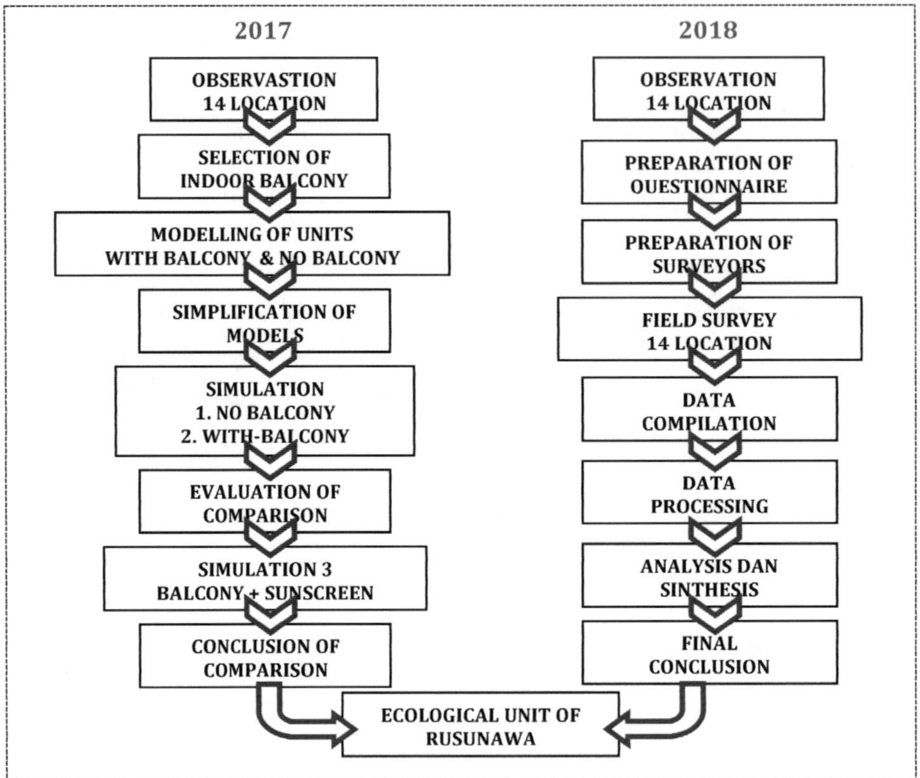

Figure 2. Steps taken in the whole research

3. Results

First of all it is necessary to explain that all questions in the questionnaires had already provided with certain answers that should be selected by the respondents. The intention was to provide certainty and ease [16] either for respondents or for analysers during the recapitulation of answers. In the choice of model or balcony design was also already given by samples in images in order to facilitate the respondents in illustrating and deciding choices. The number of respondents was 10 residents for each *rusunawa* that spreads out in 11 locations in Surabaya. The respondents were selected randomly, so the number of respondents in this study can be ascertained as many as 110.

3.1. Result on whether balcony is desired by occupants

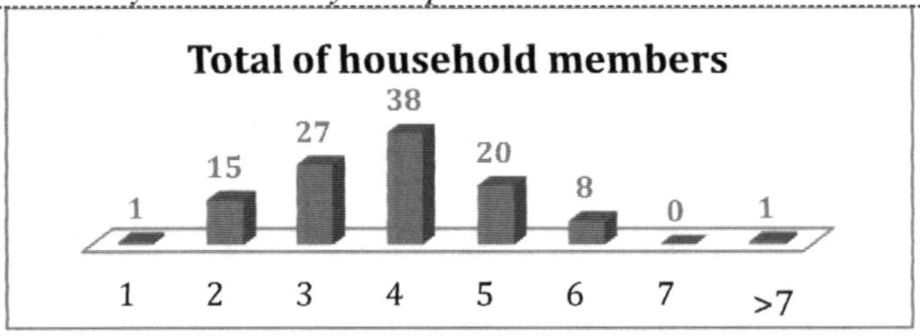

Figure 3. Household members.

Before discussing the desire on balcony, it is necessary to declare the mean or average number of household members. From the graph above it is seen that the dominant total member of household was 4 people, so from total of 110 respondents, the average amount of house' occupants was 3.81 members, if it was rounded up it became 4.

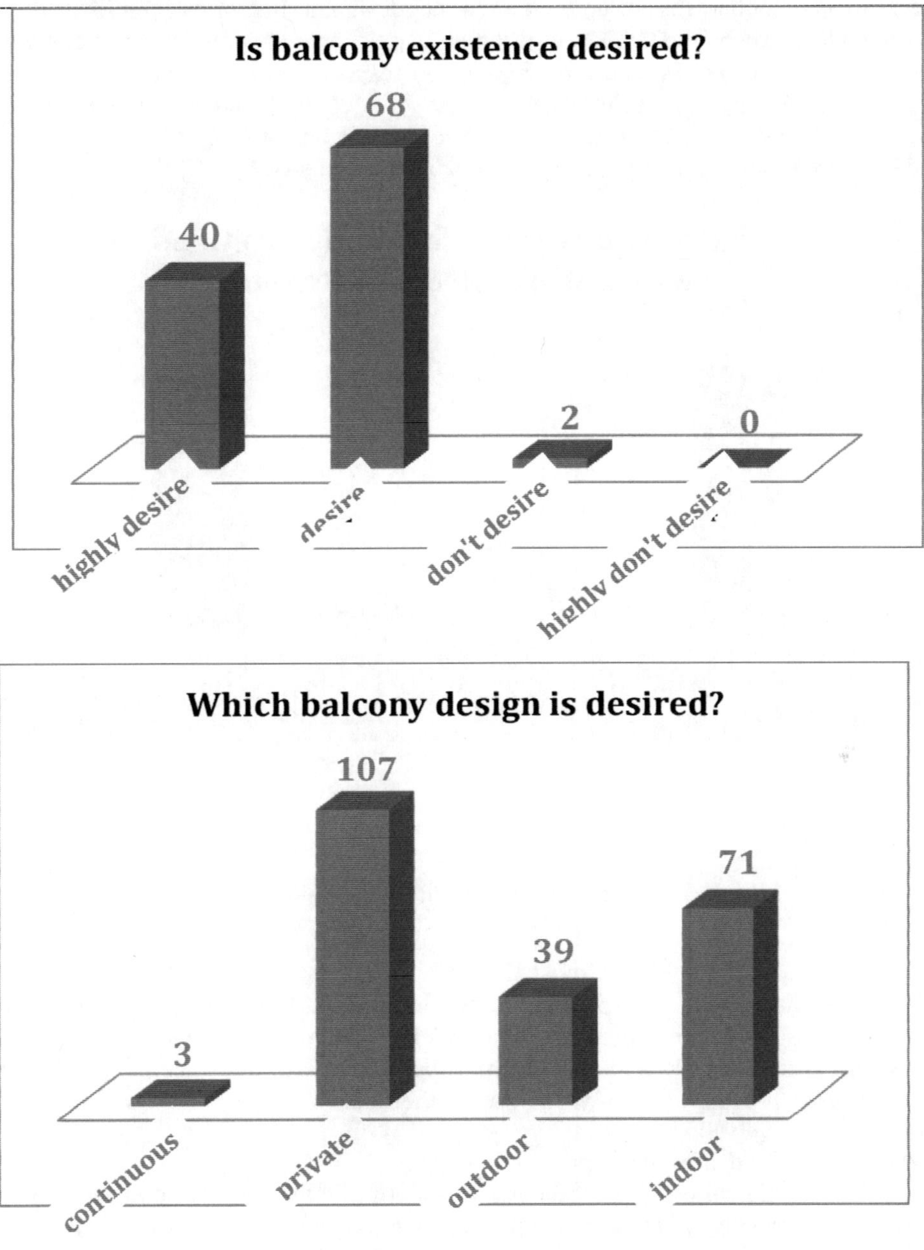

Figure 4. Is balcony existence desired; which balcony design is desired?

Figure 4 above shows that almost all respondents desired the existence of balcony. From 110 respondents only 2 of them did not need balcony on their units. Even 40 of 108 respondents who desired the availability of balcony declared that they were highly in need of the balcony. From here it can be ascertained that the presence of balcony was definitely required by the occupants of *rusunawa*.

IOP Conf. Series: Earth and Environmental Science **213** (2018) 012039 doi:10.1088/1755-1315/213/1/012039

Regarding the design, seen from the right graph, almost all respondents expressed their desire to have a private balcony instead of a continuous one that can be shared and accessed by neighbours. From all 110, only 3 respondents preferred to have their balcony designed continuously. It was proved that almost all residents wanted a balcony that separated each other from their neighbour's. Further design requirements asked whether the balcony should be designed indoor or outdoor, 65% of total respondents liked their balcony to be indoor, meaning that the balcony should be protected or bordered by walls and glasses or windows, and only 35% or 39 respondents preferred that their balcony was provided outdoor or directly open to outside fresh air. This contradicted the previous research result that required the balcony to be designed outdoor. The outdoor balcony complete with the sun devices is proven to be successful in reducing the temperature of internal space.

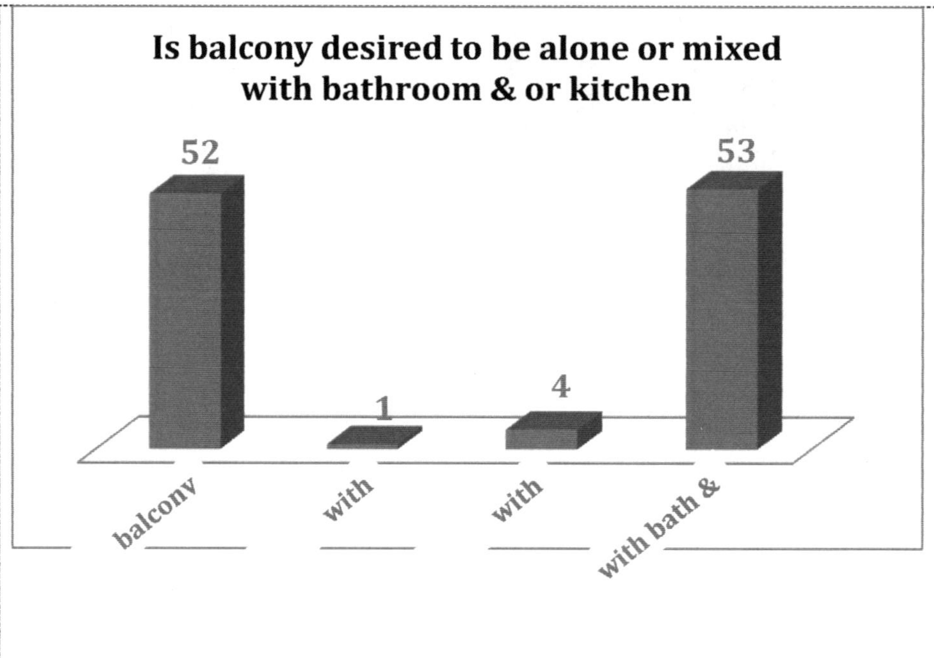

Figure 5. Whether balcony desired to be alone or mixed.

Other design requirement resulted in almost fifty-fifty calculation. From all 110 respondents, 52 of them preferred that their balcony be provided alone with no bathroom or kitchenette mixed inside. While the other 48% or 53 respondents desired to have their balcony be provided along with bathroom and pantry within. 5 families liked their balcony to be mixed with only one of them. Four of them wanted to be mixed with pantry, and only one wanted their balcony to be mixed with bathroom.

3.2. Result on whatever activities are done on the balcony

The most activity that carried out at the balcony was storing items. 101 respondents or 92% of them put their items at the balcony. The second top was drying in which 87 families or 79% respondents dried their laundry or others there. The third top was cooking. 73% or 80 respondents stated that they cooked at the balcony.

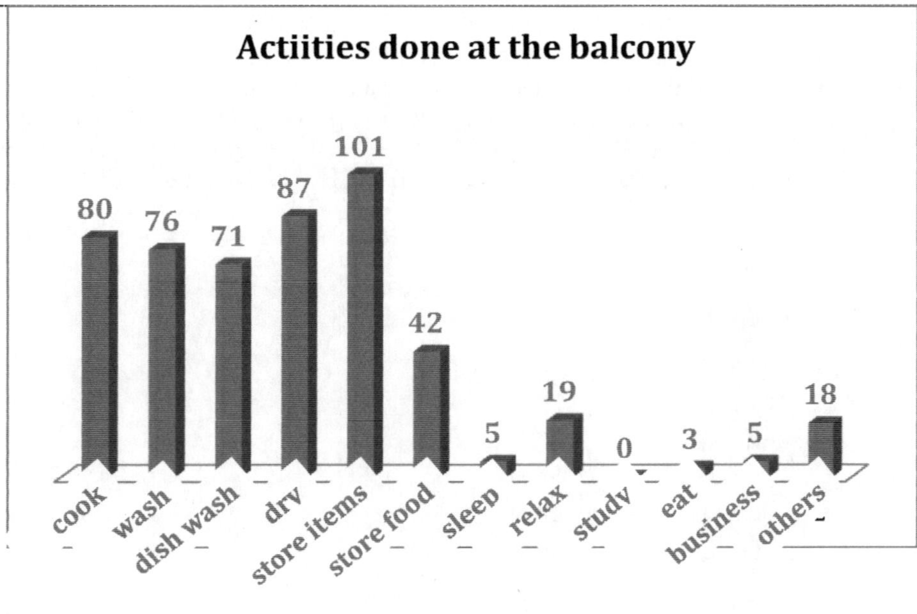

Figure 6. Whatever activities carried out at the balcony

The top four was washing, which was done by 76 households, followed by dishwashing done by 71 respondents, storing food by 42 of them, and the rest were activities conducted by smaller than 20% of total respondents e.g. relaxing which was done by 19 respondents, sleeping and home based business by 5 respondents respectively, and eating by 3 respondents. Nobody studied at the balcony.

3.3. Result on the role of balcony related to space adaptation

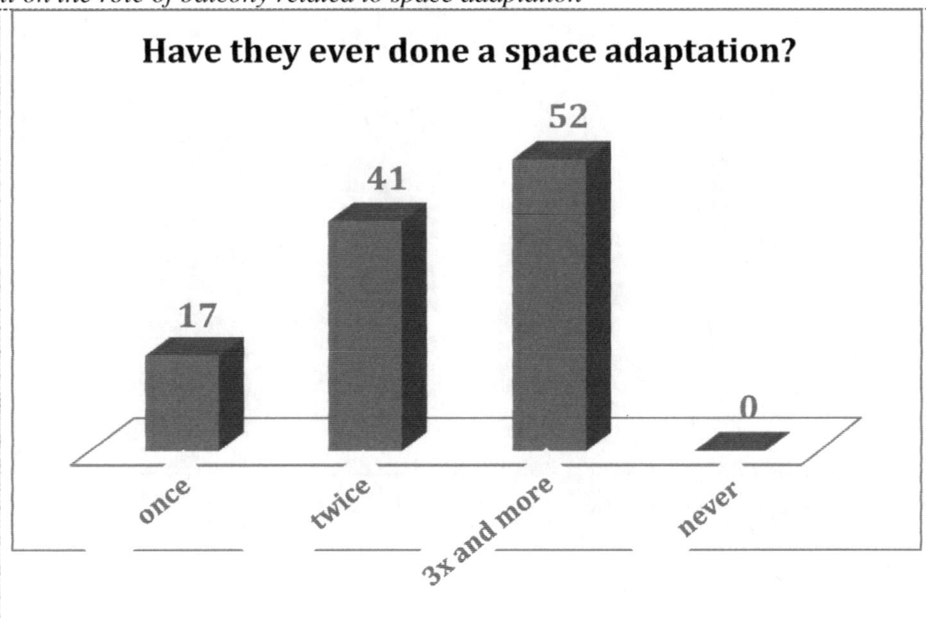

Figure 7. Have they ever made space adaptation?

From the figure it is shown that all respondents had ever adapted their space to meet their requirements, even most of them namely 52 respondents 47% did adaptation three times or more. The other 41 respondents or 37% did it twice, and the rest 17 or 15% did it once. Nobody of them had never done spatial adaptation. It can be confirmed that all occupants made adaptation as time goes by.

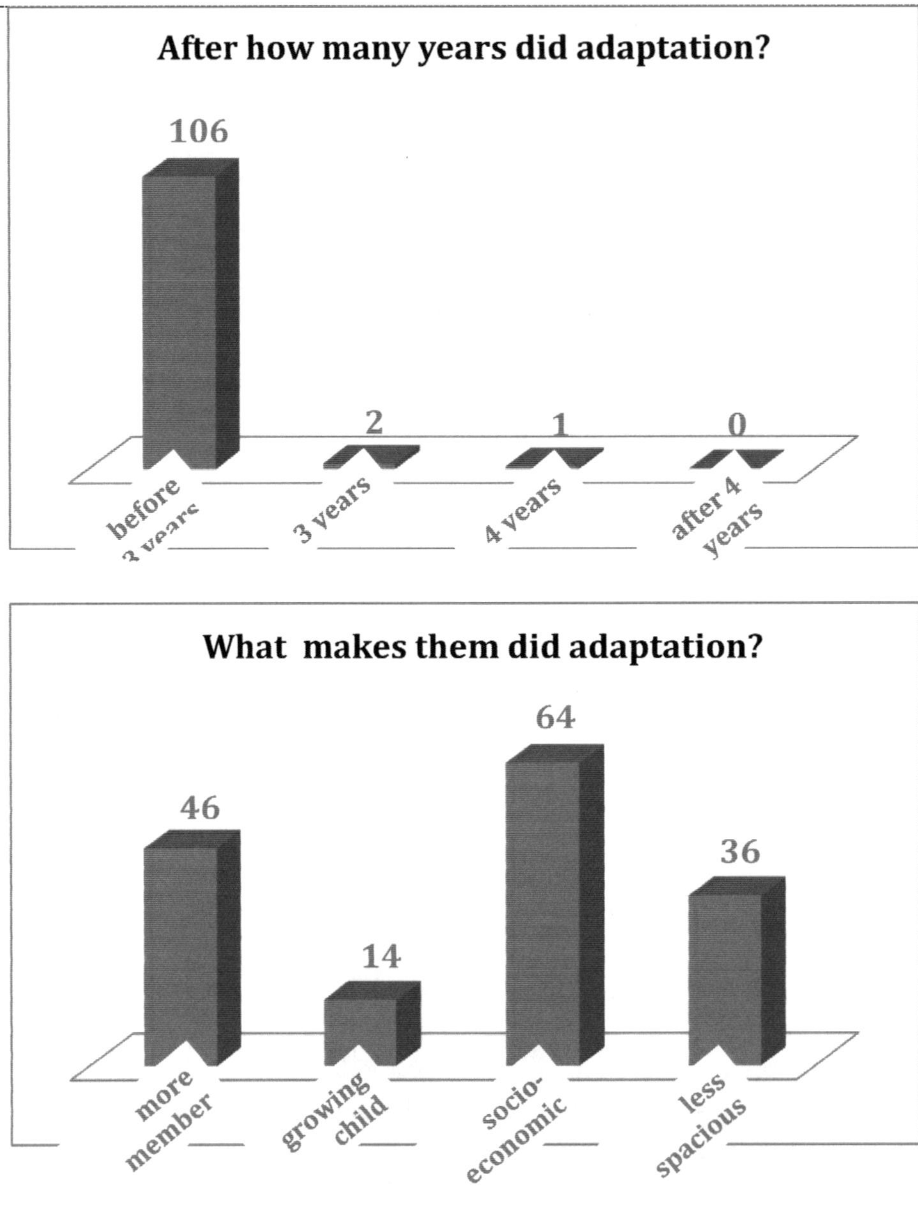

Figure 8. When space adaptation is made? What reason made them do adaptation?

The spatial adaptation even had already been made before 3 years of their living in *rusunawa* unit. As seen above, there were 106 families or 96% respondents adapted their space earlier than 3 years living in the unit. Only less than 4% respondents did adaptation after three years. This proved that occupants of *rusunawa* absolutely needed to do spatial adaptation [17]. This must be well anticipated [18].

When they were asked about what reason that encouraged them to do adaptation, most of 64 respondents said socio-economic condition made them adapt spatially. The second top reason was because of additional household members. The other 36 respondents said because of less spacious space. While the rest 14 respondents stated that growing up with children pushed them to do spatial adaptation.

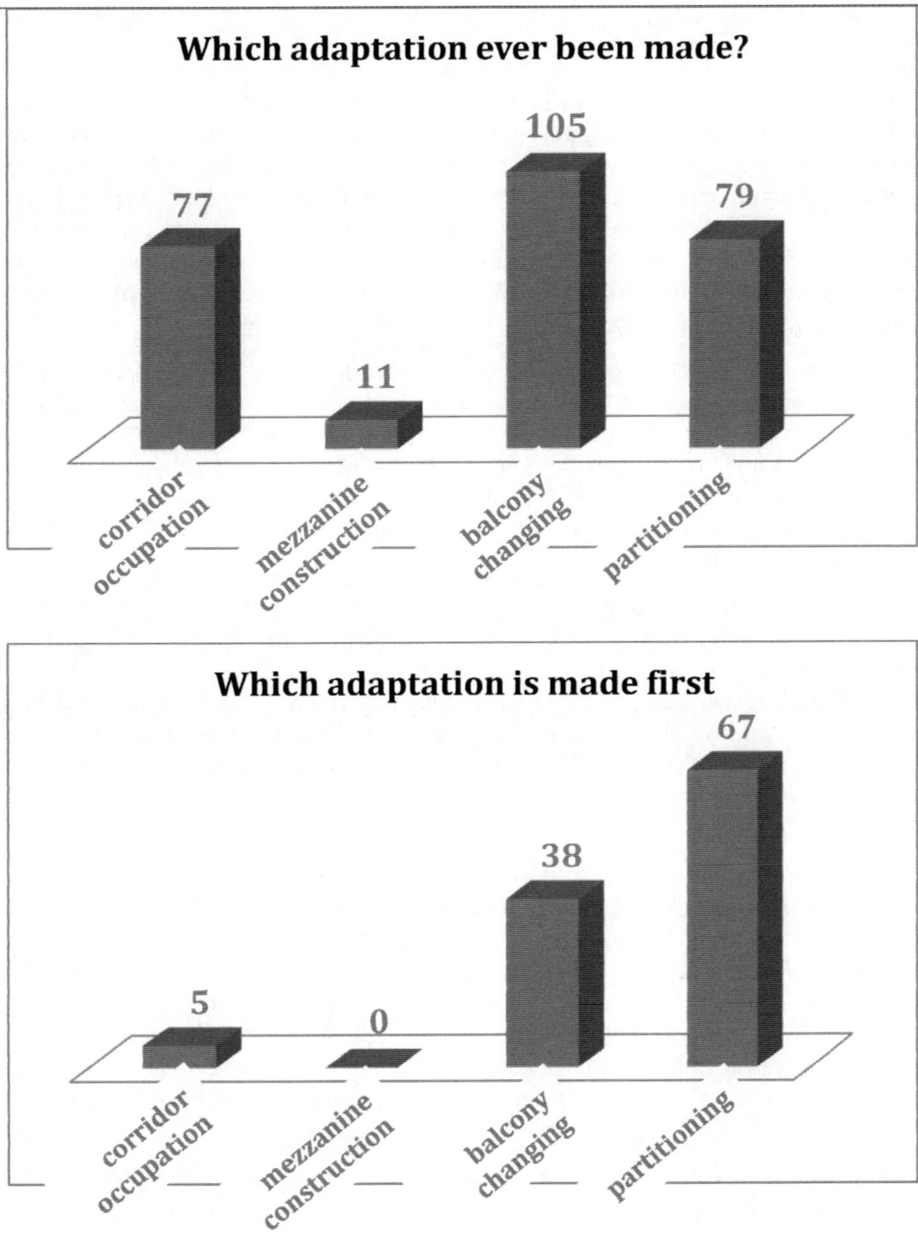

Figure 9. Whichever adaptation has been made? Which adaptation is done first?

Regarding whichever adaptation has been made? Almost all respondents did more than one type of adaptation. However, from the figure above it is shown that most type of adaptation was balcony change. 105 respondents ever changed their balcony into another functional space or a more living space. Partitioning took the second place by making 79 respondents divided their internal space into smaller ones. Corridor occupation took the third in which 77 respondents occupied parts of the corridor space. While mezzanine construction was only done by 11 respondents.

IOP Conf. Series: Earth and Environmental Science **213** (2018) 012039 doi:10.1088/1755-1315/213/1/012039

From all 4 types of adaptation, partitioning was done firstly, in which 67 respondents said that from all adaptation types they had ever made, they initially did partitioning. The other 38 respondents declared that the first adaptation they made was balcony changing, and only 5 respondents stated that the first adaptation done was corridor occupation. Nobody said that they started by constructing mezzanine.

4. Discussion

Firstly from figure 4, it could be assumed that the existence of balcony on *rusunawa* units was definitely desired and demanded by residents, and it should be a private balcony, not a continuous one. The reality that the comparison between residents preferred their balcony to be indoor and outdoor was 65:35 confused the author. I am afraid that this is possibly because their balcony is currently an indoor balcony. In fact, the research in previous year proved that balcony that was able to reduce the heat of air and fix the thermal comfort of the internal space was an outdoor balcony. However, it is an absolute requirement that this outdoor balcony be equipped with solar thermal protectors or sun devices that meet the needs, both the design and the material.

The balcony (fig. 5) can be designed to be alone outside the unit, or mixed with bathroom and pantry. Either one was fine since the result indicated the same amount between respondents' interest in balcony alone and in balcony mixed with bathroom and pantry. When the balcony was designed to be alone then the bathroom and pantry were located inside. However these two should get fresh air and healthy sunlight from the balcony, which was located outside the unit. When the design of the balcony was mixed then the bathroom and pantry were located outside at the balcony. Both choices can be made in good and healthy design.

Considering (fig.6), it is clear that *rusunawa* occupants consider that the balcony was more appropriate for storing and drying space. However, cooking, washing, dishwashing, and storing food were also in favour for them to be conducted at the balcony. Since storing items cannot be ascertained in size, so it was recommended that the size of the balcony be designed by referring to the need for drying activity. At least the distance of balcony should cover times of drying line that needed 60 cm per line. It still needed extra space for the balcony itself at least 90 cm. So the distance of balcony should be minimally 1.50 m, however, more lines for drying were indeed preferred when additional distance was possible.

From (fig.3) it is confirmed that the rate of household members should account as 4 people. However with the increase of time there should be development occurred that could be caused by the growth and addition of children or the development of socio-economic condition of the family, which could be either an improvement or a decline. Frequently in low cost apartment, there is an establishment of home-based enterprise either when a family experiences an improvement or a decline in their socio economic condition. Both of them require additional space since for this business there should be a space that cannot be mixed with the private space, as there will be consumer who is outsider instead of member of the family. The space must be considered as public or semi public space rather than a private space.

Here the space that must be released is the existing living space, so that the family needs additional substitution of space transferred into the public space, as it becomes a business room. This situation will inevitably force residents to adapt spatially in order to meet their space demand. This is evident in this study where almost all residents had adapted 3 times and the adaptation was firstly done before 3 years they inhabited the unit. If it was like this usually occupants did some action, if not occupy the corridor they changed the balcony into a living or internal space because their private space was already sacrificed as a business room.

Regarding adaptation, it is clear that compared to the other two adaptation actions, corridor occupation and mezzanine construction, changing the balcony into an inner space is the most correct and legal, as well as inexpensive action. So, it is appropriate that most widely adaptation action chosen by residents is balcony changing, although the first action is partitioning space instead of changing balcony.

From the results of research survey it is proved that residents of *rusunawa* needed the balcony. Related to artificial ventilation, [19] has proven that applying the hybrid ventilation method using balcony space in apartment housings will not only create a comforting atmosphere but also reduce the cooling energy consumption compared to the constant mechanical ventilation. In *rusunawa*, most of residents apply mixed ventilation system by using fan and natural ventilation; only a few applied air conditioner and they only turned on the AC when it was too hot. When the air was cool enough they did not use AC. Unlike the hybrid system, *rusunawa* did not have intelligent control system that can switch automatically between natural and mechanical modes in order to minimize energy consumption. The design elements of a balcony affected the solar radiation entering the building (e.g. balcony depth, window-to-wall ratio, glazing optical properties, etc. have a significant impact on the heating loads [20]. Therefore the dimensions as well as materials applied need further investigation.

5. Conclusion

It can be concluded that the existence of balcony proved not only desired and needed but also benefited the inhabitants because they could carry out several household activities. It is also potential to solve the problem of lack of space, or to anticipate space adaptation. Moreover this balcony is proven to quantitatively improve the thermal conditions of its interior space. So it must be guaranteed that the balcony is provided on the design of the upcoming *rusunawa* units. Local government as the provider of *rusunawa* must be able to assure the availability of a private outdoor balcony at each unit of future *rusunawa* with the least distance of 1.5m.

6. Acknowledgments

This article is derived from a research of a relevant topic that is financially sponsored by Institut Tekonologi Sepuluh Nopember (ITS) Surabaya, Indonesia under a scheme of Laboratory Research.

7. References

[1] Palmer Isabelle (2012). *The Balcony Gardener: Creative ideas for small spaces.* Kindle Edition. Ryland Peters & Small and CICO Books.

[2] Maslow (1970). *Motivation and Personality.* Harper & Row. New York.

[3] Lestan, Erzen, Golobic (2014). *The Role of Open Space in Urban Neighbourhoods for Health-Related Lifestyle.* International Journal of Environmental Research and Public Health Vol.11/6, pp 6547-6570. doi: 10.3390/ijerph110606547. https://www.ncbi.nlm.nih.gov/pmc/articles/PMC4078595/

[4] Canizares AG (2006). *Outdoor Spaces (Good Ideas).* Collins Design. ISBN-10: 0060893230.

[5] Givoni B (1998). *Climate Consideration in Building and Urban Design.* Copyright 1998, by Van Nonstrand Reinhold. Publisher ITP, a division of International Thomson Publishing Inc.

[6] Kennedy R, Buys L, and Miller E (2015). *Residents' Experiences of Privacy and Comfort in Multi-Storey Apartment Dwellings in Subtropical Brisbane.* Journal of Sustainability 2015, vol.7, issue 06, page 7741-7761; ISSN: 2071-1050; doi:10.3390/su7067741. file:///C:/Users/OWNER/Downloads/sustainability-07-07741-v2.pdf

[7] Kwong W.C, Siu K.W and Chung Y.Y (2004). *The value of the provision of a balcony in apartments in Hong Kong.* Journal of Property Management Vol.22 No.3, 2004 pp. 250-264 http://www.lantm.lth.se/fileadmin/fastighetsvetenskap/utbildning/Fastighetsvaerdering_VFTF05/The_value_of_balcony_in_Hong_Kong.pdf

[8] Tedjokoesoemo (2014). *Balcony and Jalousi Effectiveness Review to Promote Daylight in Interior for Tropical Country.* Journal of Dimensi Interior vol.12, no.1, 2014 pp 51-56. ISSN: 1693-3532. file:///C:/Users/OWNER/Downloads/19370-23604-1-SM.pdf

[9] Marylee Pangman (2015). *The Potted Desert Garden: Transform a Balcony into a Garden Paraduse.* CV Independent January 13, 2016. https://www.cvindependent.com/index.php/en-US/opinion/potted-desert-garden/item/1904-the-potted-desert-garden-transform-a-balcony-into-a-garden-paradise

[10] Lippsmeyer Georg (1994). *Bangunan Tropis.* Jakarta: Erlangga

[11] King P (2004). *The room is too panic, an example of film criticism and housing research.* Journal of Housing Theory and Society. Vol.21, issue 1, 2004. Page 27-35.

[12] Homedit Interior Design & Architecture (January 23, 2012). *45 Inspiring Small Balcony Design Ideas.* https://www.homedit.com/45-inspiring-decor-ideas-for-small-balconies/

[13] Kisnarini R (2015). *Functionality and Adaptability of Low Cost Apartment Space Design. A case of Surabaya Indonesia.* Tue Eindhoven, the Netherlands.

[14] Pile JF (1988). *Interior Design (book).* Library of Congress cataloguing in Publication Data. Harry and Abrams Incorporated. New York.

[15] Szokolay SV (1980). *Environmental Handbook for Architects and Builders.* The Construction Press Ltd Lancaster, England. Published in the USA by Halsted Press and division of John Wiley & Sons Inc. New York.

[16] Groat and Wang (2003). *Architectural Research Methods.* Library of Congress Cataloguing-in – publication-Data. John Wiley & Sons. Canada.

[17] Geraets Rob (2001). *Future Value of Buildings.* Faculty of Architecture. University of Technology Delft (TU Delft), the Netherlands.

[18] West BN and Emmit Stephen (2004). *Functional design and analysis of new speculative house plans in the UK.* Journal of Design Studies. Vol.25, issue 3, May 2004. Page 275-299.

[19] Won, Kim, and Leigh ( ). *An Evaluation of Energy Performance of Hybrid Ventilation System Using the Balcony Space in Apartment Housing.* CIB 8160, pp 757-762. http://www.irbnet.de/daten/iconda/CIB8160.pdf

[20] Clarke, Johnstone, Kim, Kokogianakis, Strachan, Kim, Woo, and Kang (2008). *Study of the Energy Performance of Korean Apartment Buildings with Alternative Balcony Configurations.* World Renewable Energy Congress 2008. http://www.esru.strath.ac.uk/Documents/lea70.pdf

Wall Panel Of Waste Paper Tapioca 'Perva' Answering The Ecologic Building Material Challenge

FX Bambang Suskiyatno, Krisprantono, Afriyanto Sofyan
Faculty of Architecture and Design Soegijapranata Catholic University Semarang, Indonesia

Corresponding e-mail: bambang_sus@unika.ac.id

Abstract. Building material in a building occupies more than 90% physically. In accordance with the provisions and regulations in many countries construction of buildings requires certification as 'green' as green building regulations. One of the ways to obey the rule is through the use of ecological building materials. Waste paper wall panels - tapioca "Perva" can be one of the viable alternative. Waste paper when strived as raw material exists in the realm of 'recycle' ecological. Paper and tapioca are organic materials containing organic elemental cellulose 'non toxic' building material. Procurement of paper waste raw materials can be done by collecting used paper. The formation of 2 (two) basic materials into new materials can be produced by simple technology easily done by people or 'self-help' technology. The weakness due to the use of materials with high levels of vulnerability requires the collaboration of additional raw materials to complement each other and form a favorable positive force. Composity with other ingredients can be done by excluding the ecological value if it is not found, natural ingredients with the desired criteria provided minimal and with a low non-ecological percentage value. Research of wall building materials Paper Waste - Tapioca has been done since the embodiment determines the aggregate composition, forms the demensi of certain materials, and makes the walls in a conventional way. Testing was conducted on several aspects related to the feasibility of building wall materials. Research continued on increasing the value of the shape as a composite panel. Experimental laboratory tested in architectural aspect, material feasibility aspect, its application in building and also feasibility aspects of wall construction. The research performed aggregate percentage test, the way of printing, the feasibility of wall construction to the adhesive aspect of coating material or surface protector and adhesive media. Arrangement as a wall with a certain extent used the system of persuasion at a certain dimension. The lateral load testing of wall construction resulted in effectiveness of the area of paper wall - tapioca.

Keywords: *paper, tapioca, panel, composite, wall*

1. Background

The Indonesian economy is increasing along with the strengthening of national development that must be accompanied by the spirit of 'Sustainable Development' with real application. The regulation of green building applications in Indonesia has been established at the national level; some up to the municipal government deals with the existence of buildings. The eco-friendly 'Green Building' system is already a demand in response to the worsening conditions of Global Warming and Climate Change. It responds to the country's commitment to reduce carbon emissions by up to 29% in responding to the decisions of some of the world's 'Green Conference' (COP) where almost every line of development leads to an eco-friendly footing. From the architectural aspect, the biggest influence is felt in the use of building materials. The selection and use of environmental friendly building materials are highly recommended and even controlled by regulation. Implementation in the field of building construction from residential houses to public facility buildings leads to 'Green Building' (Green Building). Building materials of more than 90% are used in a building, so it has a very important role when directed by the use of materials with ecological characteristics. Use of ecological building materials can increase the 'Green' value on a green building system. The effort of innovation and creativity is very important in the development of building materials. During this time almost all building materials seize enormous natural resources. Saving energy from non-renewable sources is very necessary as well. Many of them are 'toxic' and not hygienically safe. Perva is the name of the material we propose as the term of the new material we developed from 'paper' (paper) and 'cassava' (tapioca). The Perva composite panel classified as environmentally friendly building materials has been through a gradual experimental research review as a refinement. The problems that arise are the characteristics of raw materials on the core material of the panel products and the composite surface coating supporting material to the core panel. Other research that was carried out is the aspect of its ability to the construction application. The research of this new building material used Government and internal fund of Faculty, which is 'Multi years' and has stepped on 2nd year (two).

2. Methods

Discussion of research implemented in this research used quantitative and qualitative methods of laboratory experiments. Quantitative methods were performed using equipment with measurable indicators, while qualitative methods used laboratory means through visual and mechanical observations. The research method was done from the research of making substantial product of paper granule building - tapioca in the shape and demensi of brick and possibly other form with aesthetic character. The continuation of the research was the ability of building material product produced to construction aspects, such as compressive strength test, test of cutting success, test of surface coating with other material and screwing test. Research evolved towards the effectiveness of the area of the vertical wall against the lateral load of the wall. In the realm of production of new building materials waste paper - tapioca to the formation of wall construction was also through the research ability of the workmanship by the layman without a background of construction expertise.

Research increased on the formation of new building materials with panel products for building walls from waste paper - tapioca and composite panel which was a collaboration of paper materials - tapioca with other complementary building materials from the aspect of new material characteristics. The method discussed in this paper compared the results of research to the rules of ecological rules that could be achieved in new building material products.

3. Results and Discussion

The waste composite panel product was tapioca. This research aimed at testing effectiveness new wall building material never done before. This lead to creative innovation of easy raw material easily found but still maintained the strength and stability aspects of the wall. This initial step was to form a material that could be made from 2 (two) main raw materials of paper and tapioca. The collaboration of these two materials was very possible considering that both were derived from organic cellulosic elements of the ecological flora so that it could be incorporated adhesively. Basically organic cellulose was a natural element and not toxic, especially on tapioca raw materials. Paper raw materials were adjusted with preservatives and dyes were chemical and toxic. The process of pourage was expected to dissolve the chemical levels of preservatives and colouring.

Mixing of raw materials to form new building materials required special processing. Because the paper in units of sheets with a lightweight could be heavy when stacked, so there was almost no cavity between them. For relatively light amount and weight efficiency, the processing must lead to the end result of the number of cavities, within the paper itself and the cavities between the papers. The paper grain shape (pulpy granule) would be able to answer predicted forecasts. The paper granules were still trapped in the air inside, and the grain of paper was put together in large quantities, allowing the presence of many cavities between them. Wasted paper was mixed up by stamping manually to become pourage and then dried in the outdoor area.

In aggregate formation, the paper granules were mixed with tapioca. Flour tapioca was used as a thick liquid colloidal. Tapioca coloid was cooked up to boiling point conversion with a total cooking time of only 15 minutes using a gas stove. It was mixed and ensured that each grain of paper was coated with tapioca colloidal as a means of adhesive gluing. New building materials in trials were conducted in the form of 50 x 50 x 50 mm demension and increased in certain format dimension with estimation still within reach of pressure pressures by hand and manpower. The recommended demands were 50 x 100 x 230 mm printed press and removed. Printing was done manually relying on human weight and drying of renewable solar heating.

The product of the object formed when applied as a building wall component would have a very vulnerable surface to the aspects of the outside, heat, rain, sound waves, pressure and air humidity. Some can be anticipated, but some would cause damage and even material destruction. Therefore the material product formed requires a surface protector to anticipate. Some materials are possible to collaborate with staple materials, among others, cement, cement-sand mortar plaster, white cement plamuur - PVAC glue, emulsion wall.

Table 1. Selection of surface coating materials

SURFACE COATING & SAMPLE MOCK-UP50X50X50 MM	**CEMENT PASTE**	**MORTAR CEMENT – SAND (1 : 2)**	**PLAMUUR WALL WHITE CEMENT – PVAC**	**WALL PAINT EMULSION**
A	Sticking almost full The coating is broken a little corner Strong hard	Less sticky The coating is intact Strong hard	Sticking full The coating is intact Strong hard enough	Sticking full The coating is intact Soft soft
B	Sticking full A broken middle crack coating Strong hard	Stick full The coating is broken a bit Strong hard	Sticking full The coating is intact Strong hard enough	Sticking full The coating is intact Soft soft
C	Sticking apart The coating is intact Strong hard	Stick full The coating is intact Strong hard	Sticking almost full The coating is broken corner Strong hard enough	Sticking full The coating is intact Soft soft
EXPERIMENTAL RESULT	MIXED MIDDLE PERFECT	MIXED MIDDLE PERFECT	MIXED MIDDLE PERFECT	MIXED PERFECTLY

Those were experimental results of lateral load test of horizontal thrust against the vertical wall plane of the paper-tapioca brick structure with the same material adhesive medium. The test was performed on the area of 1.00 m2, 4.00 m2 and 9.00 m2 with edge frame.

Figure 1. Mock-up wall paper – tapioca

The experiment result showed that the feasibility of robustness, strength and stability of the recommended vertical wall was between 1.00 - 4.00 m2. This result was based on the effectiveness of the area. Every material wallpaper – tapioca 1.00 – 4.00 m2 had to apply the stiffness frame in every edges and could be categrized as partition wall.

The development of research focused on the effectiveness of wall without any reduction to the stiffness, robustness and stability of the wall. The result of optimizing the recommended area could be synchronized with the partition wall (although the shape was close to the wall filler), seen from the aspect of wall performance. Then an efficiency step was possible.

Configuring the arrangement of the shape of the paper brick - tapioca which was originally 'sleep' was changed to 'stand' so that the configuration of the order of the wall area had enough thickness, only 50 mm and forming 'wall panel'. Innovation steps led to the composity of the paper panels - tapioca form with the idea of leading to the recommended vertical demension of the area of 1.00 - 4.00 m2. It was necessary to add additional materials that strengthened the weakness factor of the wall panel, among others, the possibility of reinforcing the surface of the panel that could add the value of robustness in the ability of strength and stiffness of the field, especially lateral load strength. Stability led to the ability of the vertical panel to remain in position.

Successful addition of a surface layer of paper brick walls, one of which used 1 sand: 2 cement, showed perfect adhesiveness and added to the stiffness, robustness and stability value. The possibility of adding material strength to the surface of the field could inspire its application in wall panel research. Then it was possible to merge 2 (two) elements of materials with different characters into a form of composite material that could support each other from positive aspects that complement each other and strengthen. Therefore, the study led to the formation of new building materials in the form of Composite Panel of paper bricks - tapioca with other materials that needed to be tested before it was determined on many considerations. Of course in determining the material was also considered aspects of resistance to water and moisture.

There were various possible types of 'second' materials to be compiled with paper panels - tapioca including:

- Mortar cement - sand with composition (1: 2) (wet aggregate with ram wire)
- Calcium board (sheet)
- GRC (glass fiber reinforce concrete) (sheet)
- Acrilic thixotropic (pasta)

Table 2: Selection of composite panel composite materials

MATERIALS & ASPEK	**MORTAR PASTE CEMENT SAND (1 : 2) Thick 3 – 4 cm**	**LAYER CALSIUM CEMENT BOARD Thick 3 mm**	**LAYER GRC BOARD Glass Fibre Reinforce Concrete Thick 3 mm**	**PASTE PAINT ACRYLIC THIXOTROPIC (waterproof coating) Thick 0.5 mm**
The level of rigidity	Hard enough	Easily broken	Hard	Soft
Power level	Strong	Strong enough	Strong	Not very strong
Level of stiffness	Very stiff	Rigid	Very stiff	Bending
Waterproof level	Dainty	Simply impermeable	Very impeccable	Simply impermeable
Sticky media	Wire netting & interlocking cement paste	PVAC Glue	PVAC Glue	PVAC on the substance
Ease of workmanship	Need special skills	Easy simple	Easy simple	Quite hard & plated
DECISION	**NOT RECOMMENDED**	**RECOMMENDED** (2)	**RECOMMENDED** (1)	**NOT RECOMMENDED**

From the theoretical study conducted on 4 (four) possible materials that could be used as reinforcement of paper panel - tapioca panels, which was tested, it is concluded that the most recommended materials were GRC Sheet Board and Calsium Cement Board Sheet.

The composite panel formed was named as Composite Panel Perva - GRC and Composite Panel Perva - Calcium Cement. Composite Wall Panels Perva - GRC / Calcium Cement was formed with a demension of 1000 x 1000 x 60 mm, to meet the provisions of previous research results that the recommended effective area ranged from 1 to 2 m2, based on lateral load strength on the wall. A further recommendation is the provision of a stiffening frame and a wall holder in the form of steel or galvanish hollow mild steel. Composite Wall Panel Perva - GRC / Calcium Cement as partition wall can be collaborated with building structural system of any material such as reinforced concrete frame, profile steel frame, light galvanish steel frame for light building.

3.1. Tested Product

The cutting test of product showed the results of Composite Panel research that could be perfectly cut using manual hacksaws. Cutting with a hacksaw produced fine cut traces while using a wood saw produced pieces that were not perfect or fail because of the relative destruction. The cutting result could still be used as a component for the wall at a smaller demension.

Experimental test screwing outline using 3 types of screws each 3 sample showed that:

IOP Conf. Series: Earth and Environmental Science **213** (2018) 012040 doi:10.1088/1755-1315/213/1/012040

1. 50 mm long demension screw was firmly embedded, average scale was 6 out of 10 points
2. 40 mm long demanding screws were firmly embedded, averaging a scale of 4 out of 10 points
3. 30 mm long demanding screws were firmly embedded, average scale of 2 out of 10 points

The results read in this experimental study were only at the limit of vertical load embedded, not from the aspect of vertical tensile strength and lateral tensile strength given the softness of basic materials such as paper.

Self-help work test on experimental paper brick production - tapioca was done personally with no previous building construction expertise. At optimal working for 8 hours could produce an average of 45 - 50 pieces of brick size. The trials were also carried out to housewives and students and sixth semester students of the 3rd year on certain lectures, demonstrating the perfect results from the adhesive aspects of materials, product hardness and demension without deformation. Process production was only using simple tools with adding water and bake under the sun.

4. Conclusion

Production was carried out manually using print equipments and by hands. People involved did not require special skills and could even be done by the men and women of adolescence. People with no special skills could also do the construction of composite panels on the framework of the building. Therefore, the aspect from production to building construction could be categorized as ecological.

From the study of all aspects, among others materials, production processes and producers who involved even until the construction phase were categorized as ecological. Thus, the new building materials of Perva Composite Panel in this research could be categorized as ecological building material and have feasibility in ecological building. The Perva Composite Panel can be used as one building component to support the creation of ecological buildings and may be submitted in the proposing of 'Green Building Certification'.

5. References

[1] Anink David, Chiel Bonstra, John Mak, (1996). **Handbook Of Sustainable Building**, James and James (sciencepublisher) Limited, London.

[2] Bodgan R & S. Taylor (1992). **Pengantar Metoda Peneletian Kualitatif** *(terjemahan)*, Usaha Nasional, Surabaya.

[3] Frick, Heinz, FX. Bambang Suskiyatno. (1998). **Dasar-dasar Eko Arsitektur**, Penerbit Kanisius, Yogyakarta.

[4] Frick, Heinz, FX. Bambang Suskiyatno. (2007). **Dasar-dasar Arsitektur Ekologis**, Penerbit Kanisius, Yogyakarta.

[5] Frick, Heinz, Mulyani Tri Hesti (1998). **Arsitektur Ekologis**. Yayasan Kanisius, Yogyakarta.

[6] Frick, Heinz, Koesmartadi CH (1999). **Ilmu Bahan Bangunan**. Yayasan Kanisius Yogyakarta.

[7] Gabriel, JF (2001). **Fisika Lingkungan**, Hipocrates, Jakarta.

[8] Gate (ed), (1980), **Techniquest For Low Cost/Self-Help Houshing**, GTZ, Esdiborn.

[9] Nawawi, H & M. Martini (1994). **Penelitian Terapan**, Gajah Mada University Press, Yogyakarta

[10] Robinette, Gary O (1977). **Lanscape Planning for Energy Conservation,** Van Nostrand Reinhold Company, New York.

[11] Setiono, kuswijatri, Johan S Masyur, Anna Alisyabana (1998). **Manusia Kesehatan dan Lingkungan** Kualitas Hidup dalam Perspektif Perubahan Lingkungan Global, Penerbit Alumni, Bandung.
[12] Hidayat, d, Suparmin Sarino (1979). **Petunjuk Praktek Bahan Bangunan**. Jakarta.
[13] Illston JM, (1994), **Construction Materials Their Nature And Behavior**, Second Edition, E 7 FN Spont and imprint of Champman & Hall, London.
[14] Intep AG und Peter Steiger, (1996), **Hochbaukonstruktionen Nach Okologishen Gesichtspunten, Sia**, Scheizerischer Ingenieur-und Architekten-Verein, Zurich.
[15] Scheder, rolf. (ed). **Natural Building**. (IL 27) stuttgart: Institut Fur Leiche flachentragwerke, 1081. 310 halaman.
[16] Stulz Roland, (1981), **Appropriate Building Materials, Skat** Publication No. 12, St. Gall.

Conservation of 18th Century Java Industrial Heritage

Krisprantono
Faculty of Architecture and Design, Soegijapranata Catholic University, Semarang, Indonesia

Corresponding e-mail: kris_prantono@yahoo.com

Abstract. Industries and trading have important roles in the development of the city. Sugar industries in Java developed by the Dutch were influenced by Industrial Revolution. Sugar trade worldwide was related to transportation that included train, railways and bridges and lately came to harbor. Industrial activities, such as industrial heritage, steam machineries, locomotives and carriages, railway heritage, bridges, and anything related to industrial culture played a critical role in Indonesia's economic and social development that began in the late 18th century. The significance of study is that trade activities influenced urban development that also included architectural development. These activities generated a rich heritage of built structures, artifacts and records. But this heritage is at risk due to the growth and modernization of the Indonesian economy. Financial pressures result in the demolition of obsolete facilities located in urban areas, equipment is sold for scrap, and records are discarded. The purpose of the study is conservation of industrial building equipment and emplacement surrounded area. The method of the study was by historical method by tracing sugar history; agriculture, process of production, architectural record, trading activities and spread out of sugar industries in Java. Preservation of industrial heritage could play an important part in attracting visitors, as an alternative and complement to the island's more conventional tourists attractions.

Keywords: industrial heritage, archaeology and conservation

1. Introduction

The idea of the European continent on Enlightenment in the 17th century, industrialization in the 18th century and modernization in the 19th century simultaneously led to the colonization era. At the beginning of the 19th century European influence on trade management and the introduction of technology affected the world. This idea followed by Western European countries and also influenced the Dutch colonies in Java in developing agriculture and plantation industries, especially the sugar industry and the development of railroads for transportation to the port [1].

The ultimate creativity in the industrial revolution was the achievement of science in the form of technological change from human, animal or natural power. As a result of this technological change, the steam engine replaced human, and animal or natural power drove industrial machinery and transportation. The wider influence in the discovery of this new industry was to convert raw materials, mineral exploitation and agricultural products to produce a product for daily living needs.

Nowadays, the evidence of the above events is in the form of industrial buildings equipped with hydraulic engines and chimney burners which are symbols of the production process in 'space and time'. To transport raw materials to be processed from plantations and fields to industrial sites, manufactured materials were transported to ports for export using locomotives that pulled the carriages over long tracks.

IOP Conf. Series: Earth and Environmental Science **213** (2018) 012041 doi:10.1088/1755-1315/213/1/012041

Industry has an important role in economic, social and political development in the development of modern history of Indonesia. In the dynamics of the environment, the source of cultural inspiration certainly has economic value and shows the existence of a cultural heritage attraction. The legacy of cultural heritage presents itself as a tangible proof as an interest in the result of history relating to the object of the relic.

2. Methodology

This action research was carried out to understand historical potential agricultural land of Java sugar agriculture, Industrial agriculture of sugar, trading history of sugar and the consequence of 18 century Java Industrial architectural design in wider and more comprehensive means in the field of architecture. First, the writer as researcher defined and identified ground-theory by exploring data 'Recording and Documenting Historic Structures' archaeological research to find the original architectural shape and machineries, material of entire compound (emplacement, buildings and structure). After that, researcher reviewed architectural structure's system in a more comprehensive approach, doing a structural benchmark of industrial building. Qualitative approach with logical explanatory was taken to explore this type of buidings, machineries and structure. Quantitative approach was previously conducted by most of researchers [2]. Furthermore the value of this Industrial Building was obtained by asking 'What is the Significance of Historic Structures'.

3. Results and Discussion

The arguement will be conducted in multiple aspect-specific subjects and arranged gradually so the connecting aspects would be seen in comprehensive approach. The starting point is understanding a variety of the original building shape which supports industrial architectural design, continued by study of elementary architectural structure. Afterwards, discussion about structure will be explained and applied conservation architectural design by deepen understanding consrvation practical sense.

3.1. Industrial Archeology

The Industrial Revolution born in England in the 18th century was triggered by the discovery of the English scientist James Watt. His remarkable discovery was to convert the heated water vapor and then process it technologically into power that could drive the grinding machine, the locomotive engine and the ship's engine. It was this period that characterized the modern lifestyle, the agrarian transformation of society into an industrialist society in Western Europe. The Dutch community at that time brought the idea of the development of industrial technology steam engine to Java in the mid-18th century. They combined technology steam engine industry with the agricultural industry (agriculture) in Java. The coffee, tea, rice, tapioca and sugar industries were developed at that time. The peak of the Dutch East Indies economic development was the introduction of the system of forced cultivation.

The modern spirit of the Dutch State provided a remarkable development of industrial development, the discovery of various sources of energy for machinery, the intensification of capitalism and the intensive colonialism into modern imperialism. The most spectacular result of the industrialization was that Java, together with Cuba, became the world's largest supplier of sugar. After the abolition of forced cultivation many sugarcane and sugar industry plots fell into private hands (private). With this modern regulation the sugar industry became a seedling of capitalism [3]. This heritage which later gave prosperity to the Dutch people on the side of the Javanese society because of the System of Cultivation. Although the Cultivation System was abolished in 1870 but sugar cane as a raw material for sugar and coffee making industry was still in effect because sugar and coffee were very profitable commodities of Dutch trade.

Industrial Archeology is a branch of science that sees a period of history in which humans use technology as an important means of life. Against the cultural heritage of Java industry, there is a need for proper handling so that we can save the archaeological sites of great importance and the history of the machinery industry in the form of the development of steam engines, diesel engines and electric power machines [4]. The development of Archaeological Sciences in western Europe and the

United States today is not only concentrated on the site of medieval sites (15th century) and earlier but also oriented to the relics of the 18th century. Industrial archeology growing in Western Europe is a branch of archeology that studies the site industrial in the 18th - 19th century in the form of factory driven by steam engine. Archaeological Sciences of the industry is not only about industrial problems in the form of factory buildings but also facilities and infrastructure related to industries such as railway transportation, railway stations, irrigation systems, bridges and so on (Figure 1 and Figure 2). So Archaeological Sciences is a combination of Archaeological Sciences, Conservation, Architecture, Civil Engineering, Mechanical Engineering, History, Economics and Culture. Now even an international association has been formed for the conservation of industrial buildings as TICCIH (The International Committee for the Conservation of the Industrial Heritage) which often holds international discussions and seminars. The development of industrial archeology has had many results. In former Western Europe the former industrial area is conserved to serve as a tourist destination, the object of a former industrial building has a new function as museums, art galleries or other new functions appropriate for economic development. In the small town of Saltaire near Leeds, England many industrial buildings are still equipped with chimneys made office and apartment prestigious (Figure 1). Trains with locomotives with steam engine power are turned on again as an attraction for tourists. In France some ancient railway stations were made into museums [5].

Figure. 1. The building of a powerhouse industry built in 1830 in London (Source: Google Image)

Figure. 2. Industrial building as port infrastructure, Liverpool Dock (Source: Google Image)

3.2. Sugar Industry Cultural Heritage in Java

The development of sugar in Java was strongly supported by the climate and soil in Java which was suitable for sugarcane. Besides that, after the Industrial Revolution in Western Europe, the Dutch brought a form of steam engine technology that could move the milling industry on a large scale in Java. Sugarcane farming areas in Java are found in Central Java and East Java. The best climate for sugarcane cultivation is a hot area and enough rainfall. Noted in the map 'Suikerfabrieken op Java', sourced from Cartography. Inr. J.H. de Bussy. Amsterdam 1914 stated that in that year there were 191 sugar factories active in production, while from the map of 'Kaart der Suikerfabrieken, Spoor en Tramwegen van Java en Madoera, Bijgewerkt tot 1 June 1925 also from Cartography source. Inr. J.H. de Bussy, Amsterdam, on the map shows in Java in 1925 there were about 201 sugar factories that were actively producing. The sugar industry played an important role in changing the economic tradition in Java from the agricultural economy to the mechanization of the economy in the period

1880-1955. In the mid-19th century after the mechanization period of the sugar industry the position of sugar production in Java rose to the best compared with other agricultural and plantation products. In the mid-18th century Java became the world's largest supplier of sugar in competition with Cuba and the results were still felt during the independence period until the 1970s [6].

From the sequence of events above, the sugar factory in Java is recorded as a heritage of the Javanese, Chinese and Dutch communities that are now agreed as 'shared heritage' or 'mutual heritage' [7]. Sugar mill factory grew in line between Cirebon to Kudus area which at that time was built transportation means in the form of transportation line for the purpose of trade in the form of famous railway with the nickname 'Golden Road' to connect the area of sugar-producing areas, such as sugarcane farm and sugar mill factory so that with easily transported sugar for export through the port of Semarang. Office of administration office to operate sugar production and trade mostly exist in Semarang Old Town which was the forerunner of Semarang city development [6].

The legacy of the sugar industry in Java, which is 1 ½ centuries old now is slowly threatened with extinction. The process of decline in the sugar industry in Java began in about the 1920s when Mount Merapi erupted causing the building of many smashed sugar factories, around Yogyakarta – Surakarta. There were about 15 sugar factories collapsed due to earthquake such as PG. Medarie, PG Kedaton Pleret, PG. Gedaren, PG. Gedoeng Banteng, PG. Gesiekan, PG. Demak Idjo and PG. Ngelom that no longer existed, only rubble debris or the name of the area. While some sugar factories that still existed until now around Yogyakarta and Surakarta only PG. Mojo in Sragen, PG. Tasik Madu in Surakarta (Figure 3) and PG. Gondang Baru in Klaten while PG. Ceper near Klaten and PG. Colomadu in Surakarta no longer reproduced. Then a recession in 1930 hit the sugar industry in Java, known for the "malaise" era where many factories were closed and destroyed [1]. Since the Second World War Java was under Japanese rule and many Dutch people returned to the Netherlands the fate of an sugar mill was erratic. After the independence of the Republic of Indonesia there were roughly about 50 of the remaining sugar mills. After the rehabilitation period the machines of the industrial sugar factory began to recede in the period between 1970-1998. Since the economic crisis of 1998 many sugar industry dismissed its operation due to costly production cost.

The sugar factory is a witness to the history of economic development during the industrialization period in Java which gave many advantages not only to the Dutch East Indies colonial government in Java but also the government of the Republic of Indonesia after independence until the mid 1970s. The Dutch industrial heritage is not only 'tangible' (factory buildings that are physical buildings, machinery and locomotives and lorries) (Figure 4), but also 'intangible' (i.e. non physical system of intangible) that is cultivation system of sugarcane farming and technology the process of making sugarcane into sugar. In addition, regarding the forced cultivation system to obtain raw materials in the form of sugarcane, farmers were forced to plant the plants determined by the VOC as the ruler. Farmers as owners and processors of land were forced to plant sugarcane on 1/5 of their land. This system survived even though Indonesia became independent and the sugar factory was under the Ministry of Prosperity of the State Sugar Company Administration (*Badan Penyelenggara Perusahaan Gula Negara* - BPPGN) in Yogyakarta.

3.3. Cultural Heritage Preservation

In its process, conservation is always questioned "What is the significance of the object(s)", what is the importance of the object to be preserved [8]. In the age of more than one and a half century sugar factories had the value of importance to be preserved as a cultural heritage (Benda Cagar Budaya). The Law on Cultural Heritage 2010 states that cultural heritage objects are a cultural richness of the nation which is essential for the understanding and development of history, science and culture, so it needs to be protected and conserved for the sake of the nation's awareness of national identity and national interests. Maintaining the preservation of cultural heritage objects required regulation steps that are useful for the protection, maintenance, management, utilization, and supervision of cultural heritage objects. From historical data it can be understood that sugar has an important meaning and meets the above criteria not only the history of economic development but also the history of port development and the history of urban development (urban morphology). Sugar factories also have aesthetic value as an architectural industry equipped with office and residence with colonial architectural style. This is important for the scientific field of architecture and civil engineering to learn how to study architecture and structure of the past. In addition, sugar production used the locomotive as a means of transportation and machine engine with steam power technology which in its development was combined with diesel engine power engine and electric power. This unique and very interesting phenomenon can be developed for the benefit of not only tourism but also science in the field of machinery technology. In addition, the traces of sugar factory's history cannot be separated from the surrounding life of sugar cane farming which is an economic, social and cultural problem.

This event occurred again in economic recession of 1997-1998 where many sugar factories dismissed its operation. The 1997 economic crisis affected the shrinking productivity of sugar mills. At its peak about a dozen sugar factories were unable to operate again in Central Java. After the year 2000 even Indonesia began to import sugar to supply domestic needs. The closure of the post-crisis sugar factory had a very bad effect, firstly many people working in the factory had no jobs, including sugar cane farmers who had to think of planting other crops in the sugarcane field before. The third most serious problem was the closing production of many sugar factories because of less benefit of sugar production compared with importing sugar. As a result many facilities in sugar factories were damaged and decayed ranging from factory buildings, warehouses, office facilities, housing, and other facilities whose historical and architectural values were so high. They were not well preserved. There was much looting from elements of existing home materials, railroad trains to old iron machines that have historical value. In addition, due to heat and rain there is a lot of damage to the building because there is no treatment because there is no system to take care of it. This situation is ironic. On the one hand, many people need housing to reside and school as a place to study; on the other hand, the government let sugar factory uninhabited where the factory buildings and other facilities were gradually destroyed.

The remaining problems are about 40 more sugar factories still in production and dozens of other sugar factories that are no longer producing in Central Java and East Java under the coordination of PTP IX, X and XI. Production of sugar factories is highly dependent on the supply of sugarcane as the raw material for sugar production comes largely from sugar cane farmers. Now the quality of sugarcane is not as good as it used to be, the rendement is not good anymore because now the sugarcane farming process is not cultivated well, especially in Central Java. In addition, the land for sugar cane farming is also decreasing as farmers have moved to plant other plants that are more profitable. In East Java there are still many sugar factories in production, sugar cane is also still quite extensive and well cultivated. But in general the problem of the main sugar factory is the cost of production which often exceeds the price of the sugar product. If left unchecked this will result in the closure of the sugar factory slowly and we will lose the legacy of historical relics that are priceless. Alternative to tackle this problem should be considered together because the assets 'Industrial Archeology' in Java is already a global class. Furthermore, this sugar factory is a living museum whose existence is rare in this world. If this asset is well managed and thought together by the stakeholders then it is not impossible will give a positive outcome.

3.4. Sugar Factory Revitalization

Revitalization of Regions is an effort to revive a region by providing new functions to generate the vitality of existing and degraded areas and buildings but not yet fade through physical and non-physical interventions in the fields of economy, environmental management, heritage management, socio-cultural engineering and institutional development. According to Tiesdel in Revitalizing Historic Urban Quarters (1996) revitalization is: Improving the quality of the property to address certain dimensions of that property obsolescence. Owners and occupiers of buildings can address the dimensions of obsolescence that are within their abilities, especially the structural, functional and image dimensions. There are redevelopment, renovation and re-function. The Revitalization approach should be able to recognize and exploit the potential of the environment (history, meaning, location uniqueness and place image). With the support of revitalization plan control mechanisms, it should be able to raise the strategic issues of the region both in the form of social and cultural activities and economic and physical character of the region [9].

IOP Conf. Series: Earth and Environmental Science **213** (2018) 012041 doi:10.1088/1755-1315/213/1/012041

The existence of tourist areas for recreational facilities will provide economic sustainability to be able to support the existence of Gondang Baru Sugar Factory. The plan includes a museum of sugar that is passive and active for the benefit of education and preservation of cultural heritage in the form of sugar processing industry using 19th century technology in the form of steam engine. Revitalization of the area as an effort to save and preserve the area of Sugar Mill Tasik Madu Plant in Karang Anyar Surakarta previously owned by Mangkunegaran has been done by the PTP. The location of the emplacement is used as a tourist area while the sugar production process is still running. This activity contributes to the preservation of redevelopment, renovation and refunction but the work of this process does not involve archaeologists, so that the value of authenticity decreases and many materials that are not original are replaced with new material.

Figure 3. Tasikmadu Sugar Factory Karang Anyar Surakarta.

Figure 4. Former sugar cane machine is placed in an open area as a sculpture for children's playground.

Archeological principles in the implementation of conservation need to be introduced to the world of architectural and regional design:

Minimum intervention: As in Burra Charter Art.3, 1979, New Zealand Charter Art. 4.iii, 1992 which emphasizes that "conservation action is an act of intervention to historic buildings on the level of building authenticity. Therefore, any intervention action on the grounds of research and technical action from the beginning should be kept to a minimum.

Minimal loss of fabric: As evident in Deschambault Declaration Art. V-C, 1982, New Zealand Charter Art. 4.ii, 1992 which emphasizes that "all conservation actions should take into account the original loss of building materials to a minimum."

Riversibility: As is evident in Burra Charter Art.1.10, 1979, Appleton Charter Art.D., 1983 which emphasizes that all intervention measures to historic buildings should refer to the possibility of broad future developments or remedial action on issues not yet seen where the originality of the source may be impaired.

Legibility (clarity): As written in Venice Charter Art.12, 1964, Burra Charter Art. 19-3, 1979 which asserts that "all new reimbursement of missing elements in historic buildings must be distinguished from the original to avoid counterfeiting of historical evidence.

Buildings in the factory area (emplacement) in particular the original residential buildings and elements of machinery and architecture are conserved by referring to the rules that have been established and revived. Steps that need to be done are as follows:

- **Recording - Documenting historic structures**
  The historic original building is preserved by conducting the first historical exploration studies to show that the building has historical value so it needs to be preserved. Second, hold complete measurements and architectural reconstructions to record the original data of the building [10]. Third, reconstruct steam engine machines that still work both milling and locomotives because the machine is a rare item and priceless and will be the main attraction because of its uniqueness.
- **Improve environmental quality**
  Facilities and infrastructure in the sugar factory area need improvement such as roads, drains, exhaust system which often cause problems of pollution in the surrounding environment.
- **Rules of rules in the preservation process**
  Parties PTP and BPCB (Cultural Heritage Preservation Agency) representing the government should do conservation program with rules according to the rules of conservation right. Where necessary, any new material replacement should be done according to the correct rules [11]. The need for architecture data for the purposes of conservation for the authenticity of building buildings and machine engines that have this historical value awake authenticity. The government can also cooperate with universities and professional organizations, such as HAPBI (Association of Building Maintenance Experts) or IAI (Indonesian Architects Association).
- **Economic feasibility study**
  Physical revitalization in the form of buildings, machinery and environment will increase confidence in the area, but to keep confidence to be maintained, there must be a revitalization in the field of economy. Tim Heath Tiesdel in his Revitalizing urban quarters proposes that 'Functional restructuring' new uses or activities of replacing the former ones; while 'functional regeneration' is an alternative to existing uses. Moreover, the change in building functionality will provide new alternatives to increase revenue [9].

3.5. Conservation of 18 Century Java Industrial Heritage

Industrial heritage in Java that include sugar, tea and coffee factories are part of our heritage. Economic history also recorded this heritage in relation to world trade. In the sense of trading history, the heritage included not only factory but also train railway and bridges to transport the product of tea, coffee from the factory to the harbor. All of the 18-century factories and all of their facilities, train railway bridges and harbor still exist in Java. Some of them are still in operating system. They have value in the sense of technological development, trading system and agricultural science in Java.

According to Government Regulation related to Heritage; *Act no 11 year 2010 about* cultural heritage is something handed down from one generation to next that has cultural value. 18th century Java sugar mills have historical, scientific, educational and cultural values included as cultural heritage.

1. **Conservation Program**: BPCB (Cultural Heritage Preservation Agency) representing the government should do conservation program with rules according to the rules of conservation right.
2. **Recording and Documenting**: The need for data, history, archaeology, architecture and structure for the purposes of authenticity of building and machine that have historical value.
3. **Revitalization:** To improve the economic level of sugar mills, a competitive development is needed for emplacement of sugar factories for better community life. One of which is the means of recreational and educational activities to bring the wheels of the economy up.
4. **Authenticity:** Conservation action is an act of intervention to historic buildings on the level of building authenticity. Therefore, any intervention action on the grounds of research and technical action from the beginning should be kept to a minimum.
5. **Improve Environmental Condition:** Environmental conditions that exist in the area of sugar factories are usually slum because a lot of goods in the form of ferrous iron had no place and then littered the area. They need to be arranged and maintained.
6. **Reversibility and Legibility:** All intervention measures to historic buildings should refer to the possibility of broad future developments or remedial action on issues not yet seen where the originality of the source may be impaired. All new reimbursement of missing elements in historic buildings must be distinguished from the original to avoid counterfeiting of historical evidence.

4. Conclusion

In fact, all sugar factories in Java are highly feasible to be able to revitalize both those that are still in production and that stopped. All stakeholders must support each other. Archaeologists should play a great role in this activity to preserve the value of the authenticity of industrial cultural heritage. The surrounding community will definitely support this effort if this revitalization has a positive impact for the surrounding community. The Pecangaan Sugar Factory near Jepara is one of the oldest factories in Java now still operating with other functions as a sack factory where at least the existence of the building can still be maintained. Revitalization of sugar factories is not only for tourist sites but also other facilities. For instance, the area of Sugar Factory Colomadu is no longer operable but it can be used as other facility where its machines can be stored in the museum. The Olean Sugar Factory near Situbondo East Java is a small but very attractive sugar mill. All facilities are still complete and original from factory building, housing, machine, locomotive along with its rail and old *munggur* tree. If the revitalization is carried out correctly, it is not impossible that this area can be proposed to be 'World Heritage Industrial Site'. All can be done as long as 'no damage' in the sense of maintaining the authenticity of the area and the building of cultural heritage.

5. References:

[1] Ricklefs M . 1991 *A History of Modern Indonesia Since 1300* (London: The Macmillan Press Ltd.)
[2] Schodek D L 1999 *Strukture Edisi Kedua* (Jakarta: Erlangga)
[3] Hall D G . 1964 *A History of South East Asia* (London: Macmillan & Co Ltd)
[4] Palmer M and Neaverson P 1993 *Industrial Arcaheology* (Routledge)
[5] Stratton M and Trinder B 2000 *Twentieth Century Industrial Archaeology* (London: Spon Press)
[6] Nas P J . 1986 *The Indonesian City: Studies in Urban Development and Planning* (Amsterdam: Floris Publications)
[7] Fielden B 1994 *Conservation Historic Buildings* (London: Butterworth Henemann)
[8] Earl J 1997 *Building Conservation Philosophy* (Reading: The College of Estate Management)
[9] Tiesdell S, Oc T and Heath T 1996 *Revitalizing Historic Urban Quarters* (Oxford: Architectural Press)
[10] Alfrey Y and Putnam T 1992 *The Industrial Heritage* (London: Routldge)
[11] Cunningham P 1988 *Change of Use* (London: A&C Black Publisher)

The Impact of Building Mass Configuration Towards Wind-Driven Natural Ventilation in Apartment in Jakarta

Fathina Izmi Nugrahanti, Irma Handayani Lubis, Dibya Kusyala
Institut Teknologi Bandung, Bandung, Indonesia

Corresponding e-mail: fathinaizmi@gmail.com

Abstract. Recently, based on the data released by Ministry of Energy and Mineral Resources (ESDM), housing is the second most energy-consuming sector after industry sector in Indonesia. Furthermore, the biggest energy consumption in housing is for air conditioner. Meanwhile, the rise of temperature cause people to use artificial ventilation more often, leaning to higher energy consumption. In order to reduce that, architects need to provide a design option, so people can achieve thermal comfort through natural ventilation by utilizing wind forces. Cilincing District is selected as case study for its microclimate characteristic. Since it is located in the coastal area of North Jakarta, the prevailing wind that comes from north-east can be utilized for wind-driven natural ventilation. There are some factors related to natural ventilation, one of them is building mass configuration. The aim of this paper is to find the optimum building mass configuration inside the given site area in order to achieve the optimum air movement around the buildings. The methodology in this research was design iteration using Ansys Fluent CFD simulator of 5 design models; each consisted of 6 building blocks. The results showed that model 5 with un-linear (zig-zag) building layout and combination of various building distances (wide – narrow – wide), which narrowed building distance in the middle of the site, provided optimum air movement around the buildings because it could create a kind-of tunnel/ trap that strengthened the wind which produced a constant wind velocity along the site.

Keywords: wind-driven natural ventilation, passive thermal comfort, green design, walk-up apartment

1. Introduction

Based on the data provided by Southeast Asia Energy Outlook [1], Indonesia plays a major role in energy consumption, accounting for over 35% of the region's total energy demand. Household has become the second most energy-consuming sector after industry in this country, which mostly used for air conditioning (ventilation). Meanwhile, with the rise of temperature nowadays, the tendency of people to use active strategy to achieve thermal comfort, becomes inevitable. This energy consumption will keep rising along with the increase number of housing, especially in big city like DKI Jakarta.

According to this challange, architects need to develop a housing design solution that can save more energy especially for ventilation by reducing cooling load. Bioclimatic design, utilizing passive design strategies, have been considered as the main solution to achieve sufficient degree of thermal comfort with less energy consumption. Idham [2] mentioned that there are five principles of thermal control, i.e. thermal insulation (related to heat), air movement (related to wind), cooling and heating (related to temperature), also dehumidification and humidifaction (related to humidity). Due to high humidity and temperature levels in tropical-climate building, heat avoidance and natural ventilation cooling could be the best options to adapt comfort in indoor spaces [3].

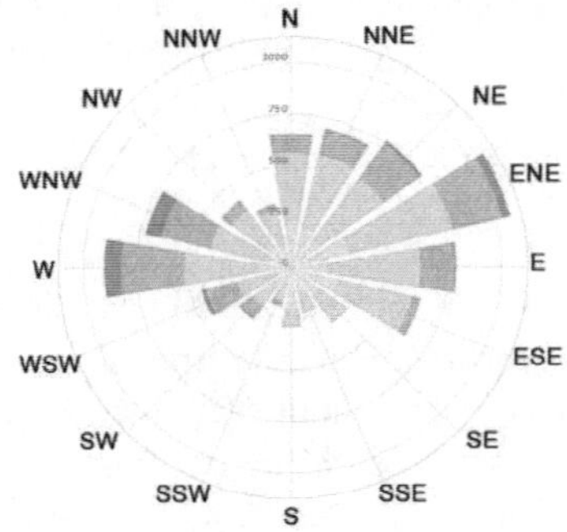

Figure 1. Wind rose of North

This research object is a walk-up apartment building located in Cilincing District, North Jakarta, only 550 meters from Marunda Beach. Adjacent to Java Sea, this location has a strong character of coastal region, where the prevailing wind is potential to use as natural ventilation that can achieve thermal comfort, thus create energy-efficient building design.
Based on local data from Indonesian Agency for Meteorology, Climatology and Geophysic (BMKG) Tanjung Priok Station, this area has monsoon climate with average wind velocity 3,3 m/s which mostly comes from north-east, average temperature 28,96°C, and average humidity 107,83% around the year.

In early design process, some design factors were used to determine building form that can affect air movement around the building. Busato [4] explained that *shape and orientation* of the building relative to direction of the wind are both directly related to the way in which air movement behaves around building. Rectangular plan alongside east-west, limiting the exposure on east and west sides, is the best *building shape* in the tropics [5]. Shallow floor plan (narrow *building depth*) is more efficient where it encourages more air-flow inside the building [6]. The optimum *building proportion* for hot-humid climate is 1 : 1,7, elasticity 1 : 3 [7] with the most effective *building orientation* should be defined based on prevailing wind and sun angle [8]. There are plenty of researches about building shape, proportion, or orientation but few have discussed about the significancy of *building configuration/ layout* towards air movement around building. The aim of this paper is to find the optimum building mass configuration on a spesific walk-up apartment complex, to achieve the optimum air movement around the buildings for natural ventilation.

2. Methods

2.1 Determine the basic mass model

Walk-Up Apartment is a vertical housing which use stair as the main vertical transportation system with the units type chosen are one bedroom 24 m^2 unit and two bedroom 36 m^2. The site has 1.4 Ha area with maximum 35% built floor area. Density that will be achieved is 460 persons/Ha, so there are six blocks of five-storeys buildings with 288 room units/block.

This research was preceded by preliminary analysis of building shape and orietation simulation, followed by the main simulation topic in this research, building layout/configuration. The first preliminary analysis was completed to find the basic mass model for further simulation, which was obtained from the adjustment between number of units needed, the desired unit area, and built area, with the following criterion as fixed variables.

Table 1. Preliminary criterion for the building

Unit	Shape	Square & Rectangular	Based on market trends analysis
	Size / Poportion	4x6 (24 m2) & 6x6 (36m2)	Based on market trends analysis
Massa	Circulation	Single-loaded	[3] & Based on market trends analysis
	Basic shape	Rectangular (Box) Atrium	[5][7][14]

2.2 Set up the simulation

The computational wind simulations were completed by Ansys Fluent Simulator to investigate the effect of building mass configuration, which received the optimum airflow for wind-driven naturally ventilated housing. In order to model the air movement of the site area, the simulations used a scaled model, scaled to 1:100. As a result, adjustment should be done in the boundary conditions, especially the wind speed. Similarity of the simulation on the scaled model to the real scale model depended of the Reynolds Number [9]. In this study, the wind speed of the real scale was 3.3 m/s, which in the scaled models should be stated as 330 m/s due to Reynolds Number formula.

The analysis of airflow behaviour around the buildings was conducted by the emphasis of qualitative analysis, which confirmed by quantitative analysis. Hence, visual analysis of wind and temperature contour images were selected. Two-plane level was selected as a sample for the velocity and temperature magnitude, where plane at 2 represent(s) the ground level and 10 represents the high level.

3. Discussion

3.1 Wind-driven natural ventilation

Generally, pressure discrepency can be generated by wind (wind-driven ventilation) and temperature/ thermal bouyancy (stack effect ventilation) or by combination of both [4]. The stack effect occurs when indoor temperature is higher than outdoor temperature where the warm indoor air will be risen up and replace the cooler, denser air from below. Wind-driven natural ventilation occurs when higher wind velocity hits the building surface and creates gher air pressure (+) on the inlet and lower pressure (-) on the other side that will initiate air to move across the building, bring the fresh air and also reduce room temperature [10].

Naghman K. et al. [11] observed that in hot-humid conditions, where the temperature differences between indoor and outdoor of buildings (diurnal variation) are small, stack effect is reducesd so stack ventilation is insufficient to achieve thermal comfort. Therefore, Lim [12] suggested that wind-induced natural ventilation design has a great potential to achieve the desirable air speed in the indoor building environment to improve thermal comfort for building occupants especially in tropical climatic conditions. Since this research located adjacent to North Jakarta sea which the prevailing wind is potential for natural ventilation, this paper will be focus on wind-induced natural ventilation.

3.2 The effect of building mass configuration on airflow around building

Besides building shape and orientation, other factors affecting the airflow pattern around the buildings are the physical relationship to surrounding volumes and obstacle such us other building mass, fences, vegetation, etc. In a building complex where there are multi – building masses, every building mass can affect each other's airflow pattern. For example, the positioning of buildings within the suction zone (negative pressure – shaded area) of other structures can impact on the velocity of air coming into the inlets of the naturally ventilated building [4].

In order to avoid this 'shadow effect', the *distance between buildings* has to be adjusted. Markus [13] proposed that the distance between each building mass should be minimum six times the height of the first obstructiong volume, which actually quite a wide (if it is assumed that the obstrcuting volume is a one-storey building). Other technique mentioned by Markus is using a *staggered layout* to organize the masses.

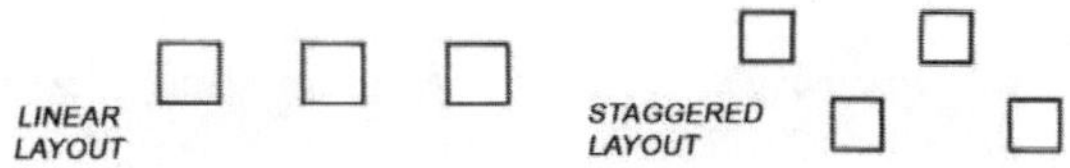

Figure 2. Proposed staggered layout

Other research conducted by Hong [14] found that trees arrangement, buildings layout patterns and their orientations with respect to wind have significant effects on the outdoor wind environment. Furthermore, long facades of building, which are parallel to the prevailing wind direction, can accelerate horizontal vortex airflow at the edges and obtain pleasant thermal comfort and wind environment at pedestrian level. In line with Hong, Olgyay [7] stated that in tropic areas the buildings are freely elongated. The houses are separated to utilize air movements which generate a scattered and loose layout.

3.3 Preliminary analysis

The preliminary criterion mentioned at **Table *1***, generated two building shape options, i.e. the rectangular plan with the size of 35x15x16 m and the square plan (25x20x16 m). However, the rectangular plan was chosen because it could reduce the building heat gains. This is in line with Konya [5] who said that it is better for tropic-climate to have rectangular plan where wind can be distributed evenly along the building surface and gives less east-west exposed units which can reduce heat gain. Then, followed by building orientation simulation to determine the chosen shape for further analysis. Burnett [16] proposed to test three different orientation 0°, 45°, 90°. From the simulation, model which is parallel to wind direction 0° is chosen because it has the least facade facing east – west.

3.4 Mass-Configuration Factor

From the previous reseaches mentioned before, it can be summarize that there are two main suggestions related to building mass configuration, *pararrel layout* [7] [14] and *staggered or zig-zag layout* [13]. Those are used as the basis of the building mass configuration simulation.

Beforhand, minimum distance between masses was determined based on technical guidelines from building permit issued by Indonesian goverment [17]. For two opposite buildings that both have large openings or transparant facades, the distance must be minimum twice of the clearance number determined.

Table 2. Minimal distance between buildings

Min.distance between buildings	Transparant facades / facades with large openings Clearence (Y) for each 5-storeys building is 5.5m	D = Ya + Yb = 5.5 + 5.5 = **11 m**

From the standard above, some model options were developed for the simulation.

IOP Conf. Series: Earth and Environmental Science **213** (2018) 012042 doi:10.1088/1755-1315/213/1/012042

Table 3. Comparison for each configuration

Case	v (m/s)	lv.	Contours	Data
Configuration A				
(A.1) 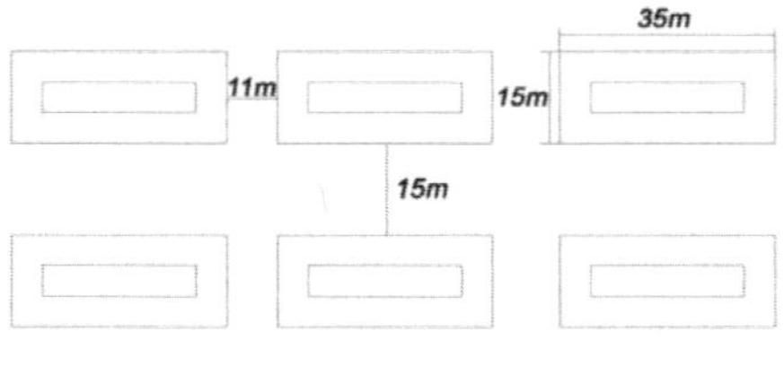 v average = 3.132 m/s	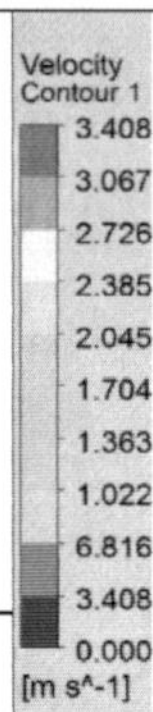	2m		3.11 m/s 27°C
		10m	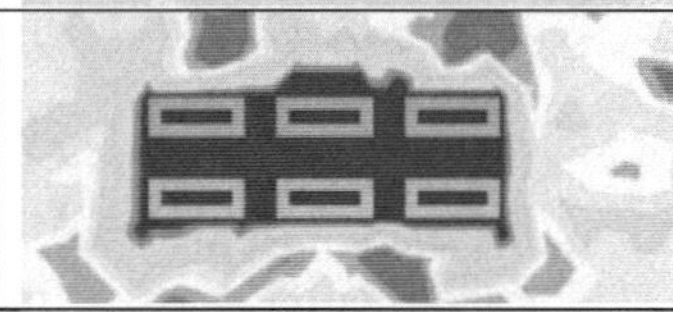	3.15 m/s 27°C
(A.2) 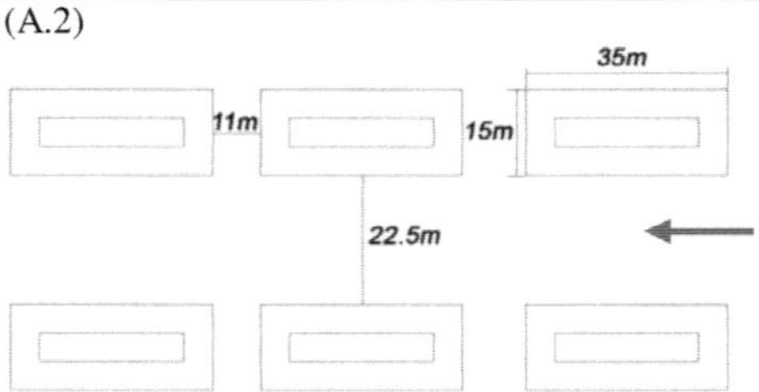 average = 3.109 m/s		2m	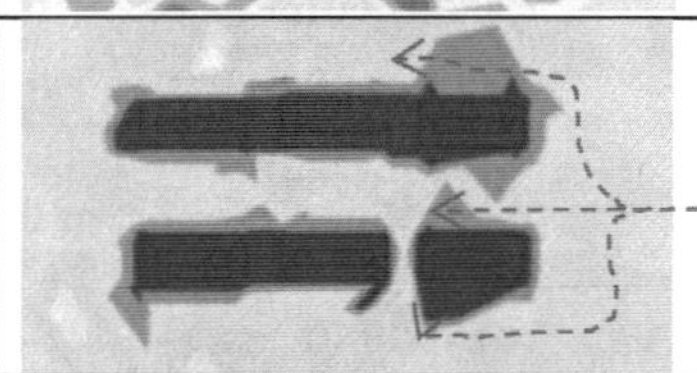	3.09 m/s 27°C
		10m	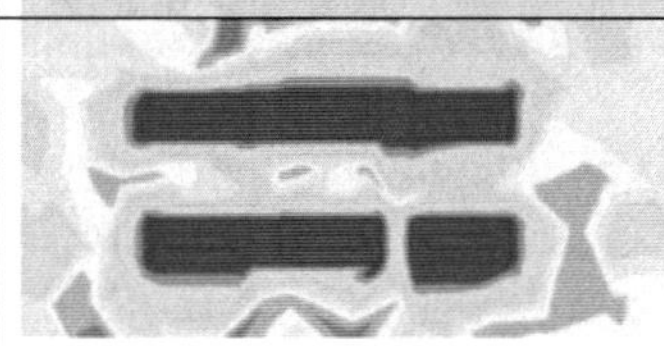	3.13 m/s 27°C
Configuration B				
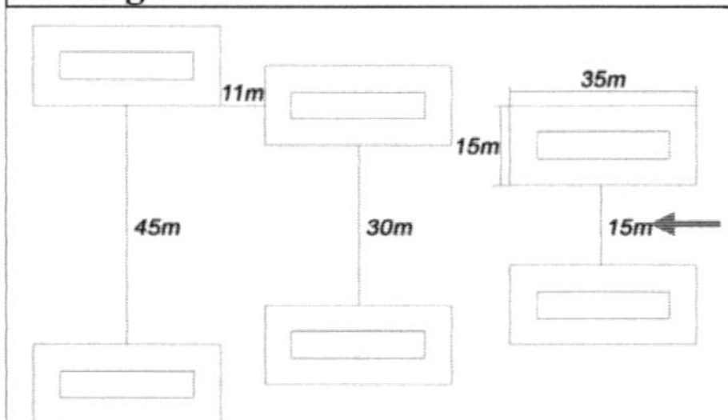 average = 3.066 m/s	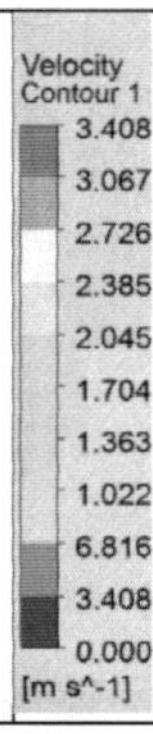	2m	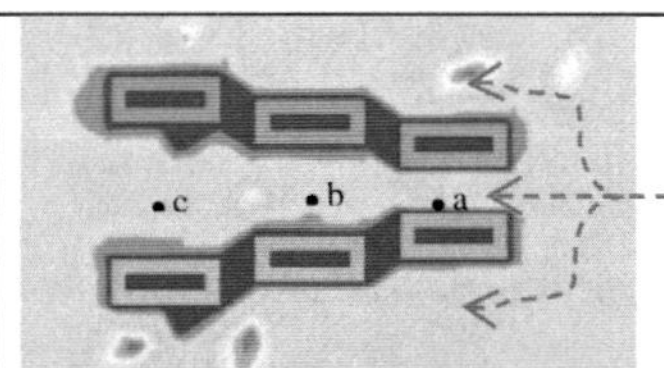	3.03 m/s 27°C
		10m	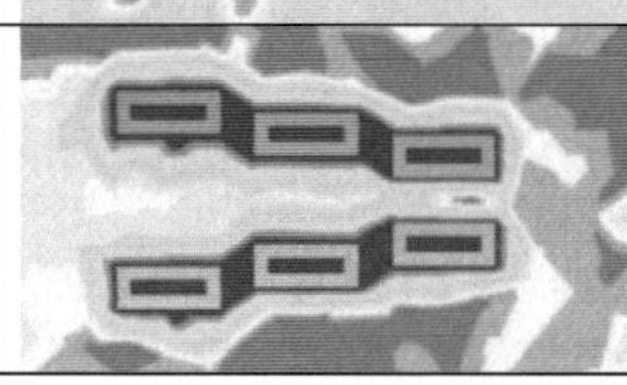	3.10 m/s 27°C
Configuration C				
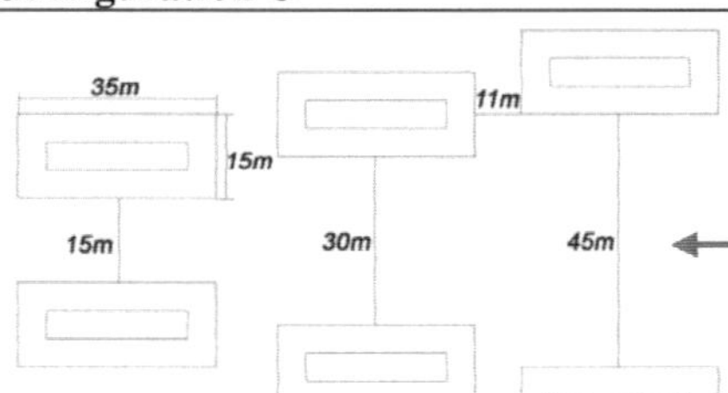 average = 3.062 m/s	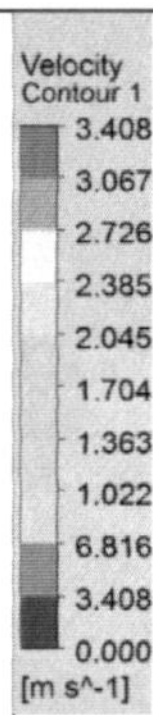	2m	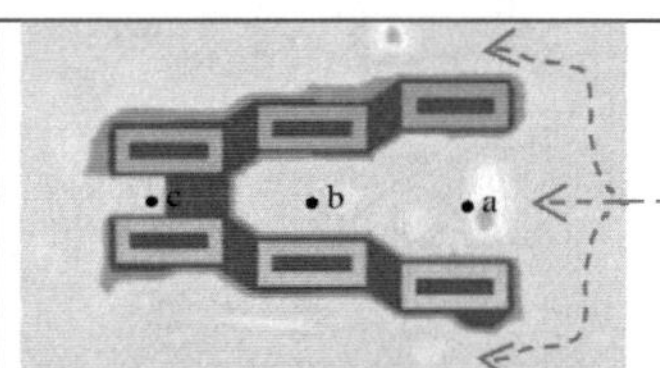	3.03 m/s 27°C
		10m	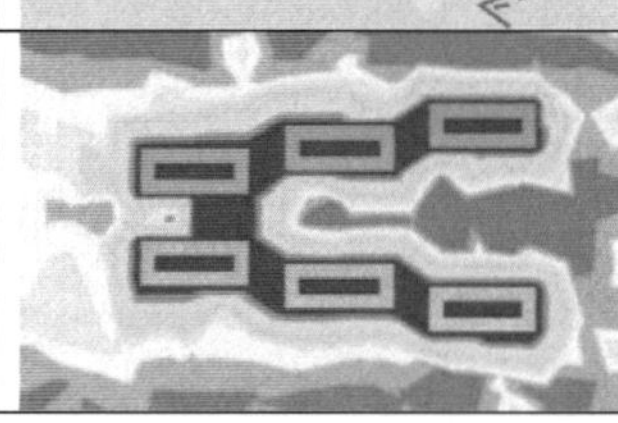	3.10 m/s 27°C

IOP Conf. Series: Earth and Environmental Science **213** (2018) 012042 doi:10.1088/1755-1315/213/1/012042

Configuration D				
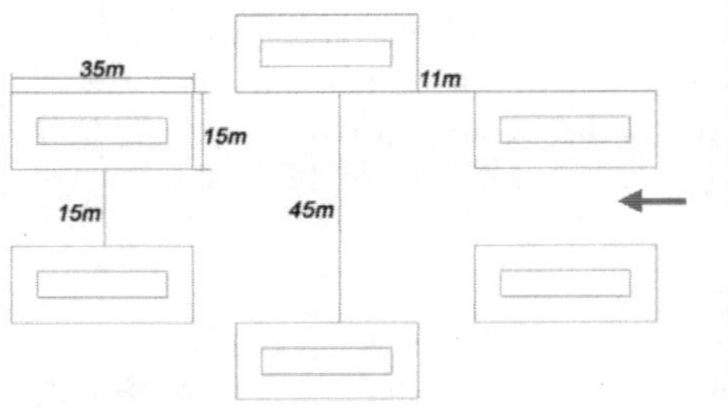 average = 3.106 m/s	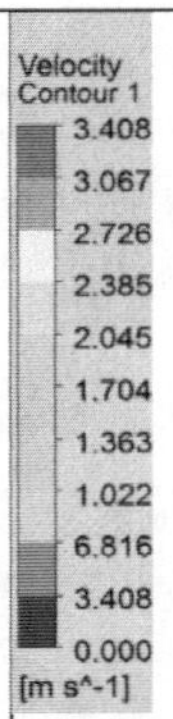	2m	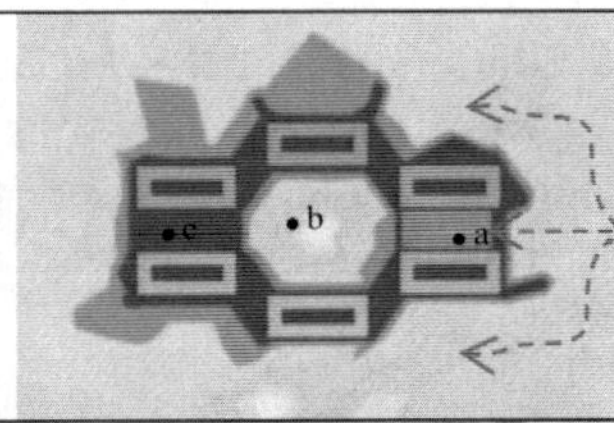	3.09 m/s 28°C
		10m	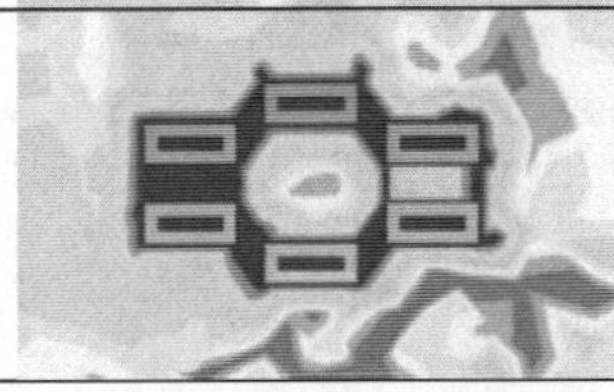	3.13 m/s 27.5°C
Configuration E				
E.1 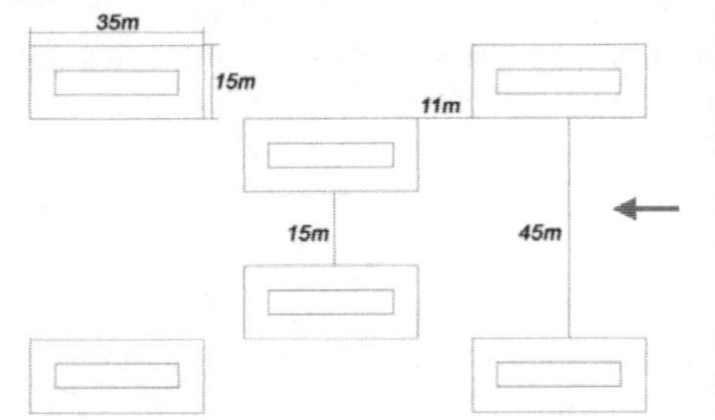 average = 3.110 m/s	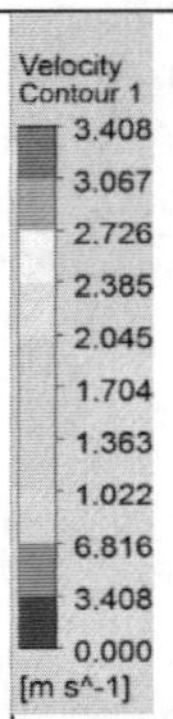	2m	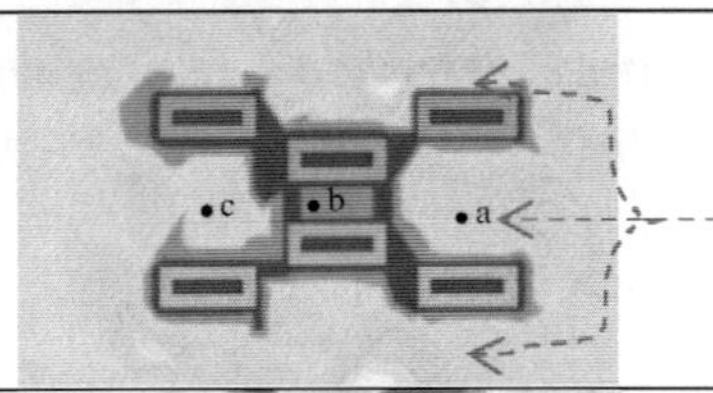	3.09 m/s 28°C
		10m	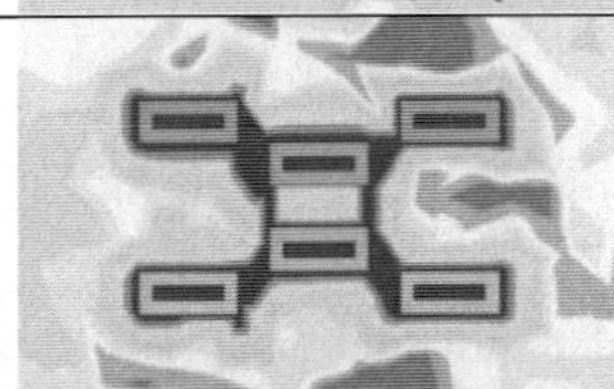	3.13 m/s 27.5°C
E.2 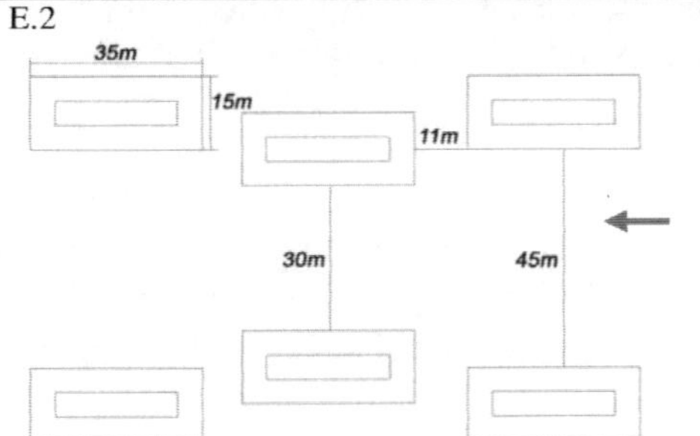 average = 3.121 m/s		2m	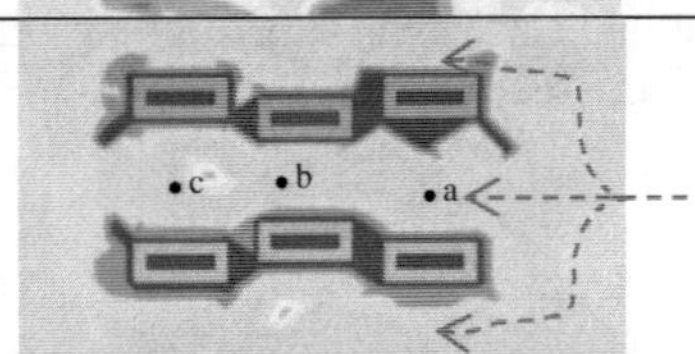	3.10 m/s 28°C
		10m	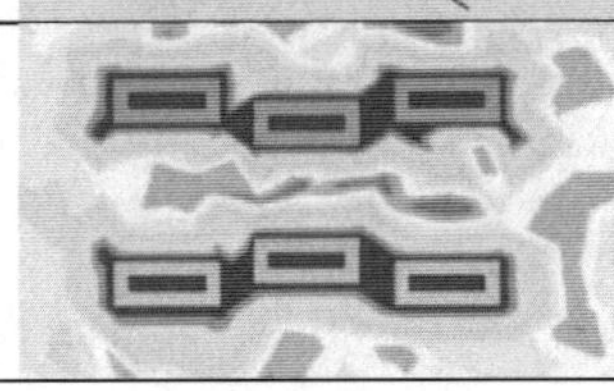	3.14 m/s 27,5°C

3.4.1 Configuration A

The first model represented the basic parallel layout where the distance between the two-row buildings was constant (15 m). The result simulation showed that the incoming wind did not occur much deflection as it was immediately split into two directions. The distance between the two-row buildings was narrow so airflow did not appear at the central space, creating wide shaded area, thus experiencing a low air movement.

Therefore, the other model alternative was conducted to investigate whether wider distance would give better velocity. In simulation A.2, distance between the long-side facade was widened to one a half of building's height (22.5 m). The result showed that wind could go across the middle area and give more evenly wind contour. However, because of this layout wind was split into 3 directions, the velocity produced was slightly lower that in A.1 (0.02 m/s difference).

IOP Conf. Series: Earth and Environmental Science **213** (2018) 012042 doi:10.1088/1755-1315/213/1/012042

3.4.2 Configuration B

Model B was a modification of model A, in which the distance between the long-side facade slowly widened from east to west. From the contour it could be seen that wind speed spread more evenly indicated by less shaded – blue area. However, the colour in the middle changed from green to light blue. It showed that wind speed decreased slightly from east to west because the distance was widened.

3.4.3 Configuration C

Model C was a contradiction of model B, in which the distance between the long-side facade slowly narrowed from east to west. However, it showed the same wind behaviour where wind speed in the middle also decreased from east to west like model B. It happened because wind lost its strenght as it met more obstructions along the site.

3.4.4 Configuration D

Model D represented a courtyard – like layout, in which the distance changed from narrow (15m) – wide (45m) – narrow (15m). Apparently it caused more shaded area in between the long-side facades except in the middle which is 45 meters wide. This model confirmed the result from model A.1 that distance of 1x height of the building (15 m) gave more shaded area which was not enough to supply airflow for natural ventilation.

3.4.5 Configuration E

Model E.1 was the opposite of model D, in which the distance changed from wide (45m) – narrow (15m) – wide (45m). It gave much better airflow distribution than model D but still had shaded area in the very middle since it was only 15 meters wide. So, model E.2 was tested. In fact, this layout eliminated the shaded area in model A.1. Interestingly, unlike the other options, wind speed in between the long-side facades was slightly increased from east to west. Presumably, this result was caused by wind tunnel effect created by distance combination of wide (45m) – narrow (30m) – wide (45m).

3.4.6 Comparison

From all the options tested above, model A.1 had the biggest value of wind speed. But it did not mean that model A.1 was the best option since visually it hasdwide shaded area and very low wind speed in the middle of the blocks. This high wind speed occured because the windward (short – side) facade was very narrow giving the least obstruction.

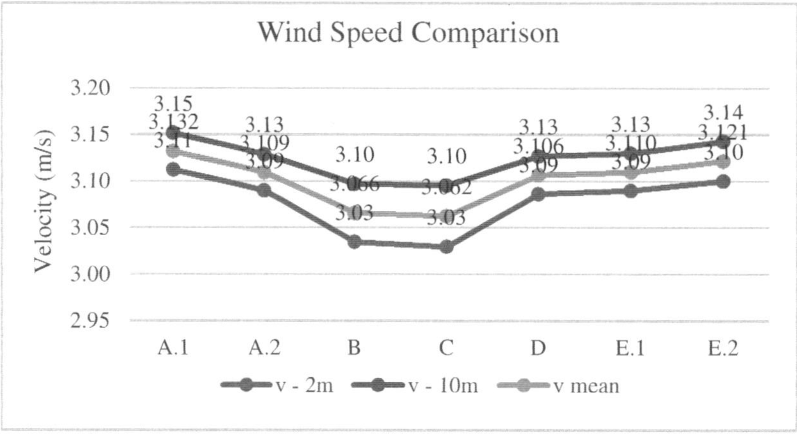

Figure 3. Wind speed comparison for each configuration

Model A.2, D, and E.1 had similiar wind speed value, but visually option D (courtyard – like layout) did not give great impact since it had more shaded area in the middle of the buildings. Meanwhile option A.2 (linear layout) and model E.1 (staggered layout) visually had a decent airflow distribution.

Model E.2, where distance changes from 15 m to 30 m wide with the combination of distance wide – narrow – wide, gave the most optimum result since it gave higher wind speed with the best airflow distribution visually.

4. Conclusion

In order to supply enough airflow inside the building for natural ventilation, 0.1 – 0.35 m/s (SNI 03-6572-2001), it required sufficient wind speed around the building. Shaded area, which was shown by dark blue in colour, indicated area with verly low wind speed. Therefore, some options were explored to find the best wind speed value and also the best wind distribution visually.

Between parallel and staggered layout which were mentioned in the literature study, for this project, model E.2 was the most optimum. Zig-zag configuration allowed wind to move in the small gaps between buildings because it created bouncing pattern of the wind since the buildings directed the wind to subsequent structures. Besides that, the configuration of building distances (wide – narrow – wide) created a kind of trap or wind tunnel – like feature that strengthened the wind in the middle of the site, so it received a constant (even bigger) wind velocity along the site.

Other than that, there were some conclusions about the effect of building configuration to the behaviour of airflow around buildings. High wind velocity did not always represent the best condition since the airflow distribution played big role in the building. The smaller winward blocks area (wind-facing facades), the less deflection would occur, thus the bigger value of velocity, and vice versa. Distance between masses should be more than one of building's height so airflow could to go through it. Widening the distance would produce less shaded area, but slightly lower velocity number and vice versa, which is the bernoulli principle in fluid. But there was also condition where wider distance gave higher velocity, which happened because wind had less obstuctions. In fluid dynamic like natural ventilation, each case would have unique airflow pattern that must be specifically simulated. To sum up, the study summarized that building mass configurations could impact air movement around the buildings to provide natural ventilation.

5. Acknowledgments

This research is part of master degree thesis in Architectural Design Master Program SAPPK ITB and funded by P3MI ITB. The local climate data assigned was obtained from Mr. Hastuardi, BMKG Tanjung Priok Station.

6. References

[1] IEA, "Southeast Asia Energy Outlook 2017," p. 149, 2017.
[2] N. C. Idham, *Arsitektur dan Kenyamanan Termal*. Yogyakarta: Penerbit ANDI, 2016.
[3] A. Aflaki and N. Mahyuddin, "Study on Efficiency of Passive Cooling Strategies on Thermal Comfort Attainment within Tropical Climate," no. November, 2015.
[4] L. Busato, "Passive cooling and energy efficient strategies for the design of a hotel on the Southern coast of Pernambuco , Brazil," London Metropolitan University, 2003.
[5] A. Konya, *Design primer for hot climates*. London: The Architectural Press, 1980.
[6] A. N. Tombazis and S. A. Preuss, "Design of passive solar buildings in urban areas," *Sol. Energy*, vol. 70, no. 3, pp. 311–318, 2001.
[7] V. Olgyay, *Design with Climate: Bioclimatic Approach to Architectural Regionalism*, New and ex. New Jersey: Princeton University Press, 2015.
[8] R. Thomas and T. Garnham, *The Environments of Architecture*. Great Britain: Routledge, 2007.
[9] T. Petrila and D. Trif, *Basics of fluid mechanics and introduction to computational fluid*

dynamics, XIV. Springer, 2005.
[10] P. La roche, C. Quirós, G. Bravo, E. González-Cruz, and M. Machado, "Keeping Cool: Principles to avoid overheating in buildings," in *PLEA notes: Passive and low energy architecture international Design tools and techniques.*, S. V. Szokolay, Ed. New South Wales: Research, Consulting and Communications (RC&C), 2001.
[11] N. Khan, Y. Su, and S. B. Riffat, "A review on wind driven ventilation techniques," *Energy Build.*, vol. 40, no. 8, pp. 1586–1604, 2008.
[12] C. H. Lim, O. Saadatian, B. Ali, M. Sulaiman, S. Mat, and K. Sopian, "Air Changes and Extraction Flow Rate Analysis of Wind-Induced Natural Ventilation Tower under hot and humid climatic conditions," *IEEE Business, Eng. Ind. Appl. Colloq.*, vol. 6, no. 5, pp. 488–495, 2012.
[13] T. A. Markus and E. M. Morris, *Buildings, climate and energy*. London: Pitman Publishing, 1980.
[14] B. Hong and B. Lin, "Numerical studies of the outdoor wind environment and thermal comfort at pedestrian level in housing blocks with different building layout patterns and trees arrangement," *Renew. Energy*, vol. 73, pp. 18–27, 2015.
[15] M. Baharvand, M. Hamdan, B. Ahmad, and T. Safikhani, "Thermal Performance of Tropical Atrium," no. November 2013, pp. 34–40.
[16] J. Burnett, M. Bojic, and F. Yik, "Wind-induced pressure at external surfaces of a high-rise residential building in Hong Kong," *Build. Environ.*, vol. 40, no. 6, pp. 765–777, 2005.
[17] P. M. P. UMUM and N. : 29/PRT/M/2006, "Pedoman Persyaratan Teknis Bangunan Gedung," 2006.

From sacred to secular: *Baruk* architecture on secular buildings in Sarawak

Atta Idrawani Zaini, Bambang Karsono, Awang Hasim Awang Sulong
Department of Architecture, Faculty of Engineering, Universiti Malaysia Sarawak, Kota Samarahan, Sarawak.

Corresponding e-mail: izatta@unimas.my

Abstract. The Bidayuh people of Sarawak are known for their *baruks*. A *baruk* is often observed as having a round form with a conical roof that has various functions such as a courthouse, a meeting place and a residence for the male adolescents. These are all secular purposes built around its religious functions, constructed based on their traditional beliefs that qualify it to be considered as a sacred building. Recently, the pursuit for fulfilling political and commercial objectives has made the architecture of *baruks* to be copied onto various secular buildings in Sarawak, ranging from a visitor centre to a governmental building. Despite much have been praised as an attempt to revive this endangered architecture, applying its language onto contemporary secular buildings sparks a debate whether the sacredness of the original *baruks* are respected or is it just a cultural commodification. This paper attempts to justify the adaptation of *baruk* architecture into secular buildings. It is discussed through a grounded theory approach, with the methods used were in the form of literature review and on-site observations. This paper suggests the extent of applying *baruk* sacred architecture into secular buildings should be observed, allowing more potential discussions in the academia pertaining adaptation of sacred indigenous architecture in contemporary buildings.

Keywords: Baruk, Bidayuh, secular building, sacred building, Neo-Vernacularism

1. Introduction

The new Sarawak State Legislative Assembly (SSLA) building is located just a few hundred meters opposite of the Kuching Waterfront, standing far enough by the river to signify the might of the Sarawakian people and their belief in the democratic system [1]. Designed by Hijjas Kasturi and completed in 2009, it is intended to be an icon - a testament to Sarawak's bold vision for the future. This building resembles what the local lay public might be familiar with – consciously echoing architectural characteristics of a *baruk*, a special community hall and a sacred place for the Bidayuh ethnic group [2]. The conspicuous *baruk* form reflects the institution as the oldest state legislative assembly in Malaysia [1]. It is one of many contemporary architectural pieces in Sarawak that echoes the image of this sacred building.

IOP Conf. Series: Earth and Environmental Science **213** (2018) 012043 doi:10.1088/1755-1315/213/1/012043

Figure 1. The SSLA building [24]

Borrowing vernacular architectural languages into modern, contemporary buildings is not an uncommon practice in Malaysia. Many have sparked debates, but this trend has constantly reappeared and seems to resonate with the local leaders' quest in creating a unique local architectural identity. Similarly, it is also a straightforward schema in formulating a vivid appearance of an identity which is also perceived can be beneficial for commercial objectives. *Baruks*' image can be seen across various building typologies such as on the visitor centre in Matang National Park, the White Baroque Restaurant near the Kuching Waterfront and the Damai Beach Resort in Santubong just to name a few. Many of these buildings are built in Kuching division in the western region of Sarawak as there are more Bidayuhs living there than the other parts of the state.

Figure 2. An earlier sketch of a *baruk* drawn by Hugh Low in 1848 [14]

In the modern days, despite some *baruks* are still being used for religious activities mostly by the traditional Bidayuhs practising the old belief, many have been demolished and become obsolete. This is a direct result from the conversion of most traditional Bidayuhs to Christianity [3], apart from other factors associated with modernization. Thus slowly, *baruks* which once were a Bidayuhs' sacred institution has slowly become just a showpiece of their historical past. Therefore, the attempts of reviving *baruks'* architectural language in modern and mostly secular buildings in Sarawak have been well received as an effort to monumentalize their identity, but there is one concern that is rarely questioned. *Baruks* are highly revered by the Bidayuhs and have played a very important role in their traditional beliefs, as the *baruks* were likened to their sacred temples [4]. As a sacred building, *baruks'* architectural image that is borrowed to mostly secular buildings may yield a question of morale pertaining to their new status. Therefore, in this paper, the rationales behind embracing *baruks'* architectural language on secular buildings are discussed.

2. Methods

This paper mainly relies on qualitative method of dredging information from our prior knowledge, literature sources and on-site observations on neo-vernacular buildings that adapted the *baruks*' architectural characteristics as case studies. It is worth to note that the Bidayuh people and their material cultures such as *baruks* do also exist in other parts of Borneo, particularly in Indonesia. However, it is only in the interest of this paper to focus on secular buildings in Sarawak, Malaysia. The literature sources are mainly associated with the research conducted on *baruks* by previous scholars and observers from as early as the 19th century to the present ones. Additionally, this paper serves as an overview, thus it is not necessarily of our intention to derive a conclusive outcome. This study is also inclined to use a grounded theory approach to arrive at the issue – a logical method to explore an alternative viewpoint of an issue.

3. Discussion

3.1. The Bidayuh People

The Bidayuhs are believed to be true indigenous inhabitants of Borneo [5], despite the origin of the Bidayuhs in Sarawak is uncertain and unknown, as many depend on legends and oral traditions of how they came thus it is difficult to prove its accuracy [6]. Their cultures and architectural knowhows may be a result of long cultural exchange and improvements, as according to [7], the vernacular process of design, procurement and implementation in building construction of indigenous community in Sarawak were usually passed down from shared community knowledge.

Geographically, Sarawak is located on the island of Borneo. It was reported in 2010, there are around 190,000 Bidayuh people in Sarawak [8] with a large number of Bidayuh people live in the western region namely Kuching and Samarahan (94% of the total Bidayuh population). Before the advent of Christianity, the Bidayuhs on the olden days believed in the omen [3]. The sort of danger from evil spirits what may have caused misfortunes and sickness, thus every action in their daily lives were an act of taking care of these bad spirits by not provoking them. Occasionally if these spirits were not pleased, they will conciliate them by means of ceremony or feast called *Gawai*. The common means of protection is also through seeking protection from *Tampa Riyuh* (God) and the good spirits of their ancestors to drive the bad spirits away. They show gratitude and seek guidance from *Tampa Riyuh* to help them in all matters in life to protect them from dangers [8]. The traditional Bidayuhs did pray and believe in the concept of God and much of their acts of tribute to the God and spirits were in the form of respecting the land and nature itself.

The traditional Bidayuhs believe the communication from the omen birds is a form of forewarning in certain circumstances. If a bird is heard calling while they were hunting, it is a sign that they should return home. Obeying these signs from the birds can prevent bad omens that it is common to see the replica of these omen birds placed at the peak of *baruks'* roof. To some extent, the traditional Bidayuhs had so much to be afraid of that they were viewed to be ideal subjects to conversion to Christianity [3]. This conversion later changed the fate of *baruks* in terms of their function as a sacred traditional belief institution.

The Bidayuhs were also once an oppressed group in which their villages were usually the target of vicious attacks and massacres by the Ibans and the Brunei Rulers. It was due to this that the Bidayuhs had decided to set up their settlements in mountain areas and far into the interior to avoid threats. The Bidayuh people have survived these hardships and modern Bidayuhs since 1963 have participated in the creative process of maintaining, discovering, restoring and creating their traditions to some extent. Such is evident in the establishment of Dayak Bidayuh National Association (DBNA) by the educated Bidayuhs that modernizes the ethnic group's values and identity which include the recognition of *baruks* as one crucial architectural heritage of the Bidayuhs [9,10].

3.2. Sacred Architecture

The 'sacred' and 'secular' are 2 classic dichotomous sides that often been discussed as being opposite to one another, divided conveniently, or sometimes naively, into a holy and a profane institution. This divide came about into the mainstream view after the dispute between the religious and the scientific institutions during the Renaissance in the 18th to 19th century, which eventually resulted to what is come to known today as the separation of the church and the state. The term sacred stemmed from the

IOP Conf. Series: Earth and Environmental Science **213** (2018) 012043 doi:10.1088/1755-1315/213/1/012043

Latin word *sacer* means 'to make holy' [11], thus being sacred is the act of aligning to the holy forces that set apart humans from the ordinary world that is already known. Sacred buildings or spaces then can be directly implied as the venue to perform the act of worshipping. Sacred spaces are for performing the ritual for the divine through the act of consecration, or sacrifice which excluded from the dominion of the profane, often determined by the architectural framework of the temple itself or by the act of sanctification [12]. It usually can be identified through a limited penetration by human agencies, in which it can only be accessed by a certain priestly caste, group of people or individuals, usually revolves around the idea of 'precinct' and 'threshold'[11]. There is a concept of 'precinct' and 'threshold' that split certain spaces in *baruks* that can be accessed for sacred occasions and by certain individuals, that some are through a granted permission from the important figures in the community. There is also the sense of connection of the building to the spiritual world, an explicit notion to suggest that *baruks* are indeed a sacred building.

3.3. Baruk – A Bidayuh Sacred Architecture

Baruks are described as a traditional headhouse, a meeting place, a bachelor residence, a ritual place and many others [7,8,13]. To some extent, it is a typology of its own that has no precedence manifested by other ethnic groups in Malaysia. The first *baruk* in Sarawak is believed to be built in the in Kampung Opar in Bau (western Sarawak region) after a disease outbreak which claimed the lives of their children. The *baruk* then was built using *belian* and *kapur* timber as a place for them to discuss solutions to the outbreak [8]. Through the process of improvements, *baruks* slowly evolved into a more complex structure to become one of the most recognizable forms that have been emulated in different sites.

The *baruk*, which is also referred to as a *pangah*, *balu* and *pangaran* in different dialects [14], is also widely known mostly to the western observers as the headhouse as old human skulls of their headhunting days are positioned above the fireplace in the centre of the hall in reverence [15]. Apart from the skulls, *baruks* are also where the traditional musical instruments particularly the drums which are called the *sebbang* (used for certain rituals) are stored [5]. The Bidayuhs believed that human skulls are a symbol to instill bravery and vigilance among the youths in defending their villages[10]. In an observation, there were a few cases where there were even no skulls in certain *baruks* thus it reinforces the fact that a 'headhouse' is purely some westerners' term because the Bidayuh people also never called their *baruks* as such [16].

Figure 3. An original *baruk* in Opar, Bau still serves as a ritual and ceremony place for the traditional Bidayuhs today [10]

In 1993, the Bidayuh Community Cultural Seminar has addressed this misconception to re-elevate *baruks'* status as a significant cultural manifestation more than just a headhouse. Admiration over *baruks* among Bidayuhs is apparent in their myth tales, epics and traditional poems or *bajo*, which contain mostly praises and reverences over this structure. The sanctity of this institution is ultimately defined by the religious activities associated with it. Without *baruks*, these rituals and ceremonies cannot be done elsewhere. The comments from several observers have mentioned *baruks* were also a gathering place [3,8], which implies that other non-religious activities were just secondary to that of the religious ones. *Baruks* as more like a temple that is considered sacred that no one could enter and sleep in it without granting the permission from the village chief [4,8]. Even the heads were taken

IOP Conf. Series: Earth and Environmental Science **213** (2018) 012043 doi:10.1088/1755-1315/213/1/012043

from battles needed to undergo elaborated ceremonies and rituals for them to be finally qualified to be displayed in *baruks*.

3.3.1. Site & Construction
The Bidayuh people put great stress on *baruks'* height and the ideal location are often on the edge of a steep terrain as to give an illusion of hanging from the sky when viewing from the inside [10,13]. It is either by this or they would prefer to build *baruks* much taller when there is no such terrain [16]. *Baruks* are often built strategically in the middle of the village, which reflects its institutional appeal and accentuates its importance. The construction of *baruks* would be by unmarried Bidayuh men through collective efforts [14]. Even if the decision to erect a *baruk* was rejected by just one member of the community, the project will be halted to avoid discontent among the villagers [8]. Dishonouring this would only summon misfortune to the village itself, which in turn might befall their yearly crop and badly impact other aspects of their life. With the culture that honours land as a source of substance and life [3], the selection of site should be done respectfully. The site must be in a strategic location with a good soil condition, which is to be selected by the shamans [8]. Certain prohibitions were imposed on every aspect throughout the process. Prayers and rituals accompanied most actions in the construction process from the early stage until completion [8]. It is through the blessings from the spirits that *baruks* can eventually be erected successfully.

3.3.2. Functions
Apart from as a defense fort, *baruks* also served as a crucial social institution such as a residence for the bachelors, a guesthouse for visitors, a site for performing rituals or ceremonies, a meeting place and a local court [8,13]. It is in *baruks* where the elderlies discuss local issues regarding their community [15]. Residing in *baruks* is an act of gratitude and honour for the visitors or guests. When issues need to be trialled, *baruks* function as a local court for the chief to decide what kind of punishment to serve. Despite these seemingly common functions, ultimately *baruks* are a manifestation of the traditional Bidayuh beliefs as there are rituals and ceremonies exclusively tied to *baruks*. During Gawai celebration, *baruks* are a venue for the traditional and religious ceremony [3]. One of the ceremonies celebrated is *Gawai Muka*, which is the returning of warriors with the enemy heads. The dried skulls with various omens had to undergo rituals to make them favourable to be installed in the *baruks*. The spirits of the old skulls are highly revered and respected among the traditional Bidayuhs that they dignified them in the highest regards [8].

A baruk is a place for interaction between humans and the hidden spirits that protect them in life [17]. It is, therefore, the best site to serve as a place for shamans to cure their male patients that without *baruks* the activities pertaining the worshipping of the spirits cannot be performed. Other religious ceremonies that had to be taken place in *baruks* are often related to their agricultural activities such as worshipping the spirits of the paddy before the farming can take place [8]. This connection between the humans and the spirits is highly significant in the traditional Bidayuhs' belief. All these rituals had to be done in *baruks* as they believed that *baruks* contain holy and mighty forces that are more powerful than mere mortals.

3.3.3. Architecture
Baruks' usually had a simple, centric layout of internal space. The floor of the main gallery is usually made out of double-layered bamboo halves just so it would not be penetrable to attacks from spears coming from below [18]. The floor structure is split into 2 levels with the higher one being approximately 2 feet higher than the one below. Circling the main gallery is a raised timber platform, or *akat*, which serves as a sleeping and seating area [8]. The lower section is split into another 2 sections, which are *akat guna* and *akat batin. Akat guna* is a space for placing offerings for the spirits of the skulls while *akat batin* is a place reserved for the spirits during ritual and ceremonies in *baruks*. These spaces are considered sacred and are restricted for human use.

IOP Conf. Series: Earth and Environmental Science **213** (2018) 012043 doi:10.1088/1755-1315/213/1/012043

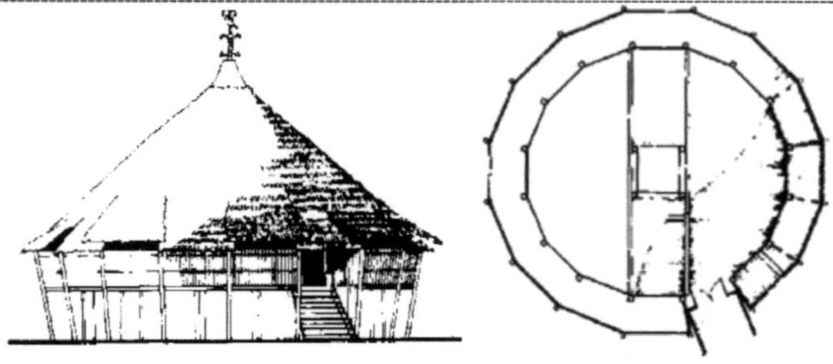

Figure 4. An elevation and a floor plan of a *baruk* [19]

In the middle part of the gallery, there are usually 4 central posts bounded by rattan and ropes that extend to the peak of the roof, which supports much of the weight of the roof itself. The roof has open skylights usually braced by a bamboo or a timber stick at the lower end of the roof to allow lights and fresh air to ventilate into it [15]. Palm-leaf thatch is greatly relied upon as the material rather than timber shingles which are scarce. The high pitch steep of the gable widened as it goes lower along the roof elevation as to quicken the withdrawal of rainwater and control any sprays from it [8]. The omen bird replica usually found at the peak of the roofs is associated with the Bidayuhs' belief that they can give early warnings of misfortunes. The omen bird replicas were also installed to dispel satans and bad spirits from perching on the roof saddle.

The Bidayuh people relied heavily on bamboo as they live in areas in which lumber trees are scarce [19]. Some observers regarded *baruks* as octagonal or round with a conical roof, which in some cases there were also in square and irregular forms [13]. He added round is inconsistent with the characteristics of other Dayak groups as evident in their traditional longhouse architecture and other artefacts in which are all rectilinear. It is still very much unknown why they favour a round shape but Simmins [11] has pointed out generally indigenous peoples often believe certain fundamental geometrical shape such as round has a certain power. It is also not difficult to understand their preference, as round floor plans are generally practical as they serve large, unobstructed area for gatherings. This is also aligned with an observation reported by Riri [19] on the *Maniamas* dance performed by Kalimantan (Indonesia) Bidayuhs. The dance is performed as a ritual of circling the gallery to symbolize unity and unobtrusive flow of current. Some *baruks* are also observed as having parallelogram forms and some also had a non-conical, high steep pitched roof [3]. There were also some inconsistencies reported based on the observations by European travellers and writers who had visited Sarawak regarding the structure of *baruks*. They reported some were built in a single storey while some had double, and some had doors on the side and some had no doors on the side but with a trap door on the floor [3]. The universal vernacular language of *baruks* is not consistent, thus this inconsistency suggests that the idea of sacredness are not associated with *baruks* physical form.

Figure 5. An earlier baruk in Kampung Taee [10]. Note the non-conical roof shape [13]

3.4. Baruks Architecture on Secular Buildings

As a reaction to the economic liberalization in the 1980s, Malaysian cities are fast becoming modernized through buildings that are directly influenced by internationalization [20]. Some architects with the support from the government officials have pursued to a more alternative style that would be able to reflect the local culture. Neo-vernacular approach as according to Rasdi [21] is a popular method in fulfilling the plea of local leaders in creating a new regional identity, despite mostly were built using modern methods of construction to create an image of 'progress'. Neo-vernacularism can be argued as a goof-off imitation, simply underplay the significant history of local architecture that came about through rigorous process.

Apart from imitating the modern language imported from the west, contemporary buildings in Sarawak also seem to be drifted along with this trend of borrowing the vernacular aesthetics from the local tribes. There is a tendency of having the 'Sarawak element' into everything. Explicably, notable neo-revivalism language applied to buildings throughout Sarawak resonate to the immediate culture of a majority group in each area or division. The institution buildings in Mukah for instance, which has about 60% of the people are predominantly of Melanau descent [22], tend to apply the metaphor of a Melanau *terendak* (traditional Melanau headwear). The Kuching South City Hall, which is in a predominantly Chinese area, borrowed its image from the pagoda, in sympathy with its Chinese neighbourhood [23].

Figure 6. Left: An institution building in Mukah [23]. Right: The Kuching South City Hall [2]

3.5. Physical Architecture

In 1988, through a meeting in celebrating Sarawak's 25th independence through Malaysia, the government had requested the representatives from various ethnic groups of Sarawak to discuss which parts in their culture should be regarded as outdated, thus to be rejected, and which one should be retained [13]. To contribute to the state and national culture, the Bidayuh community has preferred a more distinctive round *baruks* with conical roof rather than rectilinear forms as their cultural identity, taking precedence from the one in Opar as an epitome of *baruk* architecture. Since then the round form with conical roof has been recognized as a universal language of *baruks* used widely in contemporary buildings around Sarawak. The recurring architectural traits that are being repeated in most contemporary secular buildings in Sarawak echoing that of *baruks* are summarized in Table 1.

Table 1. Most recurred architectural traits of *baruks* in contemporary buildings in Sarawak.

Conditions	Elements	Illustrations
Oblogatory (primitive form)	• Rounded/ centric/ octagonal plan • Conical roof	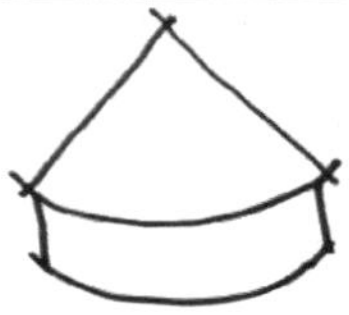
Optional (enrichment)	• Slanted outer walls; • Wider gable end & thin roof peak • On-stilts	
Omitted	• Omen bird replica • Sacred spaces (*akat guna* & *akat batin*) • Old skulls gallery • Sacred artefacts (*sebbang* etc)	

The obligatory elements applied to these buildings are mostly of its rounded form and conical roof, which at this point can be attributed as the primitive form of a *baruk*. Some examples also show the tendency to enrich the image of *baruks* by adding some elements such as the slanted outer walls with much thinner roof peak (SSLA building), and some were elevated to give an image of being built on stilts (Bau Civic Centre; Damai Beach Resort). Other buildings used either one of the elements or a combination of all. The omitted elements are mostly of religious and sacred elements. Despite some have retained the feature of the omen bird such as can be seen on Bau Civic Centre (designed by the late Marcel Wu of KDI Architect), the intention may not be a religious one. This is a gesture to pay homage to the historical past to many Bidayuh locals living in that part of the region and the patron of the project.

Based on this, the physical aspects of *baruks* adapted by secular buildings mostly have very little to do with, if not unrelated at all, to *baruks'* sacred elements. The physical sacred elements are only reflected through certain components in *baruks*, which are the spatial layout (*akat guna* & *akat batin*), the artefacts related to rituals (*sebbang* etc) and the omen bird replica on the peak of the roof. Despite the raised floor component in most Austronesian indigenous buildings as according to Ting [7] are used to represent their cosmological belief, there is no strong evidence suggesting that the raised floor in *baruks* are related to religious belief, unlike in the Iban architecture, in which the ground floor of their longhouses represents the underworld, the first floor represents the realm of humans while the attic or the roof represents the spirit world. The openings and the conical roof are more of a climatic response as to allow airflow and efficient rainwater withdrawal [8], rather than arousing spiritual presence. At a glance, it can be regarded as acceptable for secular buildings to use unsacred physical characteristics of *baruks*, as long as they are not replicating the sacred components.

IOP Conf. Series: Earth and Environmental Science **213** (2018) 012043 doi:10.1088/1755-1315/213/1/012043

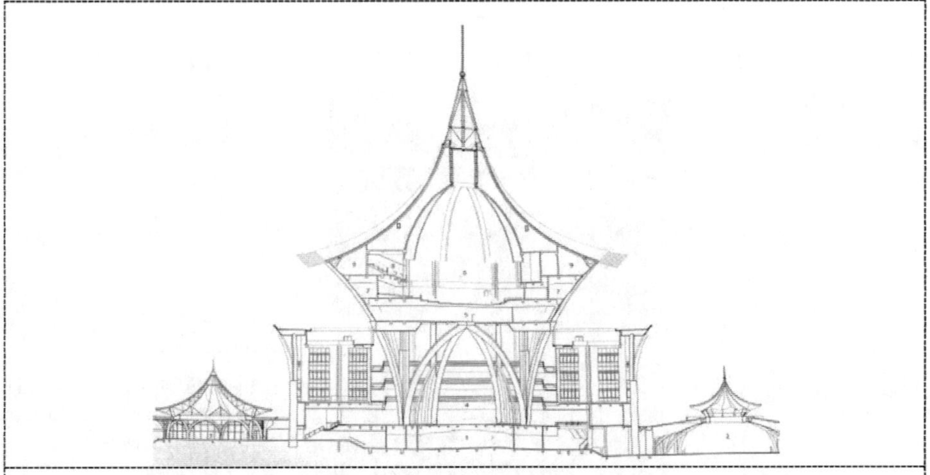

Figure 7. The cross section of SSLA Building clearly echoes the profile of *baruks* (conical roof & the illusion of being raised on stilts), but the replicated sacred components are non-existent [1].

From our on-site observations, literature sources and prior knowledge, notable physicality aspects of *baruks* adaptation on secular buildings in Sarawak can be primarily categorized into these approaches:

1. 'Museumification': Accurate replica of *baruks* built in a traditionalist way to preserve its full glory usually used for tourism and educational purposes (i.e. the Sarawak Cultural Village);
2. Hybridization: Mixing some architectural elements of *baruks* with other styles, materials, construction, and functions (i.e. Bau Civic Centre & Damai Beach Resort in Santubong);
3. Metaphor: More abstract and liberal way of interpreting and recreating *baruks'* architecture into a more sculptural or monumental structure. An advanced result of hybridization (i.e. SSLA building).

Minos stressed that the architecture of *baruks* has to be accepted and built which can keep alive the identity of Bidayuhs [6]. Ungang highlighted, if the new approach of sustaining and promoting the architectural language of *baruks* in tourism industry does not diverge to the interests of the Bidayuh people, it is most welcome if it carefully respects the sanctity or sacredness of *baruks* [8]. These secular buildings are mostly interested to replicate the form of conical roof and the round floor plan, rather than the sacred components. However, to highlight a point from [5] research, even the recording of *sebbang* played for promotional or commercial purposes are considered as non-ethical. They expanded the concern to include other Bidayuhs' cultural objects. Our paper suggests that *baruks* fall under the same category. Using this argument, the commercial use of *baruks* architectural language as exemplified in these secular buildings may have some harm to the culture, as they somehow remove the sacred meanings to create something that is less valuable. Ironically, when there are sacred components being replicated, such as the omen bird on the roof of the Civic Centre in Bau, it is rather unethical despite the good intention of being in sympathy to the local culture. Therefore, to simplify the arguments, we outlined some examples of the secular buildings with our comments built upon these concerns in Table 2.

Table 2. Some examples with comments on secular buildings with architectural references taken from *baruks*.

IOP Conf. Series: Earth and Environmental Science **213** (2018) 012043 doi:10.1088/1755-1315/213/1/012043

Building	Image	Approach/ Comments
Sarawak Cultural Village Function: Museum [27]		Museumification Respects traditional culture through cultural reenactments for educating visitors while maintaining commercial interests.
Bau Civic Centre Function: Community hall & gallery		Hybridization In sympathy with local Bidayuh community and respectful to the culture as it serves also as a gallery about Bidayuh culture. A bit careless in adapting *baruk* architecture by treating omen bird replica as just a cultural object.
Damai Beach Resort Function: Hotel [15] (*Image source: Invernizzi, 1996*)		Hybridization Aggressively using *baruks'* images that can be interpreted as cultural commodification to fulfil commercial objectives.
The White Barouk Restaurant (then Beijing Restaurant) Function: Restaurant [28]		Hybridization B*aruks'* image is used as cultural commodification to fulfil commercial objectives.
SSLA Building Function: Government institution [29]		Metaphor Paying homage to *baruks'* institutional appeal while avoiding using sacred elements. Less harmful to the sacred meaning of *baruks*.
Tebedu Community Hall Function: Community hall [31]		Hybridization Restoring *baruks'* common function as a community venue with no reference to sacred elements. Less harmful to the sacred meaning of *baruks*.

Morality of Using Baruks' Architecture in Secular Buildings

There is already consent from the Bidayuh community in the form of an approval to use *baruk* architectural language for commercial and tourism purposes. *Baruks* are essentially a sacred building but they also serve as a place of gathering, thus any new version of *baruks* in line with the spirit of gathering people may be regarded as appropriate. Despite there is already a rule indicating that it should align with the interests of the Bidayuh community and be respectful to the sanctity of *baruks* institution itself, the execution of these still very much open to interpretation. This is where the line should be drawn.

Being respectful to *baruks* can be in the form of what type of secular activities being hosted in the buildings. If it was meant for education or as an institutional building such as a legislative assembly, a tourist information centre or a community hall, despite does not in any way reflecting original *baruks'* sacred functions, this may be acceptable as the original function of *baruks* to gather the community and as a place for meeting is not fully compromised. If the building is used as a restaurant or a hotel, then there is a risk of being exploitive. Though is tolerable to the Bidayuh community, commercial activities may relegate the sanctity of *baruks* depending on the commercial owners who would decide on what type of commercial activities going to be hosted there. Commercial activities are also often ephemeral and ever changing, as there is money involved, thus it is highly probable that the activities may not be in line with the idea of sacredness. For instance, further perversion of it can happen if it was converted into a nightclub or a karaoke bar.

Suggestions

This architectural hybridization between contemporary and vernacular style is not so much the exception but the norm in Malaysia, so much so that it is quite regular to see buildings in Malaysia flexing styles taken from various ethnic references. Imposing vernacular characteristics on relatively modern buildings is considered by some as a run-of-the-mill approach, but there is also some appealing dimension to it. This is at least may be agreed from the perspective of post-modernist architects who love to spark a broader sense of appreciation by different groups of people across dichotomous levels of intelligence. The interplay between abstraction and obviosity as a result from what the American architectural historian Charles Jencks dubbed as the 'dual code' in post-modern buildings triggers a wider sense of appreciation by both learned professionals and the lay public. This has made some buildings appear to use revivalist and metaphorical approaches becoming too literal to some intellectual senses.

The reason behind of why certain architects chose to use the language of *baruks* is dubious. The format in which Hijjas Kasturi (the architect of the SSLA building) applied to most his iconic buildings in Sarawak are all similar, with round floor plan and the conical roof is set to exploit the moulding potential of concrete in giving an impression of being futuristic and give an element of visual surprise to the context [23]. This can also be found in his other works in Sarawak such as the Bintulu Development Authority (1995), the North Kuching City Hall (1994), and the South Kuching City Hall (1995). These buildings could as well be regarded as using *baruk* metaphor as they all have a similar conical roof and rounded floor plan. It seems that the justification for using the image of *baruks* could as well be secondary. Thus, removing the theme of *baruks* from the intention of architectural form making may be a more dignified approach rather than claiming one to be a thing that it is not. It is therefore possible to create something similar looking but without mentioning the intention of replicating the architecture of *baruks* itself, which may be less harmful to the culture. It is even better if the 'dual code' approach is used to let the observers to suggest for themselves that the image of *baruk* is there without even mentioning it.

IOP Conf. Series: Earth and Environmental Science **213** (2018) 012043 doi:10.1088/1755-1315/213/1/012043

Figure 8. Other Hijjas Kasturi's works in Sarawak that may resemble *baruks* but are not [31,32]

4. Conclusion

The hybridization of the language of *baruks* with some other language may hurt the original architectural language, what they represent and their meanings. For instance, using the external façade of *baruks* with modern spatial configuration does not represent *baruk* architecture in total. This is also due to the omission of sacred spaces of *akat guna* and *akat batin* that define what *baruk* architecture is and how it functions. Such can be seen in almost all secular baruk buildings in Sarawak, in which what has been retained only the physical appearance of a *baruk*. The metaphor approach of translating *baruks* language into an avant-garde result can be perceived as an artistic sculpturing of the original structure, thus itself can be considered as a form of art that uses physicality of the environment as the canvas. This approach generally may be less harmful, as it may be just an artistic rendition of *baruks* and often can be open to interpretation.

The initiatives to incorporate *baruks'* image into contemporary buildings are applauded and well received. However, as far as architectural morality and ethics are concerned, there would be a very limited way to do this, as it is always best not to jeopardize the original sacredness of *baruks*. Due to this dilemma, the most practical solution by far is by respectfully compromising the sacred elements of *baruks* just for keeping the architectural image alive. This seemingly paradoxical approach may not be the best, but it may be what keeps the modern society conscious of this architecture, thus empowering the culture itself. However, by doing this in a long run may also give birth to a new problem in the future. There is already an abundance of examples of architectural works that used sacred language that has become a new breed of secular architectural language itself. It is not impossible that someday this is going to happen to *baruks*. It may also be plausible in future, if without the intellectual interventions done in the process, *baruks* would only become just an image that can be copied and reproduced without having to know its *raison d'etre*, submerged or drifted along with the current of other architectural styles.

5. References

[1] Lee, C. W 2009 *The New DUN Sarawak (Sarawak State Legislative Assembly Building): Symbol of the People's Sovereignty.* Singapore: Editions Didier Millet

[2] Finch, G. A 2015 *The Sarawak State Legislative Assembly Building: Hybrid Architecture for National Identity.*

[3] Chang, P. F 2002 *History of Bidayuh in Kuching Division Sarawak.* Kuching: Sarawak Press.

[4] Macdonald, M 1956 *Borneo People.* Oxford University Press.

[5] Jähnichen, G., & Frank, B. (2015). The Bidayuh's Sebbang: Preservation Issues. In *UPM Book Series on Music Research* **7** 175-184

[6] Minos, P 2000 *The Future of the Dayak Bidayuhs in Malaysia.* Kuching: Lynch Media & Services.

[7] Ting, J. H 2018 *The History of Architeture in Sarawak Before Malaysia.* Kuching: Malaysian Institute of Architects Sarawak Chapter

[8] Ungang, C 2014 *Baruk: Falsafah dan Pemikiran Masyarakat Bidayuh.* Medium Publications.

[9] King, V. T. (2013). Culture and Identity : Some Borneo Comparisons. **1**1–67

[10] Ungang, C 2006 Baruk: sebuah institusi sosial dan budaya masyarakat Bidayuh. Retrieved from http://ir.unimas.my/607/
[11] Simmins, G 2008 *Sacred Spaces and Sacred Places: A Comparative Approach* (Vol. 1). VDM Verlag Dr. Müller
[12] Piatkowska, K 2017 Technical Limitations in Merging Secular and Sacred Functions in Monumental Churches. In *IOP Conference Series: Materials Science and ENgineering 245.* http://doi.org/10.1088/1757-899X/245/5/052079
[13] Winzeler, R. L 1996 Bidayuh Architecture: Tradition, Change, Revival. *The Sarawak Museum Journal.* 1-23
[14] Atar, S., Ansely, A., & Bala, B 2017 Fungsi dan Cabaran Baruk Sebagai Warisan Budaya Bidayuh. *Jurnal Borneo Arkhailogia (Heritage, Archeology and History),* **11**
[15] Invernizzi, L 1996 *Sarawak Style.* Singapore: Times Media
[16] Geddes, W. R 1957 *Nine Dayak Nights.* Singapore: Oxford University Press.
[17] Geddes, W 1954 *The Land Dayaks of Sarawak.* London: Her Majesty's Stationary Office.
[18] Sim, S 2010 *Redefining the Vernacular in the Hybrid Architecture of Malaysia Under the supervision of*
[19] Riri, L 2018 Fungsi Tari Maniamas dalam Upacara Adat Nyobeng pada Suku Dayak Bidayuh Desa Sebujit Kabupaten Bengkayang. Pontianak, Indonesia: Universitas Tanjungpura.
[20] Moser, S 2012 Circulating Visions of "High Islam": The Adoption of Fantasy Middle Eastern Architecture in Constructing Malaysian National Identity. *Urban Studies*, *49*(13), 2913–2935.
[21] Rasdi, M. M 2008 *Seni Bina: Sejarah, Teori dan Kritikan.* Skudai: Penerbit UTM
[22] Unit, S. S. P 2012 Sarawak Facts and Figures 2012. Kuching: Chief Minister's Department.
[23] McGillick, P 2006 *Concrete Metal Glass: Hijjas Kasturi Associates Selected Works 1977-2007.* Kuala Lumpur: Editions Didier Millet.
[24] naim.com.my
[25] recoda.com.my
[26] *http://photos.wikimapia.org/p/00/02/00/10/55_big.jpg*
[27] *scv.com.my*
[28] *tripadvisor.com*
[29] *theborneopost.com*
[30] *tripmondo.com*
[31] minsar.com.my
[32] kkeb.com.my

IOP Conf. Series: Earth and Environmental Science **213** (2018) 012044 doi:10.1088/1755-1315/213/1/012044

Thermal condition of passive cooling system in Bogor Cathedral Church

Augi Sekatia[1], Erni Setyowati[2], Gagoek Hardiman[2]
[1]Student at Doctoral Program of Architecture and Urbanism, Universitas Diponegoro, Semarang, Indonesia
[2]Department of Architecture, Faculty of Engineering, Universitas Diponegoro, Semarang, Indonesia

Corresponding e-mail: augisekatia@gmail.com

Abstract. Sustainable architecture recently becomes a priority design concept for all buildings in Indonesia. All designs attempt to lower the energy consumption and save more energy to achieve a sustainable building. In Dutch colonial era, passive cooling system became an effective building technology to cope with Indonesian tropical climate without wasting lot of energy. Bogor City, West Java, has a historical building named Bogor Cathedral Church that is equipped with passive cooling system. Bogor Cathedral Church has low ventilation that are adjustable (open or closed). Research was carried out three times during church services and at 15 (fifteen) measuring points where low ventilations were open and closed. Dry bulb temperature, relative humidity, air velocity, and effective temperature (ET) were measured. The purpose of this study is to investigate the thermal conditions of this passive cooling system building. The results of this study show that low ventilations made a significant change to the thermal conditions inside the building. The measuring points near the low ventilation had lower relative humidity and effective temperature, and higher air velocity compared with another points. This study can contribute to enhance understanding that passive cooling system is essential to sustainable architecture.

Keywords: sustainable architecture, passive cooling, thermal condition, historical building

1. Introduction

A building design must be sustainable with the local climate in order to achieve a sustainable design and save energy. By adjusting the local climate, thermal comfort can be achieved. But today's building designs tend to pay attention only to aesthetics without taking into account the climate where the building is built, so that it results in high-energy consumption. Indonesia is a country that has a humid tropical climate.

Tropic or from Greece "*tropikos*" means turning line which covers 40% of the Earth. This line is 23°27' South Latitude. Tropical climate has high solar radiation, very high reative humidity level, and also high rainfall [1]. Lippsmeier (1994) mentioned that to achieve a comfortable thermal condition, it must achieve the right effective temperature and the characters of wet tropic area, including the type of rainforest, the landscape on the equator line, red soil covered by grass, the high variety of plants, the same seasons all over the year, the 60-90% cloudy weather with sky lamination of 850-7000Cd/m2, the high solar radiation, the air temperature with fluctuation of 21-32°C during day and night, the high precipitation in rainy season, the humidity of 55-100%, and the low air velocity.

Thermal comfort is a state of mind that expresses one's satisfaction with its thermal environment [2]. Thermal comfort is one element of comfort that is very important, because it concerns with the condition of a comfortable room temperature. Thermal comfort can be obtained when the effective temperature is 23˚C - 27˚C, the air humidity is 30% -60%, the metabolic rate is 1-1.2, the clothing condition is 0.5 - 0.6, and the wind speed is more than 0.2 m / s [3]. If thermal comfort can be achieved, the building can reduce energy consumption and lead to sustainable development. Historical buildings are examples of buildings that apply the local climate as a design concept so that it can be an example of a building that reaches its thermal comfort and energy efficiency [4].

Bogor City is a city that has many historical building in the Dutch colonial era. One historical building from Dutch colonial era that still stands firm and used until today is Bogor Cathedral Church. Bogor Cathedral Church was build on 1905 and it is still used until now. As a public historical building, Bogor Cathedral Church uses passive cooling system in its building which becomes a perfect object for finding thermal condition. It has unique ventilations installed at the bottom area of the wall called low ventilations. Low ventilation can optimize thermal comfort [5].

Figure 1. Bogor Cathedral Chrch

This study aims is to find the thermal conditions of this building with only passive cooling and to prove that low ventilations can optimize thermal comfort. All variables measured during a worship activity. The purpose of this study was to examine the thermal condition of the building in which only passive cooling system was functioned during prayer time. Bogor Cathedral Church is equipped with air conditioner (AC) too, but when the data was collected, AC would be turned off to optimize passive cooling system performance. The hypothesis of this study is that thermal conditions of Bogor Cathedral church still enter into the thermal condition's comfort zone. The passive cooling system of this building designed at colonials era is also still capable to cope with the tropical climate and lower down the building's effective temperature.

2. Methods

There were two stages of data collections: first was site observation and second was thermal measurement inside the cathedral. Thermal measurement data collected were dry bulb temperature (˚C), air velocity (m/s), and relative humidity (%). As shown in Table 1, there were 3 (three) periods of worship time and on 15 (fifteen) spots inside the building as shown in Figure 4. There were also several gears calibrated properly as showm in Figure 3 and Figure 4. The measurement was performed in two different conditions: open and closed low ventilations (Figure 5).

This study employed quantitative method. Data was collected and then analyzed to communicate results and discussion. Results and discussion included tables and charts so that all information, validity and reliability could be evaluated by the readers.

Table 1. Time Measurement

Measuring Time	Period
Morning	8.45
Evening 1	16.30
Evening 2	18.15

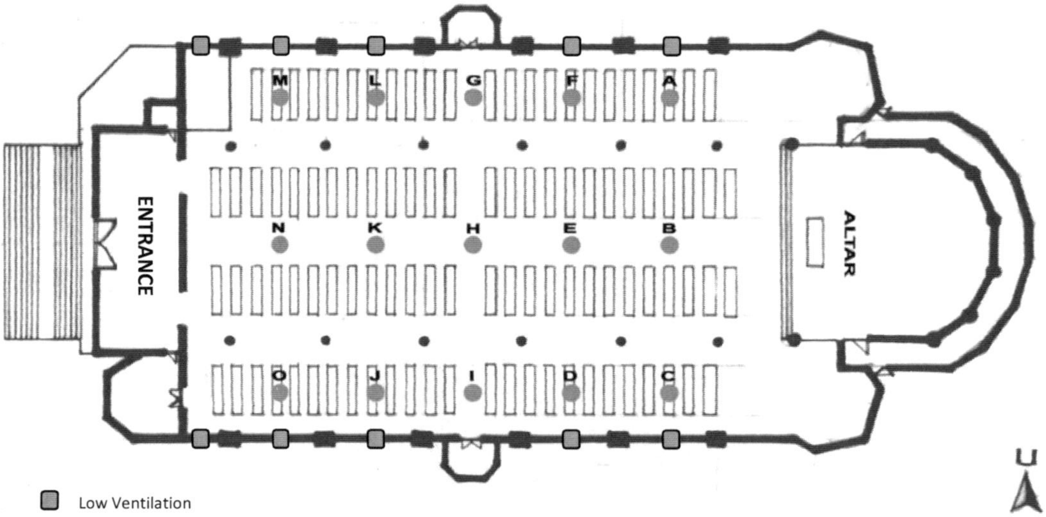

Figure 2. Measuring Spot Plan

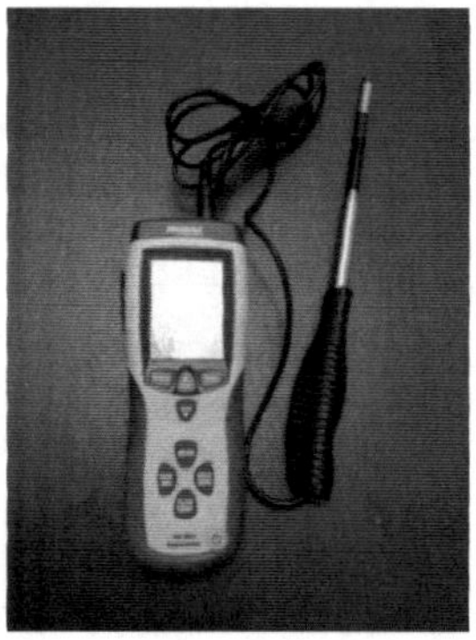

Figure 3. Hotwire Anemometer.

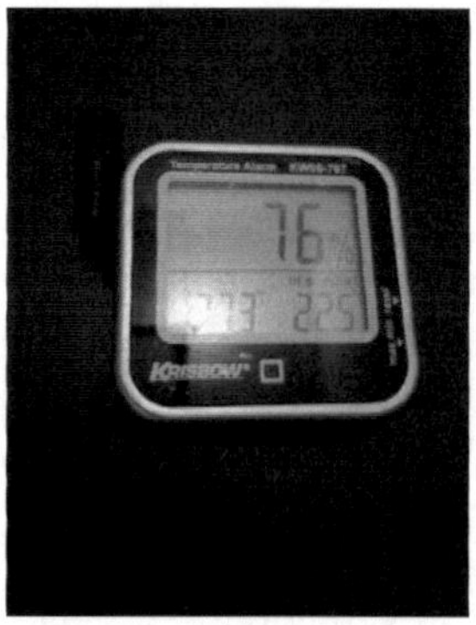

Figure 4. Digital Thermo Hygrometer

Figure 5. Low Ventilation. It is located 63cm from the floor. It can be adjustable as open or closed.

3. Measurement Results

Conditions: Low Ventilation Opened, With Prayer Activities

Measuring Time	Measuring Spots	Weather	MRT (°C)	Temperature (°C)	Relative Humidity (%)	Air Velocity (m/s)	Effective Temperature (°C)
8:45	A		26,5	27	85	0,33	25.9
	B		26,7	27,4	88	0,25	27,2
	C	S	25,95	25,9	84	0,39	24,2
	D	U	25,7	25,4	84	0,34	23,9
	E	N	26,3	26,6	85	0,22	26,3
	F	N	25,35	24,7	85	0,39	23
	G	Y	25,3	24,6	84	0,2	24,3
	H		25,35	24,7	85	0,2	24,4
	I		25,35	24,7	92	0,17	25,1
	J		25,3	24,6	84	0,48	22,5
	K		25,2	24,4	87	0,2	24,1
	L	26	24,95	23,9	83	0,49	21,8
	M	75%	25	24	81	0,53	21,8
	N		25	24	82	0,18	23,9
	O		25	24	78	0,32	22,6
16:30	A		26,25	26,2	83	0,36	24,7
	B		26,35	26,4	82	0,22	26
	C	C	26	25,7	82	0,43	23,8
	D	L	25,55	24,8	83	0,47	22,8
	E	E	25,75	25,2	83	0,28	24,2
	F	A	25	23,7	82	0,49	21,7
	G	R	24,8	23,3	81	0,24	22,7
	H		24,8	23,3	84	0,28	22,5
	I		24,85	23,4	80	0,31	22,3
	J		24,8	23,3	79	0,46	21,4
	K		24,9	23,5	80	0,26	22,7
	L	26,3	24,85	23,4	79	0,43	21,7
	M	78%	24,8	23,3	78	0,47	21,4
	N		24,9	23,5	79	0,19	23,3
	O		24,8	23,3	79	0,43	21,6
18:15	A		25,5	26	86	0,49	23,6
	B		25,5	26	87	0,23	25,3
	C	C	25,2	25,4	85	0,36	23,5
	D	L	24,8	24,6	87	0,38	22,7
	E	E	24,95	24,9	85	0,3	23,4
	F	A	24,65	24,3	85	0,43	22,2
	G	R	24,85	24,7	87	0,25	23,6
	H		24,9	24,8	88	0,23	23,9
	I		24,8	24,6	87	0,27	23,2
	J		24,15	23,3	87	0,4	21,5
	K		24,2	23,4	86	0,32	22
	L	25	24,1	23,2	84	0,39	21,4
	M	83%	23,65	22,3	84	0,44	20,4
	N		23,8	22,6	84	0,31	21,3
	O		23,5	22	85	0,46	20

Conditions: Low Ventilation Closed, With Prayer Activities

Measuring Time	Measuring Spots	Weather	MRT (°C)	Temperature (°C)	Relative Humidity (%)	Air Velocity (m/s)	Effective Temperature (°C)
8:45	A		26,65	27,3	89	0,24	27,3
	B		26,8	27,6	90	0,19	28,4
	C	S	26,25	26,5	86	0,26	25,9
	D	U	25,8	25,6	86	0,27	24,7
	E	N	25,85	25,7	88	0,17	26,1
	F	N	25,5	25	87	0,28	24
	G	Y	25,65	25,3	86	0,12	26,1
	H		25,75	25,5	87	0,1	25,3
	I		25,85	25,7	90	0,13	26,8
	J		25,3	24,6	85	0,33	23,2
	K		25,6	25,2	89	0,2	25,1
	L	26	25,15	24,3	86	0,35	22,9
	M	75%	25,25	24,5	84	0,36	22,9
	N		25,25	24,5	85	0,13	25,1
	O		25,15	24,3	83	0,21	23,8
16:30	A		26,35	26,4	84	0,25	25,8
	B		26,45	26,6	86	0,14	27,7
	C	C	26,15	26	85	0,27	25,2
	D	L	25,75	25,2	85	0,31	24
	E	E	26	25,7	87	0,19	25,8
	F	A	25,25	24,2	86	0,29	23,2
	G	R	25	23,7	84	0,18	23,7
	H		25,1	23,9	87	0,13	24,7
	I		24,95	23,6	83	0,2	23,3
	J		25,05	23,8	80	0,32	22,6
	K		25,05	23,8	81	0,23	23,2
	L	26,3	24,9	23,5	81	0,33	22,3
	M	78%	24,9	23,5	80	0,37	22
	N		24,9	23,5	82	0,11	24,1
	O		24,85	23,4	81	0,32	22,3
18:15	A		25,65	26,3	87	0,3	25
	B		25,75	26,5	89	0,19	26,5
	C	C	25,4	25,8	86	0,2	25,3
	D	L	25,1	25.2	86	0,22	24,4
	E	E	25,25	25.5	87	0,16	25.6
	F	A	24,85	24,7	87	0,28	23,4
	G	R	25	25	87	0,19	24,6
	H		25,3	25,6	89	0,17	25,6
	I		25,1	25,2	88	0,17	25,1
	J		24,3	23,6	88	0,28	22,4
	K		24,4	23,8	89	0,23	23
	L	25	24,2	23,4	87	0,29	22,2
	M	83%	23,75	22,5	86	0,31	21,3
	N		23,9	22,8	87	0,26	21,9
	O		23,65	22,3	86	0,34	20,9

Figure 6. Thermal conditions during acivities

The results of measurement, as shown in Figure 6, were as follows:

- When low ventilations were open, ET had range between 1.1 to 1.3°C lower than ET when low ventilations were closed,
- When low ventilations were open, relative humidity (RH) was 1.5 to 2.5% lower than RH when low ventilations were closed,
- When low ventilations were open, air velocity was 0.09 to 0.11 m/s higher than air velocity when low ventilations were closed.

The measurement results revealed that when low ventilations were open, ET and RH became lower, but the air velocity became higher. There were significant differences.

4. Discussion

Effective Temperature (ET) is a thermal comfort's function of radiant heat, temperature, RH and air movement or air velocity [6]. ET is defined as the temperature of saturated air in near or not moving

IOP Conf. Series: Earth and Environmental Science **213** (2018) 012044 doi:10.1088/1755-1315/213/1/012044

conditions (± 0.1 m/s) [7]. Effective temperature is categorized as comfortable when its temperature is around 23˚C - 27˚C [3]. As shown in Figure 7, the charts of effective temperatures have various dynamics. Some spots in all measuring time and in all conditions (low ventilations open/closed) were included into comfortable effective temperature (23˚C - 27˚C), but some other spots had lower ET. It shows that basically, Bogor Cathedral church has ET included into comfort to cool temperature. B spot was the only spot that entered not comfortable ET because its ET was higher than 27 ˚C. Two graphics in Figure 7 had a difference. ET became lower when low ventilations were open than when low ventilations were closed. Lowest ET was located at 6:15 pm, it can be seen at figure 7 that 8 from 15 spots had lower ET than 23˚C. Spots near the low ventilations (spots A, C, D, F, J, L, M, O) had lower ET from other spots.

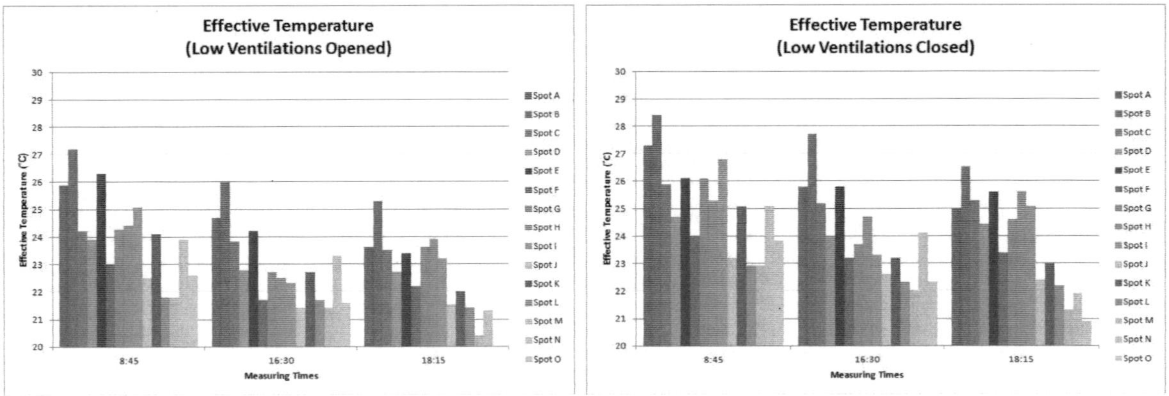

Figure 7. Effective Temperature during activities.

Relative humidity shows the comparison between the existing steam pressure to the maximum possible steam pressure (degree of saturation) under certain air temperature [1]. RH between 30 – 90% is included into comfort zone of relative humidity [8]. Figure 8 shows that most measuring spots and measuring times were included into comfortable relative humidity. It was only 1 spot at 8:45 am that had higher than 90% RH. As seen in Figure 8, relative humidity became higher when low ventilations were closed. It was caused by the lack of air velocity that entered the building so that relative humidity became higher and higher. At 8:45 am and 6:15 pm had higher relative humidity than at 4:30 pm.

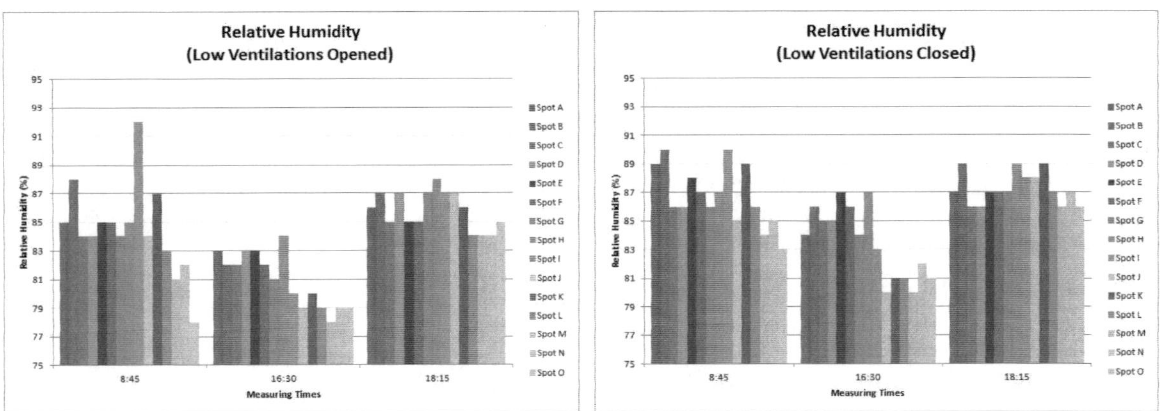

Figure 8. Relative humidity during activities.

Air movement or air velocity assists in the evaporation process which can lower the air temperature [6]. Air velocity between 0.2m/s - 0.8m/s is included into comfortable zone of air velocity [2]. As shown in figure 9, almost all spots, measuring times, and conditions were categorized as comfort zone.

When low ventilations were closed, almost half of all measuring spots had air velocity lower than 0.2 m/s. It means that when low ventilations were closed, almost half of measuring spots were categorized into not comfortable air velocity. As shown in Figure 9, spots A, C, D, F, J, L, M, and O near the low ventilations had higher air velocity in compared to the other spots.

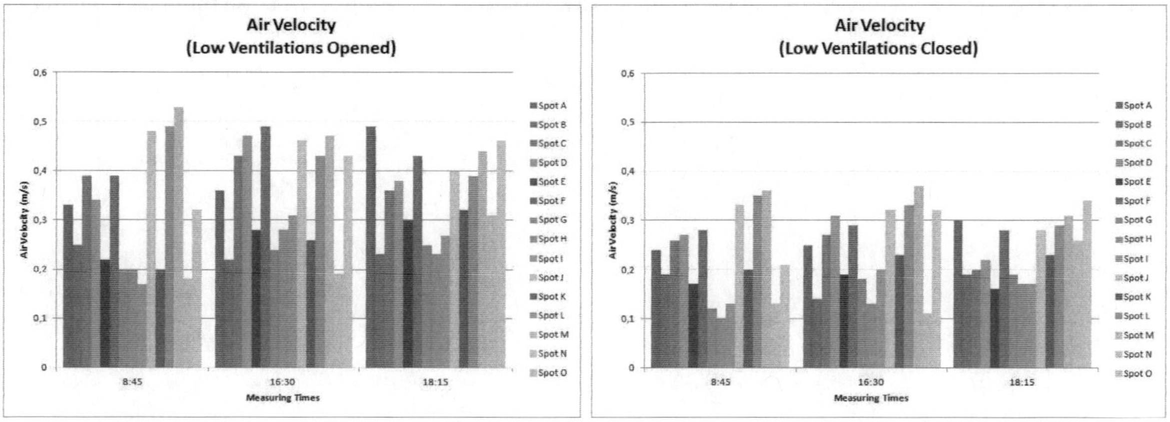

Figure 9. Air velocity during activities.

5. Conclusion

From the discussion above, we can conclude that Bogor Cathedral church is categorized as a thermal comfortable building. All measurement data and analysis were included in the comfort zone. The achievement of thermal comfort zone of this building could not be separated from the exixtence of low ventilations. Findings of this study revealed that low ventilations successfuly lowered down the ET so that the temperature of this building became cooler. The air humidity having comfort zone from 30% - 90% was also achieved. Spots near the low ventilations (A, C, D, F, J, L, M, O) had lower value of ET and air humidity than other spots. The air velocity also achieved its comfort zone (0.2m/s – 0.8m/s) while low ventilations resulted in more air velocity entering the building. Passive cooling system of this Bogor Cathedral church was proven to achieve thermal comfort, and low ventilations were also proven to optimize the thermal conditions to entered the comfort zone.

6. Acknowledgements

This paper would not exist without any guidance and assistance from several people. So that, we would like to express our deepest gratitute to; first of all thank you Pastor R.D. Tukiyo as the leader of Bogor Cathedral church for giving us permission to conduct a research in this beautiful building. Secondly, thank you Mrs. Erni Setyowati and Mr. Gagoek Hardiman as our supervisor for the guidance. Lastly, thank to all friends and colleagues for their support to finish this paper.

7. References

[1] Lippsmeier, Georg 1994 Tropical Building. 2nd edition. Syahmir Nasution, translator. Jakarta: Erlangga

[2] Szokolay 1973 Manual of Tropical Housing And Building. India: Orient Longman

[3] ASHRAE 1989 Handbook of Fundamental Chapter 8 Physiological Principles, Comfort, and Health. USA: ASHRAE

[4] Martínez-Molina, A., Tort-Ausina, I., Cho, S. and Vivancos, J. L 2016 'Energy efficiency and thermal comfort in historic buildings: A review', *Renewable and Sustainable Energy Reviews.* **61** 70–85

[5] Sekatia,Augi, Hasritanto, Bangun I.R, Setyowati, Erni and Hardiman, Gagoek 2017 Thermal Condition in Semarang Cathedral's Passive Cooling System. *Proceedings of International Conference on Engineering, Technology, and Industrial Application*

IOP Conf. Series: Earth and Environmental Science **213** (2018) 012044 doi:10.1088/1755-1315/213/1/012044

[6] Frick, Heinz and FX.Bambang Suskiyatno 2007 Dasar dasar Arsitektur Ekologis. Yogyakarta: Kanisius
[7] Houghten, F.C. and C.P. Yagloglou. 1923. Determining Lines of Equal Comfort. *Transactions of the American Society of Heating and Ventilating Engineers.* **29** 163
[8] ASHRAE STANDARD 55 2004 Thermal Environmental Conditions For Human Occupancy. USA: ASHRAE

A Study on Core House as Housing Reconstruction Program after the Central Java-Yogyakarta Earthquake 2006 in Kasongan Village Indonesia

Edward Endrianto Pandelaki[1], Resza Riskiyanto[2]
[1]Department of Architecture, Faculty of Engineering, Universitas Diponegoro, Semarang, Indonesia,
[2]Department of Architecture, Faculty of Engineering, Universitas Diponegoro, Semarang, Indonesia

Corresponding e-mail: epandelaki@arsitektur.undip.ac.id

Abstract. This paper is a result of study on core house as one type of post disaster housing reconstruction program in Indonesia. The aim is to explore the role of core house in restoring the beneficiary's condition after the earthquake disaster. Core house is a starter house provided with basic rooms that could be occupied directly and then developed incrementally by the occupants. This kind of house has a long history in housing development in Indonesia, which then implemented it as one type of post disaster housing reconstruction program. The case study is core house built in Kasongan Village after the Central Java-Yogyakarta Earthquake in 2006. Qualitative method was carried out according to the aim of the study. Long-term view is needed to conduct housing reconstruction program after disaster. Therefore, giving attention toward physical aspect needs to be balanced with nonphysical aspect.

Keywords: Core house, post-disaster housing reconstruction, Central Java-Yogyakarta earthquake 2006

1. Introduction

An earthquake of 6.3 on Richter scale struck Central Java Province and Special Region of Yogyakarta, Indonesia, in May 27th 2006. According to Bappenas [1] the earthquake took over about 5.700 lives and injured more than 40.000 people. In the housing sector approximately 154.000 houses were completely destroyed, 260.000 houses were slightly damage, and 600.000 people were homeless. There were supports and assistances from local and international institutions to conduct the post disaster activities.

Various types of housing reconstruction efforts were conducted, started from repairing assistance until donation of readily built starter house, and in-situ houses development until resettlement. One of the housing reconstruction activities was the provision of core houses in Kasongan Village, a well-known pottery craft village in Special Region of Yogyakarta, donated by the local government of Bengkulu Province, Indonesia. This activity was conducted as one of the in-situ housing reconstruction programs earlier right after the earthquake. Within this context, it is interesting to study the role of core house to its beneficiaries' post disaster life. The aim of this study is to explore the role and advantages of core houses as an in-situ housing reconstruction program for the beneficiaries in Kasongan Village, Yogyakarta. Qualitative method was used in the analysis, based on field observation and interview.

IOP Conf. Series: Earth and Environmental Science **213** (2018) 012045 doi:10.1088/1755-1315/213/1/012045

This paper is divided into several sections. The first section is introduction that describes background, aim, and method of the study. The second section describes the Kasongan Village and the provided core houses from the Bengkulu Province. The third section contains findings and discussion. And this paper ends with conclusions in the fourth section.

2. Kasongan Village and the Core House

Kasongan Village is located at the South side of Yogyakarta City (Figure 1). Administratively the village is part of Bantul Region. One of the well-known activities in this area is economic activities of its inhabitant in creating pottery handcraft. On the main street of the village there are the shops and showrooms of the pottery handcraft, while in the inner part behind the main street there are craftsman who makes the pottery in their own houses. When the Central Java-Yogyakarta Earthquake 2006 struck, Kasongan Village was damaged and many houses were affected with minor as well as major damage.

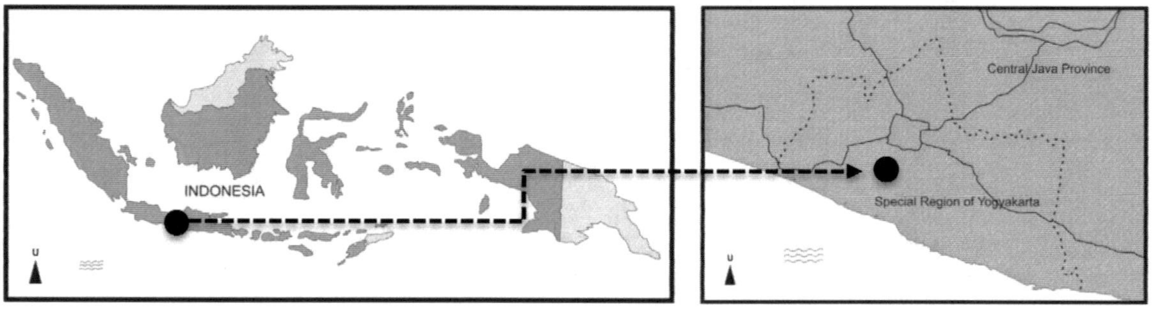

Figure 1. Location of Kasongan Village

Various post-disaster housing reconstruction programs were delivered in Kasongan Village. One of the housing reconstruction programs was in the form of donation of core houses from the Bengkulu Province. This program was one type of housing reconstruction implemented right after the disaster with the form of in-situ reconstruction (Figure 2). Core house was starter house that could be occupied immediately and then developed incrementally along with the occupants' need and capability. The core house was developed in the former lot of the beneficiary. It had 18-m^2-floor area with a gable roof style. Stone foundation and concrete steel column were used in the construction with bricks wall and roof tile from clay.

Figure 2. In-situ core house as one type of post disaster housing reconstruction program

IOP Conf. Series: Earth and Environmental Science **213** (2018) 012045 doi:10.1088/1755-1315/213/1/012045

3. Findings and Discussion

Based on field observation, the built core houses as the part of post disaster housing reconstruction program after the Central Java-Yogyakarta earthquake 2006 in Kasongan Village have been developed in various forms (Figure 3). The transformations conducted in building quantity as well as quality aspects. The quantity aspect means the floor space area of the house. The occupants have extended the current core houses with various width of floor space area. With these floor space area extensions, room structure or room completeness in the houses has become better, complete and adjusted according the occupants' need. This condition can be seen in the function of the rooms that become clearer and separated each other, such as guest room, bedroom, kitchen, production space for the craftsman, and other rooms. It seems that the occupants still emphasize the quantity aspect in developing their core houses although several core houses already did not function as the main houses of the occupants. On the quality aspect, it is found that the occupants have conducted enhancement on the building quality through improvement of the finishing materials. These quality enhancements were mostly conducted not in the whole part of the house but only toward several parts of the house, such as material of floor tile, wall surface, and finishing of the ceiling. The enhancement was also usually conducted in the certain parts of the house that had the nature of public area, such as terrace, guest room, and the front elevation of the house. According to Pandelaki [2] this kind of core house transformation can be categorized as moderate. This transformation is development emphasizing the quantity aspect of floor space extension with room structure that becomes clearer, altogether with development in quality aspect that is still implemented in several or certain parts of the house. These conditions show that the occupants have the capability to develop their houses incrementally. Although based on observation on the condition of space and its utilization in several houses, it seems that sometimes assistance needs to be given in order to improve the house to be a more comfortable place to live.

Figure 3. Core houses that have been transformed by the occupants

The core houses receive positive response from the beneficiaries. The occupants satisfy with the core houses because they can receive support after the disaster struck. However, the point of strength and fast implementation of the core houses also becomes positive factors put forward by the occupants. Its simple form makes the core house has the flexibility in space utilization and easier establishment and development. This condition is in accordance with the core house concept as a starter house that can be occupied immediately and then developed according to the need and capability of the occupants. The core houses use simple construction technology that can be conducted by the local community and local building labours. As mentioned above, the core houses from Bengkulu Province became one of early supports given to the beneficiaries in the form of in-situ reconstruction. Social bond owned by the community and kept within the community with the in-situ housing reconstruction supported the achievement of the objective. This condition is needed for post-disaster situation. Social bond becomes one capital that could support the implementation of post disaster housing reconstruction.

The development or transformation of core houses was conducted in phases. Several houses were transformed right after the occupants started to live in. Almost all of the core house had been extended or transformed after one year of development [3].The core house transformation was related to the source of funding, which might come from several sources, such as government support, family savings, and other sources. The utilization of lot and materials from former houses as remaining assets gave advantages in the development of the core houses afterward. These assets could be utilized in the housing reconstruction process. The occupants' response related to the obstacles in the core house transformation process is more toward financing the transformation.

The current development or transformation of core houses receives a positive response from the occupants. Various forms of transformed core houses reflect their various needs, capabilities, and priorities. Therefore, the transformed core houses have various roles and capabilities in fulfilling the occupants' needs. Within the framework of Maslow's hierarchy of needs, basically the occupants gave positive response toward fulfilment of the needs by the current house form. An important point in post-disaster housing reconstruction program is concern with the development of core houses that could also support the recovery in family's economic condition. There were people who worked as pottery craftsman in their houses before the earthquake in Kasongan village. Through the implementation of core houses as an in-situ housing reconstruction, the craftsman could immediately return to conduct their economic activities of creating pottery to support their livelihood (Figure 4). Therefore, their economic aspect as craftsman can also immediately be restored.

Figure 4. Inhabitant create pottery handcraft in their own houses

The findings revealed that the core houses with their current various transformation forms had enough flexibility in accommodating the spectrum of occupants' needs in the bionomic process. It also appears in the responses to the obstacle in the development of the core houses, which was more toward financing instead of physical aspects. Several advantages also could be seen from the core houses as an in-situ

IOP Conf. Series: Earth and Environmental Science **213** (2018) 012045 doi:10.1088/1755-1315/213/1/012045

housing reconstruction program, such as maintaining togetherness of the inhabitants, and supporting recovery in economic aspect through the existing assets. Therefore, the post-disaster housing reconstruction program needs to give concern on not only physical aspect of providing immediate shelter for the disaster's victims, but also long term condition. In this regard, non-physical aspect needs to be given in order to comprehensively recover the life of the beneficiaries.

4. Conclusion

The beneficiaries provided positive responses on the original core houses and the current developed or transformed core houses. Various forms of transformation reflected the occupants' needs, capabilities, and priorities. This condition also reflects the role and capability of the core houses, which is flexible enough in accommodating the spectrum of various needs, capabilities, and priorities of the occupants. Furthermore, this study revealed that as an in-situ housing reconstruction program after the earthquake, the core houses in Kasongan Village had several advantages for its occupants, such as: support in maintaining the family's and neighbourhood's togetherness and support in economic recovery by maintaining and utilizing the existing assets. Post-disaster housing reconstruction program will have a better role when given long term and balanced attention between physical and nonphysical aspects. The house should provide not only an immediate shelter but also recovery for the occupant's life comprehensively.

5. Acknowledgments

The authors would like to express their gratitude to Faculty of Engineering, Universitas Diponegoro, for the support in this study.

6. References

[1] Bappenas. 2006. Preliminary Damage and Loss Assessment, Yogyakarta and Central Java Natural Disaster. A Joint Report from BAPPENAS, the Provincial and Local Governments of D.I.Yogyakarta, the Provincial and Local Governments of Central Java, and international partners, June 2006. Jakarta.

[2] Pandelaki, Edward Endrianto, Shiozaki, Yoshimitsu. 2010. Core House Adjustments and Its Typologies during the Dwelling Process, Case Studies: Low Cost Housings by Perum Perumnas in Semarang City Indonesia. Journal of Architecture and Planning, volume 76, number 660, page 295-304.

[3] Ikaputra. 2008. Core House: A Structural Expandability for Living. Dimensi Teknik Arsitektur, volume 36, number 1, July 2008, page 10-19.

IOP Conf. Series: Earth and Environmental Science **213** (2018) 012046 doi:10.1088/1755-1315/213/1/012046

'No Fort in Fort City': Lost Heritage and The Shift of Urban Conservation in Tangerang, Indonesia

Mushab Abdu Asy Syahid, Kemas Ridwan Kurniawan
Department of Architecture Universitas Indonesia, Depok, Indonesia

Corresponding e-mail: mushab.abdu1@ui.ac.id

Abstract. The progress of modern heritage conservation has now reached a very local level. The need for researching and strengthening locals' history and identity is inevitable, yet oftentimes the physical heritage was lost before the idea of preservation really caught on. As a local city that bears the retrospective name "Benteng" (fort) reflecting its history, Tangerang City is similarly challenged with the lack of actual fort heritage. This paper attempts to explain the dynamics in how the lost heritage had shaped the city identity throughout architecture and urban structure. This research shows that the disappearance of physical heritage consequently led to the loss of the main historical identity of Tangerang. It affected how local government and communities take different conservation strategies to develop other potentials of the place, including architectural buildings and urban landscape eco-tourism surrounding the old site of the fort. This indicates a shift in the city conservation efforts to other alternative to perpetuate the reminiscence of the historic city produced in various form of commodification of the memory about "Benteng" in Tangerang, Indonesia.

Keywords: conservation, fort, heritage, landscape, Tangerang, urban

1. Introduction

The progress of modern heritage conservation has now reached into a very local level, and the need to strengthening locals' history and identity is inevitable, yet oftentimes the physical heritage they had was lost before the idea of preservation really caught on. Besides, there is much lesser attention and appreciation towards any historical areas whereas the everyday traditional activities are still living in. The dynamic evolution of conservation practices also shows the extension in which the conservationists perceive elements that are considered "heritage". The focus of urban conservation generally changes from the ranges within 1) single monument, 2) urban morphologies, 3) urban structure, and finally with 4) the natural properties of the urban landscape. This indicates an intention and transition to reach an understanding towards city heritage in a wider scope. [1]

This dynamic also occurs in the condition of conservation in local cities in Indonesia. Bearing the retrospective name "Benteng" (fort) which perhaps reflects its history, Tangerang is similarly challenged with the lack of actual fort heritage. Fort was one architectural piece that had generally shaped the modern city structure we wish could preserve rather than destroy. The fort became a historical significance since the early period of Dutch colonial enterprise (VOC) in East Indies (Indonesia). However, due to the lack of historical researches on this topic, more researches only focused on the communities of Pasar Lama or other city heritage buildings and seldom mentioned the influences of the lost fort heritage in the city conservation. [2]

Pusat Dokumentasi Arsitektur (PDA – Architectural Documentation Center) by the commission of Indonesian Government under Ministry of Culture and Tourism and co-sponsored by The Government of the Royal Kingdom of Netherlands had curated, invented and identified during 2007-2010 about 422 forts built in 16th-20th century, scattered throughout the archipelago. The fort in this paper is also referred to "Benteng", that by the PDA's definition is a range of the various types of defendable buildings or settlement formed in massive earth mound. Some of other notable demolished forts were Kasteel Batavia in Jakarta, Fort Vastenburg in Solo, and Fort Amsterdam in Manado bombed by Allies in World War II. [3]

Therefore, this paper attempts to explain the dynamics of how the lost heritage had actually shaped the identity of a city throughout architecture and urban structure. The aim of this research is to analyze the rational consequence of Tangerang as a modern city that had lost its main heritage. This condition of the loss could negatively lead to a sense of 'heritage at risk', or exactly a positive stimulus for developing any new potentials of city heritage in wider urban scale [4]. Meanwhile, the protected status of architectural heritage should also be the part of urban development for the area. The physical and functional quality layouts of the environment adjacent neighborhood are important as well [3].

Furthermore, the findings will be able to open the discussion in re-evaluating what local government and stakeholders have done in the city development in accordance with the conservation principles. It is also important to look at the efforts on communities level in preserving the reminiscence of Tangerang heritage, or at least their own related history in the city.

2. Methods

The research employed the qualitative method and descriptive approach to narrate the current urban conservation program surrounding "Pasar Lama" (Old Market) of Tangerang where the lost fort was located. Those efforts involved local government's intervention on historic urban landscape in Tangerang alongside Cisadane River, as well as the privates and local community actions toward heritage buildings. The data was collected by conducting field observations in some of built infrastructures related to the urban heritage conservation framework in the city which would be explained more detail in the following sections.

The findings later will be wrapped up by critical analysis using any relevant historical archives and documentation, as well as comparison of interviews with the community actors who have preserved the heritage buildings. The primary data will be elaborated by some supporting literature reviews, including the history of fort development during the early colonial period in 1684-1809.

3. Discussions

This section will firstly narrate the history of Tangerang as a consequence of colonial expansion. After the fort demolition, the story continues to the modern city infrastructure built by colonial government. Moving on to the next sub-section, it will describe the efforts of conservation practices regarding the recent situation of urban and architectural heritage in the city, such as the development of Museum Benteng Heritage and other program alongside Cisadane River as the urban landscape.

3.1. Historical Background of Tangerang as "Kota Benteng"

The study on historical archives helps to trace the stories on how the fort(s) during the early period of colonial enterprise of *Verenigde Oost-Indies Compagnie* (VOC) in Dutch East Indies (1602-1799) were developed in urban scale and also shaped the cities of Indonesia later. In the context of Tangerang since 1684, fort had signified a military defence purpose of the regional border formed by natural landscape and political clashes between colonial *Ommelanden* of Batavia (Jakarta) and indigenous Banten Sultanate in Java.

The etymological origin of name *Tangerang* or *Tangeran* (in Dutch archives' version) and *Tamgara* or *Tangaram* (in Tomé Pires version) is debatable to be derived from Sundanese word, *tengger*, *tanggeran* or *tatengger* that indicates a 'sign' to a border. The Cisadane River was also referred as one

IOP Conf. Series: Earth and Environmental Science **213** (2018) 012046 doi:10.1088/1755-1315/213/1/012046

of the six important trading ports in Sundanese Kingdom in Western Java amongst others, namely forts of Bantam (Banten) and Calapa (Sunda Kelapa) [5].

The fort building of Tangerang was initially a semi-permanent palisade (fences) covering amount size of land in Babakan across Cisadane River as an army camp (figure 1). Commonly, colonial forts were decidedly replaced the existing vernacular forts, and the fortification structures had begun since 16th century by the natives since the colonial arrival in 1511. Along with its development amongst the construction of many other fortresses and towers *(ronduyt)* around Batavia in 1708, VOC then designed and built in the old site a solid 20-foot-walled fort made of brickworks and stone taken from Aria Soetadilaga as the local regent of Tangerang [6]. In addition, the first construction of the riverfront *(quala)* guardhouse was built during the war with Banten Sultanate since 1682 and strengthened in 1752 after the second rebellion [7].

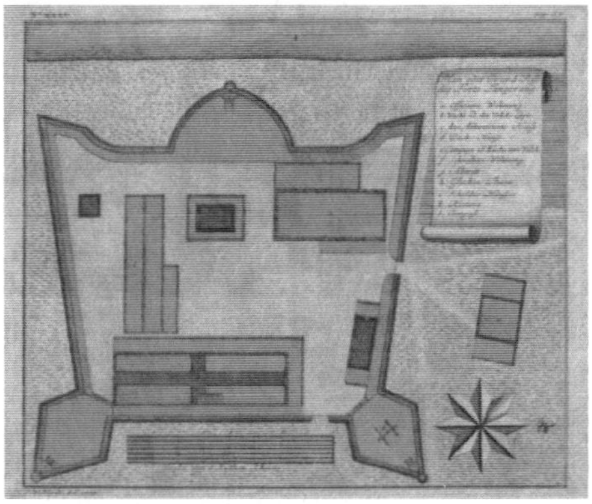

Figure 1. *Plan oder Grund-Riss des Forts Tangerang*. Source: Atlas of Mutual Heritage (Public domain)

Figure 2. The situation outside the fort Tangerang in 1763. Source: Atlas of Mutual Heritage (Public domain)

The fort building in 1779 (figure 2) was constructed with four bastions and ammunition powder storage and occupied by 38 Europeans inside and Makassar garrison outside the fortress, so it was also called as "Makassar fort". Although it was not clear how the first settlement had firstly appeared between the fortification and the natives' living environment, it is an undeniable fact that many of the villagers came to the city to live and made new settlement surrounding Makassar fort. It became a melting pot of many migrants with various ethnic backgrounds, such as Javanese, Sundanese, Malay, Arabic to Chinese community. The majority of the population in the Old Market was a mix of Sundanese-Betawi community, but there was also Chinese diaspora who influenced the early development of the city [8].

The land of Tangerang once used a cadastral and geometrical grids system, as shown in the map of living environment surrounding the fort and alongside Cisadane River dated 1770. The area consisted of Kampung Benteng Makassar, traditional market, and private land owned by Chinese landlords. It was planted with coffee, sugar cane and pepper. It also showed the residence of Aria Soetadilaga I-VII, which was the origin of Masjid Jami Kali Pasir, the first mosque in Tangerang [9]. However, in 1809 following the bankruptcy of VOC in 1799, Willem Daendels as General Governor of the Dutch East Indies (1808-1811) as well as military dictator of the Bonaparte's envoy instructed to abandon the entire fortresses in Java, although a new planning to build the fort in was constructed in 1800.

This also refers to the information in the Superintendent of Public Building and Work dated 6 March 1816, stating, "*The fortress and barracks in Tangerang are now neglected, no one wants to see it again. The doors and windows are damaged and even taken for their own benefit.*" It was also stated that the fort then was used as a prison, denoted pillboxes and bunkers of 20th century during World War II. The former site of the lost fort today is where the public mosque, shopping mall Plaza Tangerang, church, and other public buildings stand nearby the bridge over Cisadane River (figure 3).

Nevertheless, colonial government continued the history of first modern infrastructure in the city by building the water infrastructure (Bendungan Pasar Baru) and a train station of Tangerang during Ethical Policy period of Dutch East Indies Government in the early of 20th century.

3.2. Some Efforts on Heritage Conservation in Tangerang City

After the demolition of the fort, the urban heritages in the area were the complex of settlement, market and community buildings. The remaining name of "Benteng" there is only found in its street name, Jalan Bentang Raya. Later on, "Benteng" was also found in the prominent brands of home-industry products located in the Old Market, such as "Kecap Benteng Tong Giok Sen" and "Kecap SH (Shiong Hin)" made by ethnic Chinese descendants since 1882 and 1920. There is also a street called "Jalan Benteng Betawi", although it is located far enough from the Old Market area.

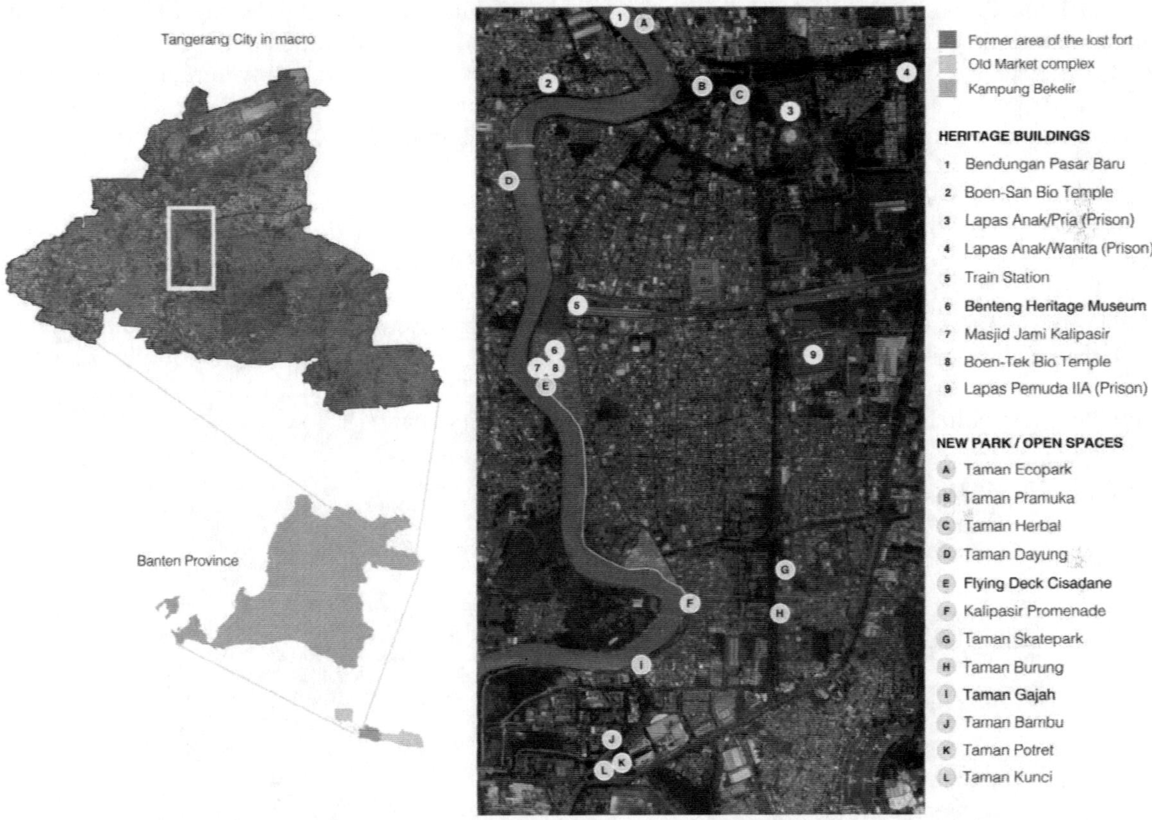

Figure 3. Map of Tangerang City, the location of architecture/urban heritage and new open spaces

The fort as the lost physical main heritage affects the locals to undeniably find other related historical buildings. In August 2011, the municipal government of Tangerang City officially published *Surat Keputusan Walikota Tangerang* (Decree of the Tangerang Mayor) after selecting nine old architectures as the formal heritage buildings ("*Bangunan Cagar Budaya*") in the city. They are 3 old buildings of prison, 2 local Chinese temples of Boen San-Bio and Boen Tek-Bio, 1 Masjid Jami Kalipasir, 1 water levee of Bendungan Pasar Baru, 1 train station of Tangerang and 1 vernacular local-Chinese house that is now functioned as a museum (Benteng Heritage). (figure 3)

The selection was held under the supervision of *Balai Pelestarian Peninggalan Purbakala Serang* (BP3S - Serang Archaeological Heritage Preservation Center) due to the absence of independent heritage conservation institution in Tangerang, and they took about a half-year during March-August 2011 for the research. The criteria given were the age of the building, the architectural quality, and other valuable considerations. The buildings were the legacies of Dutch colonial institutional

infrastructure and the public prayers buildings, which have strong relationship to the daily spiritual life and spaces of the community [10].
Ironically, the government's intervention was only considered superficial effort, yet to be done recently in their planning is the erection of 2 x 1 m formal epigraph ceramic stones in the 7 out of 9 city heritage buildings, although the similar functioned board has actually been existed in each of the sites since 2011-2012. This action, regarding the official statement from Jajat Y. Bun'Yani as the Section Head of History and Preservation in Department of Culture and Tourism of Tangerang City, surprisingly costs approximately Rp12,000,000,- for each epigraph. However, they provides no assistance when it comes to the building maintenance costs; with an exception for the functioning public facilities like the train station, prisons and the water levee.

As excerpted from the statements of some community actors like H. Syairodji as the advisory board of Masjid Jami Kalipasir and Udaya Halim as the founder of Museum Benteng Heritage, they admitted that they did the efforts for daily operation, maintenance, and even renovations of the buildings only by themselves. In case of the Kalipasir mosque, they take expenses from any funding and social charities from local Muslims living surrounding the mosque complex. For Benteng Heritage, the museum collected Rp25,000,- from the entrance ticket for each individuals.

3.2.1. The Conservation of Local Communities' Heritage
On 11 November 2011, 3 months after the release of *Surat Keputusan Walikota Tangerang* (Decree of the Tangerang Mayor), Udaya Halim as one of the ethnic Chinese descents who used to spent his childhood times in Old Market community officially opened Museum Benteng Heritage *The Pearl of Tangerang* (2012) together with KPH. Prof. Dr. HE. Harimurti Kridalaksana and Ping Sopandi. Halim took the name "Benteng" for this historical museum that was a byproduct of adaptive reuse of an old 18th-century Chinese traditional house he bought. It is located nearly behind the Boen Tek-Bio temple on Jalan Baharudin No. 8 Tangerang 15111, Banten-Indonesia (figure 4).

Figure 4. Front entrance of Benteng Heritage

Figure 5. A stone epigraph erected on the museum's front wall

However, after visiting the place, one hardly finds anything about the historical documentation of "Benteng" (fort) heritage itself, except for only a few mentions. The museum is nothing but cultural history of local Chinese in the city. Not surprisingly, many visitors claimed it as 'the first museum of Chinese communities in Indonesia', whilst an epigraph erected on the building's wall stating the museum as cultural heritage of Chinese Tangerang (*Warisan Budaya Peranakan Tionghoa Tangerang*) (figure 5).

IOP Conf. Series: Earth and Environmental Science **213** (2018) 012046 doi:10.1088/1755-1315/213/1/012046

Udaya Halim once submitted Benteng Heritage Museum to UNESCO Bangkok and Asia and Pacific Regional Bureau Education to globalize his project, and also to receive 2012 UNESCO Asia-Pacific Awards for Cultural Heritage Conservation. As learnt from Halim's mailing archive, Tim Curtis, whose office in the Head of Culture Unit in UNESCO Bangkok on 14 September 2012 positively implied his appreciation of Halim for his efforts and dedication in conserving the hybrid ethnic Chinese culture, although the juries later also stated that the project "*was difficult to fully understand...*", regarding its insufficient level of proper documentation.

Further analysis is that the extension of this conservation did not really attach to the whole story of the city (primarily the history of Tangerang as "Benteng" itself), because the collections and stories provided were more about the ethnic Chinese since the first place. It perplexes the UNESCO Jury, reflected from their impression that questioned the extent of the works, historic significance, and its cultural authenticity of the building, whereas more modest conservation interventions were actually more appreciated. This kind of conservation program done by privates (although officially recognized by the government) is still able to givs different sense and unique example of how the history of the locals was preserved.

3.2.2. Approaches in Historic Urban Landscape

The heritage of urban landscape in Tangerang is now commercialized with culinary festivals, semi-permanent outdoor food stalls, and building infrastructure for future investments to gear the economic activities in the city. According to Arief Rachadiano Wismansyah as the Mayor of Tangerang, his vision to the city is manifested through the program grand design called "*Tangerang Layak Dikunjungi*" (Tangerang Worth Visiting). This program engages cultural heritage, local tourism, and the implementation of waterfront city concept in the urban landscape, where the river is the main plot of urban development.

Historically speaking, Cisadane River has become an integrated historic urban landscape that linked two communities' sacred annuals: first is *Arakan Perahu* festival for local Muslims to celebrate the birth of Prophet Muhammad and second is *Peh Cun* festival for local Chinese community (that was held on last 18 June 2018). Besides that, the city government also has organized other annual "Cisadane Festival". The program in developing eco-tourism also happens in the settlement along the river, as it is popularly implemented in "Kampung Bekelir" program that was finished in cooperation with PT. Pacific Paint anniversary in 2017 to create a unique character of local *kampung* (figure 6).

Figure 6. Aerial view of Kampung Bekelir. Source: Fahrurozi

Figure 7. Pedestrian promenade in Cisadane Walk, as shown in figure 3, the line between E-F

Besides, the local government has also provided many open spaces of theme parks. In the last four years alone as reflected in the results of field observation by the first author, more than 10 parks have been built surrounding the city. Two of them were built alongside of the Cisadane River named Taman Gajah and Flying Deck Cisadane. The other parks are: "Kalipasir Promenade", an open space functioned as public sport field, "Taman Kunci" (an abbreviation of *Taman Kupu-Kupu & Kelinci*), Taman Pramuka (2013), Taman Pisang (2014), Taman Potret (2015), Taman Eco-Park (2015), Taman Bambu (2016), Taman Skatepark (2016), Taman Dayung (2016), Taman Gajah (2017), Taman Herbal (2017) and Taman Burung (2018). (figure 3)

Taman Gajah was an outcome of cooperation between Municipal Government of Tangerang City and Corporate Social Responsibility (CSR) of Gajah Tunggal, Tbk., the largest integrated tire industry in South-East Asia, operated in the city since 1978. The park, opened the area up to 6.066 m^2, consists of open spaces and amenities facilities for recreation, outdoor food court, bird watching, mosque, and the view of Cisadane river landscape alongside the park. Another open space related to Cisadane River is "Cisadane Walk", that was built in last 2017 and located in western of Old Market complex, and along the street of Jalan Benteng Jaya to Jalan Kalipasir and Jalan Perintis Kemerdekaan, Tangerang. These pedestrian sidewalks install culinary stalls, open stage, children playground, flying deck and benches provided in stopping point of the sidewalks (figure 7).

Nonetheless, the banality of urban rejuvenation affects many local city landscapes by using a popular big 'Hollywood sign' style as the templates. As the examples above, "Flying Deck Cisadane" (figure 8) and "Kampung Bekelir" (figure 9), indicating a colorful theme for a living kampong settlement of the natives besides Old Market complex. On the other hand, this kind of facility inevitably creates a superficial city landmark surrounding the natural urban landscape.

Figure 8. A stopping point in the flying deck of Cisadane River

Figure 9. A sign in front of the settlement of Kampung Bekelir

4. Conclusion

From the initial research findings that authors have described and analyzed, this study concludes that the loss of a major physical cultural heritage of a city will be indirectly followed by the loss of the intrinsic values of historical city's identity in the vast modern development. In this case, the lost heritage becomes only a name with an unknown physical object. It also demands different strategies of local government and local communities to preserve what remains.

This paper also shows the early eco-tourism program led by the local municipal government of Tangerang surrounding the old site of the fort. These efforts may not always refer to the main history of Tangerang as "Kota Benteng" at all, but this indicates an extension in the conservation scope, whereas Cisadane River plays a significant role to be a complementary historical urban landscape. In other words, the disappearance of "Benteng" can also be seen as a metaphorical symbol of the shift of city heritage conservation focus from one monumental object into a wider landscape.

On the other hand, the representation of the fort heritage seems to be produced in various forms of the commodification of the memory, as what Udaya Halim had attempted with Benteng Heritage Museum, that "Benteng" name was preserved as a reminder only to a particular history of the community. This commodification, either by local government or by individuals, somehow only works in particular individuals' interests. Asserting a new kind of integrated conservation program is immediately needed to overview the Old Market complex, Cisadane River and the heritage buildings in Tangerang as a whole area. Finally, further research related to this issue is still urged to deal more with the past and future of city heritage in order to harmonize between preserving historical reminiscence and optimizing the potential of the modern urban spaces.

5. Acknowledgements

Some of the material parts of this research were the revised version of the first author's unpublished thesis work to obtain Master Degree in History and Theory of Architecture, in Department of Architecture, Faculty of Engineering Universitas Indonesia, Depok, which was defended on August 2017. The first author would like to express gratitude to Prof. Kemas Ridwan Kurniawan, ST., M.Sc., PhD. as his main supervisor since 2015 and for his insightful inputs during the processes of working research. The research was also supported by 2017 UI Research Publication Grant, under the scheme of PITTA UI (*Publikasi Internasional Terindeks Tugas Akhir UI*).

6. References

[1] Corten JP, Geurts E, Meurs P & Vermeulen R 2014 *Heritage as an asset for inner-city development: an urban manager's book* (Rotterdam: nai010 publishers), 22-27.

[2] Menteri Kebudayaan dan Pariwisata Republik Indonesia 2010 *Forts in Indonesia: inventory and identification*. Jakarta: Pusat Dokumentasi Arsitektur, 280-281.

[3] O Syahrida and J Sumabrata 2018 *IOP Conf. Ser.: Earth Environ. Sci.* 126 012191.

[4] Rico, T 2014 The limits of a 'heritage at risk' framework: The construction of post-disaster cultural heritage in Banda Aceh, Indonesia, *J. Soc. Arch.* **14**, 157-176.

[5] Cortesão A 1944 *The suma oriental of Tomé Pires and the book of Francisco Rodrigues, an account of the east, from Red Sea to Japan*, **1**, London: Council of Hakluyt Society, 166-171.

[6] Knaap G 2007 *Grote atlas van de verenigde oost-indische compagnie, II Java en Madoera*, Royal Dutch Geographical Society, Utrecht.

[7] Atsushi Ota 2003 Banten rebellion, 1750-1752: Factors behind the mass participation, *Modern Asian Studies* **37**, 613-651.

[8] Sumarda H 1985 *Tangerang 1946-1949: Pemerintahan dan rakyat*. Unpublished undergraduate thesis work, Department of History Universitas Indonesia, Depok.

[9] Lukito, YN & Syahid, MAA, 'Conservation of Kalipasir jami' mosque in Tangerang, Banten', *INSIST*, 2, 52-57 (2017).

[10] Surat Keputusan Walikota Tangerang No. 430/kep.337-disporbudpar/2011 tentang Bangunan Cagar Budaya tahun 2011.

Spatial System Station NIS Yogyakarta-Bantul Railway

Masyiana Arifah Alfia Riza[1], Ikaputra[2], Dimas Wihardyanto[2]
[1]Universitas Diponegoro, Semarang, Indonesia
[2]Universitas Gadjah Mada, Yogyakarta, Indonesia

Corresponding e-mail: arifah@arsitektur.undip.ac.id

Abstract. Station is an architectural heritage that has local potential value. NIS Station Yogyakarta-Bantul railway is one of the first generation stations in Indonesia. It was established in 1985. Since some station buildings began to collapse, the purpose of this research is to know the architecture of the buildings seen from spatial space and station functions that were still in operation. Therefore, this research formulated spatial system of station building, which is found based on old station building typology that exists in Indonesia. The research method used qualitative rationalistic. The aim of this research is to know the different spatial systems of stations on the Yogyakarta-Bantul railway while operated until 1976. Therefore the study depended on the type of old station building found. There were 3 types of old stations that form spatial characters in each of Yogyakarta-Bantul railway stations. Three spatial types are is *haltegebouw* type at Dongkelan Station, *stasiongebouw* with small type station at Winongo Station, Bantul Station, and Palbapang Station, and *stasiongebouw* with medium type station at Ngabean Station. The spatial system of building stations on Yogyakarta-Bantul railway had their roles by themselves because they were influenced by setting area where the stations located.

Keywords: Architecture, Station, Spatial, Yogyakarta.

1. Introduction

The station is a building that serves as a railway stop. The railway is often called by Indonesian people *"sepur"*. *Sepur* is adopted from the Dutch *"spoor"* in which according to etymology means footprints or trail of wheels abandoned. In other words "*spoor*" is the path that has been made to pass [1]. The first station building in Indonesia was established in 1864 in Tanggung District. It was inaugurated in 1867. The number of stations in Indonesia was 1516 in 1955 and lagged 571 in 2000, down 62% in 45 years [2]. Yogyakarta to Sewugalur railway is one of the first generation train lines with 1435 mm rail width. Until now, there are 5 stations still standing from the Yogyakarta-Sewugalur railway. The five remaining stations are in operation to connect Yogyakarta to Bantul. Otherwise the railway that connected Bantul and Sewugalur (Kulonprogo) had disappeared after the Japanese occupation. The operational period of the Yogyakarta-Bantul railway reached 1976 and then closed due to road access.

Understanding of the spatial system of station building architecture is one of the efforts to maintain the sustainability and conservation of the local potential. It that has been instrumental in the establishment of the identity of the city. Building spatial system is one of the important aspects of architectural component of the building. In architecture there are 3 elements as the main formers in the theory of Vitruvius. They are in the form of *utilitas*, *venustas*, and *firmitas* [3]. This study specifically discussed

IOP Conf. Series: Earth and Environmental Science **213** (2018) 012047 doi:10.1088/1755-1315/213/1/012047

the spatial building that is part of the *utilitas* theory by Vitruvius. This study focused on the spatial aspects of spatial function and the governance of interconnected spaces.

1.1. Spatial System

Architecture is created from form and space. Space in a building determines a function and circulation to facilitate human activity. The spatial function is also explained in one of the basic architectural theories by Vitruvius that architecture is made up of *utilitas* (function), *venustas* (beauty), and *firmitas* (strenght). The emphasized *utilitas* is a good space arrangement, based on function, inter-space relationships, and building technology (lighting, airing, etc.).

> "...convenience, when the arrangement of the apartments is faultless and presents no hindrance to use, and when each class of building is assigned to its suitable and appropriate exposure...." [3]

Habraken desribed spatial system that is related to the floor plan that includes the form of layout, organization space, orientation and hierarchy of space [4]. *Utilitas* include needs, functions and floor plans. Architecture has a function that is to not only shade and accommodate human beings with all activities and all the necessary furniture in the activity, but also give the atmosphere, image, and direct the thoughts and feelings and behavior of the users [5]. Ching states that space is a component that accommodates all kinds of functions according to its purpose embodied in a certain form that can be seen and felt. Space can be distinguished on the organization of form and space, spatial relationships, spatial organization, hierarchy space, and circulation [6]. But a space is influenced by the function inside in which this function is a component that contains all human needs, especially activities that must be accommodated [6]. Components of this function have several factors, such as cultural, socioeconomic, customs, and behaviour [6]. This affects the shape of the building plan, the more complex. While the emphasis on *utilitas* is to set a good space based on the function of inter-space relations and technology [5]. Spatial functions as a container of human activity both physically and psychologically [7]. It also causes spatial pattern to be seen as the relationship between architecture, environment and culture where the spatial is located [7]. Supstandar describes spatial system as a system or a way of setting space which is able to meet the requirements to accommodate the needs of users [8]. The position of interior study is part of architectural science, as a result of the development of technology, science and art as a whole. According to that interior setting must walk together, parallel, with the exterior. So, in the other words spatial system cannot be separated from exterior design in architecture. The results of several theories regarding spatial system will be the objects of research, including the building settings, building mass orientation, building configuration, arrangement and orientation space, relationships and organization space, hierarchy of space, and the pattern of circulation space.

1.2. Station Building

The station building is taken as an object of study because it has a unique characteristic compared to other modes of transit space [9]. Station is a place where passengers can ride railway transportation. The station's architecture in Java, generally can be divided into the type of *haltegebouw* and *stationgebouw* (see Figure 1) [10]. *Haltegebouw* is a small station which has only one or two spaces. The building size is small and simple in terms of design and facilities for passengers as well as for the execution of its main railway service. *Stationgebouw* is a station that has more complete facilities than *haltegebouw*. This station consists of several spaces for passenger and train service. S*tationgebouw* types are divided into small type, medium type, and big type (main station) [10]. According to Ikaputra, the old stations in Indonesia are influenced by the indish style.

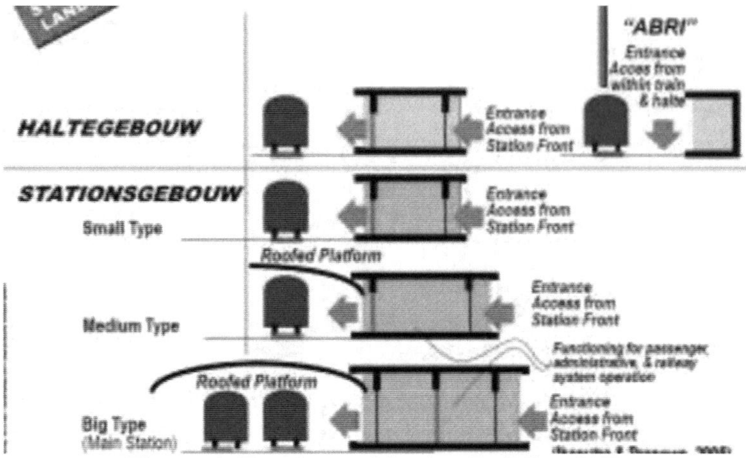

Figure 1. Typology of Station Building [10].

Broadly speaking, the existing room in the station is divided into two groups, first for the activities of passenger services as public space and second for the technical space of railway operation. When compared between the two groups, the public space in the station shows more diversity. The technical space of railway operation has the same standards for all stations. The difference is only the dimensions due to the number of emplacement [9].

According Sulistyani [9], spaces that existed in the stations of the three companies NIS-SS-SJS in the period 1864-1930 were:

- *Vestibule (hall)*
  The front room of the station that accommodates activities such as waiting for tickets, and transit passengers. Hall has a door which is the main access to the station building. This space is used to purchase tickes. *Vestibule* (hall) does not require any furniture because the activities of the visitors are only transit for a short time.
- *Kaartjis* (*counter*)
  Place or space to buy or book train travel tickets. *Kaartjis* (counter) places together with hall to accommodate the queue of people who buy tickets.
- *Watchkamer*
  Space that is functioned to await the arrival of the train.
- *Domes toilet*
  The toilets, where each station has a different standart.
- *Buffet*
  The dining room, where each station has a different standart.
- *Baggage/Magajizn*
  The store room, where each station has a different standart.

.

2. Methods

This research used qualitative research method with rationalistic paradigm approach. The analysis of descriptive research results created an overview and exposure as well as explored the existence of physical changes in the style of buildings on the station building in Yogyakarta-Bantul railway.

IOP Conf. Series: Earth and Environmental Science **213** (2018) 012047 doi:10.1088/1755-1315/213/1/012047

2.1. Data Collection Method
Data collection methods included primary data collection through surveys, observations, interviews and secondary data of various related information in the form of writings, newspapers, books, and literature studies. From the survey results 5 remaining stations were found (Figure 2) from the Yogyakarta-Sewugalur railway and became the Yogyakarta-Bantul railway. Ngabean Station (1), Dongkelan Station (2), Winongo Station (3), Bantul Station (4) and Palbapang Station (5).

Figure 2. Five Remaining Stations on Yogyakarta-Sewugalur Railway Currently, (1) Ngabean Station, (2) Dongkelan Station, (3) Winongo Station), (4) Bantul Station, (4) Palbapang Station

2.2. Data Analysis Method
Data analysis in this research with descriptive method described and gave description about character of building style at each station. Exposure was done by analyzing the history and context of the environment that occurred at the station and the picture of the building was done by modeling the building so the physical character of the station building was known. Then it would be linked to the colonial style that affected the physical character of the building. Furthermore, the overlay of the analysis served as a basis for knowing the physical character changes that occurred at stations on the Yogyakarta-Bantul railway.

3. Discussion

3.1. Building Setting
The observation of the building's structure on the station architecture in the Yogyakarta railway is seen in the type of station building. It is based on the form of the station that aims to see the activity space of passengers. The setting of this building is also intended to see the orientation of the building so it can see the trend of station for user activity.

Table 1. Building Setting Analysis

DESCRIPTION		STATIUNGEBOUW TYPOLOGY				
		HALTEGEBOUW TYPOLOGY	SMALL TYPE STATION			MEDIUM TYPE STATION
		DONGKELAN STATION	WINONGO STATION	BANTUL STATION	PALBAPANG STATION	NGABEAN STATION
Regional Setting	Location in town	■	□	■	□	■
	Direct access to sugar factory	□	■	■	■	□
	Direct acces to main road	■	□	■	■	■
Building Setting	Parallel station shapes	■	■	■	■	■
	Orientation facing railway	■	■	■	■	■

	Single Mass Building Station	■	■	■	■	■
	There is a House Service Complex	□	■	■	■	■
Station Role	As the distribution of industrial products	□	■	■	■	■

■ **: found**

□ **: not found**

Judging from the setting (Table 1), Yogyakarta-Bantul stations are parallel type based on the shape setting. The five stations have railway lines on one side only. In this stations the activity of passengers who ride train are only done on one side only. From the observation it is found that the orientation of the building does not depend on a particular direction so that the orientation of the building can face any direction. The orientation of the building is more influenced to face the train motion. In this case the train moves from Yogyakarata to Bantul and instead causes a longitudinal path north of the south, thus making the building facing the direction of the train motion. Priority orientation of the building facing the direction of the train motion is because with the building that extends along the railway lines cause the spaces to be more efficient and optimal. Building stations that accommodate both in terms of rail and passengers resulted in the building of Yogyakarta-Bantul stations that have two orientations facing the railway line and facing in the direction of coming passengers. The ease of achievement as an orientation direction is influenced by the functional needs and rational thinking of Europeans who do not base on mental spiritual and religious mythology, but rather on logical thinking [11]. The role of spatial efficiency and the role of spatial building function become the main priority in determining the setting and orientation of the building.

The stations on the Yogyakarta-Bantul railway have single building masses and their extensions are extended along the direction of the train motion. In the surrounding of the station emplacement, there is official residence of station's workers.

From the setting of this building there is a typical character of the station building found on the station that is the type of *stationgebouw* on the Yogyakarta-Bantul railway. The location of this line station did not depend on the access of the main road or its location in the center of the city but the approaches of the sugar factory to transport the production. In addition, the station on this track was the type of emplacement of goods and passengers. Its important role was the transportation of the sugar factory industry connected the industry from the sugar factories in Bantul to the area of Yogyakarta. This important role was evident from the station of urgency station that was not to pick up passengers but the proximity of access to sugar factory.

3.2. Building Spatial System

3.2.1. Dongkelan station is a haltegebouw type.

Type *haltegebouw* obtained in this study has the orientation of the building facing the railroad. Simple room completeness of *kantoor/kaartjees* and *wachtkamer* are arranged in linear with circulation patterns that disappear in space (Figure 3).

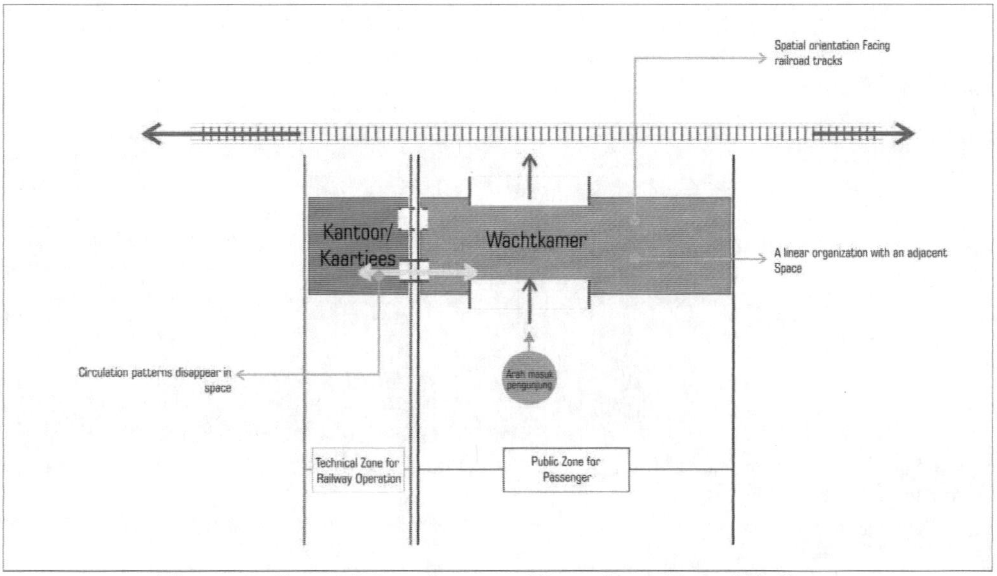

Figure 3. Type *Haltegebouw* on Yogyakarta-Bantul Railway

The *haltegebouw* type that is found on Dongkelan Station has a *wachtkamer* with 12m x 4m, and a *kantoor/kartjees* with 4m x 4m (Figure 4). The *wachtkamer* area is 3 times bigger than the *kantoor/kartjees* room in order to keep the room comfortable with the many activities inside, such as the arrival of passengers both from train and from outside, wait for train arrival, and train ticket purchase. The spatial proportions of Dongkelan Station were found to be a 1: 1 (Figure 4). The ratio was thought to be influenced by the proportion of renaissance by the Dutch community established at the time and had not changed the building style from the beginning.

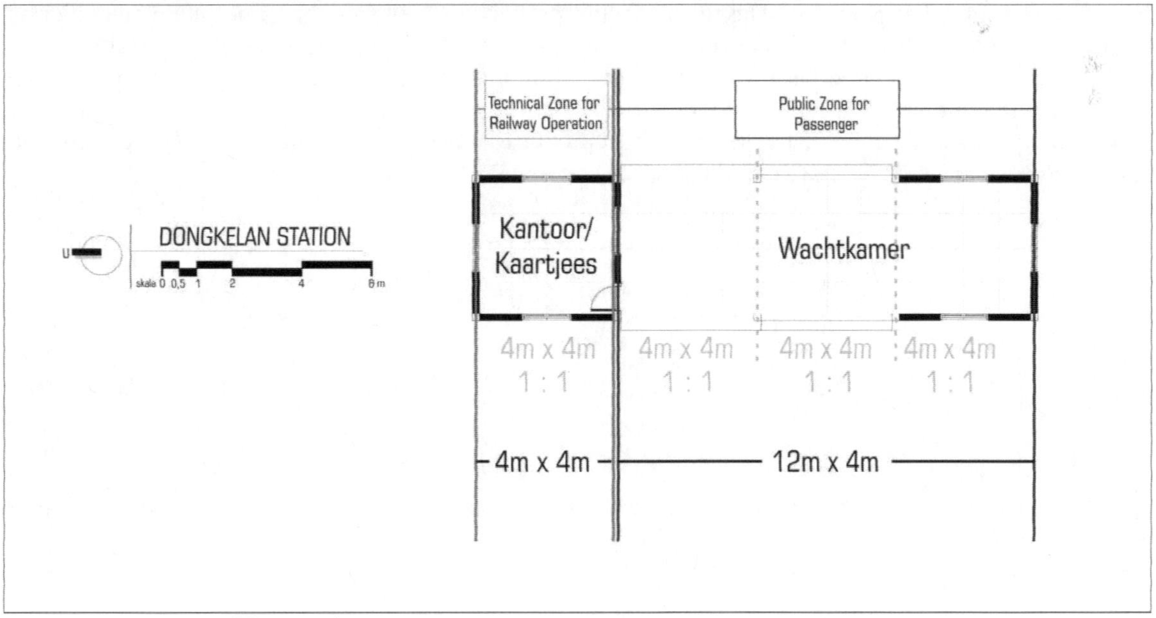

Figure 4. Type *Haltegebouw* on Yogyakarta-Bantul Railway

3.2.2. Winongo Station, Bantul Station, and Palbapang Station are small type stastiongebouw.

Winongo Station, Bantul Station, and Palbapang Station are small type of *stastiongebouw*. *Stastiongebouw* type obtained in this study has the orientation of the building facing the railroad.

Completeness of space in the form of *wachtkamer, kantoor/kaarjtjees, condecteur*, and *magajizn* is arranged in linear with circulation patterns that pass through space. The shape of the circulation is open on one side (Figure 5).

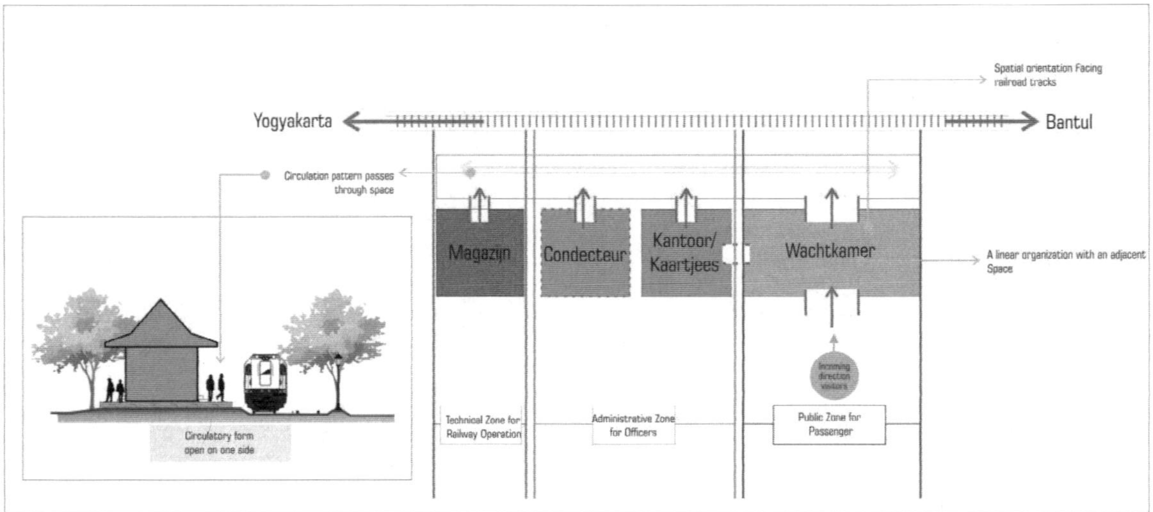

Figure 5. Stationgebouw Small Type on Yogyakarta-Bantul Railway

The *stastiongebouw* types found on Winongo Station, Bantul Station, and Palbapang Station have *wachtkamer* with 6m x 4m, *kantoor/kartjees with* 8m x 4m, and 4m x 4m (Figure 6). The *wachtkamer* area is 2 times bigger than the *kantoor / kartjees* space in order to keep the room comfortable with many activities inside, such as the arrival of passengers from both train and outside, wait for train arrival, and train ticket purchase. The result of this stations analysis has the same modular, which is formed from module 2m x 2m then developed according to space requirement at different station. This modular proportion was used because in these three buildings allegedly had undergone changes from their original building to meet the needs of the station compared to the previous one. Modular proportions is thought to replace the previous proportion because it is more easily applied and developed its shape in meeting the needs of space functions.

IOP Conf. Series: Earth and Environmental Science **213** (2018) 012047 doi:10.1088/1755-1315/213/1/012047

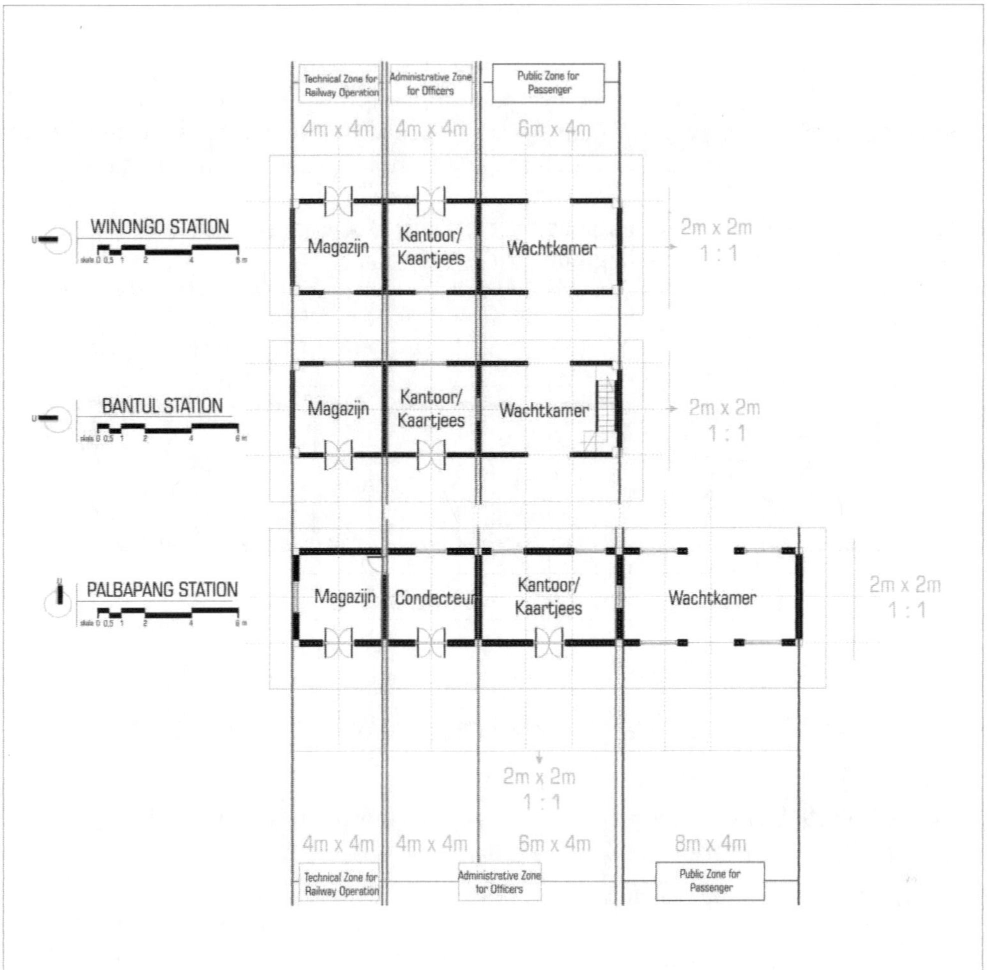

Figure 6. Stationgebouw Small Type on Yogyakarta-Bantul Railway

3.2.3. Ngabean Station is a type of stationgebouw medium type.

Type *stastiongebouw* obtained in this study has the orientation of the building facing the railroad. Completeness of space in the form of *wachtkamer, kantoor/kaarjtjees*, and *magajizn* is arranged in linear with circulation patterns that pass through space. Open circulation forms with roof cover (Figure 7).

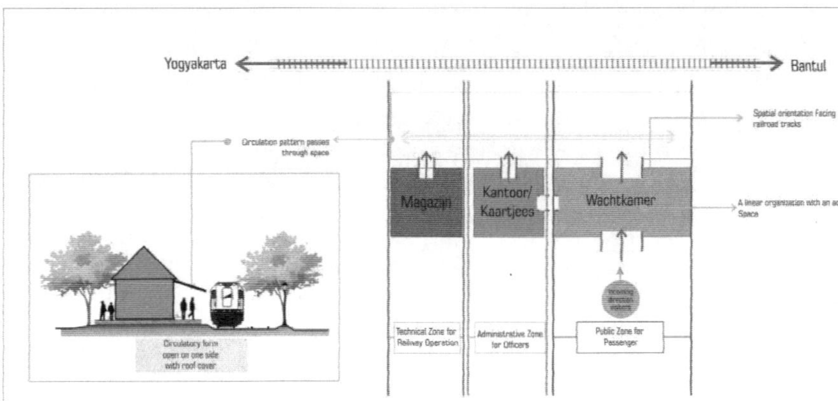

Figure 7. Stationgebouw Medium Type on Yogyakarta-Bantul Railway

IOP Conf. Series: Earth and Environmental Science **213** (2018) 012047 doi:10.1088/1755-1315/213/1/012047

The *stastiongebouw* type found on Ngabean Station has a *wachtkamer* with 8m x 5m and a *kantoor/kartjees* 5m x 5m (Figure 8). The *wachtkamer* area is 2 times bigger than the *kantoor/kartjees* space in order to keep the room comfortable with many activities inside such as the arrival of passengers from both train and outside, wait for train arrival, and train ticket purchase. Spacial Proportion of Ngabean Station is obtained 1: 1 ratio and the proportion of golden section is thought to be influenced by the proportion of renaissance by the Dutch society established at that time and has not changed the building style from the beginning of establishment.

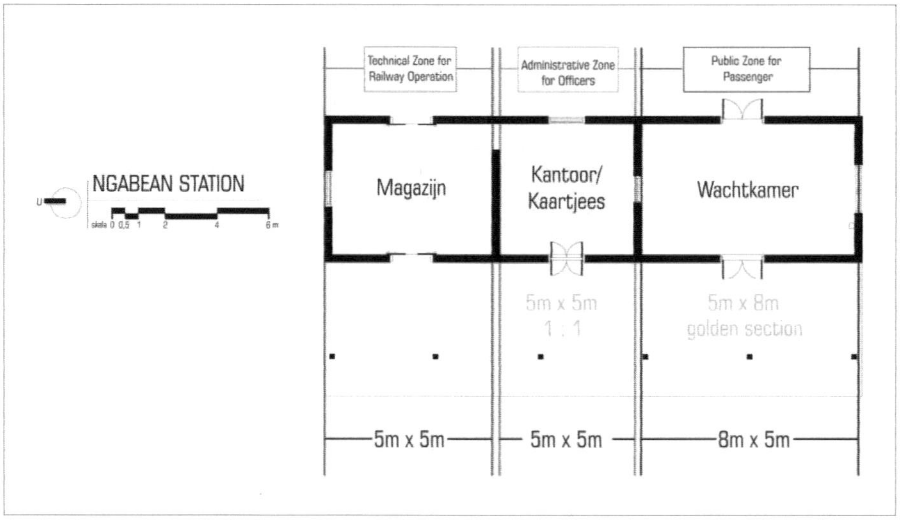

Figure 8. Stationgebouw Medium Type on Yogyakarta-Bantul Railway

Based on the spatial system disussion above, Yogyakarta-Bantul station has uniqueness with its building setting. The stations established on this line prove the role of the stations as the transportation of the factory industry so the stations depended on the setting of the sugar factory area. Small type of *stastiongebouw* accommodated sugar factory to transport produce so that the station was close to or directly adjacent to the location of the sugar factory. Because of this role, small type station did not have to be close to the highway or in the middle of the city. The train that had carriage from the factory then transported to s*tastiongebouw* medium type. The role of the medium type stations was to distribute the produce both within the city of Yogyakarta and outside Yogyakarta region. Because of that the station is located in the middle of the city and close to the main station. From that the carriage can be transmitted to other main railway. Furthermore, the *haltegebouw* type on this path has no role as a station. It only serves as a temporary halt, or for the train to pull over to give other train chances to follow (Figure 9).

IOP Conf. Series: Earth and Environmental Science **213** (2018) 012047 doi:10.1088/1755-1315/213/1/012047

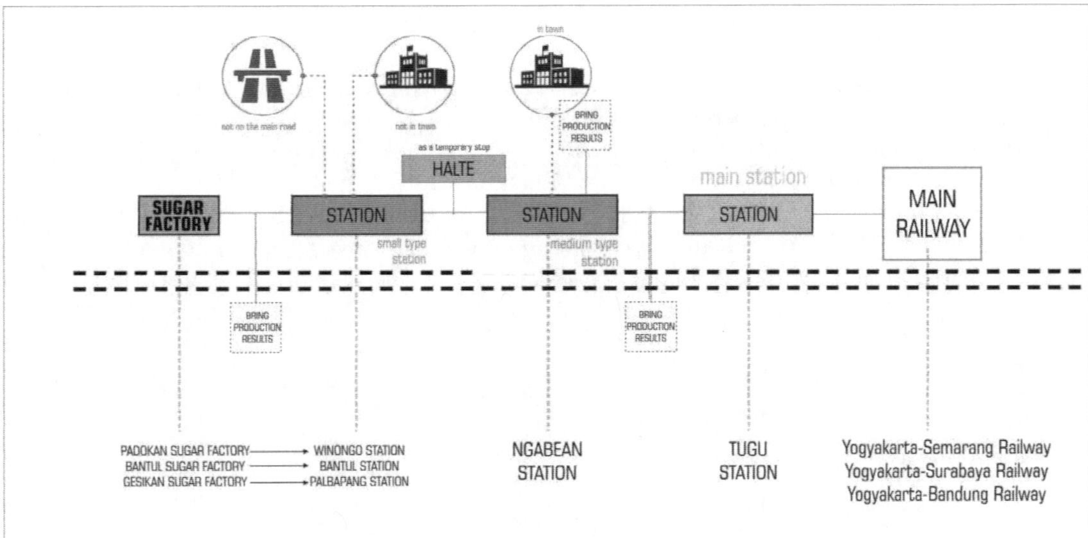

Figure 9. Setting Yogyakarta-Bantul Railway

4. Conclusion

Based on the typology of the building of the indish station, the stations on the Yogyakarta-Bantul railway are divided into 3 different typologies. Yogyakarta-Bantul railway stations are divided into *haltegebouw* type, small type of *stationgebouw*, and medium type of *stationgebouw*. The *haltegebouw* type on this railway is found on Dongkelan Station. In this type of station building consists of a simple arrangement of space that is only a *wachtkamer* and *kantoor/kaartjis*. Then, the *stationgebouw* type has more complex arrangement that is *wachtkamer, kantoor/kaartjis,* and *magazijn*. Small type station of *stationgebow* on this railway is found on Winongo Station, Bantul Station, and Palbapang Station. Next medium type station of *stationgebow* on this railway is found on Station Ngabean. The difference in the medium type with the type below is that there is a wide building terrace with a roofed platform. The medium type of building stations had a lot of passengers and its location was in the middle of the city so that the activity was busier cause the station building was bigger than the type under it despite their almost same space arrangement. Typology of the buildings of the stations on the Yogyakarta-Bantul railway has their own roles because it is influenced by the setting of the territory.

The spatial characteristics of the Yogyakarta-Bantul railway stations are unique to their setting context. The stations established on this line proved that the role of the station as the transportation of the factory industry so that it depended on the setting of the sugar factory area. Furthermore, the *haltegebouw* type on this path had no role as a station as it only served as a temporary halt, or for the train to pull over to give other train chances to follow.

5. References

[1] Olivier Johannes Raap 2017 *Sepoer Oeap di Djawa Tempo Doeloe by Olivier Johannes Raap* (Kepustakaan Populer Gramedia)

[2] Santoswana 2013 *PT Kereta Api (Persero) dan Pendirian PT Kai Commuter Jabodetabek* (Universitas Indonesia)

[3] Pollio V 1914 *Vitruvius: The Ten Books on Architecture* (Harvard university press)

[4] Habraken N J 1988 Type as a social agreement *Proceedings of the Asian Congress of Architects, Seoul*

[5] Elfendes R *Teori Arsitektur Vitruvius*

[6] Ching F D 2000 Bentuk, Ruang dan Tatanan edisi kedua *Erlangga: Jakarta*

[7] Primadewi S P N 2013 Inkulturasi Arsitektur Tradisional Bali Dalam Pola Spasial Rumah Etnis Tionghoa Di Desa Adat Carangsari, Badung *Kurva Tek.* **2** 108–25

[8] Suptandar J P 1999 Desain Interior: pengantar merencana interior untuk mahasiswa desain dan arsitektur *Jakarta: Djambatan*
[9] SULISTYANI H 2010 Karakteristik tata ruang dalam bangunan stasiun kereta api di jalur Semarang - Vorstenlanden periode 1864-1930
[10] Lehner E, Doubrawa I, Ikaputra., IVA-ICRA I for C R in A (Vienna). and Gadjah Mada University. Department of Architecture and Planning. 2013 *Insular diversity : architecture, culture, identity in Indonesia* (IVA-ICRA, Institute for Comparative Research in Architecture)
[11] Mangunwijaya Y B 2013 *Wastu citra : pengantar ke ilmu budaya bentuk arsitektur, sendi-sendi filsafatnya, beserta contoh-contoh praktis*

Mapping the socio-religious spaces of a kampong: a case of Kauman Semarang

Atiek Suprapti[1], Dhanoe Iswanto[1]
[1]Department of Architecture, Engineering Faculty, Diponegoro University, Indonesia

Corresponding e-mail: atiekbudiarto@gmail.com

Abstract. Cultural identity and creativity in diversity rise up as dominant issues in international development agenda. Identity is an important element of a nation to show who the citizens really are in the era of globalization. The existence of an urban kampong is often marginalized due to 'bad' infrastructure and social conditions. But a kampong as the residence of common people in the city center is also recognized as a successful place in reflecting human values. Kauman is a kampong where people live around the city's great mosque and demonstrate characters of a devout Muslim community in the Javanese traditional city. In Semarang Kauman is a pioneer of a city characterized by the social life of non-elite class, coastal Islamic culture, and trading activities. The role as a spiritual-religious area has been continually attached to Kauman since the Mataram Kingdom's era to the present despite different improvisation of developments. At this moment Kauman seems to experience the effects of Johar Market burning in 2016. As a kampong with stories of the history of the city's development, Kauman has many local cultures that potentially become a city identity. However, the condition of infrastructure and socio-cultural aspect appears to be in degradation due to aging and modernization-capitalization pressure from its surroundings. This research aims to explore the spatial phenomenon influenced by socio-religious aspect that makes Kauman able to survive with the identity of a 'religious spiritual' place. This research employed ethnographic methods. The results showed that socio-religious values influenced community's everyday life and encouraged the community participation in the making of place.

Keywords: urban heritage, place, social cohesion, Kauman

1. Introduction

Built environment basically is an organization among space, time, resources, and communication (Rapoport, 2005). The running of time and events has resulted changes in modernity civilization that focuses on the advancement of science, technology and capital. In the case of downtown, a heritage settlement needs resilience in order to face the growing pressure of capitalization modernization in the era of globalization. A form of resilience in the socio-ecological system is adaptive capacity related to the ability of humans to compromise with changes in their environment through observation, learning, and changing the way of interaction (Gunderson, 2000). Berkes & Folke (1998) stated that adaptive management approaches based on local knowledge, institutional learning conditions and organizations will build sustainable resilience.

Semarang as the capital of Central Java has a long history. In the 17th century the Dutch came, brought modernizm and spread Christianity. Ethnic groups lived in groups at strategic locations, such as near the smelters, near the river and near the city center. One ethnic group with a strong character is

Kauman. Generally Kauman is a settled area in "Javanese Cosmic City" located near the Great Mosque as residence for *santris* (students of traditional Islamic boarding school) and *kyais* (Masters of tradisional Islamic boarding school), and their relatives. Kauman Semarang develops as a religious kampong (a village in city centre) serving religious needs throughout the city. Occupying a strategic location right in the heart of the city close to Johar market area as a great market gives an effect on the high pressure of modernization-capitalizaation (Suprapti, 2017). In the 19th century until the early 20th century Dutch settlement was growing rapidly with various activities, such as housing, business, education, socio-cultural and religious activities. Likewise the Chinatown area was growing rapidly with different business, socio-cultural and religious activities. The politics of social segregation was used by the Dutch to suppress the local community and to make the indigenous leaders moved to fight back. Besides that, the problem of ignorance was also an internal threat. Therefore, as an educated group, the Kauman community was challenged to create education with strong values and religious beliefs for the community. So, there were a study institute and an Islamic education institution that function as a tool to carry out religious teaching.

Controllability of space in Muslim settlement is the ability of a place to manage the development of its spatial plan sustainably through socio-cultural & economic sources (Suprapti, 2017). Furthermore, the reciprocal relationship between spaces for religious activities and spaces for Muslim economic activities has created a spatial protection strategy. It will encourage the formation of spatial arrangements to form a controlled space with the support of symbiotic space. This paper aims to explore the space for socio-religious activities in the Kauman kampong of Semarang.

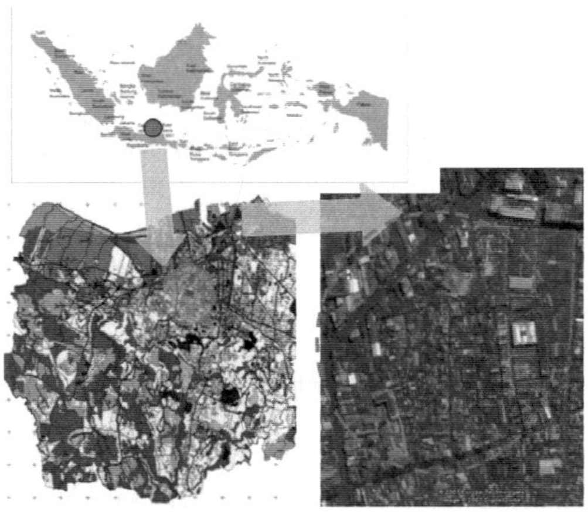

Figure 1. Location and area of Research in Kauman kampong of Semarang

2. Methods

2.1. Research Locus, Informant & Source & Quarrying Data

The location of the research covered the kampong of Kauman that consists of two parts: Bangunharjo and Kauman districts (see Figure 1). In the qualitative research, informants were interviewed in order to obtain primary data dan information (Lexy, 2007). This research used snowball-sampling technique. The technique was to dig up information from the initial informant that would show in the next informants. Extracting information was stopped after the repetition of data occured. The data could be categorized into two types: the primary data that came from the field, and secondary data derived from the documents. Field data consisted of: (1) physical data in the form of the building and its environment and the elements making up space, and (2) non-physical and data that included: worldview, ideology, spiritual outlook, and social system, especially relating to the management and arrangement of space. Information was gathered from anywhere, either in writing or verbally in order to explore the source of primary data, socio-cultural investigation. Researchers entered the filed to collect data and then conducted interviews or observations openly and unstructured. Data from this participatory observation was dug out from several sources, and crossed check with other data in order to obtain fixed data.

In the section of Mini Tour, in-depth observation was conducted to develop a spatial pattern, formed by social situations. In the section of Domains analysis, searches were first performed in the initial domain to find the names of the domain of the original space. Then exploration was carried out throughout the observation unit in order to rediscover domains in any space, which has a great influence in the lives of local people. Then, the Focused Observations were performed to deepen the observations that had been made at the earlier stage. Further taxonomic analysis led to the internal structure of each domain to discover the principles of the preparation and the symbols within the domain of the original space. The meaning of the symbols of the original space arose from the way the elements were connected to one another. Meanwhile componential analysis was the systematic search for a component of the space meaning, which was regularly associated with the symbol. Componential analysis was carried out by searching for meaning through observation on how the symbols with the other symbols were associated. The theme of space was the principle of which was repeated in a number of cognitive the domain, either hidden or open, and served as the connection between the branch systems of meaning.

3. Discussion

This research showed three themes, which are descibed further below.

3.1. Space of Waqf

The space of *Waqf* is a space with awareness raised by the owner to institutions/religious institutions/organizations. This is because traders who came from the *santri* and *ulama* communities have a big share in the establishment of the *waqf* space in Kauman. When they received a substantial amount of profits from their business, they tended to: (1) purchase buildings to increase business activities and (2) purchase property for endowments, so that the waqf space increased in number from time to time. According to the interview results, the community's motivations to do *waqf* were: (1) carrying out charity, (2) maintaining a family tradition, and (3) preparing an employment as a cleric (*ustadz*). The waqf space is spread in both the Kauman and Bangunharjo Kelurahan (districts). The possible material for *waqf* included a plot of land, buildings, furniture, money, and so on. *Waqf i*n Kauman shows a significant role in the development of Islamic education institutions and other religious institutions. In Kauman, people's awareness to participate in conducting *waqf* was driven by *Kyais* in collaboration with entrepreneurs who were oriented towards a range of local and regional services. Meanwhile around the Great Mosque, the *waqf* tradition collaborated with local entrepreneurs and from outside of Kauman had successfully supported the educational institution of *Badan Wakaf* that was developed very rapidly in the regional and global scales within five decades. *Badan Wakaf* is an educational foundation established in 1950 which managed an elementary school. Within five decades it has managed four elementary schools, two junior and senior high schools, and one big university named Islamic University of Sultan Agung with a big hospital. In this regard, *Waqf* traditions in Kauman have been going on for centuries since the establishment of the Great Mosque of Kauman, because the mosque actually was the result of endowments from the Demak Sultanate. Since then many Kauman people who have high faith attempted to validate some of their assets, lands or buildings for the continuity of religious propaganda. Family *Waqf* Space is a space as a result of a *waqf* from a particular family for religious broadcasting activities in Kauman. Based on the process of transferring rights, the *waqf* space can be categorized as below (see Figure 2 and Figure 3).

3.1.1. The traditional waqf space. The process of transferring rights of lands or buildings is based on mutual trust, sincerity, spontaneity, and with a simple process. It was used for activities included in other places. For example: Pondok Pesantren of Raudhatul Qur'an, Quran reading and writing school of Raudhatul Qur'an, Kauman Great Mosque, Astajiddin Mosque, Al-Iman Mosque, Al-Ikhsan Mosque (see Figure 4).

3.1.2. The traditional waqf space. The process of transferring rights of lands or buildings is based on mutual trust, sincerity, spontaneity, and with a simple process. It was used for activities included in other places. For example: *Pondok Pesantren* of Raudhatul Quran, Quran reading and writing school of Raudhatul Quran, Kauman Great Mosque, Astajiddin Mosque, Al-Iman Mosque, Al-Ikhsan Mosque (see Figure 5).

3.1.3. The expectation waqf space. The process of transferring rights of lands or buildings is still under a plan. The building to be represented has been prepared, but it is still used for other activities (such as: commercial activities). The location of the waqf space can be concentrated in one location or widely spread.

Figure 2. The Spread of Space of modern *waqf,* traditional *waqf, and* expectation *waqf* in Kauman Semarang

IOP Conf. Series: Earth and Environmental Science **213** (2018) 012048 doi:10.1088/1755-1315/213/1/012048

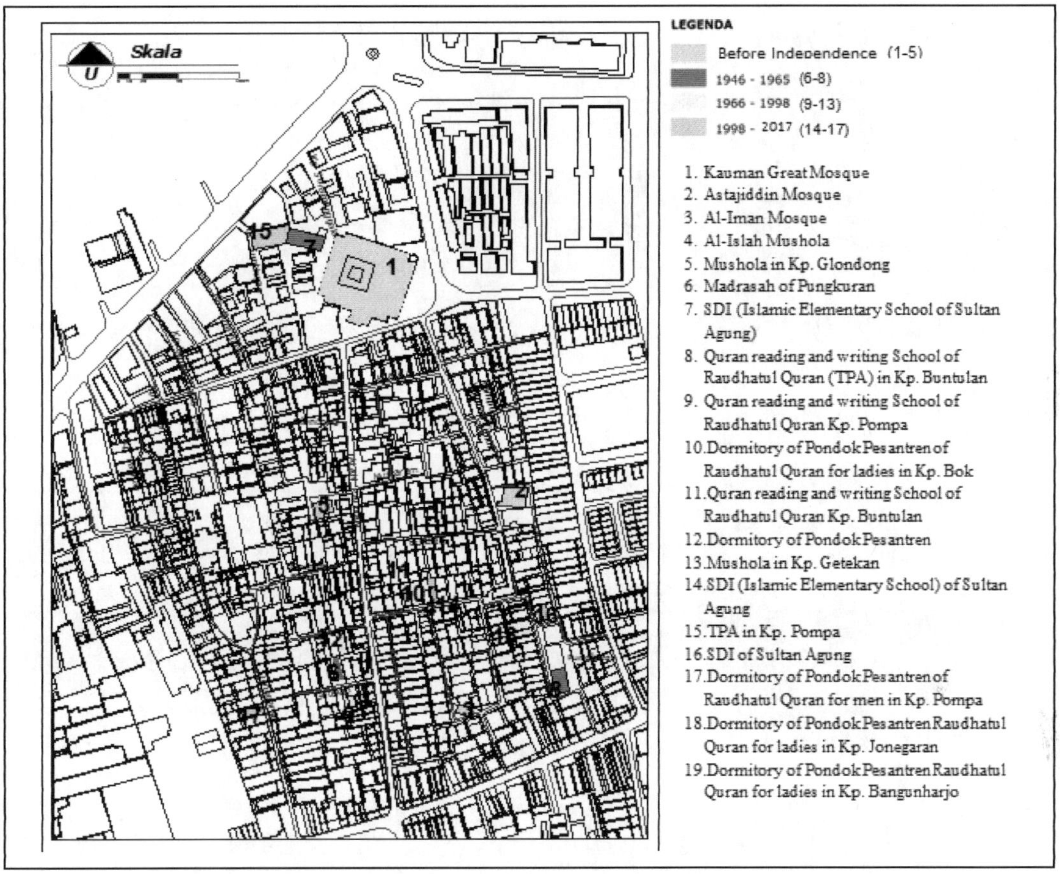

Figure 3. The Spread of Space of *Waqf* in Kauman Kampong Semarang

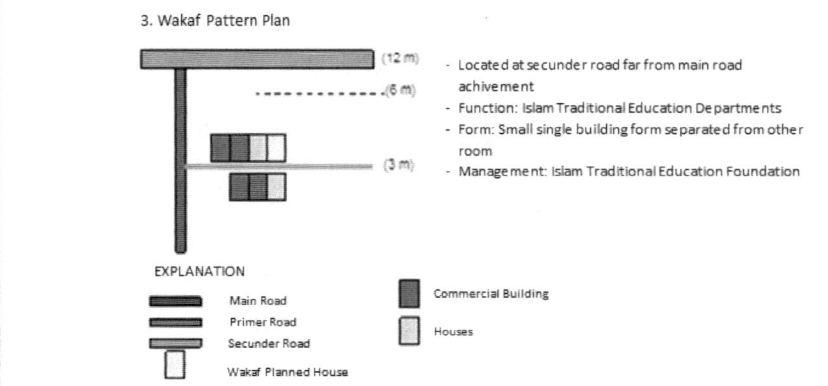

Figure 4. The Pattern of Modern *Waqf* in Kauman Semarang

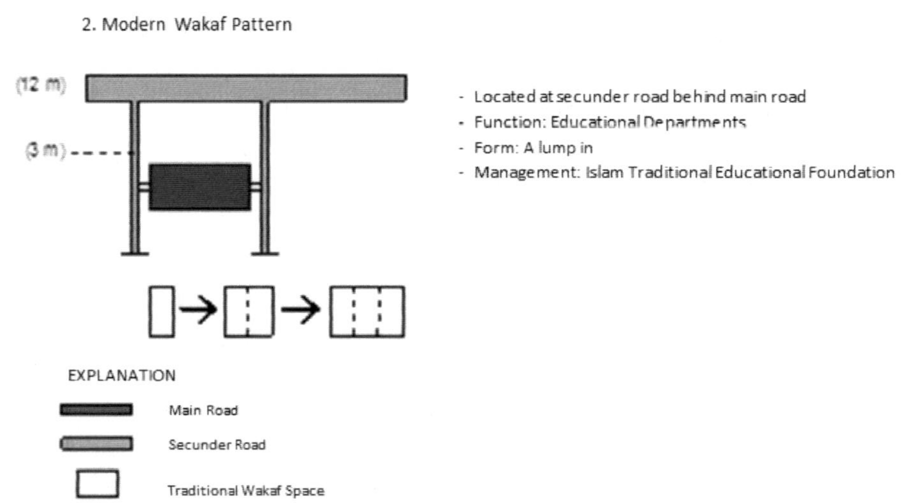

Figure 5. The Pattern of Traditional Waqf in Kauman Semarang

3.2. Space for teaching a religion knowledge (mulang).

The space for teaching religion knowledge is the space used to conduct religious knowledge teaching activities by clerics/clerics for students/students and the congregation. This is done to expect a reward from God Almighty. The physical form is a building or a part of the main building owned privately so that it can be used temporarily for repeated activities. This activity has been carried out by this community since the beginning of the 20th century and it still exists today. For actors and owners of this space, space has two meanings. The first meaning is real space in the world used for the purpose of worldly life. The second meaning is imaginary space in the afterlife or heaven where this space must be prepared by humans in the real world. According to its characteristics, the space for '*mulang*' can be categorized into three types as follows.

3.2.1. Space to mulang separately from the Kyai's house. Kyai/Santri traditionally lend space in the form of buildings/houses to teach a religion. The family of Kyai Abdullah & Kyai Turmudzi has developed a space for '*mulang*', namely *Pondok Pesantren* (the Islamic Boarding School) separated from the house, but it still uses their home to teach the religious knowledge (*mulang*).

3.2.2. Space for 'mulang' *belongs to the family of Kyai.* Besides teaching to the community, the Kyai also lent a part of their house.

3.2.3. Using a family house to support the activity of 'mulang'. They usually use the living room for '*mulang*' or other related activities. Like the late Kyai Ridwan Mujahid, Kyai Abdullah and Kyai Turmudzi used parts of their houses, including the front of their house to '*mulang*' (teach) the students. While the late Kyai Sahli deliberately designed a special space in his house to '*mulang*' his congregation, and his descendants continued this tradition nowadays. This shows the contribution of private space to the public on the basis of sincerity.

3.2.4. . Space for 'mulang' by utilizing a prayer space/mushola/ mosque. In addition to lend space in their house for '*mulang*' the *santris* (students of pondok pesantren), the Kyais also '*mulang*' in the musholla, mosque or school. The *ustadzs* have the principle that taking part in education requires sincerity and modesty to benefit the surrounding community. Although they receive only small fee, but they still feel responsible for the education of the next generation. For them, this is an investment in the afterlife, which will be a medium to enter an eternal space in heaven.

3.3. Space of "Pengajian" (Recitation).

The space of "*Pengajian*" (recitation) is a space used to discuss a religious knowledge that can be applied in everyday life of people to become a Muslim who has an *akhlakul karimah* (a good Muslim morality), and serves as a spiritual refresher for the Kauman community and surrounding. People usualy use the public religious space for various purposes, inlcuding mushola, mosque, and houses of citizens. The recitation is conducted to foster the mentality of the citizens to become a Muslim with good morality. Recitation is carried out in groups for children and adults in all days. Recitation is also carried out by kyai's relatives to strengthen the kinship and to improve the morality of the nobility. Several study groups established the "*Majlis Taklim*" that provided material focusing on specific problems to solve.

3.4. Space of Education for Ustadz Candidate's.

It is a formal space used to educate *santris* in a traditional Islamic boarding school, so they can be an *ustadz* (cleric) in the future. Raudhatul Qur'an Boarding School produces *hafidz* (people who memorize the Quran). Furthermore, the *hafidz* will usually work as *ustadz*, or work in a place that is related to the teaching of the Qur'an. The space of educational for ustadz candidate is actually a social reproduction space (Lefebvre, 1991). This space produces *ustadz* who will control the moral of youngers generations and have good moral virtues of *akhlakul karimah.*

3.5. Space of formal education.

Space of formal education is a space used to educate the younger generation with formal and modern knowledge in Islam. It is divided to several levels starting from basic level, namely PAUD (Early Childhood Education), TKI (Islamic Kindergarten), SDI (Islamic Primary Schools), or Madrasah. There are also Quran Education, such as TPA (School of reading and writing Quran), and Al-Hafidz Islamic Boarding School. The *ustadz* are from local community as well as outside of Kauman, while the students come from Kauman and its surrounding area. The oldest school is Madrasah Pungkuran that was established in 1926. The one that has been growing larger and faster is SDI (Islamic elementary school) of Sultan Agung because it received endowments from the community participation. Its buildings appear in a three-story building and some buildings spread throughout the

IOP Conf. Series: Earth and Environmental Science **213** (2018) 012048 doi:10.1088/1755-1315/213/1/012048

village mingling with residents' houses. So, there is a symbiosis between the space for developing Islamic education and the private space around it. Furthermore, the space for developing Islamic education is private where the financing matters are carried out independently and there is the participation of the community. The success of managing these educational institutions invited many outside parties to cooperate together. This is supported by the existence of good relations between religious leaders and outside community leaders. Therefore, a lot of cooperation is established by developing a network of formal education activities outside Kauman. The Islamic Elementary School of Sultan Agung as an educational institution opened a network outside Kauman covering several districts of Semarang (Seroja, Bedas, Kaligawe) and out of town, such as in Jepara. This educational institution receives *waqf* from various circles of society. The Islamic education network in Kauman had been developed long before the Independence era.

3.6. The Place Making of Kauman.

Waqf (endowments) is granting some of the assets of a Muslim to certain parties, then the assets will be used for some religious activities. The principle in *waqf* is a transfer of rights from those who give (*waqif*) to those who exercise the mandate (*nadzir*). *Nadzir* is usually a religious institution that will manage it. *Waqfs* are carried out by Muslims as a form of an obedience to religious orders, *wakif* (trustees) believe that their actions are a form of investment that will be rewarded by God with a blessing or a place of honor in Heaven (Suprapti, 2017). In their view the *waqf* space is a mediator that connects the worldly space with the imaginary space in Heaven. Heaven's imaginary space is a space constructed from three elements. Firstly it is the financial ability to provide *waqf* space that is devoted to Islam for education, recitation, and worship activities. Secondly it is the values of strong belief in religious teachings and has taken place in time. Thirdly it is the knowledge of the *waqf* space inherited from their parents. Furthermore *waqf* practitioners believe that they will have space in Heaven. This is a reflection of the view of the Kauman Muslim community cosmology.

Waqf is a form of community participation in development where the community contributes actively to take a part inside (Yuwono, 2001). This will increase social cohesiveness in society (Handayani, 2006). Ancok (2003) stated that a social capital has an important role for the nation development in the era of information technology (Ancok, 2003). Furthermore, Ancok explained that the social capital could be created from formal-informal education, character buildings and social meetings. Later Woolcock (2001) mentioned components that play important roles, as tools of achieving development goals, which are appropriate social relations. The Kauman community has a strong social relationship. Kauman has some social meetings of recitation groups, such as formal-informal, and modern-traditional Islamic education institutions, for all level of society from children to adolescents. The meetings serve as tools to control as well as develop their social capitals. The *waqf* tradition possessed by Kauman is not only social capitals, but also economic capitals. It also becomes spiritual capitals which is very important for the development of the city characteristics in the era of globalization. Those are very important capitals to be preserved for their sustainability. Planning, designing and managing the public spaces are known as the place making theory. This process includes community participation in contributing, planning-designing and managing a space collaboratively. It will produce an inherent and memorable public space for the community (Nicodemus, 2012). The next process of place making is considered as an improvement on the quality of people's lives, because it provides inspiration to the community through shared memories and builds a public space participation (Project for public spaces, 2008). It seems like Kauman's place is a form of compact residential space with mixed functions, high density and friendly for pedestrians (Jenks M., Burton E, William, K., 2005).

3.7. Concept of Spiritual Space & Islamic Education.

From the 19th century to the beginning of the 20th century, the Dutch developed the city rapidly where they established housing, business places, education places, socio-cultural facilities, and religious activities. Likewise the Chinatown area grew rapidly with different business, socio-cultural and religious activities. The policy of community segregation used by the Dutch to suppress local communities made indigenous leaders move and fight against the Dutch. These things were considered as threats to the Kauman community. On the other hand, the problem of ignorance was also an internal threat. Therefore, the Kauman community was challenged to create religious formation for the surrounding community. In this case, the Kauman community established an educational institution and an Islamic education institution that function as tools to carry out religious teaching.

4. Conclusion

The values of social religious predominantly influenced the place making in Kauman. Kyai, entrepreneurs and commons residents play active roles in the formation of social religious space by participating in development spiritually and materially through *waqf*. The number of *waqf* spaces increased in the kampong which then reduced the amount of residential space. This social capital has a role in strengthening the social cohesion of the community and consolidating the religious moral control of residents. Kauman kampong shows the relationship among the built environment settings, including religious social spaces in the form of *waqf* space, space for teaching religion knowledge, space of recitation, space of education for religious candidates, space of formal education. It also has intangible aspects in the form of socio-cultural and religious aspects over time especially within a century period. This is the city's DNA that is the core of its identity. Learning from the Kauman case, the formation of an area of a city required a city DNA to create its identity. This will strengthen the quality of public space, which will improve the quality of people's lives, including their happiness, health and welfare. This social capital is essential to the development of the nation.

5. Acknowledgments

This article is a part of research that was finnancially supported by the Faculty of Engineering, Diponegoro University Indonesia through Strategic Research Grant Program in 2018.

6. References

[1] Rapoport A 2005 Theory social and culture Tylor and Francis Metapress
[2] Gunderson L H 2000 Ecological resilience — in theory and application annual review of Ecology and Systematics **31** 425
[3] Berkes F, and Folke C 1998 Linking Social and Ecological Systems: Management practices and social mechanisms for building resilience Cambridge University Press **133** 429
[4] Suprapti A 2017 Control of spatial protection in Kauman Semarang Journal of Archutecture and Urbanism **414** 268
[5] Moleong J L. 1996. Metodologi penelitian kualitatif Bandung PT Remaja Posdakarya
[6] Lefebvre H 1998 The production of space Blackwell Oxford UK & Cambridge USA.
[7] Yuwono T. 2001 Manajemen otonomi daerah membangun daerah berdasar paradigma baru Semarang Clogapps Diponegoro University
[8] Handayani S 2006 Perlibatan masyarakat marginal dalam perencanaan dan penganggaran partisipasi Surakarta Kompip
[9] Ancok D 2003 Social capital and s ociety community Jurnal Psikhologika Jurnal Pemikiran dan Penelitian Psikhologi **1815**
[10] Woolcock M 2001 The place of social capital in understanding social and economic outcomes Canadian Journal of Policy Research **21** 11
[11] Nicodemus A G 2012 Creative placemaking 2.0 grantmakers in the arts http://wwwgiarts.org/article/creative-placemaking-20 (cited on May 18, 2018)
[12] Project for Public Spaces 2008 Street as places using streets to Rebuild Communities. New York AARP
[13] Jenk M, Burton E, and Williams K 2005 The compact city a sustainable urban f orm? Tylor and Fancis e-Library

AUTHOR INDEX

Adiprakoso, Gregorius 247
Antariksa 32
Asriningpuri, Handajani 81
Bangun, I. R. H 308
Basri, Ardi 225
Boli, Benediktus 277
Bramiana, Chely N. 308
Darmawan, Edy 262
Djonda, Ubaldus 277
Ernawati, Jenny 16
Erwindi, Collinthia 160, 168
Fauzi, Hendra 225
Felasari, Sushardjanti 288
Firmandhani, Satriya Wahyu 41
Firzal, Yohannes 59
Hamzah, Baharuddin 210
Hanan, Himasari 256
Hardiman, Gagoek 132, 380
Harsritanto, Bangun Ir 67, 118
Ikaputra 400
Indrawan, Iwan Adi 332
Intan, Kusumo Dewi Diah 295
Irvansyah 160, 168
Isnaeni, Hendrajaya 186
Istiadji, A. Djoko 132
Iswanto, Dhanoe 411
Jamaluddin, Rizqi 67
Jamila, Rona Fika 67
Junara, Nunik 318
Karsono, Bambang 367
Kim, Saehoon 9, 125, 142
Kisnarini, Rika 332
Koesmartadi, Ch. 247
Krisdianto, Johanes 332
Krisprantono 343, 350
Kurniati, Rina 295
Kurniawan, Kemas Ridwan 194, 392
Kusumawanto, Arif 148
Kusyala, Dibya 358
Lake, Reginaldo Christophori 277
Lubis, Irma Handayani 358
Lukito, Yulia Nurliani 217, 232, 239
Marasabessy, Firdawaty 148
Marlina, Avi 106
Miranda, Fildza 232
Muhammad, Asri A. 148
Nareswari, Ardhya 148
Noerwasito, Vincentius Totok 25
Nugrahanti, Fathina Izmi 358
Nuraeny, Elita 194
Nurini, Nurini 295
Pandelaki, Edward Endrianto 387
Permana, Asep Yudi 96, 177
Prabowo, Bintang N. 308
Pratama, Putri Widya 217
Prima, Listen 49
Purwanto, Edi 262
Rahim, Mustamin 210, 225
Ramandhika, Mirza 308
Riskiyanto, Resza 387
Riza, Masyiana Arifah Alfia 41, 400
Rochimah, Estuti 81
Roychansyah, M. Sani 288
Ryu, Suha 125
Samalavicius, Almantas 1
Samodra, Fx Teddy Badai 160, 168
Santosa, Herry 16
Sari, Wulani Enggar 67
Satwiko, Prasasto 132
Sekatia, Augi 380
Seo, Seungwoo 142
Setiadi, Yusup 186
Setioko, Bambang 118
Setyaningsih, Wiwik 73
Setyowati, Erni 380
Siwa, Yohanes Nurak 277
Sofyan, Afriyanto 343
Sugeng, Triyadi S 318
Sulong, Awang Hasim Awang 367
Suprapti, Atiek 411
Surjono 32
Susanti, Indah 177
Susanti, Retno 295
Suskiyatno, Fx Bambang 343
Syahid, Mushab Abdu Asy 392
Temas, Irham 59
Titik, S. Yulita 247
Titisari, Ema Yunita 32
Wardhani, Mustika K 118
Widiarso, Tulus 256
Widiastuti, Ratih 67, 308
Wihardyanto, Dimas 400
Wijaya, Karto 96, 177
Wulandari, Lisa Dwi 16, 32
Yudantini, Ni Made 202
Yuliani, Sri 73

Zahra, Rima Aisha 239
Zaini, Atta Idrawani 367

Institute of Physics
Dirac House, Temple Back
Bristol BS1 6BE UK

ISSN: 1755-1307
ISBN 978-1-5108-7955-3

CURRAN ASSOCIATES INC.
proceedings.com